Correlation Functions and Quasiparticle Interactions in Condensed Matter

NATO ADVANCED STUDY INSTITUTES SERIES

A series of edited volumes comprising multifaceted studies of contemporary scientific issues by some of the best scientific minds in the world, assembled in cooperation with NATO Scientific Affairs Division.

Series B: Physics

RECENT VOLUMES IN THIS SERIES

Volume 27 – Topics in Theoretical and Experimental Gravitation Physics
edited by V. De Sabbata and J. Weber

Volume 28 – Material Characterization Using Ion Beams
edited by J. P. Thomas and A. Cachard

Volume 29 – Electron–Phonon Interactions and Phase Transitions
edited by Tormod Riste

Volume 30 – Many Degrees of Freedom in Field Theory
edited by L. Streit

Volume 31 – Many Degrees of Freedom in Particle Theory
edited by H. Satz

Volume 32 – Electron and Ion Spectroscopy of Solids
edited by L. Fiermans, J. Vennik, and W. Dekeyser

Volume 33 – Microscopic Structure and Dynamics of Liquids
edited by J. Dupuy and A. J. Dianoux

Volume 34 – Path Integrals and Their Applications in Quantum, Statistical, and Solid State Physics
edited by G. Papadopoulos and J. T. Devreese

Volume 35 – Correlation Functions and Quasiparticle Interactions in Condensed Matter
edited by J. Woods Halley

Volume 36 – Strongly Coupled Plasmas
edited by Gabor Kalman

Volume 37 – Coherence in Spectroscopy and Modern Physics
edited by F. T. Arecchi, R. Bonifacio, and M. O. Scully

This series is published by an international board of publishers in conjunction with NATO Scientific Affairs Division

A Life Sciences **B Physics**	Plenum Publishing Corporation London and New York
C Mathematical and Physical Sciences	D. Reidel Publishing Company Dordrecht and Boston
D Behavioral and Social Sciences	Sijthoff International Publishing Company Leiden
E Applied Sciences	Noordhoff International Publishing Leiden

Correlation Functions and Quasiparticle Interactions in Condensed Matter

Edited by
J. Woods Halley
University of Minnesota

PLENUM PRESS • NEW YORK AND LONDON
Published in cooperation with NATO Scientific Affairs Division

Library of Congress Cataloging in Publication Data

Nato Advanced Study Institute on Correlation Functions and Quasiparticle Interactions in Condensed Matter, Spring Hill Conference Center, 1977.
Correlation functions and quasiparticle interactions in condensed matter.

(NATO Advanced study institutes series: Series B, Physics; v. 35)
Includes index.
1. Solid state physics–Congresses. 2. Liquids–Congresses. 3. (Quasiparticles (Physics)–Congresses. 3. Correlation (Statistics)–Congresses. I. Halley, James Woods, 1938- II. Title. III. Title: Condensed matter. IV. Series.
QC176.AlN32 1977 530.4'1 78-18837
ISBN 0-306-40018-9

Proceedings of the NATO Advanced Study Institute on Correlation Functions and Quasiparticle Interactions in Condensed Matter held at Spring Hall Center, Wayzata, Minnesota, August 17–27, 1977

A Division of Plenum Publishing Corporation
227 West 17th Street, New York, N.Y. 10011

Printed in the United States of America

PREFACE

This volume contains the proceedings of a NATO Advanced Study Institute devoted to the study of dynamical correlation functions of the form

$$C_{AB;AB}(\omega) = \frac{1}{2\pi}\int_{-\infty}^{+\infty} e^{-i\omega t}\langle A(o)B(o)A(t)B(t)\rangle dt \tag{1}$$

where A and B are physical operations in the Heisenberg representation and

$$\langle\ldots\rangle = \frac{Tr(e^{-\beta\mathcal{H}}\ldots)}{Tr\, e^{-\beta\mathcal{H}}}$$

is an equilibrium average. In equation (1) it is useful to regard the product AB as the product of two operators in cases in which A and B refer to different spatial points in a condensed matter system and/or in which A and B behave dynamically in a quasiharmonic way. In the second case, one has a two quasiparticle correlation function and $C_{AB;AB}(\omega)$ gives information about quasiparticle interactions.

Condensed matter physics has increasingly turned its attention to correlation functions of this type during the last 15 years, partly because the two point and/or one-particle correlation functions have by now been very thoroughly studied in many cases. The study of four point and/or two quasiparticle correlations has proceeded somewhat independently in several diverse fields of condensed matter physics and it was one purpose of the institute to bring experts from these different fields together to describe the current state of their art to each other and to advanced students.

The fields represented included solid state physics of phonons and magnons and superconductivity, dynamical critical phenomena and the physics of classical and superfluid liquids. In this volume, the first section contains four lectures on basic theoretical relationships and techniques which underlie much of the interpretation of of the physics of the systems discussed in the following sections. These lectures describe basic theory of response functions (R. Stinchcombe), many-body perturbation theory (J.W. Halley), theory of dynamical critical phenomena (G. Mazenko) and the foundations of hydrodynamics (I. Oppenheim). In the next three sections of this volume, studies of correlation functions in magnetic systems, liquids and phonons in solids are described. The identity of the operators A and B in these various fields is indicated in Table 1. Lecturers on these topics were H. Bilz (phonons), R. Cowley (liquids, magnons and phonons), P. Fleury (liquids and magnons), W. Gelbart (liquids), A. Rahman (liquids), I. Silvera (quantum solids) and M. Thorpe (magnons).

Twelve people contributed brief manuscripts on special aspects of the subject on which they gave seminars at the institute. These were M.G. Cottam, U. Balucani and V. Tognetti, and P.D. Loly (magnons), A. Ben-Reuven, D. Frenkel, H. Metiu and R. Kapral (classical liquids), C. Murray, R. Hastings and F. Pinski (superfluid ^{4}He), A. Goldman (superconductivity) and W. Buyers (phonons). Seminars were also given at the institute on superionic conductors (S. Ushioda), phonon bond states (J. Ruvalds) and long time behavior of spin correlation function (D. Huber).

In reviewing the contents of this volume certain impressions emerge. The problem of analyzing experiments which give information on correlation functions of the sort considered here, falls into two parts: 1) The computation of the coupling between the probe (usually electromagnetic radiation or neutrons) and the many-body system and 2) the computation of the correlation function itself.

TABLE 1. Identity of the operators A,B of equation (1) in various systems

STATE	A,B	QUASIPARTICLE
Solid (non-magnetic)	$\rho^{\underset{\sim}{k}}, \rho^{-\underset{\sim}{k}}$	phonons
Solid (magnetic)	$S^{\mu}_{\underset{\sim}{k}}, S^{\nu}_{-\underset{\sim}{k}}$	spin waves (magnons)
Critical Regions	both above	none
Liquids	$\rho^{\underset{\sim}{k}}, \rho^{-\underset{\sim}{k}}$	phonons and rotons (^{4}He) ——— None (classical)

In magnetic systems for T below the ordering temperature, both aspects of the problem are now under reasonably good control except in the critical region where the techniques described by G. Mazenko have not yet been seriously applied to the sort of correlation functions discussed at the institute. Above the ordering temperature little has been done; the work described by Balucani and Tognetti in this volume is a start. (Work by P. Fedders should also be mentioned in this connection.)

In liquid physics, W. Gelbart describes the problem of computing second order coupling to light in this volume. Large uncertainties remain in this area. It is possible that x-ray scattering experiments could contribute to a resolution of this problem. The problem of the computation of the higher order correlation function in classical liquids is probably amenable to solution by molecular dynamics. Preliminary work is described by A. Rahmann in this volume. (The HNC techniques described by F. Pinski in this volume are useful for integrated scattering but not for the dynamics.) The talk by R. Hastings shows that attempts to calculate dynamical four point response functions in liquid ^{4}He remain at the phenomenological level.

In phonon systems, the reviews by R. Cowley and H. Bilz indicate good qualitative understanding of most higher order phonon spectra and couplings though quantitative microscopic (as opposed to phenomenological) models are not yet available. In quantum solids, as described by Silvera, the basic dispersion relations have not yet been obtained from neutron scattering and this remains a major obstacle to a quantitative understanding of the higher order optical spectra. It remains unclear whether bound states of phonons have been seen in solids. People interested in this question might wish to be more systematic about seeking optimal materials in which to see this phenomenon unambiguously.

Finally, in general response theory it appears that a large area of theoretical development in the area of non-linear optics in condensed matter has only begun to be discussed.

Many people contributed to the success of this institute. I will mention first Professor Charles Campbell who, as local committee chairman, was indispensable and who also contributed in a very important way scientifically. The organizing committee provided indispensable assistance over a period of more than a year. The committee members were Heinz Bilz, Charles Enz, Paul Fleury, William Gelbart, Satoru Sugano, Michael Thorpe. The following local companies contributed financially: Deluxe Check Printers Foundation, Magnetic Controls Company and Northern States Power Company. We also received some aid from the University of Minnesota Graduate School. Indispensable administrative assistance from the office of Associate Dean of the University of

Minnesota, Institute of Technology, Walter Johnson and particularly from Louise Shea of that office is gratefully acknowledged. Sybil Leuriot is thanked for typing parts of the manuscript and other conference materials. Finally, of course, we are grateful to the Scientific Affairs Division of NATO which made the entire project possible.

J. Woods Halley
Minneapolis, Minnesota 1978

CONTENTS

I. General Theory

II. Magnetic Systems

III. Liquids

IV. Phonons

Part I
General Theory

KUBO AND ZUBAREV FORMULATIONS OF RESPONSE THEORY

R.B. Stinchcombe

Theoretical Physics Department
Oxford University, U.K.

ABSTRACT

The theory of the linear response of systems to external adiabatic and isothermal perturbations is reviewed. The relationship of the response coefficient to correlation functions and to Green functions is developed following the work of Kubo and Zubarev. The Green function method is illustrated with examples, and brief treatments are given of sum rules and of the fluctuation-dissipation theorem. It is shown how the framework may be used to discuss quasiparticle and collective modes, equilibrium thermal properties, and cross sections for neutron and electromagnetic scattering.

1. INTRODUCTION

The underlying theme of these lectures is that most properties of systems are closely related to the response of the system to appropriate external 'fields'.

This statement applies to properties ranging from bulk thermal, mechanical and magnetic properties (such as specific heat, compressibility and susceptibility) to those properties, such as phonon and spin wave energies and lifetimes which are characteristic of individual modes of the system.

The relationship is rather obvious for the bulk properties, which are in most cases observed by measuring the effect on the system of temperature changes, or the response to mechanical forces or electromagnetic fields.

It is less obvious for the single mode properties, such as excitation spectra which are usually obtained from the cross section for inelastic scattering of neutrons, electrons or photons, in which the change of state of the scattered particle is associated with a definite energy transfer ω and momentum transfer q to the system. In that case a specific normal mode at frequency ω and wave vector q is probed, so long as the normal mode couples to the scattered particle, and so long as q is a good quantum number for the modes, as is the case for e.g. a perfect crystal*. Whether or not a single mode, or many, are involved, it will be shown that such experiments are closely related to the (adiabatic) response of the system to wave-vector and frequency-dependent external fields, i.e. to "generalised susceptibilities".

Another point to be stressed is the relationship of such "susceptibilities" (response functions) to correlation functions.

The usual bulk isothermal magnetic susceptibility is the ratio of the magnetisation $\partial F/\partial h$ to the applied magnetic field h and is therefore a second derivative of the free energy F with respect to field. Similarly the isothermal compressibility is a second derivative with respect to pressure. As is well known and will be illustrated later such second derivatives are identical to static correlation functions between "macroscopic"operators (the total magnetization operator in the case of magnetic susceptibility.

The bulk adiabatic susceptibility or compressibility is the ratio of the response to the adiabatically applied field or pressure. We shall show, following the original development by Kubo [1], that this type of response coefficient can again be calculated in terms of correlation functions between the same "macroscopic" operators; but in this case a time-dependent correlation function is required. This type of description also applies to transport coefficients such as the electrical conductivity, and generalises immediately to frequency-dependent processes.

---

*

In the case of imperfect crystals, the momentum transfer to the system q will be shared between all the normal modes at frequency ω. A sum over normal modes at a particular energy is again involved in, for example, optical absorption, in which only the energy dependence is measured. In other cases, such as X-ray scattering, the wave vector change is identified but all contributing normal modes, whatever their energies, are summed over, yielding information on the structure but not on the dynamics.

For the generalised susceptibilities involving wave vector as well as frequency dependence, such as those related to scattering by particular normal modes, or to the response to non-uniform time-dependent fields, the related time-dependent correlation function is between operators which carry particular momentum labels.

Sometimes it is convenient to work directly in terms of the generalised susceptibility or the related correlation function. However a compact formalism exists for the calculation of static or time-dependent correlation functions by means of Green function equation of motion methods as described by Zubarev [2] and others [3]. The Green functions depend on two time labels and, like the correlation functions they are designed to lead to, they normally involve also thermal aspects since averaging over some (e.g. canonical) ensemble is involved. Even where the required correlation functions relate to macroscopic properties of the system the Green function method does not require complete normal mode solutions for the system.

We proceed at once (Section 2) to a statement of Kubo's theory of adiabatic linear response. This is followed by a brief section (§3) on isothermal response. In these two sections the relationship between response functions and correlation functions appears. The two-time Green function is then introduced and related to the correlation function (Section 4). In Section 5, simple examples are given to illustrate the main points previously discussed. After a short section (§6) on other types of Green function, the direct relationship between Green functions and normal mode properties is then exhibited (§7). Section 8 contains formal properties, such as sum rules, while Section 9 derives and discusses the importance of the fluctuation-dissipation theorem. The following section (§10) treats the various scattering processes which allow the normal modes to be probed, and the relationship of cross-sections to the correlation functions. The concluding section (§11)gives some generalisations of the previous discussions, and refers to situations where a breakdown of the linear response approach could occur.

2. KUBO FORMULA FOR LINEAR RESPONSE [1,4]

(i) Response function

The linear response of a system to an adiabatically applied external field F(t) may be easily derived as follows when the field can be introduced by means of an additional term in the Hamiltonian, of the form

$$H_1 = - A F(t) \tag{2.1}$$

Here A is the operator to which the field linearly couples (e.g. a dipole operator if F is a uniform electric field, or a magnetic moment if F is a magnetic field).

It is supposed that the response to the field is observed by measuring the change ΔB it produces in the mean value of a particular operator B (e.g. a current operator). Such mean values can be obtained from the density matrix $\rho(F)$ for the system in the applied field

$$\rho(F) = \rho + \delta + O(F^2). \tag{2.2}$$

ρ and δ are here the zeroth and first order terms respectively in the expansion of $\rho(F)$ in powers of F. The linear response is then

$$\Delta B \equiv \mathrm{Tr}\, \delta B \tag{2.3}$$

If H is the Hamiltonian of the system in zero field, $\rho(F)$ satisfies (taking $\hbar = 1$)

$$\dot{\rho}(F) = -i\,[(H+H_1), \rho(F)] \tag{2.4}$$

from which the terms linear in the applied field are

$$\dot{\delta} = -i\,[H, \delta] + i\,[A, \rho]\, F(t). \tag{2.5}$$

This has the formal solution

$$\delta(t) = i\int_{-\infty}^{t} dt'\, e^{iH(t'-t)}\, [A,\rho]\, F(t')\, e^{-iH(t'-t)} \tag{2.6}$$

where use has been made of $\delta(-\infty) = 0$, since the field was switched on adiabatically from zero in the distant past. Inserting (2.6) into (2.3) and using cylic invariance within the trace and the usual definition of Heisenberg operators gives the linear response at time t

$$\Delta B \equiv \mathrm{Tr}\, \delta B = \int_{-\infty}^{t} \phi_{BA}(t-t')\, F(t')\, dt' \tag{2.7}$$

where ϕ_{BA} is the "response" or "after-effect" function

$$\phi_{BA}(t) = i\,\mathrm{Tr}\,[A,\rho]\, B(t) = i\,\langle [B(t), A] \rangle \tag{2.8}$$

where

$$B(t) = e^{iHt}\, B\, e^{-iHt}, \tag{2.9}$$

and

$$\langle \ldots \rangle = \mathrm{Tr}\,\{\rho \ldots\} \tag{2.10}$$

is the usual zero field average.

The response function ϕ is a measure of the ratio of the response to the applied field. It would be an adiabatic susceptibility if A and B were both total magnetic moment operators, or a conductivity for another choice of A, B. The calculation of such response and transport coefficients has therefore been reduced to a zero field evaluation.

(ii) Generalised susceptibility

In the case of an adiabatically applied periodic force

$$F(t) = F_0 \cos\omega t \, e^{\varepsilon t} \qquad \varepsilon \to 0^+ . \tag{2.11}$$

The linear response at time t can then be written

$$\mathrm{Tr}\, \delta B = \mathrm{Re}\, \chi_{BA}(\omega) F_0 e^{i\omega t} \tag{2.12}$$

where $\chi_{BA}(\omega)$ is the "complex admittance" or "generalised susceptibility"

$$\chi_{BA}(\omega) = \lim_{\varepsilon \to 0^+} \int_0^\infty \phi_{BA}(t)\, e^{-i\omega t - \varepsilon t}\, dt . \tag{2.13}$$

An alternative form of this which will be important later is

$$\chi_{BA}(\omega) = \lim_{\varepsilon \to 0^+} \int_{-\infty}^{\infty} i\,\theta(t) \langle [B(t), A] \rangle\, e^{-i\omega t - \varepsilon t}\, dt \tag{2.14}$$

where $\theta(t)$ is the usual theta-function, which takes the value 1 for positive t and is zero otherwise. In this form the generalised susceptibility is the fourier transform of a type of Green function to be discussed in Section 4.

(iii) Kubo formula for canonical ensemble

The first expression of (2.8) for the response function can be reduced to a useful alternative form when the (zero field) density matrix takes the value appropriate to a canonical ensemble:

$$\rho = \exp(-\beta H) / \mathrm{Tr} \exp(-\beta H) . \tag{2.15}$$

In that case,

$$[A, \rho] = \rho \{ e^{\beta H} A e^{-\beta H} - A \} = \rho \int_0^\beta d\lambda\, e^{\lambda H} [H, A]\, e^{-\lambda H} = \frac{\rho}{i} \int_0^\beta d\lambda\, \dot{A}(-i\lambda) \tag{2.16}$$

where

$$\dot{A} \equiv i [H, A] . \tag{2.17}$$

Thus

$$\phi_{BA}(t) = \int_0^\beta d\lambda \langle \dot{A}(-i\lambda) B(t) \rangle = -\int_0^\beta d\lambda \langle A(-i\lambda) \dot{B}(t) \rangle . \qquad (2.18)$$

The generalised susceptibility for a system described by a canonical ensemble can be obtained by substituting (2.18) into (2.13). One form of the result is

$$\chi_{BA}(\omega) = \lim_{\varepsilon \to 0^+} \int_0^\beta d\lambda \int_0^\infty dt \langle \dot{A}(-i\lambda) B(t) \rangle e^{-i\omega t - \varepsilon t} . \qquad (2.19)$$

This expression is known as the Kubo formula.

For the case of $\omega = 0$, an alternative form for the resulting (adiabatic) susceptibility is

$$\chi_{BA}(0) = \int_0^\beta \langle (A(-i\lambda) - A^\circ)(B - B^\circ) \rangle \, d\lambda \qquad (2.20)$$

where A^O and B^O are the diagonal parts of A and B with respect to H.

The results (2.18), (2.19) and (2.20) may be easily generalised to the case of a system described by the grand canonical ensemble by making the replacement

$$H \to \mathcal{H} = H - \mu \hat{N} \qquad (2.21)$$

(where $\hat{N}$ is the number operator and μ the chemical potential) in all the preceding steps of this subsection.

Similarly the classical versions of any of the results so far derived can be obtained by replacing (1/i) [,] by the Poisson bracket, ρ by the distribution function f (p,q) in phase space and Tr by an integral over phase space.

To conclude this development, we emphasise that the expressions (2.8), (2.14), (2.18), (2.19) and (2.20) all involve "correlation functions" of the type

$$\mathrm{Tr}\, \rho A B = \langle A B \rangle . \qquad (2.22)$$

(iv) Example: Electrical conductivity

As an example of the above linear response results, consider the electrical conductivity $\sigma_{\mu\nu}$, defined by

$$\langle j_\mu \rangle = \sigma_{\mu\nu} \mathcal{E}_\nu \qquad (2.23)$$

where $\langle j_\mu \rangle$ is the average current flowing in the μ-direction when an electric field is applied in the ν-direction. This special case corresponds to

$$B = j_\mu , \quad H_1 = -e x_\nu \mathcal{E}_\nu(t) , \quad F = \mathcal{E}_\nu , \tag{2.24}$$

so that

$$A = e x_\nu , \quad \dot{A} = e \dot{x}_\nu = j_\nu . \tag{2.25}$$

Thus, using the Kubo formula (2.19),

$$\sigma_{\mu\nu}(\omega) \equiv \chi_{j_\mu, e x_\nu}(\omega) = \lim_{\varepsilon \to 0^+} \int_0^\infty dt \int_0^\beta d\lambda \, e^{-i\omega t - \varepsilon t} \langle j_\nu(-i\lambda) j_\mu(t) \rangle . \tag{2.26}$$

Expressions equivalent to this were first given by Kubo[5] and by Greenwood and Peierls[6]. The usual static case has $\omega = 0$.

This result, like all the above linear response results, is exact. However it is abstract and the current-current correlation it involves is normally very difficult to evaluate because of the central rôle played by collisions in limiting σ. Methods (normally approximate) can be adopted for the evaluation of the correlation function which do not suffer from the limitations of the earlier Boltzmann equation approach for the conductivity (see for example Sections 4 and 5). For example quantum effects, such as the de Haas-Schubnikov oscillations, and magneto phonon oscillations, which occur in the conductivity in very high magnetic fields[7] can be extracted [8,9,10] from the above expression (2.26) but not from the usual classical Boltzmann equation. Nevertheless, in the appropriate approximation the evaluation of σ by way of (2.26) can be shown to be equivalent to the Boltzman equation evaluation[11,12,13].

We also remark that an expression for the resistivity involving a force-force correlation function, has been proposed [14,15] but it seems to be of limited value [16,17].

3. ISOTHERMAL RESPONSE

(i) Isothermal generalised susceptibility

The preceding section dealt with the effect of an adiabatically applied perturbing field. In the case of isothermal response, the field is applied in such a way that the system remains in equilibrium at the initial temperature. For the canonical case, and considering only a time-independent field F, the perturbed density matrix is then proportional to

$$e^{-\beta(H-AF)} = e^{-\beta H}\,[1 + \int_0^\beta A(-i\lambda)F\,d\lambda + O(F^2)]. \tag{3.1}$$

The change in the average of B is therefore

$$\Delta B = \frac{\langle [1+\int_0^\beta A(-i\lambda)F\,d\lambda + o(F^2)]\; B\rangle}{\langle [1+\int_0^\beta A(-i\lambda)F\,d\lambda + o(F^2)]\rangle} - \langle B\rangle$$

$$= F\int_0^\beta d\lambda\,[\langle A(-i\lambda)B\rangle - \langle A(-i\lambda)\rangle\langle B\rangle] + o(F^2). \tag{3.2}$$

The angular bracket once more denotes the zero-field (canonical) average defined by (2.10), (2.15). The static isothermal susceptibility is therefore

$$\chi_{BA}^{T} = \int_0^\beta d\lambda\,[\langle A(-i\lambda)B\rangle - \langle A\rangle\langle B\rangle]. \tag{3.3}$$

By comparing with (2.20) it can be seen that this is not in general the same as the static adiabitic susceptibility: in the process described by the adiabatic susceptibility the occupation probabilities of levels remain unchanged, which is not the situation in the isothermal case. The difference between (3.3) and (2.20) is also related to points which will be discussed in section 8.

(ii) Example: magnetic susceptibilities

The susceptibility describing magnetic resonance is the response function for the magnetization $M_\mu(t)$ produced by a magnetic field $h_\nu(t)$. Since

$$H_1 = -M_\nu\, h_\nu(t) \tag{3.4}$$

the response function is, from (2.8), (2.18),

$$\phi_{\mu\nu}(t) = i\langle [M_\mu(t), M_\nu]\rangle = \int_0^\beta \langle \dot{M}_\nu(-i\lambda)M_\mu(t)\rangle\, d\lambda. \tag{3.5}$$

The static limit (2.20) of the resulting adiabatic susceptibility is

$$\chi_{\mu\nu}(0) = \int_0^\beta \langle (M_\nu(-i\lambda) - M_\nu^\circ)(M_\mu - M_\mu^\circ)\rangle\; d\lambda \tag{3.6}$$

while the static isothermal susceptibility (3.3) is

$$\chi^T_{\mu\nu}(0) = \int_0^\beta [\langle M_\nu(-i\lambda) M_\mu \rangle - \langle M_\nu \rangle \langle M_\mu \rangle]\, d\lambda . \qquad (3.7)$$

In the classical case, or when M_ν commutes with H, $M_\nu(-i\lambda) \to M_\nu$ and, for example,

$$\chi^T_{\mu\nu}(0) \to \frac{1}{K_B T} [\langle M_\nu M_\mu \rangle - \langle M_\nu \rangle \langle M_\mu \rangle] . \qquad (3.8)$$

4. CORRELATION FUNCTIONS AND GREEN FUNCTIONS [2]

(i) Correlation functions

The time-dependent correlation function between two operators A and B in the Heisenberg representation (2.9) is

$$\mathrm{Tr}\, \rho A(t) B(t') \equiv \langle A(t) B(t') \rangle = \langle A(t-t') B \rangle . \qquad (4.1)$$

The last expression follows from cyclic invariance of the trace provided the Hamiltonian commutes with ρ. Hereafter we consider only canonical or grand canonical cases, where

$$\begin{aligned} \rho &= e^{-\beta\mathcal{H}} / \mathrm{Tr}\, [e^{-\beta\mathcal{H}}] \\ \mathcal{H} &= H \text{ (canonical)} \\ &= H - \mu\hat{N} \text{ (grand canonical)}. \end{aligned} \qquad (4.2)$$

As well as giving the linear response functions, correlation functions between appropriate operators also give the usual thermodynamic averages of, for example, kinetic and potential energy, and occur in the form of pair distribution functions in the interpretation of X-ray and neutron scattering.

The correlation functions can in principle be obtained by integrating the Heisenberg equation of motion

$$i \frac{d}{dt} \langle A(t) B(t') \rangle = \langle [A(t), \mathcal{H}] B(t') \rangle \qquad (4.3)$$

and imposing the boundary condition

$$\langle A(t) B(t') \rangle = \langle B(t') A(t+i\beta) \rangle . \qquad (4.4)$$

(4.4) follows from a cyclic permutation inside the averaging trace.

It is usually simpler to consider, instead of the correlation functions, the related two-time Green functions, these satisfy a similar equation of motion but with an inhomogeneous term. The boundary conditions are then automatically incorporated through spectral theorems.

(ii) Green functions [2,3,18,19]

For any operators A, B, retarded and advanced Green functions are defined by

$$G_{\substack{r\\a}}(t,t') \equiv \langle\langle A(t)\,;\,B(t')\rangle\rangle_{\substack{r\\a}} \tag{4.5}$$

$$= \left\{\begin{matrix} -i\theta(t-t') \\ i\theta(t'-t) \end{matrix}\right\} \langle [A(t), B(t')]_\eta \rangle \, . \tag{4.6}$$

The theta function is as defined after (2.14): the upper theta function applies to the retarded Green function and the lower one to the advanced Green function. The average is over the appropriate ensemble (c.f. (4.2)), and the operators are Heisenberg operators as previously. One point requiring care is that either a commutator (denoted by η =-1) or an anticommutator (η = +1) of the Heisenberg operators A, B can be used in the definition: the symmetry of the problem often indicates that one will be more convenient than another. For example in fermion (boson) problems it is usually convenient to take the anticommutator (commutator), but there are exceptions, and spin problems are often but not exclusively, dealt with using the commutator.

By combining exp $(i\mathcal{H}t)$ and exp $(-i\mathcal{H}t')$ factors it follows that

$$G_{\substack{r\\a}}(t,t') = G_{\substack{r\\a}}(t-t'). \tag{4.7}$$

An important point is that, if the commutator is used $\langle\langle B(t);A\rangle\rangle_r$ is apart from a factor of (-2π) the "retarded commutator" whose fourier transform gives the generalised susceptibility of (2.14).

(iii) Green function equation of motion

Since the derivative of a theta function is a Dirac delta function, and the time derivative of a Heisenberg operator is its commutator with the Hamiltonian, the Green function equation of motion is (c.f. (4.3))

$$i\frac{d}{dt}\langle\langle A(t);B(t')\rangle\rangle = \delta(t-t')\langle[A,B]_\eta\rangle + \langle\langle[A(t),\mathcal{H}];B(t')\rangle\rangle. \tag{4.8}$$

The subscripts r, a have here been omitted since the same equation applies to both types of Green function. The right hand side consists of an inhomogeneous term and a term involving another, usually more complicated, double-time Green function. The equation of motion for this new Green function can then be written and the continuation of this process leads to a chain of equations (never-ending if the Hamiltonian contains interaction terms).

(iv) Spectral representation and connection between Green and correlation functions

The fourier transform $\mathcal{J}$ of the <BA> correlation function is defined by

$$\langle B(t')A(t)\rangle = \int_{-\infty}^{\infty} \mathcal{J}(\omega)\, e^{-i\omega\tau}\, d\omega, \qquad \tau \equiv t-t'. \tag{4.9}$$

Then, using (4.4),

$$\langle A(t)B(t')\rangle = \int_{-\infty}^{\infty} \mathcal{J}(\omega)\, e^{-i\omega\tau}\, e^{\beta\omega}\, d\omega. \tag{4.10}$$

Using the definition (4.6) and (4.9), (4.10), the fourier transform of the retarded Green function is therefore

$$\begin{aligned} G_r(E) &\equiv \frac{1}{2\pi}\int_{-\infty}^{\infty} G_r(\tau)\, e^{iE\tau}\, d\tau \\ &= \frac{1}{2\pi i}\int_{-\infty}^{\infty} e^{iE\tau}\, \theta(\tau)\, \langle A(t)B(t') + \eta\, B(t')A(t)\rangle\, d\tau \\ &= \frac{1}{2\pi i}\int_{-\infty}^{\infty} d\omega\, \mathcal{J}(\omega)\, (e^{\beta\omega}+\eta)\int_{-\infty}^{\infty} e^{iE\tau}\, e^{-i\omega\tau}\, \theta(\tau)\, d\tau. \end{aligned} \tag{4.11}$$

This can be reduced by using the integral representation for the theta-function:

$$\theta(\tau) = \frac{i}{2\pi}\int_{-\infty}^{\infty} \frac{e^{-ix\tau}}{x+i\varepsilon}\, dx \qquad \varepsilon \to 0^+ \tag{4.12}$$

(which can be verified by closing the contour of integration by a large semicircle in the lower half of the complex x-plane if $\tau > 0$, or in the upper half if $\tau < 0$: the pole at $x = -i\varepsilon$ in the lower half plane gives a contribution only if $\tau > 0$). Inserting into (4.11) the integral over τ can then be performed to yield $2\pi\delta(E-\omega-x)$. Carrying out the x-integral then leads to

$$G_r(E) = \frac{1}{2\pi}\int_{-\infty}^{\infty} d\omega\, \frac{\mathcal{J}(\omega)\,(e^{\beta\omega}+\eta)}{E-\omega+i\varepsilon}\;. \tag{4.13}$$

The corresponding result for the advanced Green function has $i\varepsilon$ replaced by $-i\varepsilon$. These results suggest the definition of the following function, where E is now regarded as a complex variable:

$$G(E) \equiv \frac{1}{2\pi}\int_{-\infty}^{\infty}(e^{\beta\omega}+\eta)\,\mathcal{J}(\omega)\,\frac{d\omega}{E-\omega} = \begin{cases} G_r(E) & \text{Im } E>0 \\ G_a(E) & \text{Im } E<0 \end{cases} \tag{4.14}$$

$\mathcal{J}(\omega)$ here plays the role of a spectral distribution function, and G(E) is the analytic function made by combining G_r and G_a through their analytic continuations into the upper and lower half planes, respectively. In general G(E) has a discontinuity across the real axis. The size of this discontinuity is (with E now real again)

$$G(E-i\varepsilon) - G(E+i\varepsilon) = i\mathcal{J}(E)\,(e^{\beta E}+\eta)\,, \tag{4.15}$$

which can be obtained from (4.14) with the use of

$$\frac{1}{E-\omega-i\varepsilon} - \frac{1}{E-\omega+i\varepsilon} = 2\pi i\,\delta(E-\omega). \tag{4.16}$$

(4.15) is a very important result: it is closely connected to the fluctuation-dissipation theorem to be discussed in §9; moreover it gives the correlation function, through its fourier transform $\mathcal{J}(E)$, once the Green function has been obtained, e.g. by solving its equation of motion (4.8) (G_r and G_a, and therefore G, share the same equation of motion). Examples of this procedure where the Green function equations can be easily solved are given in the first parts of Section 5.

The technique in more complicated cases is as follows: where the chain of Green function equations referred to in §4(iii) does not terminate, and cannot be exactly solved, the usual procedure is to uncouple the hierarchy of equations, which relate Green functions to higher order Green functions, by introducing some physically motivated approximation relating a higher order Green function to a lower order one. The Green functions can then be found in terms of the inhomogeneous functions of the equations. These inhomogeneous functions are in general correlation functions. These can be obtained, through the use of (4.15), from the Green functions, so that a set of implicit equations for the correlation functions results. This technique is illustrated in §5 (iv).

Finally, we recall that, if a commutator Green function is being used, the generalised susceptibility (2.14) is directly obtained from the retarded Green function:

$$\chi_{AB}(\omega) = -2\pi\, G_r(\omega) = -2\pi\, G(\omega+i\varepsilon)\ . \tag{4.17}$$

5. EXAMPLES

(i) Model calculation

The calculation of the Green function G = <<A;B>> and related correlation functions is straightforward when

$$[A, \mathcal{H}] = \Omega A \tag{5.1}$$

where Ω is a c-number. This is essentially a non-interacting situation of which particular cases are given below. With G(E) denoting the fourier transform of << A;B>>, and γ defined by

$$\gamma = \langle [A, B]_\eta \rangle \tag{5.2}$$

the fourier transform of the equation of motion (4.8), and its solution, are

$$EG = \frac{\gamma}{2\pi} + \Omega G \tag{5.3}$$

$$G(E) = \frac{(\gamma/2\pi)}{E - \Omega} . \tag{5.4}$$

(4.15) then yields, with the use of (4.16)

$$g(E) = \frac{\gamma\, \delta(E - \Omega)}{e^{\beta E} + \eta} . \tag{5.5}$$

The correlation function (4.9) is therefore

$$\langle B(t')A(t) \rangle = \frac{\gamma}{e^{\beta\Omega} + \eta}\, e^{-i\Omega(t - t')} . \tag{5.6}$$

The corresponding result for < A(t)B(t')> is obtained from (4.10) and has an extra factor of exp ($\beta\Omega$).

(ii) Bosons and fermions

As a particular case of the above results consider

$$A = a_K , \qquad B = a_{K'}^+ , \tag{5.7}$$

$$\mathcal{H} = \sum_K (\varepsilon_K - \mu)\, a_K^+ a_K \tag{5.8}$$

where a_K and $a_{K'}^+$, are boson annihilation and creation operators, and $\mathcal{H}$ is then the 'Hamiltonian' for a system of non-interacting bosons described by the grand canonical ensemble. (5.1) then follows, with

$$\Omega = (\varepsilon_K - \mu) . \tag{5.9}$$

Moreover, if the commutator (η =- 1) is used in the definition of the Green function ($\langle\langle a_K ; a_{K'}^+ \rangle\rangle$),

$$\gamma = \delta_{K,K'} . \tag{5.10}$$

Thus, from (5.4) and (5.6)

$$G = \frac{(1/2\pi)}{E - (\varepsilon_K - \mu)}\, \delta_{KK'} , \tag{5.11}$$

$$\langle a_\kappa^+ a_\kappa \rangle = \frac{1}{e^{\beta(\varepsilon_\kappa - \mu)} - 1} \quad . \tag{5.12}$$

(5.12) is the usual result for the boson occupation number: the Bose statistics are implicit in the commutator relationship between a and a^+ which gave rise to (5.10).

The non-interacting Fermi system can be treated analogously using the anticommutator form of Green function ($\eta = +1$). The Fermi distribution function is recovered, and the Green function is again given by (5.11).

The density of states for both systems can therefore be written as

$$\rho(E) \equiv \sum_\kappa \delta(E - (\varepsilon_\kappa - \mu)) = -\frac{1}{\pi} \, \mathrm{Im} \sum_{\kappa\kappa'} G(E + i\varepsilon). \tag{5.13}$$

An expression of this form applies generally for non-interacting systems, and can in principle be evaluated in any representation, not just the above κ-representation in which G was diagonal: this is important for disorder problems[20].

(5.11) illustrates another important point: the (unperturbed) Green function has poles at the single particle energies. The generalisation of this result for interacting systems will be discussed in §7.

(iii) Paramagnets

The spin-$\frac{1}{2}$ paramagnet with Hamiltonian

$$\mathcal{H} = -\sum_i h S_i^z \tag{5.14}$$

(where S_i^z is the z-component of the ith spin) can also be treated using the results of §5(i). The spin commutation relations are

$$[S_i^+, S_j^z] = \delta_{ij} S_i^+ \tag{5.15}$$

$$[S_i^+, S_j^-] = \delta_{ij} S_i^z \tag{5.16}$$

$$\{S_i^+, S_i^-\}_+ = 1 \ . \tag{5.17}$$

Thus taking

$$A = S_i^+ , \quad B = S_i^- \tag{5.18}$$

and using (5.15) to evaluate the commutator of S_i^+ with the Hamiltonian (5.14), the result (5.1) again applies with

$$\Omega = h \ . \tag{5.19}$$

One may then use the commutator form of Green function, together with (5.6),(5.2) and (5.16), to arrive at

$$\langle S_i^- S_i^+ \rangle = \frac{\langle S_i^z \rangle}{e^{\beta h} - 1} \quad , \quad \langle S_i^+ S_i^- \rangle = \frac{\langle S_i^z \rangle e^{\beta h}}{e^{\beta h} - 1} \tag{5.20}$$

or the anticommutator form, with (5.17), to obtain

$$\langle S_i^- S_i^+ \rangle = \frac{1}{e^{\beta h} + 1} \quad , \quad \langle S_i^+ S_i^- \rangle = \frac{e^{\beta h}}{e^{\beta h} + 1} \quad . \tag{5.21}$$

These results together imply

$$\langle S_i^z \rangle = \tanh \tfrac{1}{2} \beta h \tag{5.22}$$

(which could alternatively have been obtained by adding (5.17) to (5.20), or (5.16) to (5.21)).

This example, and those in the preceding subsections, were clearly non-interacting, which is the reason why higher order Green functions did not appear in the equations of motion. The results could have been obtained by using elementary matrix methods in a basis diagonalising $\mathcal{H}$.

(iv) Interactions in boson, fermion or spin systems

If a two-particle boson (or fermion) interaction term of the form

$$\Sigma\, v\, a^+ a^+ a a \tag{5.23}$$

is added to the Hamiltonian (5.8), the equation of motion for $\langle\langle a, a^+ \rangle\rangle$ will produce a higher order Green function of the type $\langle\langle a^+ a a ; a^+ \rangle\rangle$. This is a case where an infinite hierarchy of equations would be produced and more sophisticated considerations would be needed, for example the approximate uncouplings referred to earlier.

The spin-$\frac{1}{2}$ Heisenberg ferromagnet is an example of a spin system in which interactions lead to similar complications. The Hamiltonian is

$$\mathcal{H} = -\sum_i h S_i^z - \tfrac{1}{2} \sum_{ij} v_{ij}\, \underline{S}_i \cdot \underline{S}_j \tag{5.24}$$

where v is the exchange interaction and the ith spin is at lattice site $\underline{r}_i$. With

$$A = S_q^+ \equiv \sum_i S_i^+ e^{i\underline{q}\cdot\underline{r}_i} \quad , \quad B = S_{q'}^- \quad , \tag{5.25}$$

the Green function equation of motion involves, from $\langle\langle [A,\mathcal{H}]; B \rangle\rangle$, a higher order Green function of the form $\langle\langle S^+ S^z ; S^- \rangle\rangle$. This arises from the exchange interaction which makes the system of equations insoluble. But using the so-called "random-phase" approximation[21,22]

$$\langle\langle S^+ S^z ; S^- \rangle\rangle \sim \langle S^z \rangle \langle\langle S^+ ; S^- \rangle\rangle \tag{5.26}$$

the equation of motion uncouples and becomes of the form (5.3) with

$$\Omega = \varepsilon_q \equiv h + [v(0) - v(q)] \langle S^z \rangle \tag{5.27}$$

where v(q) is the fourier transform of the exchange interaction v_{ij}. The generalised spin wave energy (5.27) then appears as the pole of the Green functions. The use of (for example) the commutator Green function $\langle\langle S_q^+ ; S_q^- \rangle\rangle$ then yields

$$\langle S_q^- S_q^+ \rangle = \frac{\langle S^z \rangle}{e^{\beta \varepsilon_q} - 1} \quad , \quad \langle S_q^+ S_q^- \rangle = \frac{\langle S^z \rangle e^{\beta \varepsilon_q}}{e^{\beta \varepsilon_q} - 1} \tag{5.28}$$

and imposing the sum rule (5.17) gives the "random phase" equation of state

$$1 = \sum_q (\langle S_q^- S_q^+ \rangle + \langle S_q^+ S_q^- \rangle) = \sum_q \langle S^z \rangle \coth \tfrac{1}{2} \beta \varepsilon_q \tag{5.29}$$

from which $\langle S^z \rangle$ can be obtained as a function of field and temperature.

(v) Hall coefficient for interacting electron system [10,23]

This section gives an example of an interacting system for which a response coefficient can be obtained exactly using the Green function method.

The Hamiltonian for a system of N electrons interacting only with each other, in the presence of a uniform magnetic field, is

$$\mathcal{H} = \sum_{i=1}^{N} \frac{m}{2e^2} \underline{j}_i^2 + U \, . \tag{5.30}$$

$\underline{j}_i$ is the current operator of the ith electron in the absence of interactions, and the electron electron interaction is taken to depend only on the separation of the interacting pair:

$$\underline{j}_i = \frac{e}{m} (\underline{p}_i - \frac{e}{c} \underline{A} (\underline{r}_i)) \quad , \tag{5.31}$$

$$U = \sum_{ij} u(r_i - r_j) \, . \tag{5.32}$$

The field $\underline{H}$ = curl $\underline{A}$ is taken to be in the z-direction.

Because V is a function only of coordinates the total current operator is unchanged by the interaction

$$\underline{J} = \sum_i \underline{j}_i \, . \tag{5.33}$$

Also, since the scatterings represented by U conserve total momentum, $\underline{J}$ commutes with U so that the equation of motion of $\underline{J}$ is the same as when U=0. Hence the combinations

$$J_{\pm} = J_x \pm i J_y \tag{5.34}$$

evolve harmonically with frequency $\pm\,\omega_c$:

$$[J_{\pm}, \mathcal{H}] = \pm \omega_c J_{\pm} \quad , \qquad \omega_c \equiv \frac{eH}{mc} . \tag{5.35}$$

The transverse component $\sigma_{xy}(\omega)$ of the conductivity is, from (2.26), given by

$$\sigma_{xy}(\omega) = \chi_{J_x, eY}(\omega) = \frac{e}{2}\left(\chi_{J_+, Y}(\omega) + \chi_{J_-, Y}(\omega)\right) \tag{5.36}$$

where Y is the y-component of the total position operator $\Sigma \underline{r}_i$ for the system. But from (4.17) the generalised susceptibility $\chi_{J_{\pm}, Y}$ is given by the retarded commutator Green function $G_{\pm}(\omega + i\varepsilon)$, where

$$G_{\pm} \equiv \langle\langle J_{\pm} ; Y \rangle\rangle. \tag{5.37}$$

To obtain the transverse conductivity it is therefore appropriate to take

$$A = J_{\pm} \quad , \qquad B = Y . \tag{5.38}$$

Then

$$[J_{\pm}, Y] = \pm \frac{Ne}{m} \tag{5.39}$$

and so, from (5.35), (5.39), the results of §5(i) again apply, with

$$\Omega = \pm \omega_c \quad , \qquad \gamma = \pm \frac{Ne}{m} . \tag{5.40}$$

Thus using (5.4),

$$\chi_{J_{\pm}, Y}(\omega) = -2\pi G_{\pm}(\omega + i\varepsilon) = \frac{\mp \frac{Ne}{m}}{\omega \mp \omega_c + i\varepsilon} . \tag{5.41}$$

A suitable experiment will therefore detect a resonance at the cyclotron frequency ω_c for the non-interacting system[24]. And the Hall coefficient is, from (5.36), (5.41),

$$[H\sigma_{xy}(0)]^{-1} = (1/Nec) \quad , \tag{5.42}$$

also unaffected by the interaction[23] (and also independent of statistics).

6. THERMAL GREEN FUNCTIONS

Another type of Green function commonly used is the thermal Green function[25,26]. It is actually closely related to the two-time Green functions described above, despite the fact that it is defined apparently quite differently, and calculated using quite separate methods.

The thermal Green function is usually of the form

$$\begin{aligned}\tilde{G}(\tau) &= \langle \tilde{A}(\tau) B \rangle \qquad \tau > 0 \\ &= -\eta \langle B \tilde{A}(\tau) \rangle \qquad \tau \leq 0\end{aligned} \tag{6.1}$$

where $\eta = \pm 1$, whichever is the more convenient choice, and

$$\tilde{A}(\tau) = e^{\mathcal{H}\tau} A \, e^{-\mathcal{H}\tau}. \tag{6.2}$$

τ is essentially an inverse temperature, rather than a time, and for purely thermodynamic results need only lie in $[-\beta,\beta]$. $\tilde{G}(\tau)$ has, for $\tau < \alpha \tau + \beta$, the 'periodicity' property (c.f. (4.4))

$$\tilde{G}(\tau+\beta) = -\eta \, \tilde{G}(\tau). \tag{6.3}$$

It can therefore be written as a fourier series

$$\tilde{G}(\tau) = -\frac{1}{\beta}\sum_{\ell} \tilde{G}(i\omega_\ell)\, e^{-i\omega_\ell \tau}, \qquad \begin{aligned}\omega_\ell &= 2\ell\pi/\beta \quad &\eta = 1 \\ &= (2\ell+1)\pi/\beta \quad &\eta = -1.\end{aligned} \tag{6.4}$$

where the fourier coefficient $\tilde{G}(i\omega_\ell)$ is

$$\tilde{G}(i\omega_\ell) = -\int_0^\beta d\tau \, \tilde{G}(\tau)\, e^{i\omega_\ell \tau}. \tag{6.5}$$

An analytic function $\tilde{G}(\xi)$ can be constructed which coincides with the fourier coefficients at $\xi = i\omega_\ell$ and falls off like $|\xi|^{-1}$ at infinity. By evaluating the following integral it can be shown that[26]

$$\frac{1}{2\pi}\int_{-\infty \pm i0^+}^{\infty \pm i0^+} \tilde{G}(\xi)\, e^{-i\xi t}\, d\xi = G_{\substack{r\\a}}(t) \tag{6.6}$$

so that

$$2\pi G_{\substack{r\\a}}(\omega) = \tilde{G}(\omega \pm i0^+), \tag{6.7}$$

which shows that $\tilde{G}(\xi)$ is the same as the analytic function $G(E)$ introduced in §4(iv). Thus, in addition to thermodynamic properties, response coefficients and dynamical quantities can be obtained from the thermal Green functions (by analytically continuing the fourier coefficients), just as thermodynamic properties as well as time-dependent ones are given by the two-time Green functions. The techniques used to obtain the two types of Green functions are different: as has been described, the two-time Green functions are usually found from their equations of motion. The thermal Green functions are usually obtained by diagrammatic methods, well known for boson and fermion systems[25,26,27] but generalisable also to spin systems[28,29,30].

7. NORMAL MODE FREQUENCIES; QUASI PARTICLES AND THEIR LIFE TIMES

(i) Energy representation

For a discussion of excitation frequencies, it is convenient to have the following development of the spectral function $\mathcal{J}(\omega)$ and the Green function G(E) in terms of exact energy eigenstates defined by

$$\mathcal{H}|\alpha\rangle = E_\alpha|\alpha\rangle. \tag{7.1}$$

Using (4.9)

$$\begin{aligned}\mathcal{J}(\omega) &= \frac{1}{2\pi}\int_{-\infty}^{\infty} e^{i\omega t}\langle B\,A(t)\rangle\,dt \\ &= \frac{1}{2\pi}\int_{-\infty}^{\infty} e^{i\omega t}\sum_{\alpha\alpha'}(e^{-\beta E_\alpha}/Q)\langle\alpha|B|\alpha'\rangle\langle\alpha'|A|\alpha\rangle e^{it(E_{\alpha'}-E_\alpha)} \\ &= \sum_{\alpha\alpha'}(e^{-\beta E_\alpha}/Q)\langle\alpha|B|\alpha'\rangle\langle\alpha'|A|\alpha\rangle\,\delta(E_{\alpha'}-E_\alpha+\omega)\end{aligned} \tag{7.2}$$

where Q is the partition function. Therefore, inserting (7.2) into (4.14),

$$G(E) = \frac{1}{2\pi}\sum_{\alpha\alpha'}\frac{1}{Q}\left(e^{-\beta E_{\alpha'}}+\eta\, e^{-\beta E_\alpha}\right)\frac{\langle\alpha|B|\alpha'\rangle\langle\alpha'|A|\alpha\rangle}{E-(E_\alpha-E_{\alpha'})}\,. \tag{7.3}$$

A useful corollary of these results is that, if $A = B^+$, $\mathcal{J}(\omega)$ is real and $(G(E))^* = G(E^*)$. It follows that for this case (4.14) implies dispersion relations for $G(\omega \pm i\varepsilon)$ and that (4.15) becomes

$$\frac{1}{2}(e^{\beta\omega}+\eta)\mathcal{J}(\omega) = \mathrm{Im}\; G(\omega - i\varepsilon). \tag{7.4}$$

(ii) Normal mode frequencies

In the examples considered in §5(ii) the unperturbed "single particle" Green functions were seen to have poles at the excitation energies. Reference to (7.3) shows that in the general case the Green function G(E) has poles at the excitation frequencies

$E_\alpha - E_{\alpha'}$ corresponding to the energy difference of exact eigenstates linked by the operators A, B. For the case $A = B^+ = a$, these states have particle numbers differing by 1 so that the energy differences in that case correspond to single particle excitation energies.

It can be instructive to consider this from a "response" point of view: from (4.17), the fourier transform $G_r(\omega)$ of the commutator Green function $\langle\langle A; B\rangle\rangle_r$ (or equivalently $G_r(\omega+i\varepsilon)$) is the generalised susceptibility $\chi_{AB}(\omega)$ giving the linear influence of a perturbation of the form $- BF_o \cos\omega t \; e^{\varepsilon t}$ on the average value of A. If the forcing frequency ω coincides with a normal mode frequency of the system, resonant response can be expected, i.e. $\chi_{AB}(\omega)$ should diverge. That corresponds to a pole in $G_r(\omega)$ (and likewise in its analytic generalisation G(E)) at the normal mode frequency. The pole should occur only so long as B couples to the normal mode in question, and A has some content relating to that mode.

When A and B are density operators the Green function $\langle\langle A;B\rangle\rangle$ will have poles at the frequencies of normal modes (sound waves) corresponding to density fluctuations. Similarly, for a charged gas the charge density - charge density Green function will have a pole at the plasma frequency. As response theory would suggest, the latter Green function is directly proportional to $(\varepsilon(q,\omega)^{-1} - 1)$ where $\varepsilon(q,\omega)$ is the frequency and wave vector dependent dielectric constant, so that the plasma mode alternatively appears as a zero in $\varepsilon(q,\omega)$. The density and charge density operators only link states with the same particle number, so the sound and plasma frequencies would not be expected to occur as poles in the single particle propagator $\langle\langle a;a^+\rangle\rangle$.

Normal modes which have occurred in previous examples are the spin wave (as the pole of $\langle\langle S^+; S^-\rangle\rangle$) and the single particle mode. The forcing terms which give rise to these are respectively a rotating transverse magnetic field (which couples to S^-) or a particle-conservation - breaking field in a forcing Hamiltonian of the form $\ldots a^+ \cos\omega t$.

(iii) Quasiparticle lifetimes

Consider, for definiteness the single particle Green function $G_s(E) \equiv \langle\langle a_s ; a_s^+ \rangle\rangle$ where s is a convenient label, for example wave vector. When written in the form (7.3), a_s and a_s^+ will in general link many $|\alpha\rangle$ to many $|\alpha'\rangle$. Thus many energy differences $E_\alpha - E_{\alpha'}$ will in general occur, and when the sum over α and α' is carried out the resulting Green function will have non-vanishing response for a 'band' of energies centred on $E = \Delta_s$, say, and with width Γ_s. If the band happens to be Lorentzian (not usually the case)

$$G_s(E) \propto \frac{1}{E-(\Delta_s + i\Gamma_s)} \tag{7.5}$$

where the spectral properties require that

$$\Gamma_s \gtrless 0 \quad , \quad \mathrm{Im}\, E \lessgtr 0 \, . \tag{7.6}$$

In this simple case the resulting time-dependent retarded Green function is proportional to[18,31]

$$e^{-i\Delta_s \tau - |\Gamma_s|\tau} \, . \tag{7.7}$$

The time $\tau_s = 1/|\Gamma_s|$ can be interpreted as the 'lifetime' of quasi particles labelled by s. Near the non-interacting limit we would expect τ_s to be large. That this is so can be seen from the fact that near this limit (i.e. when $\mathcal{H}$ approaches the form (5.8)) $E_\alpha - E_{\alpha'} = \varepsilon_s$ has the greatest weight of the O(N) possible values of the energy difference, and a narrow band centred around $\Delta_s = \varepsilon_s$ results.

8. SUM RULES; ZERO FREQUENCY COMPONENTS

(i) Sum rules

The following sum rule can be obtained from (2.13) and (2.8)

$$\frac{1}{2\pi}\int_{-\infty}^{\infty} d\omega\, \chi_{BA}(\omega) = \phi_{BA}(0) = i\langle[B,A]\rangle \, . \tag{8.1}$$

Thus for the particular case of electrical conductivity (§2(iv)),

$$\frac{1}{2\pi}\int_{-\infty}^{\infty} d\omega\, \sigma_{\mu\nu}(\omega) = ie\langle[j_\mu, x_\nu]\rangle = \frac{Ne^2}{m}\delta_{\mu\nu} \tag{8.2}$$

for a system with N charge carriers of (true) mass m and charge e.

This type of statement can be extended to the moments of χ since from (2.13), (2.8)

$$\frac{1}{2\pi}\int_{-\infty}^{\infty} d\omega\, \omega^n \chi_{BA}(\omega) = \left(i\frac{\partial}{\partial t}\right)^n \phi_{BA}(t)\Big|_{t=0}$$
$$= i\langle[[\mathcal{H},[\mathcal{H},\ldots[\mathcal{H},B]\ldots]],A]\rangle \, . \tag{8.3}$$

For the system of carriers in a magnetic field along the z-direction, with Hamiltonian (5.30),

$$e[[\mathcal{H},J_x],Y] = ie\omega_c\langle[J_y, Y]\rangle = \omega_c Ne^2/m \tag{8.4}$$

so that

$$\frac{1}{2\pi}\int_{-\infty}^{\infty} d\omega \ \omega \ \sigma_{xy}(\omega) = i\,\omega_c\, Ne^2/m \ , \tag{8.5}$$

as could have been obtained from (5.41). The multiple commutators involved in higher moments can also be obtained from (5.35) for this particular case, but in other types of systems with interactions would usually be intractable.

A similar type of discussion applies to the correlation functions. From (4.9),

$$\int_{-\infty}^{\infty} d\omega \ \omega^n \ \mathcal{J}(\omega) = \langle B[\mathcal{H},[\ldots[\mathcal{H},A]\ldots]]\rangle . \tag{8.6}$$

The case n = 1 is important in for example neutron scattering[32] where the appropriate correlation function involves the fourier components ρ_q of the density operator which, for a free gas of N nuclei of mass M has the property that

$$\langle \rho_q[\mathcal{H},\rho_{-q}]\rangle = \tfrac{1}{2}\langle[\rho_q,[\mathcal{H},\rho_{-q}]]\rangle = \frac{Nq^2}{2M} \ . \tag{8.7}$$

For the Green functions G(E), expansions in inverse powers of E are often useful[33,26]. The coefficients in these expansions can be obtained from the preceding discussion by expanding (4.14) and then using (8.6), (4.10), or alternatively by expanding (7.3). Perhaps the most direct method is however to use the fourier transformed chain of Green function equations of motion beginning with (4.8). The result is

$$G(E) = \frac{1}{2\pi}\frac{\langle[A,B]_\eta\rangle}{E} + \frac{1}{2\pi}\frac{\langle[[A,\mathcal{H}],B]_\eta\rangle}{E^2} + \cdots \ . \tag{8.8}$$

(ii) Green and correlation functions at ω = o

The zero-frequency component $\mathcal{J}(o)$ of the fourier transformed correlation function is not determined[34] by the Green function equation of motion method using commutator Green functions because of the factor $(e^{\beta\omega} - 1)$ in (4.15).

It has been suggested[35] that $\mathcal{J}(o)$ should be chosen so that the correlation function factors for infinite time separation:

$$\lim_{t\to\infty} \langle B(t)A\rangle = \langle B\rangle\langle A\rangle \ . \tag{8.9}$$

Care has to be exercised in the use of this since it requires there to be irreversibility (no Poincaré cycling) which would not be the case for isolated systems. Indeed, the difference between isolated and isothermal behaviour[1] can be associated with the zero frequency part of the Green or correlation function[36].

Anticommutator Green functions do not suffer from the problem regarding $\mathcal{G}(o)$, but their use is not always convenient. $\mathcal{G}(o)$ can also be recovered from the thermal Green function.

A controlled method of dealing with $\mathcal{G}(o)$ and G(o) within the commutator Green function method is to introduce a symmetry-breaking term into the Hamiltonian in such a way that the operator B does not link degenerate eigenstates of the new Hamiltonian[37]. Then (c.f. (7.2)) $\mathcal{G}(o)$ is zero and if the value of $\mathcal{G}(\omega)$ at finite ω is calculated, its limit as the symmetry-breaking term is turned off includes the correct behaviour at $\omega = o$. Other methods of approaching the problem have been discussed[38,39].

The problem is not an academic one since the zero frequency modes often have physical significance. One example where this is so is in structural phase transitions of for example the Jahn-Teller type[40,41] where in some circumstances the 'soft mode' of the problem corresponds to transitions between degenerate electronic states[42].

9. FLUCTUATION-DISSIPATION THEOREM

(i) Fluctuation-dissipation theorem

The response of a system to an external perturbation is related by the fluctuation-dissipation theorem to the fluctuation properties of the system[1,43-47]. The theorem relates the dissipative part of the generalised susceptibility $\chi_{BA}(\omega)$ to fluctuations which, as will be shown, are contained in the following fourier transform of the symmetrised correlation function:

$$\mathcal{F}_{AB}(\omega) = \frac{1}{2\pi}\int_{-\infty}^{\infty} dt\; e^{i\omega t} \langle \{A(o)\,B(t)\}\rangle \tag{9.1}$$

where $\{AB\} \equiv \frac{1}{2}(AB+BA)$. By (4.9), (4.10),

$$\mathcal{F}_{AB}(\omega) = \frac{1}{2}\,\mathcal{G}(\omega)\,(1+e^{\beta\omega}). \tag{9.2}$$

The theorem in its general form can be obtained as follows from some previous results. From the relation (4.17) between the generalised susceptibility and the retarded commutator Green function, (4.14) allows us to write

$$\chi_{BA}(E) = -\int_{-\infty}^{\infty} d\omega\, \frac{(e^{\beta\omega}-1)\,\mathcal{G}(\omega)}{E-\omega+i\varepsilon} = -2\int_{-\infty}^{\infty} d\omega\, \frac{\mathcal{F}_{AB}(\omega)\tanh\frac{\beta\omega}{2}}{E-\omega+i\varepsilon}. \tag{9.3}$$

Where $\mathcal{G}(\omega)$ is real, which is for instance the case when $B^+ = A$ (§7(i)), $\mathcal{F}(\omega)$ is also real and then

$$\mathrm{Im}\,\chi_{BA}(\omega) = 2\pi\,\mathcal{F}_{AB}(\omega)\tanh\frac{\beta\omega}{2}\,. \qquad (9.4)$$

If $\mathcal{G}(\omega)$ is purely imaginary the corresponding result is

$$\mathrm{Re}\,\chi_{BA}(\omega) = 2\pi i\,\mathcal{F}_{AB}(\omega)\tanh\frac{\beta\omega}{2}\,. \qquad (9.5)$$

(9.4), (9.5) are the statements of the fluctuation-dissipation theorem for the two usual cases. The symmetric combination $\mathcal{F}$ is not necessarily the most convenient one in all cases: in for example the theory of Raman scattering[47,48] it is more convenient to keep to $\mathcal{G}$.

A particularly important case is when A and B are the same Hermitian operator. Then (9.4) applies, $\langle\{A(o)\,A(t)\}\rangle$ is real, and $\mathcal{F}$ is the "power spectrum"[49]

$$\mathcal{F}_{AA}(\omega) = \tfrac{1}{2}\langle |A(\omega)|^2\rangle. \qquad (9.6)$$

This expression, which involves the mean square deviation of the fourier components of the variable A, emphasizes the relationship of $\mathcal{F}$ to the fluctuations in A. In this case the dissipative part of the response is

$$\mathrm{Im}\,\chi_{AA}(\omega) = \pi\tanh\frac{\beta\omega}{2}\,\langle |A(\omega)|^2\rangle. \qquad (9.7)$$

An example of this is Nyquist's relationship[43] between the frequency-dependent conductivity and the fourier components of the fluctuating current

$$\sigma_{\mu\mu}(\omega) = \frac{2\pi}{\omega}\tanh\frac{\beta\omega}{2}\,\langle |J_\mu(\omega)|^2\rangle. \qquad (9.8)$$

(Here the extra factor of ω arises because $\sigma_{\mu\nu} = \chi_{j_\mu,\,ex_\nu}$ is related to χ_{j_μ,j_ν} by a factor ω, since $e\dot{x}_\nu = j_\nu$).

(ii) Dissipation rate

To illustrate the dissipative aspects, consider the rate at which energy is absorbed ("dissipated") from an oscillating field $F(t) \equiv F_o \cos\omega t$ which couples to the system through an interaction Hamiltonian $-\,AF(t)$. The rate of energy absorption is[45]

$$\frac{\partial\langle\mathcal{H}\rangle}{\partial t} = \left\langle\frac{\partial\mathcal{H}}{\partial t}\right\rangle = -\langle A\rangle\frac{\partial F}{\partial t} \qquad (9.9)$$

where the first equality is the usual statement for adiabatic processes. From (2.12),

$$\langle A\rangle = \mathrm{Re}\,[\chi_{AA}(\omega)\,F_o\,e^{i\omega t}]\,. \qquad (9.10)$$

Thus, when averaged over the period $2\pi/\omega$ of the external field, the dissipation rate is

$$\frac{\partial\langle\mathcal{H}\rangle}{\partial t} = \frac{1}{2}\,\omega F_o^2 \;\mathrm{Im}\; \chi_{AA}(\omega). \tag{9.11}$$

This argument shows that, in this case ($B = A = A^+$), where (9.4) applies, Im χ_{AA} can been identified as the dissipative part of the generalised susceptibility.

The fluctuation dissipation theorem (9.4) for this case can be obtained without the use of (4.17). For an alternative form for the rate of energy dissipation is

$$\frac{\partial\langle\mathcal{H}\rangle}{\partial t} = \frac{1}{Q}\sum_{\alpha\alpha'} e^{-\beta E_\alpha}\; W_{\alpha\alpha'}\,(E_\alpha - E_{\alpha'}) \tag{9.12}$$

where the transition rate $W_{\alpha\alpha'}$ is, from perturbation theory,

$$W_{\alpha\alpha'} = \frac{\pi}{2} F_o^2\, |\langle\alpha|A|\alpha'\rangle|^2\, [\delta(\omega+E_\alpha-E_{\alpha'})+\delta(\omega+E_{\alpha'}-E_\alpha)]. \tag{9.13}$$

Thus using (7.2),

$$\frac{\partial\langle\mathcal{H}\rangle}{\partial t} = \frac{\pi\omega}{2}\, F_o^2\, (e^{\beta\omega}-1)\, g_{AA}(\omega) \tag{9.14}$$

so that

$$\mathrm{Im}\;\chi_{AA}(\omega) = \pi\,(e^{\beta\omega}-1)\, g_{AA}(\omega). \tag{9.15}$$

With (9.2), this leads at once to the appropriate form of (9.4).

One important context in which the theorem arises is in connection with differential scattering cross sections, which will be treated in the next section.

10. PHYSICAL QUANTITIES

(i) Response functions and equilibrium thermal properties

The preceding sections show how to construct the Green functions and how to obtain directly from them the response functions (generalised susceptibilities), and the correlation functions which give the isothermal susceptibilities and all other ordinary equilibrium thermodynamic properties.

That correlation functions also give the differential scattering cross sections is not so obvious, and is treated in the remainder of this section.

(ii) Neutron scattering[32,50-54]

The probability of a transition from an initial state corresponding to neutron with wave vector K, spin σ, energy E_K and target system in state $|\alpha\rangle$ with energy E_α, to a final state specified by K', σ', $E_{K'}$, $|\alpha'\rangle, E_{\alpha'}$ is, according to Fermi's golden rule, proportional to

$$K' \, |\langle K'\alpha'\sigma'|V|K\alpha\sigma\rangle|^2 \, \delta(\omega + E_\alpha - E_{\alpha'}) \tag{10.1}$$

where V is the interaction potential between the incident neutron and the target system, $\omega = E_K - E_{K'}$, and the K' factor comes from the densities of final (neutron) states per unit energy range. To arrive at the partial differential scattering cross section for all scattering processes from E to E' it is necessary to divide by the incident flux (proportional to K), sum over all final states, and average over all initial states, associating probability distributions p_α with the initial target states and p_σ with the polarisation of the incident neutron. The result is

$$\frac{d^2\sigma}{d\Omega dE'} = \frac{K'}{K}\left(\frac{m}{2\pi}\right)^2 \sum_{\alpha\sigma} p_\alpha p_\sigma \sum_{\alpha'\sigma'} |\langle K'\alpha'\sigma'|V|K\alpha\sigma\rangle|^2 \delta(\omega + E_\alpha - E_{\alpha'}) \tag{10.2}$$

which is similar in form to the last expression in (7.2).

For simplicity we consider hereafter only unpolarised neutrons (dropping p_σ and the σ, σ' labels).

For purely nuclear scattering the potential seen by a neutron at $\underline{r}$ is a sum of contributions from the individual nuclei at positions $\underline{R}_j$:

$$V(\underline{r}) = (2\pi/m) \sum_j V_j(\underline{r} - \underline{R}_j). \tag{10.3}$$

The interaction V_j in general has range very much less than the wavelength of thermal neutrons, so that V_j can be replaced by the Fermi pseudo-potential $b_j \, \delta(\underline{r} - \underline{R}_j)$ corresponding to s-wave scattering.

(10.2) can be simplified by completing the integrals over neutron coordinates in the matrix elements of V:

$$\langle K'\alpha'|V|K\alpha\rangle = \langle\alpha'|\sum_j V_j(\underline{q}) e^{i\underline{q}\cdot\underline{R}_j}|\alpha\rangle , \quad \underline{q} \equiv \underline{K} - \underline{K}' \tag{10.4}$$

where $V_j(\underline{q})$ is the Fourier transform of $V_j(\underline{r})$. It is then convenient to write

$$\delta(\omega+E_\alpha-E_{\alpha'}) = \frac{1}{2\pi}\int_{-\infty}^{\infty} dt\; e^{-it(\omega+E_\alpha-E_{\alpha'})} \tag{10.5}$$

and incorporate the $\exp(-itE_\alpha)$, $\exp(itE_{\alpha'})$ factors so introduced as Heisenberg time evolution operators $\exp(it\mathcal{H})$, $\exp(-it\mathcal{H})$ on either side of the operator $\sum_{j'} V_{j'} \exp(-i\underline{q}.\underline{R}_{j'})$ inside the second matrix element of the square modulus. The sum $\sum_{\alpha'}$ can then be carried out by closure with the result

$$\frac{d^2\sigma}{d\Omega dE'} = \frac{K'}{K}\frac{1}{2\pi}\int_{-\infty}^{\infty} dt\; e^{-i\omega t}\sum_{\alpha jj'} p_\alpha V_j^+(q)V_{j'}(q)\langle\alpha|e^{-i\underline{q}.\underline{R}_j(0)}e^{i\underline{q}.\underline{R}_{j'}(t)}|\alpha\rangle. \tag{10.6}$$

$\sum_\alpha p_\alpha \ldots$ now involves an average over nuclear spin orientations and isotope distributions of V^+V, which we denote by $\overline{V_j^+V_{j'}}$, and a thermal average over initial target states. Thus

$$\frac{d^2\sigma}{d\Omega dE'} = \frac{1}{2\pi}\frac{K'}{K}\int_{-\infty}^{\infty} dt\; e^{-i\omega t}\sum_{jj'}\overline{V_j^+(q)V_{j'}(q)}\,\langle e^{-i\underline{q}.\underline{R}_j(0)}\, e^{i\underline{q}.\underline{R}_{j'}(t)}\rangle \tag{10.7}$$

where

$$\langle\cdots\rangle = \frac{\mathrm{Tr}\, e^{-\beta\mathcal{H}}\cdots}{\mathrm{Tr}\, e^{-\beta\mathcal{H}}}, \tag{10.8}$$

$\mathcal{H}$ being the target system Hamiltonian. (10.7) exhibits the relationship of the cross section to a time-dependent correlation function.

The steps taken to get from (10.2) to (10.7) are completely analogous to the reverse of those used in developing $\mathcal{G}(\omega)$ in (7.2). Thus by virtue of the relationship (7.4) of $\mathcal{G}(\omega)$ to the commutator Green function,

$$\frac{d^2\sigma}{d\Omega dE'} = \frac{K'}{K}\left(\frac{m}{2\pi}\right)^2\frac{2}{e^{\beta\omega}-1}\quad \overline{\mathrm{Im}\; G(\omega-i\varepsilon)} \tag{10.9}$$

where $G = \langle\langle V, V^+\rangle\rangle$. This corresponds to a statement of the fluctuation-dissipation theorem in which the dissipated energy related to $d^2\sigma/d\Omega dE'$ is that lost by the incident particle (the "differential retardation"[32]).

Writing

$$\overline{|V_j^+V_{j'}|} = |\overline{V}|^2 + \delta_{jj'}\left[\,\overline{|V|^2} - |\overline{V}|^2\right] \tag{10.10}$$

leads to the separation of the cross-section into coherent and incoherent parts:

$$\left(\frac{d^2\sigma}{d\Omega dE'}\right)_{coh.} = N\frac{K'}{K}|\overline{V}|^2 S(q,\omega) \tag{10.11}$$

$$\left(\frac{d^2\sigma}{d\Omega dE'}\right)_{incoh.} = N\frac{K'}{K}[\overline{|V|^2} - |\overline{V}|^2]S_i(q,\omega) \tag{10.12}$$

where $S(q,\omega)$ and $S_i(q,\omega)$ are the frequency transforms of the following thermally averaged time-dependent correlation functions

$$S(q,t) = \sum_{jj'} \langle e^{-i\underline{q}\cdot\underline{R}_j(0)}\, e^{i\underline{q}\cdot\underline{R}_{j'}(t)}\rangle = \langle \rho_q(0)\rho_{-q}(t)\rangle \tag{10.13}$$

$$S_i(q,t) = \sum_j \langle e^{-i\underline{q}\cdot\underline{R}_j(0)}\, e^{i\underline{q}\cdot\underline{R}_j(t)}\rangle . \tag{10.14}$$

Both of these functions are determined solely by the properties of the target system. In the second expression for $S(q,t)$, $\rho_q(t)$ is the fourier transform of the target system density operator

$$\rho(\underline{r},t) = \sum_j \delta(\underline{r} - \underline{R}_j(t)) \tag{10.15}$$

so that $S(q,t)$ is the density-density correlation function; that results from the fact that the neutron's interaction with the system, when averaged over nuclear spin and isotope distributions, is a coupling to the target system density, through

$$\overline{\sum_j V_j(\underline{r}-\underline{R}_j)} = \int d^3\underline{r}'\, \rho(\underline{r}',0)\, \overline{V}(\underline{r}-\underline{r}'). \tag{10.16}$$

The fourier transforms of $S(q,t)$, $S_i(q,t)$ are respectively the pair correlation function $S(\underline{r},t)$ and the self pair correlation function $S_i(\underline{r},t)$. For example

$$S(\underline{r},t) = \frac{1}{N}\int d^3\underline{r}'\, \langle \rho(\underline{r}-\underline{r}',0)\,\rho(\underline{r}',t)\rangle . \tag{10.17}$$

Its imaginary part is

$$\mathrm{Im}\, S(\underline{r},t) = \frac{1}{2N}\int d^3\underline{r}'\, \langle [\rho(\underline{r}',t),\rho(\underline{r}'-\underline{r},0)]\rangle \tag{10.18}$$

which, comparing with (2.8), has the immediate interpretation[51] as the linear response function ϕ measuring the disturbance produced by the neutron in the density of the target system, and therefore contains poles at the frequencies of the corresponding collective modes. The above results apply equally to solids, liquids, or gases.

In a classical approximation in which the non-commutativity of $\underline{R}_j(o)$, $\underline{R}_{j'}(t)$ is ignored,

$$S(\underline{r},t) \sim \frac{1}{N}\sum_{jj'} \langle \delta(\underline{r} - (\underline{R}_{j'}(t) - \underline{R}_j(0)) \rangle$$

$$S_i(\underline{r},t) \sim \frac{1}{N}\sum_{j} \langle \delta(\underline{r} - (\underline{R}_j(t) - \underline{R}_j(0)) \rangle . \qquad (10.19)$$

These pair correlation functions can be interpreted respectively as the probability that a particle is to be found a vector distance $\underline{r}$ from the position of any or the same particle a time t earlier.

Returning to the general case (not necessarily classical), because $S(q,\omega)$ satisfies sum rules of the type discussed in §8, the leading moments ($\int_{-\infty}^{\infty} \omega^n S(q,\omega)d\omega$) of the scattering law can often be evaluated exactly. An example is the case n = 1 for a gas of free nuclei, for which the result is provided by (8.7).

As $t \to \infty$, subject to the qualifications expressed below (8.9),

$$\langle e^{-i\underline{q}\cdot\underline{R}_j(0)} e^{i\underline{q}\cdot\underline{R}_{j'}(t)} \rangle \to \langle e^{-i\underline{q}\cdot\underline{R}_j(0)} \rangle \langle e^{i\underline{q}\cdot\underline{R}_{j'}(0)} \rangle . \qquad (10.20)$$

Thus $S(q,t)$ can be separated into such a product of averages, and the remainder; these yield respectively the coherent elastic cross section (from a resulting contribution to $S(q,\omega)$ proportional to $\delta(\omega)$) and the coherent inelastic cross section. $S_i(q,t)$ can be similarly treated, giving rise to the incoherent elastic and inelastic cross sections. In what follows we consider only the inelastic cross sections, which can give information about the dynamics of the target system.

(iii) Inelastic scattering of neutrons by phonons [32,55]

In most crystals the actual nuclear position is close to the equilibrium position, and it is convenient to write $\underline{R}_\ell$ in terms of the displacements $\underline{u}(\ell)$ from the equilibrium:

$$\underline{R}_\ell = \underline{\ell} + \underline{u}(\ell) . \qquad (10.21)$$

The inelastic cross sections therefore involve averages of the form

$$\langle e^A e^B\rangle - \langle e^A\rangle\langle e^B\rangle \;, \quad A = -i\underline{q}.\underline{u}(\ell,0) \;, \; B = i\underline{q}.\underline{u}(\ell',t) \tag{10.22}$$

($\ell = \ell'$ for the incoherent case). In the harmonic approximation, the commutator of A and B is a c-number and the Bloch identity[56] applies:

$$\langle e^{A+B}\rangle = e^{\frac{1}{2}\langle (A+B)^2\rangle} . \tag{10.23}$$

Thus

$$\begin{aligned}\langle e^A e^B\rangle - \langle e^A\rangle\langle e^B\rangle &= e^{\frac{1}{2}\langle (A+B)^2\rangle + \frac{1}{2}[A,B]} - e^{\frac{1}{2}\langle A^2\rangle + \frac{1}{2}\langle B^2\rangle} \\ &= e^{-2W(q)}\left[e^{\langle (\underline{q}.\underline{u}(\ell,0))(\underline{q}.\underline{u}(\ell',t))\rangle} - 1 \right]\end{aligned} \tag{10.24}$$

where $e^{-W(q)}$ $(= e^{\frac{1}{2}\langle A^2\rangle} = e^{\frac{1}{2}\langle B^2\rangle})$ is the Debye-Waller factor. Expanding the exponential to first non-vanishing order in the $\langle u\ u\rangle$ correlation function gives the 'one-phonon' approximation for the cross sections

$$\left(\frac{d^2\sigma}{d\Omega dE'}\right)^{inel.}_{coh.} = \frac{Nk'}{k}|\bar{V}|^2 \frac{1}{2\pi}\int dt\, e^{-i\omega t} \sum_{\ell\ell'} e^{i\underline{q}.(\underline{\ell}'-\underline{\ell})} e^{-2W(q)} \langle (\underline{q}.\underline{u}(\ell,0))(\underline{q}.\underline{u}(\ell',t))\rangle . \tag{10.25}$$

$$\left(\frac{d^2\sigma}{d\Omega dE'}\right)^{inel.}_{incoh.} = \frac{Nk'}{k}\left[\overline{|V|^2} - |\bar{V}|^2\right] \frac{1}{2\pi}\int dt\, e^{-i\omega t} \sum_{\ell} e^{-2W(q)} \langle (\underline{q}.\underline{u}(\ell,0))(\underline{q}.\underline{u}(\ell,t))\rangle . \tag{10.26}$$

If the $\langle u\ u\rangle$ correlation functions are evaluated in the harmonic approximation and the time integration and lattice sums performed, the resulting coherent inelastic cross section involves a sum over k of terms each corresponding to creation or absorption of a free phonon of wave vector k and frequency ω_k and exhibiting energy and wave vector conservation (up to a reciprocal lattice vector τ) through a factor of the form

$$\{ n(k)\,\delta(\omega+\omega_k)\,\delta(q+k-\tau) + (n(k)+1)\,\delta(\omega-\omega_k)\,\delta(q-k-\tau)\} . \tag{10.27}$$

The corresponding incoherent inelastic cross-section has no momentum conservation δ-functions, and so cannot give such detailed information about the dynamics of the target system as the coherent cross-section, which can give the energies of individual modes labelled by a particular wave-vector. In an imperfect crystal, where q is not a good quantum number, even the coherent inelastic cross-section will be continuous in frequency.

For liquids, the coherent scattering, which depends on the relative motion of the nuclei, can show up collective excitations: for small $\underline{q}$, ω, it can be described by the hydrodynamic approaches[52,53] briefly referred to in §11.

(iv) Inelastic magnetic scattering of neutrons [32]

Where scattering is caused by the magnetic interaction of the neutron with unpaired electrons, the result (10.2) again applies, but now V involves the neutron-electron interaction resulting from the electron's motion and its dipole moment. Both can be written in terms of an effective spin operator and to the extent to which the electrons are localised at ionic sites $\underline{R}_\ell$, the differential cross section will be of the form[32]

$$\frac{d^2\sigma}{d\Omega dE'} = \frac{K'}{K}\sum_{\alpha\alpha'} \sum_{\substack{\mu\nu \\ =x,y,z}} F_{\mu\nu}(q)\, P_\alpha \sum_{\ell\ell'} \langle\alpha| e^{-i\underline{q}\cdot\underline{R}_\ell} S_\ell^\mu |\alpha'\rangle \langle\alpha'| e^{i\underline{q}\cdot\underline{R}_{\ell'}} S_{\ell'}^\nu |\alpha\rangle\, \delta(\omega + E_\alpha - E_{\alpha'}). \tag{10.28}$$

S_ℓ^μ is the μ-component of the effective spin at site $\underline{R}_\ell$ and the state labels α,α' include a specification of the state of each effective spin. $F_{\mu\nu}(q)$ involves the square of an ion form factor, a dipole tensor, and the square of a product of nuclear and Bohr magnetons. Using its integral representation (10.5) the δ-function can again be written in terms of Heisenberg operators and the sums over states α,α' carried out. This gives rise to a correlation function again related only to the target system:

$$\begin{aligned} C_{\ell\ell'}^{\mu\nu}(q,t) &\equiv \langle e^{-i\underline{q}\cdot\underline{R}_\ell(0)} S_\ell^\mu(0)\, e^{i\underline{q}\cdot\underline{R}_{\ell'}(t)} S_{\ell'}^\nu(t)\rangle \\ &\sim \langle e^{-i\underline{q}\cdot\underline{R}_\ell(0)} e^{i\underline{q}\cdot\underline{R}_{\ell'}(t)}\rangle \langle S_\ell^\mu(0)\, S_{\ell'}^\nu(t)\rangle \\ &\equiv S_{\ell\ell'}(q,t)\ \Gamma_{\ell\ell'}^{\mu\nu}(t) \end{aligned} \tag{10.29}$$

where the factorization into spin and nuclear correlation functions

Γ and S can be made because the spins only very weakly affect the nuclear motion. Various types of elastic and inelastic scattering can be separated by taking out the $t \to \infty$ part of either or both correlation functions. The scattering contribution coming from

$$S_{\ell\ell'}(q,\infty) = e^{i\underline{q}.(\underline{\ell}'-\underline{\ell})}\, e^{-2W(q)} \tag{10.30}$$

is elastic in the phonon system but contains both elastic and inelastic spin contributions:

$$\left(\frac{d^2\sigma}{d\Omega dE'}\right)_{mag.} = \frac{K'}{K}\sum_{\mu\nu} F_{\mu\nu}(q)\, e^{-2W(q)} \frac{1}{2\pi}\int_{-\infty}^{\infty} dt\, e^{-i\omega t}\, \Gamma_q^{\mu\nu}(t) \tag{10.31}$$

where

$$\Gamma_q^{\mu\nu}(t) = \langle S_q^{\mu}(0)\, S_{-q}^{\nu}(t)\rangle \tag{10.32}$$

and

$$S_q^{\mu} = \sum_{\ell} e^{i\underline{q}.\underline{\ell}}\, S_{\ell}^{\mu}\,. \tag{10.33}$$

When the total z-component of the spin is a constant of the motion (e.g. in the Heisenberg model),

$$\Gamma_q^{\mu\nu} = \Gamma_q^{\mu\mu}\, \delta_{\mu\nu} \tag{10.34}$$

and the cross section (10.31) becomes a sum of longitudinal terms, from Γ^{zz}, and transverse terms, from $\Gamma^{xx} = \Gamma^{yy}$. At low temperature these correlation functions can be evaluated in, for example, a linear spin wave approximation to yield from the transverse terms a one-magnon inelastic cross-section similar in form and interpretation to the one-phonon coherent inelastic cross section (10.27). The first inelastic contributions to the longitudinal cross section are two-magnon terms.

For itinerant magnetic systems a result similar to (10.31) again applies, with $\Gamma_q^{\mu\nu}$ now involving the fourier components $\mathcal{s}_q$ of the itinerant electron spin. Even at low temperatures this correlation function is difficult to treat and is not discussed further here.

(v) Electron scattering [52]

The same arguments as were used to derive the neutron scattering cross-section clearly apply to electron scattering from a system having a charge distribution. In that case the interaction is of the form (10.3) with $V_j(q) \propto ee_j/q^2$, and incorporating the charge factors e_j into the correlation function gives a coherent differential cross-section proportional to $(K'/K)\ (e^2/q^4)\ \tilde{S}\ (q,\omega)$ where $\tilde{S}(q,\ \omega)$ is the charge density correlation function.

(vi) Electromagnetic scattering

Scattering of electromagnetic radiation arises in neutral systems from the polarization fluctuations produced in the atomic charge distribution by the incident photon. The interaction of the electromagnetic wave with the system may be represented by

$$H_1 = \sum_{\mu\nu} \sum_{\underline{r},\underline{r}'} \mathcal{E}^{\mu}(\underline{r})\ P^{\mu\nu}(\underline{r},\underline{r}')\ \mathcal{E}'^{\nu}(\underline{r}')$$

where $P^{\mu\nu}$ is a polarizability and $\underline{\mathcal{E}}$, $\underline{\mathcal{E}}'$ denote the electric fields of the incident and scattered photon. The differential cross-section for electromagnetic scattering is the electromagnetic energy scattered into $d\Omega\ dE'$ per unit time divided by the incident power flow ($\propto c|\mathcal{E}|^2$). Similarly to (10.2) this cross-section is proportional to

$$\sum_{\alpha\alpha'} P_\alpha \frac{|\langle K'\alpha'|H_1|K\alpha\rangle|^2}{|\mathcal{E}|^2\ |\mathcal{E}'|^2}\ \delta(\omega + E_\alpha - E_{\alpha'}) \tag{10.35}$$

where as before α,α' label the states of the scattering system and the photon is scattered with wave vector change $\underline{q} = \underline{K} - \underline{K}'$ and energy change ω. It follows from (10.35) that

$$\frac{d^2\sigma}{d\Omega dE'} \propto \sum_{\substack{\mu\nu\\ \gamma\delta}} \frac{\mathcal{E}^{\mu}\mathcal{E}'^{\nu}\mathcal{E}^{\gamma}\mathcal{E}'^{\delta}}{|\mathcal{E}|^2\ |\mathcal{E}'|^2} \int_{-\infty}^{\infty} dt\ e^{-i\omega t} \langle P^{\mu\nu}(K,K';0)\, P^{\gamma\delta *}(K,K';t)\rangle \tag{10.36}$$

where

$$P(K,K') = \int e^{i\underline{K}\cdot\underline{r} - i\underline{K}'\cdot\underline{r}'}\ P(\underline{r},\underline{r}')\ d^3\underline{r}\ d^3\underline{r}' \tag{10.37}$$

and $P(K,K';\ t)$ is again a Heisenberg operator. In this case the scattering is given by the frequency transform of the polarization-polarization correlation function, and can be written in terms of the imaginary part of the retarded commutator Green function $\langle\langle V;V^+\rangle\rangle$ where

$$V = \sum_{\mu\nu} \mathcal{E}^{\mu}\mathcal{E}'^{\nu}\ P^{\mu\nu}(K,K')\ . \tag{10.38}$$

For the case of X-rays, the energy differences $E_\alpha - E_{\alpha'}$

associated with low-lying collective modes of the system are, for given K,K', usually very much smaller than the X-ray energies and therefore unresolvable. One must then, in effect, integrate (10.36) over all final energies, which yields a cross-section $d\sigma/d\Omega$ proportional to the instantaneous correlation function $\langle P^{\mu\nu}(K,K';0)\, P^{\gamma\delta *}(K,K';0)\rangle$. Thus X-ray scattering measures the structure, rather than the dynamics, of the system.

(vii) Light scattering [47,57]

For light scattering, laser techniques provide the resolution required to probe typical collective mode energies. K,K' are effectively zero compared to typical inverse atomic spacings so that for Raman or Brillouin inelastic scattering the cross-section involves

$$\int dt\, e^{-i\omega t} \langle P^{\mu\nu}(0,0;0)\, P^{\gamma\delta *}(0,0;t)\rangle. \tag{10.39}$$

A correlation function of a similar type is also sufficient to describe infrared absorption.

To the extent to which the polarizability is a local quantity depending on the density,

$$\begin{aligned} P(r,r') &= \sum_j c_j\, \delta(r-R_j)\, \delta(r'-R_j) \\ P(K,K') &= \sum_j c_j\, e^{i\underline{q}\cdot\underline{R}_j} \end{aligned} \quad , \tag{10.40}$$

the photon scattering cross-section involves the same density - density correlation function as in nuclear neutron scattering. Light scattering, like neutron scattering, can probe other features, such as for example electronic levels or any spin dynamics the scattering system may have [58-60]. Leaving aside electronic excitations, in general the polarization P involves contributions from phonon and possibly spin wave sources. Suppose Q_j is some coordinate (e.g. a lattice displacement u_j, or a spin operator S_j) associated with site j and on which the polarization depends. Making the expansion

$$P^{\mu\nu}(0,0) = P_0^{\mu\nu}(0,0) + \sum_j P_1^{\mu\nu\lambda} Q_j^\lambda + \sum_{jj'} P_2^{\mu\nu\lambda\delta} Q_j^\lambda Q_{j'}^\delta + \dots \tag{10.41}$$

and retaining only linear terms gives a response proportional to

$$\sum_{jj'} \int dt\, e^{-i\omega t} \langle Q_j(0)\, Q_{j'}(t)\rangle. \tag{10.42}$$

This response measures the phonon displacement-displacement or the spin-spin correlation function. For the phonon case this is the analogue of the one-phonon term in the neutron cross section and will result in response at the frequencies corresponding to phonon creation (the Stokes component) or absorption (the Anti-Stokes component), the former dominating at low temperatures because of the thermal factors.

Any multiphonon terms arise from the non-linear terms in (10.41), i.e. from non-linearities in the distortion of the ions when displaced in the crystal. For neutron scattering, multi-phonon terms arise in any case from the expansion of the exponentials in the correlation functions (see for example (10.24)) so the multiphonon terms might be expected to be less important in the light scattering case. There are however situations where they are very important in, for example, Raman scattering: in some (perfect) crystals, e.g. those with O_h symmetry, the symmetry implies zero wave-vector selection rules which make the first order (one-phonon) Raman scattering forbidden; the observed scattering then arises from the second order term in (10.41). Similar considerations apply for the spin wave scattering. In the two cases phonon and spin wave interaction effects can show up very directly in the observed spectra, which in the non-interacting case could be simply analysed as a convolution of single phonon (or spin wave) bands[61,62].

The discussion in this and the preceding subsection was for the case where the electromagnetic field interacted with the system through the induced polarization. In the case of metals the appropriate Hamiltonian for the effect of an electromagnetic probe is

$$H_1 = \underline{J}\cdot\underline{A} \qquad (10.43)$$

where $\underline{A}$ is the vector potential and $\underline{J}$ the total current operator of the system. A correlation function like $\langle J^{\mu}(K,K';0)\, J^{\nu *}(KK';t)\rangle$ then arises (c.f.§2(iv)) but otherwise the treatment is similar to that above.

11. CONCLUDING REMARKS

(i) Response to gradients in temperature, chemical potential or velocity fields

One important situation which has not been discussed above is the response to perturbations not obviously expressible in terms of an interaction Hamiltonian. Examples include gradients in temperature and in chemical potential, where the resulting flow of heat, charge or mass is related through linear response coefficients such as thermal conductivity, and thermoelectric and diffusion coefficients.

Kubo formulae have been constructed for perturbations of this type[63-67] by using a local Hamiltonian density H(r) and a local number density operator $\hat{N}$ (r). To the extent in which the perturbed system can be described by an equilibrium grand canonical distribution with density function

$$\rho = \exp\left[-\int d^3\underline{r}\ \beta(r)\left[H(r)-\mu(r)\hat{N}(r)\right]\right], \tag{11.1}$$

the effective interaction Hamiltonian is

$$H_1 = \int d^3\underline{r}\left\{\frac{\delta\beta(r)}{\beta}\left[H(r)-\mu\hat{N}(r)\right] - \delta\mu(r)\hat{N}(r)\right\} \tag{11.2}$$

where $\delta\beta(r)$ is the deviation of $\beta(r)$ from its mean β, etc. Since $-\dot{H}$ and $-\dot{\hat{N}}$ are the divergences of the energy current density and number current density the thermal conductivity coefficient is, by (2.19), related to the energy current density correlation function and the diffusion coefficient to the mass current density correlation function.

Difficulties inherent in the above approach[4] can be avoided for certain types of response to thermal or chemical potential gradients. For example, the charge current response to a thermal gradient is the inverse process to the thermal current produced by an electric field. The latter can therefore be obtained directly from (2.19) using the same interaction Hamiltonian as in §2(iv) ($\dot{A} = \underline{j}$), and the thermal current operator for B. A second example is mass diffusion which can arise from the term $\delta\mu(r)$ above, or equivalently from the effect of external (e.g. gravitational or centrifugal) fields. A Kubo formula for mass diffusion can therefore be derived using mechanical forces[68,69].

Viscosity is related to the pressure change resulting from velocity gradients. This also can be treated by using a local Hamiltonian density: since[45] a system with fixed velocity $\underline{V}$ has an additional term $\underline{V}.\ \underline{P}$ in the Hamiltonian, where $\underline{P}$ is the total momentum operator for the system, an impressed velocity field $V(\underline{r})$ simply results in an interaction Hamiltonian of the form

$$\int d^3\underline{r}\ \ \underline{V}(r).\underline{P}(r) \tag{11.3}$$

where $\underline{P}$ (r) is the local momentum operator. Alternatively, the flow can be accommodated through boundary conditions (moving walls) on the system[68].

(ii) Macroscopic approaches

In these lectures a microscopic view of dynamical response has been adopted. However the more important aspects are also normally incorporated in macroscopic approaches: in particular the idea of resonant response (§7(ii)) at the normal mode frequencies is also central in the macroscopic discussions.

The usual phenomenological methods often applied for the description of, for example, fluids, make use of conservation laws and constitutive relations. The macroscopic dynamics is then obtained since the conservation laws relate densities (of number, momentum and energy) to the corresponding fluxes and the constitutive relations relate the fluxes and stress tensor to the velocity and temperature gradients through the linear response coefficients (compressibility, viscosity, thermal conductivity, etc.)[52,53,70]. Any external mechanical force enters into the momentum density conservation law and hence affects the resulting behaviour of the system. In particular the resulting density change will be large when the frequency of the applied force coincides with, for example, the adiabatic sound wave frequency if the dissipative parameters (viscosity, thermal conductivity,...) are neglected. In the low frequency (hydrodynamic) regime where it is necessary to include these, modes like the thermal diffusion mode then also appear, again as the poles of the response function. The hydrodynamic approach implies limiting forms for the frequency and wave-vector dependence of the macroscopic response functions[53,70].

Critical dynamics can also be treated by theories of the above type using mode-coupling[71,72] or renormalisation group techniques[73].

(iii) Non-linear response

The usual non-linear dissipative effects (e.g. Ohmic heating in the case of electrical conduction which is of second order in the electric field) are normally related directly through the fluctuation dissipation theorem to linear response. In general the response itself may involve second and higher order terms in the forcing field. In the electrical conduction case the terms of order $\mathcal{E}^2$ are normally very small but there are situations where they cannot be ignored. An extreme case is a system without scattering, in zero magnetic field, where all coefficients in the expansion of the current in powers of $\mathcal{E}$ diverge. This sort of structure always occurs when the flux produced commutes with the system Hamiltonian, as it does in the above case or, for example, as the total spin components transverse to the magnetic field do in an isotropic Heisenberg magnet, causing the long wave-length transverse susceptibility to diverge.

The polarisability of metals, which is essentially σ/ω, is divergent at zero frequency even where σ is not. This is analogous to the conductivity of superconductors, in which the linear response coefficient giving the ratio of 'London' current to vector potential is finite[74] although the conductivity is not.

One of the most obvious breakdowns of linear response is shown in the behaviour of the isothermal magnetic susceptibility near a magnetic phase transition: the zero field susceptibility diverges like[75]

$$\chi \propto |T-T_c|^{-\gamma} \quad , \quad \gamma > 0 \tag{11.4}$$

near the transition temperature T_C so that the response is intrinsically non-linear at T_C. Again the behaviour can be clarified by retaining frequency-dependences. Divergent susceptibilities may always occur if fluctuations are unbounded, and this is not limited to the neighbourhood of phase transitions: another example of a divergent susceptibility is the low field, low temperature, longitudinal isothermal susceptibility of Heisenberg ferromagnets in spin wave theory, which is proportional[76] to $Th^{-\frac{1}{2}}$ when $h \ll K_BT$ and thus diverges in the zero field limit $h \to o$. In this case an expansion of the response in integral powers of the forcing field is not applicable.

These examples have been given to show that care is sometimes needed in identifying the appropriate response coefficient. Some approach may need to be adopted in which such special cases can be identified (they are usually obvious on physical grounds, or certain types can be identified by considering whether or not $[\mathcal{H}, B] = 0$ where B is the operator for the response). Subject to this, the methods of response theory, linear and non-linear,[77] have extremely wide application.

REFERENCES

1. R. Kubo, 1957, J. Phys. Soc. Japan 12, 570.

2. D.N. Zubarev, 1960, Soviet Phys. - Uspekhi 3, 320.

3. V.L. Bonch-Bruevich and S.V. Tyablikov, 1962, The Green function method in Statistical Mechanics (North-Holland, Amsterdam).

4. G.V. Chester, 1963, Rept. Prog. Phys. 26, 411.

5. R. Kubo, 1956, Canad. J. Phys. 34, 1274.

6. D.A. Greenwood, 1958, Proc. Phys. Soc. 71, 585.

7. P.G. Harper, J.W. Hodby, and R.A. Stradling, 1973, Rept. Prog. Phys. 36, 1.

8. R. Kubo, H. Hasegawa, and N. Hashitsume, 1959, J. Phys. Soc. Japan 14, 56.

9. R.J. Palmer, 1970, D. Phil. Thesis, Oxford University.

10. R.B. Stinchcombe, 1974, Lecture Notes in Physics: Transport Phenomena, ed. G. Kirczenow and J. Marro (Springer, Berlin) 368.

11. S.F. Edwards, 1958, Phil. Mag. 3, 1020.

12. G.V. Chester and A. Thellung, 1959, Proc. Phys. Soc. 73, 745.

13. D.J. Thouless, 1975, Phil. Mag. 32, 877.

14. S.F. Edwards, 1965, Proc. Phys. Soc. 86, 977.

15. N. Szabo, 1972, J. Phys. C5, L241; 1973, J. Phys. C6, L437.

16. R. Kubo, 1974, Lecture Notes in Physics: Transport Phenomena, ed. G. Kirczenow and J. Marro (Springer, Berlin) 74.

17. M. Huberman and G.V. Chester, 1975, Adv. Phys. 24, 489.

18. V.M. Galitskii and A.B. Migdal, 1958, Sov. Phys. - JETP 7, 96.

19. P.C. Martin and J. Schwinger, 1959, Phys. Rev. 115, 1342.

20. R.J. Elliott, J.A. Krumhansl and P.L. Leath, 1974, Rev. Mod. Phys. 46, 465.

21. N.N. Bogoliubov and S.V. Tyablikov, 1959, Soviet Phys. - Doklady 4, 589.

22. F. Englert, 1960, Phys. Rev. Lett. 5, 102.

23. R.B. Stinchcombe, 1974, J. Phys. C7, 4277.

24. W. Kohn, 1961, Phys. Rev. 123, 1242.

25. W.E. Parry and R.E. Turner, 1964, Rept. Prog. Phys. 27, 23.

26. W.E. Parry, 1973, The Many-Body Problem (Clarendon Press, Oxford).

27. A.A. Abrikosov, L.P. Gor'kov, and I.E. Dzyaloshinskii, 1963, Methods of Quantum Field Theory in Statistical Physics (Prentice-Hall, Englewood Cliffs).

28. R.B. Stinchcombe, G. Horwitz, F. Englert and R.Brout, 1963, Phys. Rev. 130, 155.

29. V.G. Vaks, A.l. Larkin, and S.A. Pikin, 1968, Soviet Phys.- JETP 26, 188, 647.

30. H.J. Spencer, 1968, Phys. Rev. 167, 434.

31. A.L. Fetter and J.D. Walecka, 1971, Quantum Theory of Many-Particle Systems (McGraw Hill, New York).

32. W. Marshall and S.W. Lovesey, 1971, Theory of Thermal Neutron Scattering (Clarendon Press, Oxford).

33, R.D. Puff, 1965, Phys. Rev. 137A, 406.

34. K.W.H. Stevens and G.A. Toombs, 1965, Proc. Phys. Soc. 85, 1307.

35. H.B. Callen, R.H. Swendsen, and R.A. Tahir-Kheli, 1967, Phys. Lett. 25A, 505.

36. P.C. Kwok and T.P. Schultz, 1969, J. Phys. C2, 1196.

37. J.F. Fernandez and H.A. Gersch, 1967, Proc. Phys. Soc. 91, 505.

38. G.L. Lucas and G. Horwitz, 1969, J. Phys. A2, 503.

39. J.G. Ramos and A.A. Gomes, 1971, Il Nuovo Cimento 3A, 441.

40. R.J. Elliott, R.T. Harley, W. Hayes and S.R.P. Smith, 1972, Proc. Roy. Soc. A328, 217.

41. R.B. Stinchcombe, 1977, Electron-Phonon Interactions and Phase Transitions, ed. T. Riste (Plenum, New York).

42. A.P. Young, 1975, J. Phys. C8, 3158.

43. H. Nyquist, 1928, Phys. Rev. 32, 110.

44. H.B. Callen and T.A. Welton, 1951, Phys. Rev. 83, 34.

45. L.D. Landau and E.M. Lifshitz, 1969, Statistical Physics (Pergamon, London) 2nd Edition.

46. H.J. Benson and D.L. Mills, 1970, Phys. Rev. B1, 4835.

47. A.S. Barker, Jr., and R. Loudon, 1972, Rev. Mod. Phys. 44, 18.

48. P.N. Butcher and N.R. Ogg, 1965, Proc. Phys. Soc. 86, 699.

49. D.K.C. MacDonald, 1962, Noise and Fluctuations (Wiley, New York).

50. L. Van Hove, 1954, Phys. Rev. 95, 249.

51. L. Van Hove, 1958, Physica's Grav. 24, 404.

52. P. Martin, 1968, Problème à N Corps (Many-Body Physics), ed. C. De Witt and R. Balian (Gordon and Breach, New York),37.

53. R. Puff, 1969, Lectures in Theoretical Physics 11B, ed. K.T. Mahanthappa and W.E. Brittin (Gordon and Breach, New York) 297.

54. R.A. Cowley, 1968, Rept. Prog. Phys. 31, 123.

55. A.A. Maradudin, E.W. Montroll and G.H. Weiss, 1963, Theory of Lattice Dynamics in the Harmonic Approximation (Academic Press, New York).

56. N.D. Mermin, 1966, J. Math. Phys. 7, 1038.

57. P.A. Fleury, 1972, Comments on Solid St. Phys. 4, 167.

58. R.L. Wadseck, J.L. Lewis, B.E. Argyle and R.K. Chang, 1971, Phys. Rev. B3, 4342.

59. P.A. Fleury and R. Loudon, 1968, Phys. Rev. 166, 574.

60. D.L. Mills, R.F. Wallis, and E. Burstein, 1971, Proc. Int. Conf. Light Scattering in Solids, ed. M. Balkanski (Flammarion, Paris), 107.

61. R.J. Elliott, M.F. Thorpe, G.F. Imbusch, R. Loudon and J.B. Parkinson, 1968, Phys. Rev. Lett. 21, 147.

62. R.J. Elliott and M.F. Thorpe, 1969, J. Phys. C2, 1630.

63. H. Mori, 1956, J. Phys. Soc. Japan 11, 1029; 1958, Phys. Rev. 112, 1829; 1959, Phys. Rev. 115, 298.

64. D.N. Zubarev, 1961, Soviet Phys. - Doklady, 6, 776.

65. R. Kubo, 1966, Rept. Prog. Phys. 29, 255.

66. K. Kawasaki, 1963, Prog. Theor. Phys. 29, 80.

67. J.K. Flicker and P.L. Leath, 1973, Phys. Rev. B7, 2296.

68. E.W. Montroll, 1959, Rendiconti della Scuola Instituto di Fisica "Enrico Fermi", Varenna.

69. E.W. Montroll, 1961, Lectures in Theoretical Physics, 3, ed. W.E. Brittin, B.W. Downs and J. Downs (Interscience, New York), 221.

70. J.R.D. Copley and S.W. Lovesey, 1975, Rept. Prog. Phys. 38, 461.

71. K. Kawasaki, 1976, Phase Transitions and Critical Phenomena 5A, ed. C. Domb and M.S. Green (Academic Press, London), 166.

72. Y. Pomeau and P. Résibois, 1975, Physics Reports 19C, 64.

73. P.C. Hohenberg and B.I. Halperin, 1977, Rev. Mod. Phys. 49, to appear; G.F. Mazenko, this volume.

74. F. London, 1950, Superfluids (Wiley, New York).

75. M.E. Fisher, 1967, Rept. Prog. Phys. 30, 615.

76. D.M. Edwards and E.P. Wohlfarth, 1968, Proc. Roy. Soc. A303, 127.

77. I. Oppenheim, this volume.

PERTURBATION THEORY OF RESPONSE FUNCTIONS

J. W. Halley

University of Minnesota

Minneapolis, Minnesota 55455

1. Introduction

The chapter by Dr. Stinchcombe has established connections between functions defined as

$$G_r(t,t') = \langle\langle A(t);\, B(t')\rangle\rangle_r$$

$$= -i\Theta(t-t')\,\langle\left[A(t),\, B(t')\right]_\eta\rangle$$

and scattering cross-sections and susceptibilities observed directly in experiments on condensed matter. For example, the coherent part of the neutron scattering cross-section for inelastic scattering from a monatomic liquid or solid can be written as

$$\frac{d^2\sigma}{d\Omega dE'} \propto S(\underset{\sim}{k},\omega) \equiv \frac{1}{2\pi}\int e^{-i\omega t} \langle \rho_{\underset{\sim}{k}}\rho_{-\underset{\sim}{k}}(t)\rangle dt$$

$$= \frac{-2\,\mathrm{Im}\;G_r(\omega)}{(1-e^{-\omega\beta})} \tag{1-1}$$

where

$$G(\omega) = \langle\langle\rho_{\underset{\sim}{k}}\,;\,\rho_{-\underset{\sim}{k}}\rangle\rangle_r(\omega) \tag{1-2}$$

and

$$\rho_{\underset{\sim}{k}} = \sum_i e^{-i\underset{\sim}{k}\cdot\underset{\sim}{x}_i} \tag{1-3}$$

Writing G as

$$G_r(\omega) = \frac{1}{2\pi}\int_{-\infty}^{+\infty} e^{i\omega t} \langle\langle \sum_i e^{-i\underset{\sim}{k}\cdot\underset{\sim}{r}_i(t)} ; \sum_{i'} e^{i\underset{\sim}{k}\cdot\underset{\sim}{r}_{i'}(0)}\rangle\rangle_r dt \tag{1-4}$$

one sees that the neutron scattering experiment involves two space points and two time points. On the other hand, turning to a second quantized representation for a system of fermions or bosons we have

$$\rho_{\underset{\sim}{k}} = \sum_{\underset{\sim}{q},\lambda} c^{\dagger}_{q'\lambda}\, c_{\underset{\sim}{k}+\underset{\sim}{q},\lambda} \tag{1-5}$$

where λ is a spin index so that the neutron scattering cross-section involves

$$G_r(\omega) = \frac{1}{2\pi}\int_{-\infty}^{\infty} e^{i\omega t} \sum_{\substack{\underset{\sim}{q},\lambda \\ \underset{\sim}{q};\lambda'}} \langle\langle c^{\dagger}_{\underset{\sim}{q}\lambda}(t)\, c_{\underset{\sim}{q}+\underset{\sim}{k}\lambda}(t) ;$$

$$c^{\dagger}_{\underset{\sim}{q}'\lambda'}(0)\, c_{\underset{\sim}{q}'-\underset{\sim}{k},\lambda'}(0)\rangle\rangle_r \tag{1-6}$$

and the cross-section is proportional to a correlation function involving four particle destruction and annihilation operators. Still another way to look at this cross-section is obtained if, in the case of a solid, one expands the definition

$$\rho_{\underset{\sim}{k}} = \sum_{\ell,b} e^{-i\underset{\sim}{k}\cdot\underset{\sim}{x}_{\ell b}} \tag{1-7}$$

about the lattice points $\underset{\sim}{x}^{(0)}_{\ell b}$, as was done in Chapter 1.

A similar result follows in the semi-phenomenological quasi-particle model of liquid helium, in which one writes[1]

$$\rho_{\underset{\sim}{k}} = \sqrt{S(\underset{\sim}{k})}\,(\alpha^{\dagger}_{\underset{\sim}{k}} + \alpha_{-\underset{\sim}{k}}) \tag{1-8}$$

where the operators $\alpha^{\dagger}_k$ create excitations of the phonon-roton branch and $S(\underset{\sim}{k}) = N S(\underset{\sim}{k})$ where $S(\underset{\sim}{k}) = \langle\rho_{\underset{\sim}{k}}\,\rho_{-\underset{\sim}{k}}\rangle$. ($(\underset{\sim}{k}\to\infty) = 1$)

Optical processes exist which give information about correlation functions involving four powers of ρ_k. In liquid helium the extinction coefficient associated with second order light scattering is given by [2] ($\underset{\sim}{k}_n$ and $\underset{\sim}{k}_0$ are the wave vectors of the incident and scattered light, respectively).

$$h^{(1)}_{\hat{\varepsilon}_n}(\omega_n) = (\rho^3/2\pi)\,(\omega_n/c)^4 \sum_{\underset{\sim}{q},\underset{\sim}{q}'} F(-\underset{\sim}{q})\,F(-\underset{\sim}{q}')$$

$$\cdot\, S_4\,(\underset{\sim}{q}-\underset{\sim}{k}_n,\ \underset{\sim}{k}_0 - q,\ \underset{\sim}{k}_n - \underset{\sim}{q}',\ \omega_n-\omega_o) \qquad (1\text{-}9)$$

in which

$$F(\underset{\sim}{q}) = \sum_{p,p'} \alpha_p(\omega_o)\,\alpha_{p'}(\omega_o)\,\hat{\varepsilon}_n \cdot \overleftrightarrow{T}_{pp'}(q) \cdot \hat{\varepsilon}_o \qquad (1\text{-}10)$$

$$\overleftrightarrow{T}_{pp'}(\underset{\sim}{q}) = \int d_3\underset{\sim}{r}\; e^{-iq\cdot\underset{\sim}{r}}\,\overleftrightarrow{T}_{pp'}(\underset{\sim}{r})$$

$$\overleftrightarrow{T}_{pp'}(r) = \frac{C^{pp'}(r)}{r^3}\,(3\overrightarrow{r}\overrightarrow{r} - \overleftrightarrow{1}) + t^3(r)\overleftrightarrow{1} \qquad (1\text{-}11)$$

and

$$S_4(\underset{\sim}{k}_1,\ \underset{\sim}{k}_2,\ \underset{\sim}{k}_3,\ \omega)$$
$$= \frac{1}{N^3}\int_{-\infty}^{+\infty} dt e^{i\omega t} \langle \rho_{\underset{\sim}{k}_1}(t)\,\rho_{\underset{\sim}{k}_2}(t)\,\rho_{\underset{\sim}{k}_3}(0)\ \rho_{(-\underset{\sim}{k}_1+\underset{\sim}{k}_2+\underset{\sim}{k}_3)}(0)\rangle \qquad (1\text{-}12)$$

In these equations

$$\alpha_p(\omega_o) = \frac{E_p\,|x_p|^2}{E_p^2 - (\hbar\omega_o)^2}$$

where x_p is the dipole matrix element between the electronic ground state and the electronic excited state p with energy E_p of the helium atoms. Using (1-8) in (1-12) gives an expression involving 4 quasi-particle creation and annihilation operators analogous to the situation in solids.[3] Kleban[4] has shown, however, that (1-8) is particularly bad for computation of S_4 so that the Raman scattering problem in liquid helium is more difficult than the corresponding problem in solids. More recent work on this problem[5] has not used equation (1-8).

From this brief review, one sees that in various problems of interest in the present series of lectures, one is interested in response functions of two types

$$G^{(1)}_{\underset{\sim}{k}} = \langle\langle \alpha_{\underset{\sim}{k}}(t);\ \alpha^{\dagger}_{\underset{\sim}{k}}(o)\rangle\rangle_r$$

$$G^{(2)}(t) = \langle\langle A_{\underset{\sim}{k}_1}(t)A_{\underset{\sim}{k}_2}(t);\ A_{\underset{\sim}{k}_3}(0)\ A_{\underset{\sim}{k}_4}(0)\rangle\rangle_r \qquad (1\text{-}13)$$

where $\alpha_{\underset{\sim}{k}}$ is a boson or fermion annihilation operator and $\alpha^{\dagger}_{\underset{\sim}{k}}$ is a boson or fermion creation operator. $A_{\underset{\sim}{k}}$ is either a boson or a fermion or annihilation operator. We summarize some of the possibilities in the Table. Notice that the G needed for a given system and experiment depends on the model being used. For example, in a microscopic description of He, $G^{(2)}$ would be needed to describe neutron scattering.

In each case, one also needs an appropriate Hamiltonian for computation of the quantities $G^{(1)}(t)$ and $G^{(2)}(t)$. The models usually used are described qualitatively in the table. In general one is dealing with a Hamiltonian of the form

$$H = \sum_{\underset{\sim}{k}} \varepsilon_{\underset{\sim}{k}}\, \alpha^{\dagger}_{\underset{\sim}{k}}\, \alpha_{\underset{\sim}{k}} + \sum_{\underset{\sim}{k}_1,\underset{\sim}{k}_2,\underset{\sim}{k}_3} [\, g_3(\underset{\sim}{k}_1,\underset{\sim}{k}_2,\underset{\sim}{k}_3)\ \alpha^{\dagger}_{\underset{\sim}{k}_1}\alpha_{\underset{\sim}{k}_2}\alpha_{\underset{\sim}{k}_3} + \text{h.c.}]$$

$$+ \sum_{\substack{\underset{\sim}{k}_1,\underset{\sim}{k}_2\\ \underset{\sim}{k}_3,\underset{\sim}{k}_4}} g_4(\underset{\sim}{k}_1,\underset{\sim}{k}_2,\underset{\sim}{k}_3,\underset{\sim}{k}_4)\ \alpha^{\dagger}_{\underset{\sim}{k}_1}\alpha^{\dagger}_{\underset{\sim}{k}_2}\alpha_{\underset{\sim}{k}_3}\alpha_{\underset{\sim}{k}_4} + \ldots \qquad (1\text{-}14)$$

where the coupling constants g_3 and g_4 are different for each of the models listed in the Table. $\alpha^{\dagger}_k$ and α_k are boson or fermion annihilation or creation operators. In (1-14) we refer to a simplified model of scalar quasiparticles. This is, for example, appropriate to the quasiparticle model of liquid helium.

Several methods are used for including the interaction terms of Eq. (1-14) in the computation of $G^{(1)}$ and $G^{(2)}$. The first of these is an "equation of motion" method discussed in the chapter by R. Stinchcombe and also to be utilized by Thorpe in discussing spin problems. A second set of methods appropriate to computation of correlation functions near critical points will be treated by Mazenko. Here we review methods for treating the interaction terms in Eq. (1-14) by many-body perturbation theory. In the next section, perturbation theory for a function related to $G^{(1)}$ is reviewed.

The standard diagrammatic expansion is reviewed in Section 3. Section 4 gives the Dyson equation for $G^{(1)}$. In Section 5 we develop the corresponding expansion for $G^{(2)}$. The next section gives a zero order approximation for $G^{(2)}$. In Section 7 we write a "Dyson" equation for $G^{(2)}$ (which is called the Bethe-Saltpeter equation). Section 8 applies the first approximation to a solution to the Bethe-Saltpeter equation to the two examples: the helium quasiparticle model and the electron gas, which we treat as examples throughout these lectures. Finally, in Section 9, we briefly discuss application and extension of these techniques to other problems.

2. Perturbation Expansion for $G^{(1)}$ [6,7,8]

To compute $G^{(1)}(t)$ as defined by Eq. (1-13) by perturbation theory it is useful to first compute $\tilde{G}(\tau)$, the "thermal Green function" defined in Section 6 of the chapter by Stinchcombe. We define

$$\tilde{G}^{(1)}_{\underset{\sim}{k}}(\tau) = \langle T_\tau(\alpha_{\underset{\sim}{k}}(\tau)\ \alpha^\dagger_{\underset{\sim}{k}}\ (o)\rangle \tag{2-1}$$

in which

$$\alpha_{\underset{\sim}{k}}(\tau) = e^{(H-\mu N)\tau}\ \alpha_{\underset{\sim}{k}} e^{-(H-\mu N)\tau} \tag{2-2}$$

and

$$T_\tau\left[\alpha_{\underset{\sim}{k}}(\tau)\alpha^\dagger_{\underset{\sim}{k}}(0)\right] = \begin{cases} \alpha_{\underset{\sim}{k}}(\tau)\alpha^\dagger_{\underset{\sim}{k}}(0) & \tau>0 \\ -\eta\alpha^\dagger_{\underset{\sim}{k}}(0)\alpha_{\underset{\sim}{k}}(\tau) & \tau>0 \end{cases} \tag{2-3}$$

in which $\eta=-1$ for bosons and $+1$ for fermions. (This notation differs from that of reference 9.) As discussed in the same section $G^{(1)}_r(\omega\underset{\sim}{k})$ is obtained from $\tilde{G}^\omega_k(\tau)$ by Fourier transform and analytic continuation:

$$G^{(1)}_r(\omega,\underset{\sim}{k}) = \frac{1}{(2\pi)}\ \tilde{G}^{(1)}_{\underset{\sim}{k}}(i\omega_\ell \to \omega + i0^\dagger) \tag{2-4}$$

$$\tilde{G}^{(1)}_{\underset{\sim}{k}}(i\omega_\ell) = -\int_0^\beta d\tau\ \tilde{G}^{(1)}_{\underset{\sim}{k}}(\tau)e^{i\omega_\ell\tau} \tag{2-5}$$

Table: Experimental Systems and Experiments Requiring Computation of G Functions of Type $G^{(1)}$ and $G^{(2)}$.

SYSTEM	EXPERIMENT	MODEL	NEEDED RESPONSE	COMMENT
Electron gas	Electron scattering	Fermions interacting through Coulomb interaction	$G^{(2)} = \langle\langle\rho_{\underset{\sim}{k}};\rho_{-\underset{\sim}{k}}\rangle\rangle$	A's are fermion operators
Solid	Neutron scattering	Phonons with anharmonicity	$G^{(1)} = \langle\langle\rho_{k};\rho_{-\underset{\sim}{k}}\rangle\rangle$	A's are boson operators
Solid	2nd order Raman scattering on IR	Phonons with anharmonicity	$G^{(2)} = \langle\langle\rho_{\underset{\sim}{k}_1}\rho_{\underset{\sim}{k}_2};\rho_{\underset{\sim}{k}_3}\rho_{-\underset{\sim}{k}_1-\underset{\sim}{k}_2-\underset{\sim}{k}_3}\rangle\rangle$	Bound states
Liquid helium	Neutron scattering	Quasi-particle model (Eqs. (1-8)+(1-14))	$G^{(1)} = \langle\langle\rho_{\underset{\sim}{k}};\rho_{-\underset{\sim}{k}}\rangle\rangle$	Boson operators

in which

$$\omega_\ell = \begin{cases} 2\ell\pi/\beta \;, & \eta = -1 \text{ (bosons)} \\ (2\ell+1)\pi/\beta \;, & \eta = +1 \text{ (fermions)} \end{cases} \qquad (2\text{-}6)$$

In Eq. (2-4) the result has been written in a way which includes the possibility that the thermal average in (2-1) is taken in the grand canonical ensemble. It is essential to do the thermal averaging in this way for models in which $\alpha_{\underset{\sim}{k}}$ annihilates number-conserved particles (such as electrons or helium atoms) rather than quasiparticles whose number is not conserved (such as phonons, plasmons or spin waves). The formula (2-4) is valid for either case as long as one takes $\mu=0$ in the second (quasiparticle) case. (For clarity, one should add that in interacting electron systems some literature[8] refers to electron quasiparticles whose number _is_ conserved and for which $\mu \neq 0$.)

To write a perturbation expansion for $\tilde{G}^{(1)}_{k}(i\omega_\ell)$, one introduces an interaction representation in the imaginary time variable. Writing

$$H = H_o + H' \qquad (2\text{-}7)$$

where

$$H_o = \sum_{\underset{\sim}{k}} \varepsilon_{\underset{\sim}{k}}\, \alpha^\dagger_{\underset{\sim}{k}}\, \alpha_{\underset{\sim}{k}}$$

and H' is defined by (2-7) and (1-14). We define an operator $S(\tau)$ by the relation

$$S\, e^{-(H-\mu N)\tau} \equiv e^{-(H_o-\mu N)\tau} S(\tau) \qquad (2\text{-}8)$$

Interaction representation operators are defined by

$$\tilde{\alpha}_k(\tau) = e^{(H_o-\mu N)\tau}\, \alpha_{\underset{\sim}{k}} e^{-(H_o-\mu N)\tau}$$

$$\tilde{\tilde{\alpha}}_{\underset{\sim}{k}}(\tau) = e^{(H_o-\mu N)\tau}\, \alpha^\dagger_{\underset{\sim}{k}}\, e^{-(H_o-\mu N)\tau} \qquad (2\text{-}9)$$

(Note that $\tilde{\tilde{\alpha}}_k(\tau)^\dagger \neq \tilde{\alpha}_k(\tau)$.) More generally for any operator O

$$\tilde{O}(\tau) = e^{(H_o-\mu N)\tau}\, O\, e^{-(H_o-\mu N)\tau} \qquad (2\text{-}10)$$

An equation of motion for $S(\tau)$ follows by differentiating with respect to τ

$$\frac{\partial S(\tau)}{\partial \tau} = \frac{-\tilde{H}'(\tau)}{\hbar} S(\tau)$$

for which the solution is

$$S(\tau) = 1 - \frac{1}{\hbar} \int_0^{\tau} \tilde{H}'(\tau')d\tau'$$

$$+ \frac{1}{\hbar^2} \int_0^{\tau} \tilde{H}'(\tau') \int_0^{\tau'} \tilde{H}'(\tau'')d\tau' \, d\tau'' + \ldots$$

$$= T_\tau \left[\exp \left\{ - \frac{1}{\hbar} \int_0^{\tau} d\tau_1 \tilde{H}'(\tau_1) \right\} \right]$$

where T_τ is a time ordering operator defined by generalization of (2-3). We also define

$$S(\tau_1,\tau_2) = T_\tau \left[\exp \left\{ \frac{-1}{\hbar} \int_{\tau_1}^{\tau_2} d\tau' \, \tilde{H}'(\tau') \right\} \right]$$

and have

$$S(\tau_1,\tau_2) \, S(\tau_2,\tau_3) = S(\tau_1,\tau_3)$$

$$S(\tau_1,\tau_2) = S(\tau_1) \, S^{-1}(\tau_2)$$

$$S^{-1}(\tau) = T_\tau \left[\exp \left\{ \frac{1}{\hbar} \int_{\tau_1}^{\tau_2} d\tau' \, \tilde{H}'(\tau') \right\} \right]$$

Using these results

$$\tilde{G}^{(1)}_{\underset{\sim}{k}}(\tau>0) =$$

$$\mathrm{Tr} \left[e^{\Omega\beta} e^{-(H-\mu N)\beta} e^{(H-\mu N)\tau} \alpha_{\underset{\sim}{k}} e^{-(H-\mu N)\tau} \alpha^{\dagger}_{\underset{\sim}{k}} \right]$$

$$= \mathrm{Tr} \left[e^{\Omega\beta} e^{-(H_o-\mu N)\beta} S(\beta) S^{-1}(\tau) \right.$$

$$e^{(H_o-\mu N)\tau}\ \alpha_{\underset{\sim}{k}}\ e^{-(H_o-\mu N)\tau}\ S(\tau)\ \alpha^{\dagger}_{\underset{\sim}{k}}\Big]$$

Here we have used

$$e^{-(H-\mu N)\tau} = e^{-(H_o-\mu N)\tau}S(\tau)$$

which is the definition of $S(\tau)$ and

$$e^{(H-\mu N)\tau} = S^{-1}(\tau)\ e^{(H_o-\mu N)\tau}$$

which follows directly from it. Using the group property and the definition of the "interaction picture" operators:

$$\tilde{G}^{(1)}_{\underset{\sim}{k}}(\tau>0) = \mathrm{Tr}\Big[e^{\Omega\beta}e^{-(H_o-\mu N)\beta}S(\beta,\tau)$$

$$\tilde{\alpha}_{\underset{\sim}{k}}(\tau)S(\tau)\bar{\tilde{\alpha}}_{\underset{\sim}{k}}(0)\Big]$$

Now $-\beta<\tau<\beta$ so that all the operators in $S(\beta,\tau)$ contain operators with $\tau'>\tau$. Similarly $S(\tau)$ contains only operators with $\tau'<\tau$. Thus we can write

$$\tilde{G}^{(1)}_{\underset{\sim}{k}}(\tau>0) =$$

$$\mathrm{Tr}\Big[e^{\Omega\beta}\ e^{-(H_o-\mu N)\beta}\ T_\tau\{S(\beta,\tau)S(\tau)\ \tilde{\alpha}_{\underset{\sim}{k}}(\tau)\ \bar{\tilde{\alpha}}_{\underset{\sim}{k}}(0)\}\Big]$$

$$= \mathrm{Tr}\Big[e^{\Omega\beta}\ e^{-(H_o-\mu N)\beta}\ T_\tau\big(\tilde{\alpha}_{\underset{\sim}{k}}(\tau)\ \bar{\tilde{\alpha}}_{\underset{\sim}{k}}(0)\ S(\beta)\big)\Big]$$

$$= \frac{\mathrm{Tr}\Big[e^{-(H_o-\mu N)\beta}\ T_\tau\Big(\tilde{\alpha}_{\underset{\sim}{k}}(\tau)\ \bar{\tilde{\alpha}}_{\underset{\sim}{k}}(0)\Big)S(\beta)\Big]}{\mathrm{Tr}\Big(e^{-(H_o-\mu N)\beta}S(\beta)\Big)}$$

If we define

$$\langle \ldots \rangle_o = \frac{\mathrm{Tr}\, e^{-(H_o-\mu N)\beta}\,(\ldots)}{\mathrm{Tr}\, e^{-(H_o-\mu N)\beta}}$$

then this is

$$\tilde{G}_k^{(1)}(\tau>0) = \frac{\langle T_\tau\left[\tilde{\alpha}_k(\tau)\ \bar{\tilde{\alpha}}_k(0)S(\beta)\right]\rangle_o}{\langle S(\beta)\rangle_o} \tag{2-11}$$

If we now use the expression

$$S(\beta) = T_\tau\left[\exp\left\{-\frac{1}{\hbar}\int_0^\beta d\tau'\ \tilde{H}'(\tau')\right\}\right]$$

$$= \sum_{h=0}^{\infty}\left(\frac{-1}{\hbar}\right)^n \frac{1}{n!}\int d\tau_1 \ldots d\tau_n\ T_\tau\{\tilde{H}'(\tau_1)\ldots\tilde{H}'(\tau_n)\} \tag{2-12}$$

in this, then it can be used to generate a perturbation series for $\tilde{G}_{\underset{\sim}{k}}^{(1)}(\tau>0)$.

Finally we note that we also have a perturbation theory expression for the thermodynamic properties through the relation

$$e^{-\Omega\beta} \equiv \mathrm{Tr}e^{-(H-\mu N)\beta} =$$

$$\mathrm{Tr}\left[e^{-(H_o-\mu N)\beta}S(\beta)\right] = \frac{\mathrm{Tr}\ e^{-(H_o-\mu N)\beta}S(\beta)}{\mathrm{Tr}\ e^{-(H_o-\mu N)\beta}}\ \mathrm{Tr}e^{-(H_o-\mu N)\beta}$$

so that

$$\Omega = -K_BT\ \ln\langle S(\beta)\rangle_o + \Omega_o$$

Here Ω_o is the thermodynamic potential for the unperturbed system.

The next step is to simplify the series for $G_k^{(1)}(\tau>0)$ by use of Wick's Theorem. A useful form of the theorem states that thermal expectation values of the form

$$< T_\tau \left(\tilde{\alpha}_{\underset{\sim}{k}}(\tau)\bar{\tilde{\alpha}}_{\underset{\sim}{k}}(\tau')\tilde{H}'(\tau_1)\ldots\tilde{H}'(\tau_n) \right) >_0$$

in which an equal number of factors $\tilde{\alpha}$ and $\bar{\tilde{\alpha}}$ appear can be evaluated according to the following rule: Take all possible pairings of $\tilde{\alpha}$ and $\bar{\tilde{\alpha}}$ such that each operator is paired. For each of these, rearrange so that the pairs occur together, with appropriate changes of sign. Finally replace each pair of $\tilde{\alpha}\ \tilde{\alpha}$ by non-interacting

$$\tilde{G}_{\underset{\sim}{k}}^{(0,1)}(\tau) \equiv < T_\tau \left(\alpha_{\underset{\sim}{k}} \bar{\tilde{\alpha}}_{\underset{\sim}{k}} \right) >_0$$

Notice that the theorem makes a statement here about thermal averages, and not about operators alone as it does[6] at T = 0.

Proof of the Theorem: We write a general term with an even number of $\bar{\tilde{\alpha}}$ and $\tilde{\alpha}$ in (1-21) as

$$\tilde{H}'_{2m} = \sum_{\underset{\sim}{k}_1,\ldots \underset{\sim}{k}_{2m-1}} g_{2m}(\underset{\sim}{k}_1, \ldots, \underset{\sim}{k}_{m-1}) \bar{\tilde{\alpha}}_{\underset{\sim}{k}_1} \ldots \bar{\tilde{\alpha}}_{\underset{\sim}{k}_m}$$

$$\tilde{\alpha}_{\underset{\sim}{k}_{m+1}} \ldots \tilde{\alpha}_{-\underset{\sim}{k}_1 - \ldots - \underset{\sim}{k}_{2m-1}}$$

This will have no effect in the thermodynamic limit unless its matrix elements in the number representation for the αs is proportional to the sample volume V in the limit $V \to \infty$. Using the fact that, in the limit of large V

$$\sum_{\underset{\sim}{k}} (\ldots) = \frac{V}{(2\pi)^3} \int d_3 \underset{\sim}{k} (\ldots)$$

it follows that

$$g\ (\underset{\sim}{k}_1, \ldots, \underset{\sim}{k}_{2m-1}) = \frac{\lambda(\underset{\sim}{k}_1, \ldots, \underset{\sim}{k}_{2m-1})}{V^{m-1}}$$

where $\lambda(\underset{\sim}{k}_1, \ldots, k_{2m-1})$ is independent of V. $\tilde{H}'_{2m}$ can then be rewritten as

$$\tilde{H}'_{2m} = V\sum_{\underset{\sim}{k}_1,\ldots \underset{\sim}{k}_{2m-1}} \lambda_{2m}(\underset{\sim}{k}_1, \ldots, \underset{\sim}{k}_{2m-1}) \frac{\bar{\tilde{\alpha}}_{\underset{\sim}{k}_1}}{\sqrt{V}} \cdots \frac{\tilde{\alpha}_{k_{2m-1}}}{\sqrt{V}}$$

For a term with an odd number of $\bar{\tilde{\alpha}}_k$ and $\bar{\alpha}_{\underset{\sim}{k}}$, one similarly concludes, by requiring that expressions of form

$$<N|\tilde{H}_{2m+1}|N+\ell><N+\ell|\tilde{H}_{2m+1}|N>$$

be of order V that one can write

$$\tilde{H}'_{2m+1} = V\sum_{\underset{\sim}{k}_1 \ldots \underset{\sim}{k}_{2m}} \lambda_{2m+1}(\underset{\sim}{k}_1, \ldots k_{2m}) \frac{\bar{\tilde{\alpha}}_{\underset{\sim}{k}_1}}{\sqrt{V}} \cdots \frac{\tilde{\alpha}_{k_{2m}}}{\sqrt{V}}$$

where λ_{2m+1} is independent of V. We note that

$$\tilde{\alpha}_{\underset{\sim}{k}}(\tau) = \alpha_{\underset{\sim}{k}}(\tau) e^{-(\varepsilon_{\underset{\sim}{k}}-\mu)\tau/\hbar}$$

$$\bar{\tilde{\alpha}}_{\underset{\sim}{k}}(\tau) = \alpha^{\dagger}_{\underset{\sim}{k}}(\tau) e^{(\varepsilon_k-\mu)\tau/\hbar} \quad (2\text{-}13)$$

Here we label the $\alpha_{\underset{\sim}{k}}$, $\alpha^{\dagger}_{\underset{\sim}{k}}$ (which do not actually depend on τ) with τ to keep track of time order. The product of interest becomes, supposing there are ℓ factors $H(\tau_1)$ and considering interaction terms with 2m factors $\tilde{\alpha}$ and $\bar{\tilde{\alpha}}$,

$$\frac{V^{\ell}}{V^{m\ell}} \sum_{\underset{\sim}{p}_1} \cdots \sum_{\underset{\sim}{p}_{2m\ell}} <T_{\tau}\{\alpha_{\underset{\sim}{k}}(\tau)\alpha^{\dagger}_{\underset{\sim}{k}}(0) \quad (2\text{-}14)$$

$$\alpha^{\dagger}_{\underset{\sim}{p}_1}(\tau_1)\alpha^{\dagger}_{\underset{\sim}{p}_2}(\tau_2)\ldots \alpha_{\underset{\sim}{p}_{2m\ell}}\}>_0 \exp\{\ldots\}$$

Here the exponential factor involves the factors $e^{\pm(\varepsilon_p-\mu)\tau/\hbar}$.

We now consider what values the momenta $\underset{\sim}{p}_1$, $\underset{\sim}{p}_2$, $\underset{\sim}{p}_3$, ... may take.

We take an arbitrary value of $\underset{\sim}{p}_1$. Then there must be at least one $\alpha^\dagger_{\underset{\sim}{p}_i}$ (or $\alpha_{\underset{\sim}{p}_i}$) with $p_i = \underset{\sim}{p}_1$ or $\underset{\sim}{k} = \underset{\sim}{p}_1$ or else the expectation value vanishes. Similarly to each $\underset{\sim}{p}_i$ associated with an operator $\alpha_{\underset{\sim}{p}_j}$ there must correspond an operator $\alpha^\dagger_{\underset{\sim}{p}_k}$ with $\underset{\sim}{p}_k = \underset{\sim}{p}_j$ in order to get a non-zero expectation value. (This is because number is conserved in the states involved in the trace.) Now there are several possibilities:

1) For each $\alpha_{\underset{\sim}{p}_i}$ there is only one $\alpha_{\underset{\sim}{p}_i}$ with that value of $\underset{\sim}{p}_i$ (and only one $\alpha^\dagger_{\underset{\sim}{p}_i}$).

2) For each $\alpha_{\underset{\sim}{p}_i}$ there is more than one $\alpha_{\underset{\sim}{p}_i}$ with the same value of $\underset{\sim}{p}_i$ (and several corresponding $\alpha^\dagger_{\underset{\sim}{p}_i}$).

Terms of type 1) sum up in the expression for

$$\langle T\tau(\tilde{\alpha}_k(\tau)\bar{\tilde{\alpha}}_k(\tau')\ldots)\rangle_0$$

to give something of the form

$$\frac{V^\ell}{V^{m\ell}}\sum_{\underset{\sim}{p}'}\ldots\sum_{\underset{\sim}{p}}\langle T_\tau\{\alpha_{\underset{\sim}{k}}(\tau)\alpha^\dagger_{\underset{\sim}{k}}(\tau')$$

$$\alpha^\dagger_{\underset{\sim}{p}_1}(\tau_1)\alpha_{p_1}(\tau_2)\ldots\}\rangle_0 \exp\{\ldots\}$$

where the particular pairing of the first few factors is chosen for illustrative purposes. Changing the sums $\sum_{\underset{\sim}{p}_1}$ to integrals (only valid in the thermodynamic limit) we have

$$\frac{1}{V}\sum_{\underset{\sim}{p}_1}(\ldots)\rightarrow\frac{1}{V}\frac{V}{(2\pi)^3}\int d^3\underset{\sim}{p}_1(\ldots)$$

for each summation on $\underset{\sim}{p}_i$. For terms of this type 1) one gets something of order V^ℓ in the thermodynamic limit.

Now consider the second case. For terms of this type there is a larger number $m\ell$ of factors $1/V$ than there is of sums $\Sigma(\ldots)$. Therefore, the transformation of the sums $\sum_{\underset{\sim}{p}_1}(\ldots)$ to integrals results in something of the form $(k<m\ell)$

$$\frac{V^\ell}{V^{m\ell}}\sum_{\underset{\sim}{p}_1}\sum_{\underset{\sim}{p}_2}(\ldots) \rightarrow \frac{V^\ell}{V^{m\ell-k}}\frac{1}{(2\pi)^{3k}}\int d^3\underset{\sim}{p}_1\ldots d^3\underset{\sim}{k}\,(\ldots)$$

$$= O(V^{\ell-(m\ell-k)})<O(V^\ell)$$

where k is the number of sums on $\underset{\sim}{p}$'s. Thus the terms of type 2 are all vanishingly small compared to those of type 1 in the thermodynamic limit.

To complete the proof, we return to the surviving terms of type 1). Since the times in the $\alpha_{\underset{\sim}{p}}(\tau)$, $\alpha^\dagger_{\underset{\sim}{p}}(\tau')$ are in fact only labels, we can evaluate the expectation values after time ordering, for example

$$<T\tau(\alpha_{\underset{\sim}{k}}(\tau)\alpha^\dagger_{\underset{\sim}{k}}(0)\alpha^\dagger_{\underset{\sim}{p}_1}(\tau_1)\alpha_{\underset{\sim}{p}_1}(\tau_2)\ldots)>_o$$

$$= <\begin{pmatrix} n_{\underset{\sim}{k}} \\ 1\pm n_{\underset{\sim}{k}}\end{pmatrix}\begin{pmatrix} n_{\underset{\sim}{p}_1} \\ 1\pm n_{\underset{\sim}{p}_1}\end{pmatrix}\ldots>_o \tag{2-15}$$

where one takes the upper or lower choice in each parenthesis depending on how the time ordering comes out. Now it is a fact that in non-interacting thermal average, all products of the type

$$<n_{\underset{\sim}{p}_1}n_{\underset{\sim}{p}_2}n_{\underset{\sim}{p}_3}\ldots>_o$$

factor as

$$<n_{\underset{\sim}{p}_1}n_{\underset{\sim}{p}_2}n_{\underset{\sim}{p}_3}\ldots>_o$$

$$= <n_{\underset{\sim}{p}_1}>_o<n_{\underset{\sim}{p}_2}>_o<n_{\underset{\sim}{p}_3}>_o\ldots \tag{2-16}$$

as long as $\underset{\sim}{p}_1 \neq \underset{\sim}{p}_2 \neq \underset{\sim}{p}_3$ as in this case 1). (The factorization occurs because the states in the trace are products of eigenstates of the various number operators $n_{\underset{\sim}{p}}$). Finally, for each type 1) pairing of type (2-15) we make the factorization (2-16) and put the result back in the original equation (2-14) doing the sums on $\underset{\sim}{p}_1, \ldots, \underset{\sim}{p}_k$ we get, putting the exponential factors back in, a factor $\langle T_\tau(\tilde{\alpha}_{\underset{\sim}{p}}\bar{\tilde{\alpha}}_{\underset{\sim}{p}})\rangle_0$ corresponding to each pairing $\alpha_p\alpha_p^\dagger$. Making all possible pairings of type 1) we arrive at all terms contributing to the thermal expectation value of the product and prove the theorem.

3. Diagrams for $G^{(1)}$ with Four Point Interactions ($g_4 \neq 0$)

For the noninteracting system, from (2-13)

$$\tilde{G}_{\underset{\sim}{k}}^{(1,0)}(\tau>0) = e^{-(\varepsilon_{\underset{\sim}{k}}-\mu)\tau/h} \langle T_\tau(\tilde{\alpha}_k(\tau)\tilde{\alpha}_{\underset{\sim}{k}}(0)\rangle_0$$

$$= e^{-(\varepsilon_{\underset{\sim}{k}}-\mu)\tau/h}(1 \pm n_{\underset{\sim}{k}}) \tag{3-1}$$

where

$$n_{\underset{\sim}{k}} = \frac{1}{e^{(\varepsilon_{\underset{\sim}{k}}-\mu)\beta} + \eta}$$

By use of (2-5)

$$\tilde{G}_k^{(1,0)}(i\omega\ell) = -\int_0^\beta e^{-(\varepsilon_{\underset{\sim}{k}}-\mu)\tau/h} e^{i\omega\tau}(1 \pm n_{\underset{\sim}{k}})\frac{d\tau}{h}$$

$$= \frac{1}{i\hbar\omega\ell - (\varepsilon_{\underset{\sim}{k}}-\mu)} \tag{3-2}$$

We apply (2-11) to the case in which only g_4 terms contribute in (1-14). We use (2-5) to write a corresponding series for $G_{\underset{\sim}{k}}^{(1)}(i\omega\ell)$, using

$$\tilde{G}_{\underset{\sim}{k}}^{(1,0)}(\tau) = \frac{-1}{\beta}\sum_\ell \tilde{G}_{\underset{\sim}{k}}^{(1,0)}(i\omega_\ell)^{\ell^{-i\omega_\ell\tau}} \tag{3-3}$$

to rewrite the products of $G^{(1,0)}_{\underset{\sim}{k}}(\tau)$ which appear upon application of Wick's Theorem. The leading terms in the numerator of (2-11) are

$$\tilde{G}^{(1)}_{\underset{\sim}{k}}(i\omega_\ell) = \frac{1}{<S(\beta)>_o}\left[\tilde{G}^{(1,0)}_{\underset{\sim}{k}}(i\omega_\ell)\right.$$

$$+\frac{1}{2h}\int_0^\beta d\tau_1\int_0^\beta d\tau\, \ell^{i\omega_\ell\tau}<T_\tau\left(\tilde{\alpha}_{\underset{\sim}{k}}(\tau)\bar{\tilde{\alpha}}_{\underset{\sim}{k}}(0)\tilde{H}'(\tau')\right)>_o$$

$$\left. + \ldots\right] \tag{3-4}$$

The second term is

$$\frac{1}{2h}\int_0^\beta d\tau_1\int_0^\beta d\tau\, \ell^{i\omega_\ell\tau}$$

$$\sum_{\underset{\sim}{k}_1,\underset{\sim}{k}_2,\underset{\sim}{k}_3} g_4(\underset{\sim}{k}_3-\underset{\sim}{k}_1)<T_\tau\left(\tilde{\alpha}_{\underset{\sim}{k}}(\tau)\bar{\tilde{\alpha}}_{\underset{\sim}{k}}(0)\right.$$

$$\left.\bar{\tilde{\alpha}}_{\underset{\sim}{k}_1}(\tau_1)\bar{\tilde{\alpha}}_{\underset{\sim}{k}_2}(\tau_1)\tilde{\alpha}_{\underset{\sim}{k}_3}(\tau_1)\tilde{\alpha}_{\underset{\sim}{k}_1+\underset{\sim}{k}_2-\underset{\sim}{k}_3}(\tau_1)\right)>_o \tag{3-5}$$

We illustrate the possible pairings for the case of bosons below

$$<T_\tau\left(\tilde{\alpha}_{\underset{\sim}{k}}(\tau)\bar{\tilde{\alpha}}_{\underset{\sim}{k}}(0)\bar{\tilde{\alpha}}_{\underset{\sim}{k}_1}(\tau_1)\bar{\tilde{\alpha}}_{\underset{\sim}{k}_2}(\tau_1)\tilde{\alpha}_{\underset{\sim}{k}_3}(\tau_1)\tilde{\alpha}_{\underset{\sim}{k}_1+\underset{\sim}{k}_2-\underset{\sim}{k}_3}(\ _1)\right)>_o$$

$$= <T_\tau(\tilde{\alpha}_{\underset{\sim}{k}}\bar{\tilde{\alpha}}_{\underset{\sim}{k}})>_o<T_\tau(\bar{\tilde{\alpha}}_{\underset{\sim}{k}_1}\tilde{\alpha}_{\underset{\sim}{k}_3})>_o<T_\tau(\bar{\tilde{\alpha}}_{\underset{\sim}{k}_2}\tilde{\alpha}_{\underset{\sim}{k}_1+\underset{\sim}{k}_2-\underset{\sim}{k}_3})>_o$$

$$+ \langle T_\tau(\tilde{\alpha}_{\mathbf{k}}\bar{\tilde{\alpha}}_{\mathbf{k}})\rangle_0 \langle T_\tau(\bar{\tilde{\alpha}}_{\mathbf{k}_1}\alpha_{\mathbf{k}_1+\mathbf{k}_2-\mathbf{k}_3})\rangle_0 \langle T_\tau(\bar{\tilde{\alpha}}_{\mathbf{k}_2}\tilde{\alpha}_{\mathbf{k}_3})\rangle_0$$

$$+ \langle T_\tau(\tilde{\alpha}_{\mathbf{k}}\bar{\tilde{\alpha}}_{\mathbf{k}_1})\rangle_0 \langle T_\tau(\bar{\tilde{\alpha}}_{\mathbf{k}}\tilde{\alpha}_{\mathbf{k}_3})\rangle_0 \langle T_\tau(\bar{\tilde{\alpha}}_{\mathbf{k}_1}\tilde{\alpha}_{\mathbf{k}_1+\mathbf{k}_2-\mathbf{k}_3})\rangle_0$$

$$+ \langle T_\tau(\tilde{\alpha}_{\mathbf{k}}\bar{\tilde{\alpha}}_{\mathbf{k}_1})\rangle_0 \langle T_\tau(\bar{\tilde{\alpha}}_{\mathbf{k}}\tilde{\alpha}_{\mathbf{k}_1+\mathbf{k}_2-\mathbf{k}_3})\rangle_0 \langle T_\tau(\bar{\tilde{\alpha}}_{\mathbf{k}_2}\tilde{\alpha}_{\mathbf{k}_3})\rangle_0 \qquad (3\text{-}6)$$

$$+ \langle T_\tau(\tilde{\alpha}_{\mathbf{k}}\bar{\tilde{\alpha}}_{\mathbf{k}_2})\rangle_0 \langle T_\tau(\bar{\tilde{\alpha}}_{\mathbf{k}}\tilde{\alpha}_{\mathbf{k}_3})\rangle_0 \langle T_\tau(\bar{\tilde{\alpha}}_{\mathbf{k}_1}\tilde{\alpha}_{\mathbf{k}_1+\mathbf{k}_2-\mathbf{k}_3})\rangle_0$$

$$+ \langle T_\tau(\tilde{\alpha}_{\mathbf{k}}\bar{\tilde{\alpha}}_{\mathbf{k}_2})\rangle_0 \langle T_\tau(\bar{\tilde{\alpha}}_{\mathbf{k}}\tilde{\alpha}_{\mathbf{k}_1+\mathbf{k}_2-\mathbf{k}_3})\rangle_0 \langle T_\tau(\bar{\tilde{\alpha}}_{\mathbf{k}_1}\tilde{\alpha}_{\mathbf{k}_3})\rangle_0$$

Where the imaginary time arguments of the α's have been surpressed. There are various ways to indicate the possible pairings.

One can write diagrams for the various terms. In particular the first term in (3-6) is denoted

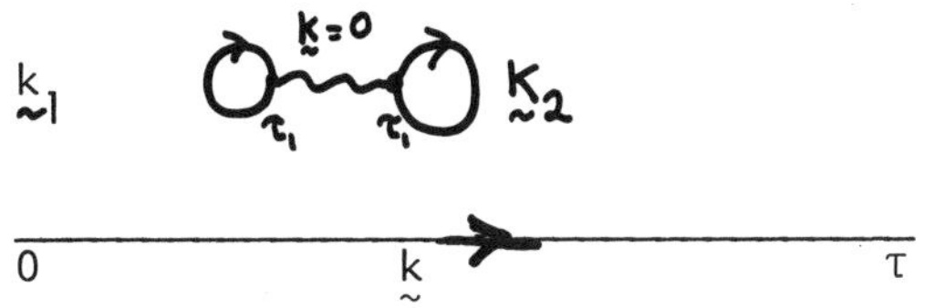

where the notations mean

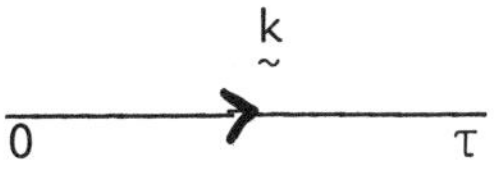

$\equiv \quad \tilde{G}^{(1,0)}_{\mathbf{k}}(\tau)$

$\equiv \quad g_4(\mathbf{k})$

Full sets of rules for writing diagrams are given in many books.[6,7,8] We cite them below for the time Fourier transform of the series. Note that, applying (3-3), a condition on the frequencies results. Thus the contribution of the first term in (3-6) to (3-5) is

$$\frac{-g_4(\underset{\sim}{k}=0)}{2\,h\,\beta}\,\tilde{G}^{(1,0)}_{\underset{\sim}{k}}(i\omega_\ell)\sum_{\substack{\ell_1,\ell_2\\ \underset{\sim}{k}_1\,\underset{\sim}{k}_2}}\tilde{G}^{(1,0)}_{\underset{\sim}{k}_1}(i\omega_{\ell_1})\,\tilde{G}^{(1,0)}_{\underset{\sim}{k}_2}(i\omega_{\ell_2})$$

This term is diagramatically denoted

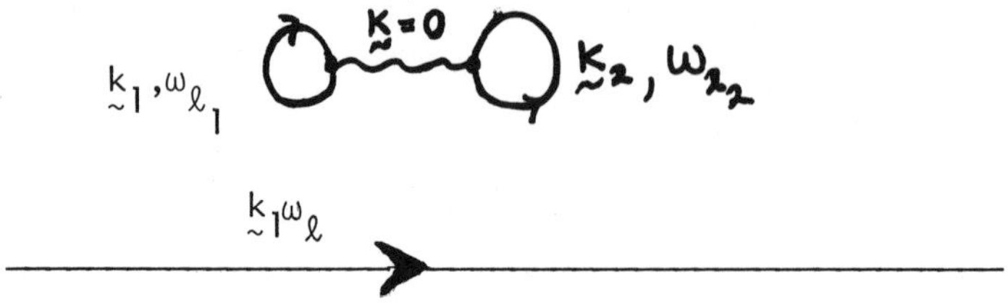

For general rules for associating terms with diagrams we adopt one of the notations used in reference 6:

$$\equiv \quad \tilde{G}^{(1,0)}_{\underset{\sim}{k}}(i\omega_\ell)$$

$$\equiv \quad g_4(\underset{\sim}{k})$$

To apply Wick's Theorem in a given order ℓ diagramatically, form all possible diagrams with ℓ wiggly lines and vertices of type  with one straight line entering and one emerging. To evaluate the resulting diagrams, make the associations (3-7) and (3-8) and require the conservation law $\sum_{\underset{\sim}{i}} k_i = 0$, $\sum_n \omega_n = 0$ at vertices. In applying the rule $\sum_n \omega_n = 0$ one formally assigns a frequency to $g_4(\underset{\sim}{k})$. Finally, sum over internal values of ω_n and $\underset{\sim}{k}$ and multiply by $(-\frac{1}{h})^\ell \frac{1}{\ell!}\frac{1}{\beta^\ell}$. A sign must be adjusted in the fermion case to take count of the anti-commutation relations of the operators.[6] The factor $1/\ell!$ disappears from contributing terms as we will see below. The diagrams associated with the other terms in (3-7) are shown below.

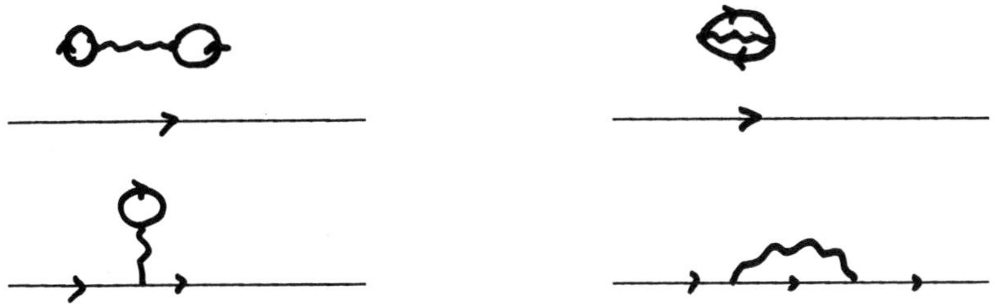

The last two appear twice.

4. Linked Cluster Theorem and Dyson Equation for $G_k^{(1)}(i\omega_\ell)$

We next review the linked cluster theorem, which shows that diagrams which are disconnected (like the first two in the example just cited) cancel with the denominator in Eq. (2-11), leaving a result which can be written as

$$\tilde{G}_{\underset{\sim}{k}}(i\omega_\ell) = \langle T_\tau\left\{\tilde{\alpha}_{\underset{\sim}{k}}(\tau)\bar{\tilde{\alpha}}_k(0)S(\beta)\right\}\rangle_0^{(\text{conn.})} \tag{4-1}$$

where (conn.) means "connected." Analytically, in disconnected diagrams the time integrals or frequency sums factor. The proof is sketched here. Details appear in many books.[6-8]

Consider a disconnected diagram of the general form

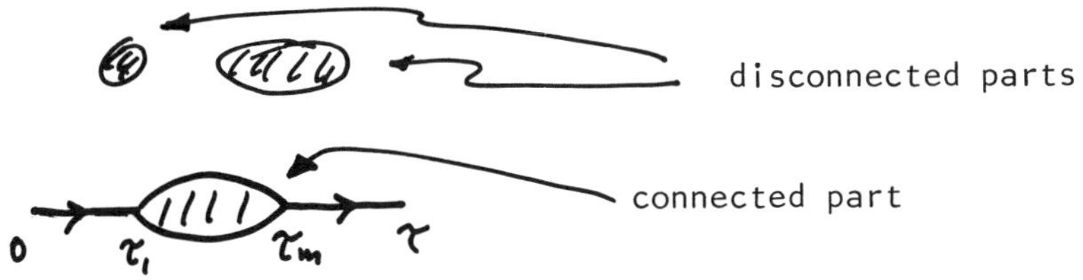

Suppose that the connected part of this contains m vertices and that the disconnected parts contain n-m vertices so that the term is of order n. Let the contribution from the disconnected part be K(n, m). We will sum this contribution over all disconnected parts while leaving the connected part fixed. There are n!/m! (n-m)! ways of picking the m imaginary times τ from the n times appearing in the integral for the contribution of the connected part and each gives an equivalent contribution. We therefore take the sum

$$\sum_{\substack{\text{all disconnected}\\ \text{parts}}} \frac{n!}{m!\,(n-m)!} K(n,m)$$

We claim that this sum is given by

$$\sum_{\substack{\text{all disconnected}\\ \text{parts}}} \frac{n!}{m!\,(n-m)!} K\ (n,m)$$

$$= \langle S(\beta)\rangle_0 \ \ X(\text{contribution of the connected part})$$

If this is true, then we can indeed drop the disconnected parts and drop the $\langle S(\beta)\rangle_0$ in the denominator as claimed.

To show this, we write out the contribution $K(n,m)$ in terms of $\tilde{H}'(\tau)$:

$$\frac{n!}{m!\,(n-m)!} K(n,m) = \frac{n!}{m!\,(n-m)!} \left(\frac{-1}{\hbar}\right)^n \frac{1}{n!} \frac{1}{\langle S(\beta)\rangle_0}$$

$$\times \int_0^\beta \cdots \int_0^\beta d\tau_1 \ldots d\tau_m \left\langle T_\tau \left(\tilde{\alpha}_k(\tau)\bar{\tilde{\alpha}}_k(0)\tilde{H}'(\tau_1) \ \ldots\ \tilde{H}'(\tau_m)\right)\right\rangle \ \text{conn.}$$

$$\times \int_0^\beta \cdots \int_0^\beta d\tau_{m+1} \ldots d\tau_n \left\langle T_\tau \left(\tilde{H}'(\tau_{m+1} \ \ldots\ \tilde{H}'\ (\tau_n)\right)\right\rangle \ \text{disconn.}$$

where the subscripts conn and disconn mean that the part of the expectation value corresponding to the contractions designated in the diagram is to be kept. Taking the sum on disconnected parts we have

$$\sum \frac{n!}{m!\ (n-m)!} K(n,m)$$

$$= \left(\frac{-1}{h}\right)^m \frac{1}{m!} \frac{1}{\langle S(\beta)\rangle_0}$$

$$\times \int_0^\beta \cdots \int_0^\beta d\tau_1 \ldots d\tau_m \left\langle T_\tau \left(\tilde{\alpha}_{\underset{\sim}{k}}(\tau)\bar{\tilde{\alpha}}_{\underset{\sim}{k}}(0)\tilde{H}'(\tau_1) \ \ldots\ \tilde{H}'(\tau_m)\right)\right\rangle \quad \text{conn.}$$

$$\times \sum_{\{\text{disconnected parts}\}} \left(\frac{-1}{\hbar}\right)^{n-m} \frac{1}{(n-m)!} \int_0^\beta \cdots \int_0^\beta d\tau_{m+1} \cdots d\tau_n$$

$$\times \langle T_\tau \left[\tilde{H}'(\tau_{m+1}) \ \ldots \ \tilde{H}'(\tau_n) \right] \rangle \quad \text{disconn.}$$

But the factor in the last two lines is just $\langle S(\beta)\rangle_0$, while the rest is the contribution of the connected part. This proves the statement. Finally, by restricting attention only to topologically inequivalent diagrams, we can drop the $1/m!$.

To derive the Dyson equation, we use the fact that a factor $\tilde{G}^{(1,0)}_{\underset{\sim}{k}}(i\omega_\ell)$ factors from all the terms in the series for $\tilde{G}^{(1,0)}_{k}(i\omega_\ell)$ for a system with spatially homogeneous Hamiltonian. The series for $\tilde{G}^{(1)}_{\underset{\sim}{k}}(i\omega_\ell)$ looks like this:

$\tilde{G}^{(1)}_{\underset{\sim}{k}}(i\omega_\ell) =$ 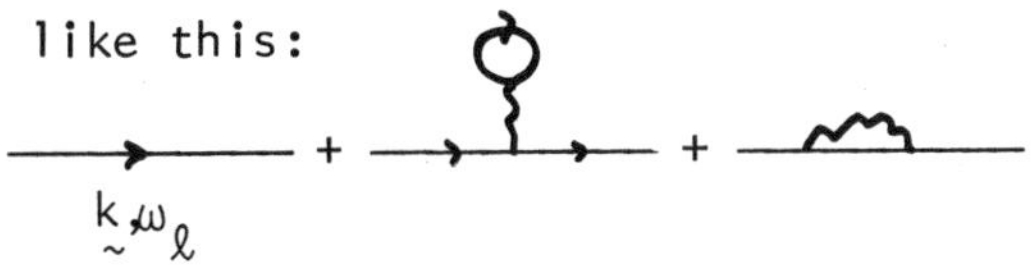

We factor off $\tilde{G}^{(1,0)}_{\underset{\sim}{k}}(i\omega_\ell)$:

$\tilde{G}^{(1)}_{\underset{\sim}{k}}(i\omega_\ell) = \tilde{G}^{(1,0)}_{\underset{\sim}{k}}(i\omega_\ell) + \tilde{G}^{(1,0)}_{\underset{\sim}{k}}(i\omega_\ell)$ (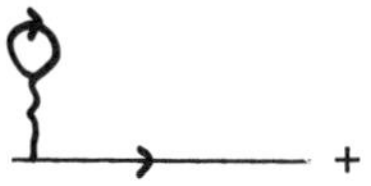

Now, in each term inside the parenthesis, we move from left to right until we find a $\tilde{G}^{(1,0)}_{\underset{\sim}{k}}(i\omega_\ell)$ line which, when cut, splits the diagram in two. We can again factor the diagram at this point. Making this factorization we have

$$\tilde{G}^{(1)}_{\underset{\sim}{k}}(i\omega_\ell) = \tilde{G}^{(1,0)}_{\underset{\sim}{k}}(i\omega_\ell) + \tilde{G}^{(1,0)}_{\underset{\sim}{k}}(i\omega_\ell)$$

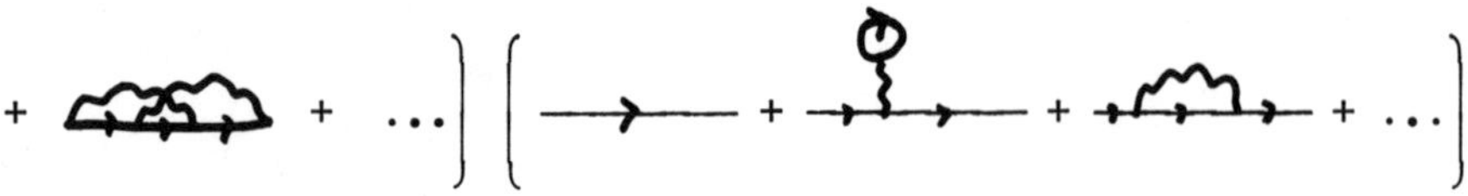

Some reflection shows that, in the right-hand parenthesis, we have everything that starts with a $\tilde{G}^{(1,0)}_{\underset{\sim}{k}}(i\omega_\ell)$ line --that is, we have the complete series for $\tilde{G}^{(1)}_{\sim}$. Denoting the stuff in the first parenthesis by Σ we have shown that

$$\tilde{G}^{(1)}_{\underset{\sim}{k}}(i\omega_\ell) = \tilde{G}^{(1,0)}_{\underset{\sim}{k}}(i\omega_\ell) + \tilde{G}^{(1,0)}_{\underset{\sim}{k}}(i\omega_\ell)\Sigma_{\underset{\sim}{k}}(i\omega_\ell)\tilde{G}^{(1)}_{\underset{\sim}{k}}(i\omega_\ell) \qquad (4\text{-}2)$$

where

$$\Sigma_{\underset{\sim}{k}}(i\omega_\ell) = \left\{\begin{matrix}\text{sum of all diagrams that}\\ \text{cannot be cut by cutting}\\ \text{one } \tilde{G}^{(1,0)}_{\underset{\sim}{k}}(i\omega_\ell) \text{ line}\end{matrix}\right\} \frac{1}{[\tilde{G}^{(1,0)}_{\underset{\sim}{k}}(i\omega_\ell)]^2}$$

(4-2) is one form of Dyson's equation. (For fermions we can take account of spin by making (4-2) a 2 x 2 matrix equation.)

Solving (4-2) for $\tilde{G}^{(1)}_{\underset{\sim}{k}}(i\omega_\ell)$ we have

$$\tilde{G}^{(1)}_{\underset{\sim}{k}}(i\omega_\ell) = \frac{1}{[\tilde{G}^{(1,0)}_{\underset{\sim}{k}}(i\omega_\ell)]^{-1} - \Sigma_{\underset{\sim}{k}}(i\omega_\ell)}$$

From Eq. (3-2):

$$\tilde{G}^{(1)}_{\underset{\sim}{k}}(i\omega_\ell) = \frac{1}{i\omega_\ell - \xi(\underset{\sim}{k}) - \Sigma_{\underset{\sim}{k}}(i\omega_\ell)}$$

$(\xi(\underset{\sim}{k}) = (\varepsilon(\underset{\sim}{k}) - \mu)/\hbar)$ Referring to Eq. (2-4), one sees that Σ shifts the energy of the pole in the noninteracting Green function which was at $\hbar\omega = \varepsilon(\underset{\sim}{k})$. For this reason $\Sigma_{\underset{\sim}{k}}(i\omega_\ell)$ is called the self-energy. The great advantage in looking at the self-energy is not only that $\Sigma_{\underset{\sim}{k}}(i\omega_\ell)$ contains only irreducible diagrams. Another advantage is that the pole in $\tilde{G}^{(1)}_{\underset{\sim}{k}}(i\omega_\ell)$ can be shifted from $\xi(\underset{\sim}{k})$ without going to very high orders in a series for Σ, while one has to go to infinite order in the series for $\tilde{G}^{(1)}_{\underset{\sim}{k}}(i\omega_\ell)$. This is because each term in the analytically continued function has a singularity at $\omega = \xi(\underset{\sim}{k})$ which can't be removed by a finite number of terms.

In the case in which the only term in (1-14) is the g_4 term, the first order terms in $\Sigma_{\underset{\sim}{k}}(i\omega_\ell)$ are given by the following diagrams:

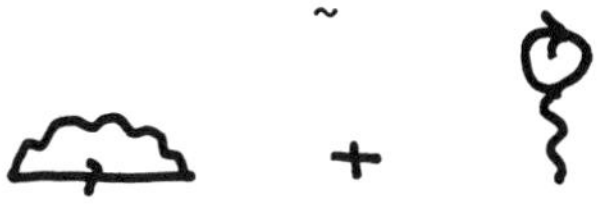

It is instructive to evaluate these and illustrate the technique for doing frequency sums. One has

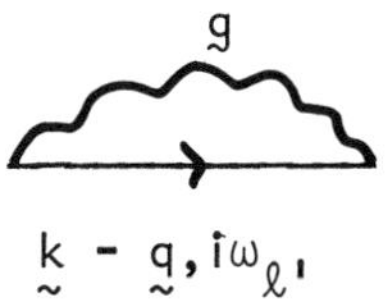

$$= \left(\frac{-1}{\hbar\beta}\right) \sum_{\underset{\sim}{q}} \sum_{\ell'} g_4(\underset{\sim}{q}) \frac{1}{[i\hbar\omega_{\ell'} - (\varepsilon_{\underset{\sim}{k}-\underset{\sim}{q}} - \mu)]} \qquad (4\text{-}3a)$$

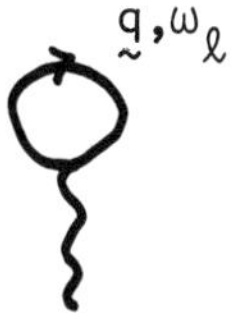

$$= \left(\frac{-1}{\hbar\beta}\right) g_4(\underset{\sim}{q}=0) \sum_{\underset{\sim}{q}} \sum_{\ell'} \frac{1}{[i\hbar\omega_{\ell'} - (\varepsilon_{\underset{\sim}{q}} - \mu)]} \qquad (4\text{-}3b)$$

The frequency sum is done as follows (we do it for bosons here):

$$\frac{1}{\beta} \sum_{\ell} \frac{1}{(i\hbar\omega_\ell - (\varepsilon - \mu))} = \frac{1}{2\pi i} \oint_C f(z) \frac{dz}{(z - (\varepsilon - \mu))}$$

where C is sketched below

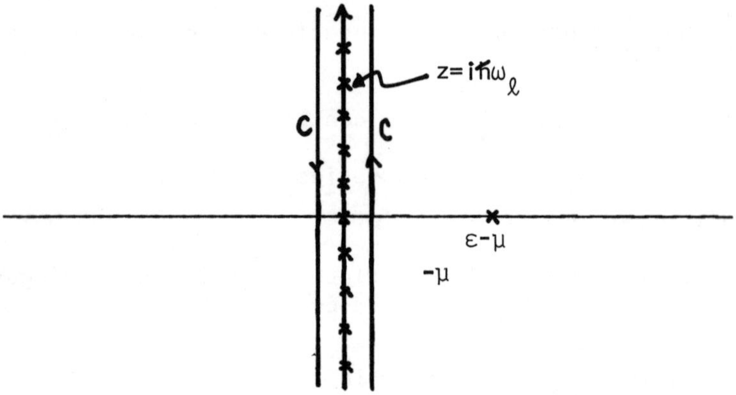

and $f(z) = \dfrac{1}{(e^{z\beta}-1)}$

$$\oint_C \frac{f(z)\,dz}{z-(\varepsilon-\mu)} = \oint_{C'} \frac{f(z)\,dz}{(z-(\varepsilon-\mu))} = -2\pi i\; f(\varepsilon-\mu)$$

where C' is

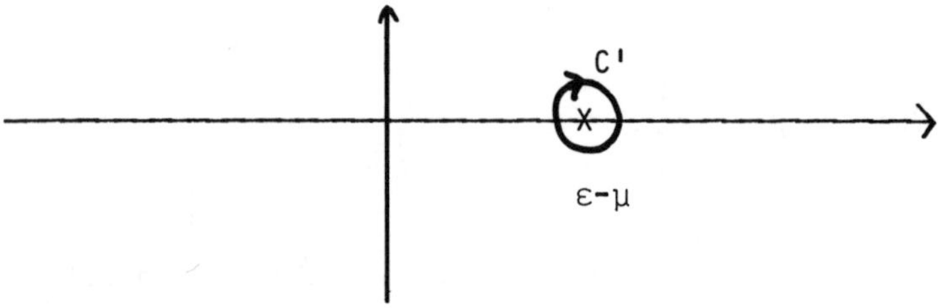

Thus

$$= \frac{1}{\hbar} \sum_{\underset{\sim}{q}} g_4(\underset{\sim}{q})\, f(\varepsilon_{\underset{\sim}{k}-\underset{\sim}{q}}-\mu)$$

 $= \frac{1}{\hbar} g_4(\underset{\sim}{q}=0) \sum_{\underset{\sim}{q}'} f(\varepsilon_{\underset{\sim}{q}'}-\mu)$

The poles in $G_r^{(1)}(\omega)$ are shifted to

$$\hbar\omega = \varepsilon_{\underset{\sim}{k}} + \sum_{\underset{\sim}{q}'} \left[g_4(0) + g_4(\underset{\sim}{k}-\underset{\sim}{q}') \right] f(\varepsilon_{q'}-\mu)$$

in this approximation.

The same example can be used to illustrate how the perturbation series expressions are used to generate uncontrolled, mean-field like approximations for $\tilde{G}_{\underset{\sim}{k}}^{(1)}(i\omega_\ell)$ in the strongly interacting case. The self-energy is

$\sum(i\omega_\ell) =$  $+$  $+ \ . \ . \ .$

The additional terms clearly include

 $+$  $+$  $+ \ . \ . \ .$

That is, all the terms which give a full $\tilde{G}_{\underset{\sim}{k}}^{(1)}(i\omega_\ell)$ (instead of a non-interacting $\tilde{G}_{\underset{\sim}{k}}^{(1,0)}(i\omega_\ell)$) as a replacement for the line 
in the diagram . We denote

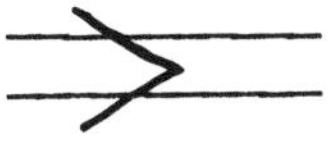 $\equiv \tilde{G}_{\underset{\sim}{k}}^{(1)}(i\omega_\ell)$

and approximate the series for $\sum_{\underset{\sim}{k}}(i\omega_\ell)$ by

$\sum_{\underset{\sim}{k}}(i\omega_\ell) \approx$  $+$  $\equiv \sum_{\underset{\sim}{k}}^{HF}(i\omega_\ell)$

making a similar replacement in  . $\Sigma^{HF}_{\underset{\sim}{k}}(i\omega_\ell)$ is so denoted because it leads a generalization of the Hartree Fock approximation. Substituting (4-4) into (4-2) one has, by use of (4-3) that

$$\tilde{G}^{(1)}_{\underset{\sim}{k}}(i\omega_\ell) = \tilde{G}^{(1,0)}_{\underset{\sim}{k}}(i\omega_\ell)$$

$$- \left(\frac{1}{\beta}\right) \tilde{G}^{(1,0)}(i\omega_\ell) \sum_{\underset{\sim}{q}',\ell'} \left[g_4(0) + g_4(\underset{\sim}{k} - \underset{\sim}{q}')\right] \tilde{G}^{(1)}_{\underset{\sim}{q}'}(i\omega_{\ell}')\tilde{G}^{(1)}_{\underset{\sim}{k}}(i\omega_\ell) \tag{4-5a}$$

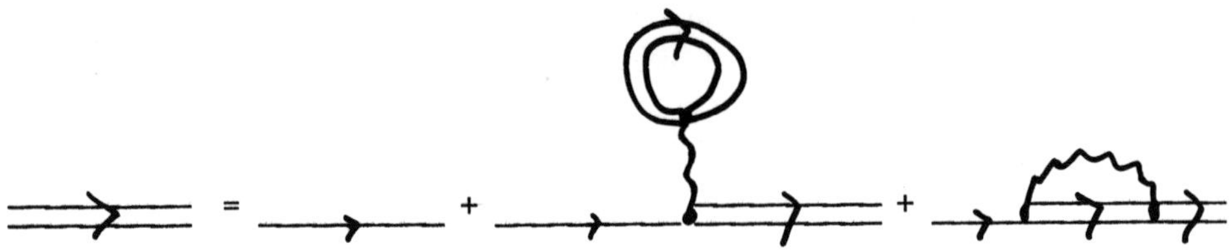

(4-5b)

Equation (4-5a) can be solved[10] self-consistently for $\tilde{G}^{(1)}$.

5. Perturbation Expansion of $G^{(2)}(\omega)$

We now turn to perturbation theory for $G^{(2)}$. As in Section 3, we define

$$\tilde{G}^{(2)}(\tau_1,\tau_2,\tau_3,\tau_4) = \langle T_\tau\left[A_{\underset{\sim}{k}_1}(\tau_1)A_{\underset{\sim}{k}_2}(\tau_2)A_{\underset{\sim}{k}_3}(\tau_3)A_{\underset{\sim}{k}_4}(\tau_4)\right]\rangle \tag{5-1}$$

The operators $A_{\underset{\sim}{k}}(\tau)$ are defined by

$$A_{\underset{\sim}{k}}(\tau) = e^{H\tau} A_{\underset{\sim}{k}} e^{-H\tau} \tag{5-2}$$

Recall that $A_{\underset{\sim}{k}}$ is either a fermion or a boson creation or annihilation operator. T_τ is the imaginary time ordering operator, which orders later times to the left in a generalization of Eq. (2-3). $\tilde{G}^{(2)}(\tau_1,\tau_2,\tau_3,\tau_4)$ is related to $G^{(2)}(t)$ of Eq. (1-20) as follows. Define

$$\tilde{G}_{\underset{\sim}{k}_1\underset{\sim}{k}_2\underset{\sim}{k}_3\underset{\sim}{k}_4}(\tau) \equiv \tilde{G}^{(2)}_{\underset{\sim}{k}_1\underset{\sim}{k}_2\underset{\sim}{k}_3\underset{\sim}{k}_4}(\tau+0^+,\tau,\ 0^+,\ 0) \tag{5-3}$$

Then one can define

$$\tilde{G}^{(2)}_{\underset{\sim}{k}_1\underset{\sim}{k}_2\underset{\sim}{k}_3\underset{\sim}{k}_4}(i\omega_\ell) = -\frac{1}{2}\int_{-\beta}^{\beta} d\tau\ \tilde{G}^{(2)}_{\underset{\sim}{k}_1\underset{\sim}{k}_2\underset{\sim}{k}_3\underset{\sim}{k}_4}(\tau)e^{i\omega_\ell\tau} \tag{5-4}$$

and can prove

$$\tilde{G}^{(2)}(\tau<0) = \tilde{G}^{(2)}(\tau+\beta) \tag{5-5}$$

with a plus sign for <u>either</u> fermions or bosons. Then (5-4) becomes

$$\tilde{G}^{(2)}_{\underset{\sim}{k}_1\underset{\sim}{k}_2\underset{\sim}{k}_3\underset{\sim}{k}_4}(i\omega_\ell) = -\int_{0}^{\beta} d\tau e^{i\omega_\ell\tau}\ \tilde{G}^{(2)}_{\underset{\sim}{k}_1\underset{\sim}{k}_2\underset{\sim}{k}_3\underset{\sim}{k}_4}(\tau)$$

with $\omega_\ell = 2\pi\ell/\beta$.

Defining

$$\tilde{A}(\tau) = A_{\underset{\sim}{k}_1}(\tau+0^+)\ A_{\underset{\sim}{k}_2}(\tau)$$

$$B(0) = A_{\underset{\sim}{k}_3}(0^+)\ A_{\underset{\sim}{k}_4}(0)$$

one sees from (6-7) of Stinchcombe's lectures and (1-13) of the present chapter that

$$\tilde{G}^{(2)}_{\underset{\sim}{k}_1 \cdots \underset{\sim}{k}_4}(\omega) = \frac{1}{2\pi}\ \tilde{G}^{(2)}_{\underset{\sim}{k}_1\underset{\sim}{k}_2\underset{\sim}{k}_3\underset{\sim}{k}_4}(i\omega_\ell \to \omega-2\sigma\mu + i0^+) \tag{5-6}$$

$$\sigma = \begin{cases} +1 \text{ if } A_{\underset{\sim}{k}_3} \text{ and } A_{\underset{\sim}{k}_4} \text{ are both creation operators} \\ 0 \text{ if } A_{\underset{\sim}{k}_{3 \atop 4}} \text{ creates and } A_{\underset{\sim}{k}_{4 \atop 3}} \text{ destroys} \\ -1 \text{ if } A_{\underset{\sim}{k}_3} \text{ and } A_{\underset{\sim}{k}_4} \text{ are destruction operators.} \end{cases} \tag{5-7}$$

Here

$$\tilde{G}^{(2)}_{\underset{\sim}{k}_1\underset{\sim}{k}_2\underset{\sim}{k}_3\underset{\sim}{k}_4}(\omega) = \frac{1}{2\pi}\int_{-\infty}^{\infty} e^{i\omega t}\tilde{G}^{(2)}_{\underset{\sim}{k}_1\underset{\sim}{k}_2\underset{\sim}{k}_3\underset{\sim}{k}_4}(t)\,dt \tag{5-8}$$

and $G^{(2)}$ is defined in (1-13).

Returning to Eqs. (5-1) and (5-3) it is straightforward to extend the derivation of Eq. (2-11) with the result

$$\tilde{G}^{(2)}_{\underset{\sim}{k}_1\underset{\sim}{k}_2\underset{\sim}{k}_3\underset{\sim}{k}_4}(\tau>0) = \frac{\left\langle T_\tau\left(\tilde{A}_{\underset{\sim}{k}_1}(\tau+0^+)\tilde{A}_{\underset{\sim}{k}_2}(\tau)\tilde{A}_{\underset{\sim}{k}_3}(0^+)A_{\underset{\sim}{k}_4}(0)S(\beta)\right)\right\rangle_0}{\langle S(\beta)\rangle_0} \tag{5-9}$$

where $S(\beta)$ has the definition (2-12). Eq. (5-9) gives the desired perturbation expansion for $\tilde{G}^{(2)}_{\underset{\sim}{k}_1\underset{\sim}{k}_2\underset{\sim}{k}_3\underset{\sim}{k}_4}(\tau>0)$. Finally, the proof of Wick's Theorem is easily extended to this case so that the terms in the series can be expressed in terms of non interacting single particle thermal green functions $\tilde{G}^{(1,0)}(\tau>0)$ or $\tilde{G}^{(1,0)}(i\omega_\ell)$ given in Eqs. (3-1) and (3-2). To introduce diagrams for this case, we again specialize to the case in which only the coupling g_4 of Eq. (1-14) is non zero. Terms in the perturbation series are defined as before except that each term is associated with 4 external lines, rather than two. Representing the terms in the series before Fourier transforms by diagrams, one sees that they have a somewhat different form in the cases indicated in Eq. (5-7). In particular, one finds for the zero order terms the following diagrams if $A_{\underset{\sim}{k}_3}$ and $A_{\underset{\sim}{k}_4}$ are creation operators

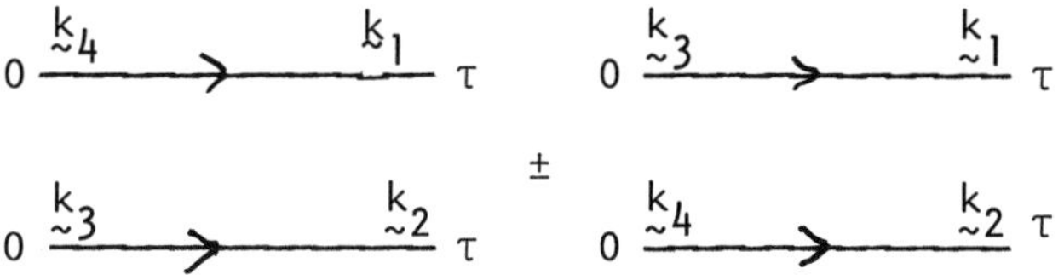

(5-10)

In general, all terms have two lines entering at time 0 and two lines leaving at time τ. For the case in which $A_{\underset{\sim}{k}_4}$ creates and $A_{\underset{\sim}{k}_3}$ destroys, the zero order diagrams are

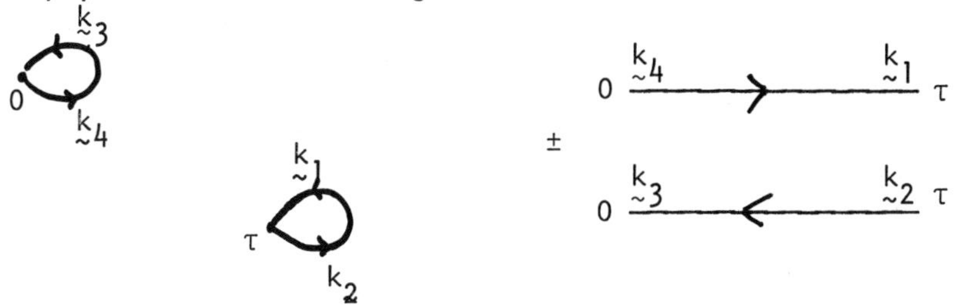

(5-11)

Here the arrows pointing to the left denote a zero order $G^{(1,0)}$ with negative time argument:

$$\equiv \langle T_\tau\left(\tilde{\alpha}_{\underset{\sim}{k}_3}(0)\,\bar{\tilde{\alpha}}_{\underset{\sim}{k}_2}(\tau)\right)\rangle_0$$

$$= \langle T_\tau\left(\tilde{\alpha}_{\underset{\sim}{k}_3}(-\tau)\tilde{\alpha}_{\underset{\sim}{k}_2}(0)\right)\rangle_0$$

$$\equiv \tilde{G}^{(1,0)}_{\underset{\sim}{k}_3,\underset{\sim}{k}_2}(-\tau) = \tilde{G}^{(1,0)}_{\underset{\sim}{k}_2}(-\tau)\,\delta_{\underset{\sim}{k}_3,\underset{\sim}{k}_2}$$

where we also indicate the meaning of the two k's on the lines here.

It is not hard to see that, in the Fourier transformed series, one associates diagrams with arrows pointing to the left with negative frequencies. The rules for constructing diagrams then require that all diagrams be formed with 2 lines entering from the left and two leaving to the right (if $A_{\underset{\sim}{k}_3}$ and $A_{\underset{\sim}{k}_4}$ both create) or, alternatively, with one line entering and one leaving from the left (if $A_{\underset{\sim}{k}_3}$ creates and $A_{\underset{\sim}{k}_4}$ destroys or vice versa). The two cases will be illustrated below with the examples of two phonon spectra in the quasi particle model of helium and the electron gas respectively.

Factors $g_4(\underset{\sim}{k})$ are denoted by wiggly lines as in Eq. (3-8).

It is now straightforward to extend the linked cluster theorem of Section 4 to the series for $G^{(2)}$ so that (5-9) becomes

$$\tilde{G}^{(2)}_{\underset{\sim}{k}_1,\underset{\sim}{k}_2,\underset{\sim}{k}_3,\underset{\sim}{k}_4}(\tau>0) = T_\tau\left\{\tilde{A}_{\underset{\sim}{k}_1}(\tau+0^+)\tilde{A}_{\underset{\sim}{k}_2}(\tau)\tilde{A}_{\underset{\sim}{k}_3}(0^+)\tilde{A}_{\underset{\sim}{k}_4}(0)S(\beta)\right\}>_0^{(conn)}$$

Here connected diagrams are defined as follows. We draw all diagrams taking care to keep dots referring to the same time exactly above each other. (This means that all wiggly lines are vertical.) A connected diagram is one which cannot be cut in two by a vertical line when the diagram is drawn this way. The next step is the derivation of a Dyson equation for $G^{(2)}$. Before doing this, we discuss some approximations for $G^{(2)}$ based on the zero terms shown in Eqs. (5-10) and (5-11).

6. Zero-Order Approximations for $G^{(2)}$

We now evaluate the zero order approximations to $G^{(2)}$ in the two cases in which two particles are created (Eq. (5-10) and in which one particle is created and another destroyed (Eq. (5-11). For the Fourier transform in time one has in the first case

$$\tilde{G}^{(2,0)}_{\underset{\sim}{k}_1\underset{\sim}{k}_2\underset{\sim}{k}_3\underset{\sim}{k}_4}(i\omega_\ell)$$

$$= -\int_0^\beta e^{i\omega_\ell\tau}\Big[<T_\tau\left\{\tilde{\alpha}_{\underset{\sim}{k}_1}(\tau)\bar{\tilde{\alpha}}_{\underset{\sim}{k}_2}(0)\right\}>_0 <T_\tau\left\{\tilde{\alpha}_{\underset{\sim}{k}_2}(\tau)\bar{\tilde{\alpha}}_{\underset{\sim}{k}_3}(0)\right\}>_0$$

$$\pm <T_\tau\left\{\tilde{\alpha}_{\underset{\sim}{k}_1}(\tau)\bar{\tilde{\alpha}}_{\underset{\sim}{k}_3}(0)\right\}>_0 <T_\tau\left\{\tilde{\alpha}_{\underset{\sim}{k}_2}(\tau)\bar{\tilde{\alpha}}_{\underset{\sim}{k}_4}(0)\right\}>_0\Big]d\tau$$

$$= -\Big\{\delta_{\underset{\sim}{k}_1,\underset{\sim}{k}_4}\delta_{\underset{\sim}{k}_2,\underset{\sim}{k}_3}$$

$$\pm \delta_{\underset{\sim}{k}_1,\underset{\sim}{k}_3}\delta_{\underset{\sim}{k}_2,\underset{\sim}{k}_4}\Big\}\frac{1}{\beta}\sum_{\ell'}\tilde{G}^{(1,0)}_{\underset{\sim}{k}_1}(i\omega_{\ell'})\tilde{G}^{(1,0)}_{\underset{\sim}{k}_2}(i\omega_\ell - i\omega_{\ell'})$$

$$= \left[\delta_{\underset{\sim}{k}_1,\underset{\sim}{k}_4}\delta_{\underset{\sim}{k}_2,\underset{\sim}{k}_3} \pm \delta_{\underset{\sim}{k}_1,\underset{\sim}{k}_3}\delta_{\underset{\sim}{k}_2,\underset{\sim}{k}_4}\right] \frac{f(\varepsilon_{\underset{\sim}{k}_1}-\mu)-f(-\varepsilon_{\underset{\sim}{k}_2}+\mu)}{(i\omega_\ell-\varepsilon_{\underset{\sim}{k}_1}-\varepsilon_{\underset{\sim}{k}_2}-2\mu)} \qquad (6\text{-}1)$$

Here

$$f(\varepsilon_{\underset{\sim}{k}}-\mu) = \frac{1}{e^{\beta(\varepsilon_{\underset{\sim}{k}}-\mu)} \pm 1}$$

and we have kept signs for bosons (upper sign) and fermions (lower sign). We note that $-f(-x) = f(x)\pm 1$ so that

$$\tilde{G}^{(2,0)}_{\underset{\sim}{k}_1\underset{\sim}{k}_2\underset{\sim}{k}_3\underset{\sim}{k}_4}(i\omega_\ell) =$$

$$(\delta_{\underset{\sim}{k}_1,\underset{\sim}{k}_4}\delta_{\underset{\sim}{k}_2,\underset{\sim}{k}_3} \pm \delta_{\underset{\sim}{k}_1,\underset{\sim}{k}_3}\delta_{\underset{\sim}{k}_2,\underset{\sim}{k}_4}) \frac{\left[f(\varepsilon_{\underset{\sim}{k}_1}-\mu) + f(\varepsilon_{\underset{\sim}{k}_2}-\mu) \pm 1\right]}{(i\omega_\ell-\varepsilon_{\underset{\sim}{k}_1}-\varepsilon_{\underset{\sim}{k}_2}-2\mu)} \qquad (6\text{-}2)$$

in this case. The sum on ℓ' in (6-1) has been done by an extension of the method used in Section 5. In the same way, we get the following result for $\tilde{G}^{(2,0)}_{\underset{\sim}{k}_1\underset{\sim}{k}_2\underset{\sim}{k}_3\underset{\sim}{k}_4}(i\omega_\ell)$ in the second case, using the diagrams shown in (5-11):

$$\tilde{G}^{(2,0)}_{\underset{\sim}{k}_1,\underset{\sim}{k}_2,\underset{\sim}{k}_3,\underset{\sim}{k}_4}(i\omega_\ell) =$$

$$- \int_0^\beta e^{i\omega_\ell\tau}\left[\langle T_\tau\left(\tilde{\alpha}_{\underset{\sim}{k}_1}(\tau)\bar{\tilde{\alpha}}_{\underset{\sim}{k}_2}(\tau+0^+)\right)\rangle_0 \langle T_\tau\left(\tilde{\alpha}_{\underset{\sim}{k}_3}(0)\bar{\tilde{\alpha}}_{\underset{\sim}{k}_4}(0^+)\right)\rangle_0 \right.$$

$$\left. \pm \langle T_\tau\left(\tilde{\alpha}_{\underset{\sim}{k}_2}(\tau)\bar{\tilde{\alpha}}_{\underset{\sim}{k}_4}(0)\right)\rangle_0 \langle T_\tau\left(\tilde{\alpha}_{\underset{\sim}{k}_3}(0)\bar{\tilde{\alpha}}_{\underset{\sim}{k}_4}(\tau)\right)\rangle_0 \right] d\tau$$

$$= -\frac{1}{\beta}\left[\delta_{\underset{\sim}{k}_1,\underset{\sim}{k}_2}\delta_{\underset{\sim}{k}_3,\underset{\sim}{k}_4} \sum_{\ell'} \tilde{G}^{(1,0)}_{\underset{\sim}{k}_1}(i\omega_{\ell'}) \sum_{\ell''} \tilde{G}^{(1,0)}_{\underset{\sim}{k}_3}(i\omega_{\ell''})\right.$$

$$\pm\ \delta_{\underset{\sim}{k}_1,\underset{\sim}{k}_4}\ \delta_{\underset{\sim}{k}_2,\underset{\sim}{k}_3} \sum_{\ell'} \tilde{G}^{(1,0)}_{\underset{\sim}{k}_1}(i\omega_{\ell'})\tilde{G}^{(1,0)}_{\underset{\sim}{k}_3}(i\omega_\ell+i\omega_{\ell'})\Bigg]$$

$$= -\beta\ \delta_{\underset{\sim}{k}_1,\underset{\sim}{k}_2}\delta_{\underset{\sim}{k}_3,\underset{\sim}{k}_4}\ f(\varepsilon_{\underset{\sim}{k}_1})f(\varepsilon_{\underset{\sim}{k}_3})$$

$$\pm\ \delta_{\underset{\sim}{k}_1,\underset{\sim}{k}_4}\delta_{\underset{\sim}{k}_2,\underset{\sim}{k}_3}\ \frac{\left[f(\varepsilon_{\underset{\sim}{k}_3}-\mu)\ -\ f(\varepsilon_{\underset{\sim}{k}_1}-\mu)\right]}{(-i\omega_\ell+\varepsilon_{\underset{\sim}{k}_1}-\varepsilon_{\underset{\sim}{k}_3})} \qquad (6\text{-}3)$$

We illustrate the significance of Eq. (6-2) and (6-3) by applying them to two simple models for well-known problems. We apply (6-2) to the calculation of the two roton Raman scattering in the quasiparticle model for liquid helium. As emphasized in the introduction, the quasiparticle model itself is inadequate to this problem and the zero order approximation is even worse so this calculation is of only pedagogical significance. Using Eq. (1-12) and (4-15) of Stinchcombe's lectures, one finds that the function $S_4(\underset{\sim}{k}_1,\underset{\sim}{k}_2,\underset{\sim}{k}_3,\underset{\sim}{k}_4,\omega)$ entering the extinction coefficient is

$$S_4(\underset{\sim}{k}_1,\underset{\sim}{k}_2,\underset{\sim}{k}_3,\omega) =$$

$$\frac{1}{N^3}\ \frac{\langle\langle A;B\rangle\rangle(-\omega+i0^+)\ -\langle\langle A;B\rangle\rangle(-\omega-i0^+)}{(e^{-\omega\beta}-1)} \qquad (6\text{-}4)$$

in which

$$A = \rho_{\underset{\sim}{k}_3}\ \rho_{-(\underset{\sim}{k}_1+\underset{\sim}{k}_2+\underset{\sim}{k}_3)}$$

$$B = \rho_{\underset{\sim}{k}_1}\ \rho_{\underset{\sim}{k}_2}$$

Using (1-8) one then finds

$$\langle\langle A;B\rangle\rangle(-\omega+i0^+) \equiv \langle\langle \rho_{\underset{\sim}{k}_3}\rho_{-\underset{\sim}{k}_1-\underset{\sim}{k}_2-\underset{\sim}{k}_3};\rho_{\underset{\sim}{k}_1}\rho_{\underset{\sim}{k}_2}\rangle\rangle(-\omega+i0^+)$$

$$= N^2\left[S(\underset{\sim}{k}_3)S(-\underset{\sim}{k}_1-\underset{\sim}{k}_2-\underset{\sim}{k}_3)S(\underset{\sim}{k}_1)S(\underset{\sim}{k}_2)\right]^{1/2}$$

$$\times \Big\{ \langle\langle \alpha^\dagger_{k_3} \alpha^\dagger_{-k_1-k_2-k_3} ; \alpha_{-k_1} \alpha_{-k_2} \rangle\rangle(-\omega+i0^+)$$

$$+ \langle\langle \alpha^\dagger_{k_3} \alpha_{k_1+k_2+k_3} ; \alpha^\dagger_{k_1} \alpha_{-k_2} \rangle\rangle(-\omega+i0^+)$$

$$+ \langle\langle \alpha^\dagger_{k_3} \alpha_{k_1+k_2+k_3} ; \alpha_{-k_1} \alpha^\dagger_{k_2} \rangle\rangle(-\omega+i0^+)$$

$$+ \langle\langle \alpha_{-k_3} \alpha^\dagger_{-k_1-k_2-k_3} ; \alpha_{-k_1} \alpha^\dagger_{k_2} \rangle\rangle(-\omega+i0^+) \qquad (6\text{-}5)$$

$$+ \langle\langle \alpha_{-k_3} \alpha^\dagger_{-k_1-k_2-k_3} ; \alpha^\dagger_{k_1} \alpha_{-k_2} \rangle\rangle(-\omega+i0^+)$$

$$+ \langle\langle \alpha_{-k_3} \alpha_{k_1+k_2+k_3} ; \alpha^\dagger_{k_1} \alpha^\dagger_{k_2} \rangle\rangle(-\omega+i0^+) \Big\}$$

The last term in (6-5) described light scattering with the production of two quasiparticles. Keeping only this last term, we can evaluate the zero order expression for the Green function by use of (6-2). Using (1-9), one finds the following for the contributions to the extinction coefficient arising from two quasiparticle creation. (μ=0 here. In the following equation, $\omega=\omega_o-\omega_n$).

$$h^{(2,0)}(\omega_n) = \rho^3\left(\frac{\omega_n}{c}\right)^4 \frac{1}{2\pi N}\sum_{q} S(q-k_n)\,S(k_o-q)$$

$$\times \frac{\left[f(\varepsilon_{q-k_n}) + f(\varepsilon_{k_o-q}) + 1\right]}{e^{-(\varepsilon_{q-k_n}+\varepsilon_{k_o-q})\beta}-1}\;\delta(\omega-\varepsilon_{q-k_n}-\varepsilon_{k_o-q})$$

$$\times \left[F(q)\,F(k_n+k_o-q) + F^2(q)\right]$$

In the limit that $k_o, k_n \to 0$ we use the fact that $F(q) = F(-q)$, and obtain

$$h^{(2,0)}(\omega_n) = \rho^3 \left(\frac{\omega_n}{c}\right)^4 \frac{1}{\pi N} \sum_{\underset{\sim}{q}} \left|S(\underset{\sim}{q})\right|^2 \left|F(\underset{\sim}{q})\right|^2$$

$$\times\ (2f(\varepsilon \underset{\sim}{q})+1)\ (f(2\varepsilon \underset{\sim}{q})+1)\ \delta(\omega - 2\varepsilon \underset{\sim}{q})$$

Finally we note that

$$(2f(x)+1)\ (f(2x)+1) = (1+f(x))^2$$

if $f(x) = \frac{1}{e^x - 1}$ as it is here so that

$$h^{(2,0)}(\omega_n) = \rho^3 \left(\frac{\omega_n}{c}\right)^4 \frac{1}{\pi N} \sum_{\underset{\sim}{q}} \left|S(\underset{\sim}{q})\right|^2 \left|F(\underset{\sim}{q})\right|^2 \left(1+f(\varepsilon \underset{\sim}{q})\right)^2 \delta(\omega - 2\varepsilon \underset{\sim}{q}) \quad (6\text{-}6)$$

Equation (6-6) is identical to the corresponding term in Eq. (5-4) of reference 11, describing a noninteracting quasiparticle theory of the Raman scattering. This is the only term which survives at zero temperature.

We next consider the electron gas[12] as an application of Eq. (6-3). The neutron scattering cross-section can be computed by calculating the density-density correlation function of Eq. (1-1). One is more interested in the longitudinal dielectric constant. Assuming an external charge density

$$z\,\rho\ \text{ext}\ (\underset{\sim}{r},t) = z\,\rho\ \text{ext}\ (\underset{\sim}{k},\omega)\ e^{i(\underset{\sim}{k}\,\underset{\sim}{r} - \mathcal{E}\omega t) + \varepsilon t} + \text{h.c.}$$

one has a Hamiltonian

$$\mathcal{H} = \mathcal{H}_o + \mathcal{H}'$$

where $\mathcal{H}_o$ describes the electron gas and $\mathcal{H}'$ the interaction with the external charge. The latter is

$$\mathcal{H}' = -\int d^3\underset{\sim}{r} \int d^3\underset{\sim}{r}'\ \rho(\underset{\sim}{r})\ \frac{ze}{|\underset{\sim}{r}-\underset{\sim}{r}'|}\ \rho_{ext}(\underset{\sim}{r}',t)$$

$$\frac{-4\,ez}{\underset{\sim}{k}^2}\ \rho_{-\underset{\sim}{k}}\ \rho_{ext}\ (\underset{\sim}{k},\omega) e^{-i\omega t + \varepsilon t} + \text{h.c.}$$

If one writes the response $\langle\rho_{-\underset{\sim}{k}}\rangle(t)$ in the presence of external charge (without the Hermitian conjugate) as

$$<\rho_{\underset{\sim}{k}}>(t) = \chi(\underset{\sim}{k},\omega)\frac{4\pi ez}{k^2}\rho_{ext}(\underset{\sim}{k},\omega)e^{i\omega t + \varepsilon t}$$

then it follows easily from the chapter by Stinchcombe that

$$\chi(\underset{\sim}{k},\omega) = -\frac{2\pi}{\hbar} <<\rho_{\underset{\sim}{k}} ; \rho_{-\underset{\sim}{k}}>>_r(\omega) \tag{6-7}$$

To relate $\chi(\underset{\sim}{k},\omega)$ to the longitudinal dielectric constant one writes the macroscopic Maxwell equation as

$$i\underset{\sim}{k}\cdot\underset{\sim}{D} = 4\pi z\,\rho_{ext}(\underset{\sim}{k},\omega)$$

$$i\underset{\sim}{k}\cdot\underset{\sim}{E} = 4\pi\left[-e<\rho>(\underset{\sim}{k},\omega) + z\,\rho_{ext}(\underset{\sim}{k},\omega)\right]$$

From which it follows easily that the longitudinal dielectric function $\varepsilon_{\parallel}(k,\omega)$ is

$$\varepsilon_{\parallel}(\underset{\sim}{k},\omega) = \frac{D(\underset{\sim}{k},\omega)}{E(\underset{\sim}{k},\omega)} = \frac{1}{1 - \frac{4\pi e^2}{k^2}\chi(\underset{\sim}{k},\omega)} \tag{6-8}$$

To compute $\chi(\underset{\sim}{k},\omega)$ in zero th order using (6-3) we employ (6-7) and (1-4) and (1-5) to give (assuming $|\underset{\sim}{k}|>0$)

$$\chi_0(\underset{\sim}{k},\omega) = \sum_{\underset{\sim}{q},\sigma} \frac{\left[f(\varepsilon_{\underset{\sim}{k}+\underset{\sim}{q},\sigma}-\mu) - f(\varepsilon_{\underset{\sim}{q},\sigma}-\mu)\right]}{\hbar\omega + i\,0^+ - \varepsilon_{\underset{\sim}{k}+\underset{\sim}{q},\sigma} + \varepsilon_{\underset{\sim}{q},\sigma}} \tag{6-9}$$

where

$$f(x) = \frac{1}{e^{x\beta} + 1}$$

Combining (6-9) and (6-8) gives the well known Lindhard dielectric function.[13]

In each of the problems just cited, one can also use the zero-th order approximation to generate an uncontrolled but presumably more accurate approximation to $G^{(2)}$ by replacing the noninteracting factors $G^{(1,0)}$ in the zero order expression for $G^{(2)}$ by the exact $G^{(1)}$. The latter can sometimes be taken from another experiment or calculated by another approximation scheme. In diagramatic language, this approach is equivalent to inserting full lines $\Longrightarrow$ for the $\longrightarrow$ in the expressions (5-10) and (5-11). We illustrate this approximation by looking at the last term of Eq. (6-5), which with (6-4) and (1-19) gives a contribution to the extinction coefficient for second order Raman scattering in the quasiparticle model for liquid helium. The resulting expression for the Green function is, using the third expression in Eq. (6-1):

$$<<\alpha_{\mathbf{k}_3}\,\alpha_{-\mathbf{k}_1-\mathbf{k}_2-\mathbf{k}_3}\,;\,\alpha^{\dagger}_{-\mathbf{k}_1}\,\alpha^{\dagger}_{-\mathbf{k}_2}>>(i\omega_\ell)$$

$$= \left(\delta_{\mathbf{k}_3,-\mathbf{k}_1}+\delta_{\mathbf{k}_3,-\mathbf{k}_2}\right)\left(\frac{-1}{\beta}\sum_{\ell'}\tilde{G}^{(1,qp)}_{\mathbf{k}_3}(-i\omega_{\ell'})\,\tilde{G}^{(1,qp)}_{\mathbf{k}_4}(+i\omega_\ell+i\omega_{\ell'})\right)$$

$$= \left(\delta_{\mathbf{k}_3,-\mathbf{k}_1}+\delta_{\mathbf{k}_3-\mathbf{k}_2}\right)\left(-\frac{1}{\beta}\sum_{\ell'}\tilde{G}^{(1,qp)}_{\mathbf{k}_3}(i\omega_{\ell'})\,\tilde{G}^{(1,qp)}_{\mathbf{k}_4}(+i\omega_\ell-i\omega_{\ell'})\right)$$

$$(6\text{-}10)$$

where

$$\tilde{G}^{(1,qp)}_{\mathbf{k}}(\omega) = <<\alpha_{\mathbf{k}}\,;\,\alpha^{\dagger}_{\mathbf{k}}>>(\omega) \qquad (6\text{-}11)$$

is related to the corresponding contribution to $S(\mathbf{k},\omega)$ as defined by Eq. (1-1) by the relation

$$S^{(qp)}(\mathbf{k},\omega) = \frac{2\ \mathrm{Im}\ G^{(1,qp)}_{\mathbf{k}}(\omega)}{(e^{-\hbar\omega\beta}-1)} \qquad (6\text{-}12)$$

In (6-10), $\mathbf{k}_4 = -\mathbf{k}_1-\mathbf{k}_2-\mathbf{k}_3$. One rewrites (6-10) using the fact that the analytically continued $G^{(1,qp)}_{\mathbf{k}}(z)$ which has the properties that

$$\tilde{G}^{(1,qp)}_{\mathbf{k}}(i\omega_\ell) = G^{(1,qp)}_{\mathbf{k}}(z = i\omega_\ell)$$

$$G^{(1,qp)}_{\mathbf{k}}(\omega) = G^{(1,qp)}_{\mathbf{k}}(z = \omega+i0^{+})$$

is analytic everywhere except on the real axis of the complex plane. Using this and the contour shown in Figure 1, one can rewrite the second factor in the last line of (6-10) as

$$-\frac{1}{\beta}\sum_{\ell'} \tilde{G}^{(1,qp)}_{\underset{\sim}{k}_3}(i\omega_{\ell'})\ \tilde{G}^{(1,qp)}_{\underset{\sim}{k}_4}(+i\omega_\ell - i\omega_{\ell'})$$

$$= -\frac{1}{\pi}\int dx\ f(x)\ \Big[\mathrm{Im}\ G^{(1,qp)}_{\underset{\sim}{k}_3}(x-i0^+)\ G^{(1,qp)}_{\underset{\sim}{k}_4}(+i\omega_\ell - x)$$

$$+ G^{(1,qp)}_{\underset{\sim}{k}_3}(-x+i\omega_\ell)\ \mathrm{Im}\ G^{(1,qp)}_{\underset{\sim}{k}_4}(-x+i0^+)$$

Analytically continuing $i\omega_\ell \rightarrow +\omega+i0^+$ one has

$$-\frac{1}{\beta}\sum_{\ell'} \tilde{G}^{(1,qp)}_{\underset{\sim}{k}_3}(i\omega_{\ell'})\ \tilde{G}^{(1,qp)}_{\underset{\sim}{k}_4}(+i\omega_\ell - i\omega_{\ell'})$$

$$\longrightarrow\quad -\frac{1}{\pi}\int dx\ \Big[f(x)-f(x-\omega)\Big]\ \mathrm{Im}\ G^{(1,qp)}_{\underset{\sim}{k}_3}(x+i0^+)\ \mathrm{Im}\ G^{(1,qp)}_{\underset{\sim}{k}}(\omega-x+i0^+)$$

Finally from (6-4) one has that the contribution of this term to $S^4(\underset{\sim}{k}_1,\underset{\sim}{k}_2,\underset{\sim}{k}_3,\omega)$ is

$$S_4^{(2\underset{\sim}{q}p)}(\underset{\sim}{k}_1,\underset{\sim}{k}_2,\underset{\sim}{k}_3,\omega)$$

$$\simeq \frac{2\,\mathrm{Im}\left[-\frac{1}{\beta}\sum_{\ell'} \tilde{G}^{(1,\underset{\sim}{q}p)}_{\underset{\sim}{k}_3}(i\omega_{\ell'})\ \tilde{G}^{(1,\underset{\sim}{q}p)}_{\underset{\sim}{k}_4}(+i\omega_\ell - i\omega_{\ell'})\right](\delta_{\underset{\sim}{k}_3,-\underset{\sim}{k}_1}+\delta_{\underset{\sim}{k}_3,-\underset{\sim}{k}_2})}{(e^{-\hbar\omega\beta}-1)}$$

$$= -\frac{2}{\pi}\ \frac{1}{(e^{-\hbar\omega\beta}-1)}\int dx\ \ \mathrm{Im}\ G^{(1,qp)}_{\underset{\sim}{k}_3}(x+i0^+)\ \mathrm{Im}\ G^{(1,qp)}_{\underset{\sim}{k}_4}(\omega-x+i0^+)$$

$$\times\Big[f(x)-f(x-\omega)\Big]\ (\delta_{\underset{\sim}{k}_3,-\underset{\sim}{k}_1}+\delta_{\underset{\sim}{k}_3,-\underset{\sim}{k}_2})$$

$$= \frac{-1}{2\pi} \frac{1}{(e^{-\hbar\omega\beta}-1)} \int dx\, S^{(qp)}_{\underset{\sim}{k}_3}(-x)\, S^{(qp)}_{\underset{\sim}{k}_4}(x-\omega) \left[e^{\beta(\omega-x)}-1\right]$$

$$\times \left[f(x) - f(x-\omega)\right]\left[\delta_{\underset{\sim}{k}_3,-\underset{\sim}{k}_1} + \delta_{\underset{\sim}{k}_3,-\underset{\sim}{k}_2}\right]$$

where in the last line (6-12) has been used. Finally we use

$$S^{(qp)}(-x) = e^{-x\beta}\, S^{(qp)}(x)$$

$$S^{(qp)}(x-\omega) = e^{(x-\omega)\beta}\, S^{(qp)}(\omega-x)$$

together with

$$e^{(\omega-x)\beta}(e^{x\beta}-1)\left[f(x)-f(x-\omega)\right] = (e^{\beta\omega}-1)$$

(which is easily shown) to obtain

$$S^{(2qp)}_4(\underset{\sim}{k}_1,\underset{\sim}{k}_2,\underset{\sim}{k}_3,\omega)$$

$$= \int \frac{dx}{2\pi}\, S^{(qp)}_{\underset{\sim}{k}_3}(x)\, S^{(qp)}_{\underset{\sim}{k}_4}(\omega-x)\,(\delta_{\underset{\sim}{k}_3,-\underset{\sim}{k}_1} + \delta_{\underset{\sim}{k}_3,-\underset{\sim}{k}_2})$$

This is identical to another approximation due to Stephen[11] for the 2 roton Raman scattering extinction coefficient. Similar approximations have been employed for spin problems.[14]

7. Higher Order Approximations for $\tilde{G}^{(2)}$.

To write a Dyson-like equation for $\tilde{G}^{(2)}$, we proceed by rewriting $\tilde{G}^{(2)}$ as (for the case of $g_4 \neq 0$ only).

$$\tilde{G}^{(2)}_{\underset{\sim}{k}_1,\underset{\sim}{k}_2,\underset{\sim}{k}_3,\underset{\sim}{k}_4}(i\omega_\ell) = \tilde{G}^{(20)}_{\underset{\sim}{k}_1\underset{\sim}{k}_2\underset{\sim}{k}_3\underset{\sim}{k}_4}(i\omega_\ell)$$

$$+ \chi \sum_{\substack{\underset{\sim}{q}_1,\underset{\sim}{q}_2\\ \underset{\sim}{q}_3,\underset{\sim}{q}_4}} \tilde{G}^{(2,0)}_{\underset{\sim}{k}_1\underset{\sim}{k}_2\,\underset{\sim}{q}_1\underset{\sim}{q}_2}(i\omega_\ell)\, V_{\underset{\sim}{q}_1\underset{\sim}{q}_2\underset{\sim}{q}_3\underset{\sim}{q}_4}(i\omega_\ell)\, \tilde{G}^{(2)}_{\underset{\sim}{q}_3\underset{\sim}{q}_4\,\underset{\sim}{k}_3\underset{\sim}{k}_4}(i\omega_\ell) \tag{7-1}$$

Here $G^{(2,0)}_{k_1,k_2,k_3,k_4}$ was evaluated in Section 5 and

$V_{q_1q_2q_3q_4}(i\omega_\ell)$ is defined below.

The factor χ is 1/2 in the case that A_3 and A_4 are both creation (or destruction) operators. For the case in which A_3 is a creation and A_4 a destruction operator (or vice versa) $\chi=1$. The factor 1/2 enters in the first case because the two contributions in the sum in (7-1) which arise from the interchange $\underset{\sim}{g}_3 \leftrightarrow \underset{\sim}{g}_4$ should only be counted once.[15] The argument making it possible to write (7-1) is closely similar to the argument leading to Eq. (4-2). The function $V_{\underset{\sim}{q}_1\underset{\sim}{q}_2\underset{\sim}{q}_3\underset{\sim}{q}_4}(i\omega_\ell)$ is an effective interaction. It is evidently defined as all diagrams (parentheses refer to the particle-hole case)
1) starting with two lines going in to the left (or one in and one out) and two going out to the right (or one in and one out);
2) involving at least one power of g_4;
3) not separable by cutting, with a vertical line, two parallel (anti-parallel) $G^{(0)}$ lines. In applying rule 3, one must count cuts of "external" lines on the V diagrams as illustrated below.

Thus, the leading diagrams in V are as follows. For the case in which A_3 and A_4 are creation operators:

$$V_{\underset{\sim}{q_1}\underset{\sim}{q_2}\underset{\sim}{q_3}\underset{\sim}{q_4}} = \quad + \quad \left[\ \cdots \ \right]$$

$$+ \quad + \quad + \quad \begin{pmatrix} \text{2nd order} \\ \text{self energy} \\ \text{insertions} \end{pmatrix} + \cdots$$

$$(7\text{-}2)$$

In second order, the diagram   is excluded by rule 3.

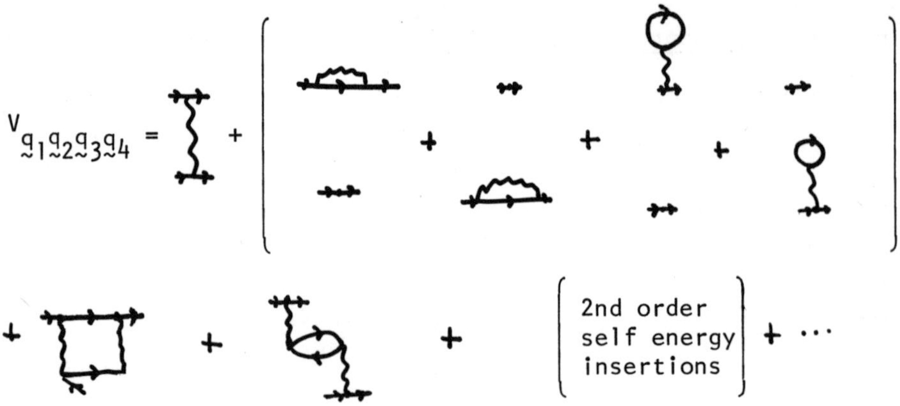

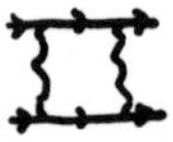

is not excluded because the dotted vertical cuts an "external" line as well as two parallel lines. For the case in which A_3 is a creation and A_4 a destruction operator

$$V_{q_1 q_2 q_3 q_4} = \quad + \quad \begin{pmatrix} \text{1st order self} \\ \text{energy insertions} \end{pmatrix}$$

$$+ \quad + \quad + \quad \begin{pmatrix} \text{2nd order self} \\ \text{energy insertions} \end{pmatrix}$$

$$(7\text{-}3)$$

In 2nd order the diagram   is omitted by rule 3. The iso-

lated dots in Eq. (7-2) and (7-3) denote Kronecker delta's of the form $\delta_{\underset{\sim}{q_1},\underset{\sim}{q_3}}$. Eq. (7-1) is easy to solve formally

$$\tilde{G}^{(2)}_{\underset{\sim}{k_1},\underset{\sim}{k_2},\underset{\sim}{k_3},\underset{\sim}{k_4}}(i\omega_\ell)$$

$$= \sum_{\underset{\sim}{q_1},\underset{\sim}{q_2}} \left[\overleftrightarrow{1} - \overset{\longleftrightarrow}{\chi \tilde{G}}^{(2,0)} V \right]^{-1}_{\underset{\sim}{k_1}\underset{\sim}{k_2};\underset{\sim}{q_1}\underset{\sim}{q_2}} \tilde{G}^{(2,0)}_{\underset{\sim}{q_1}\underset{\sim}{q_2}\underset{\sim}{k_3}\underset{\sim}{k_4}} \qquad (7\text{-}4)$$

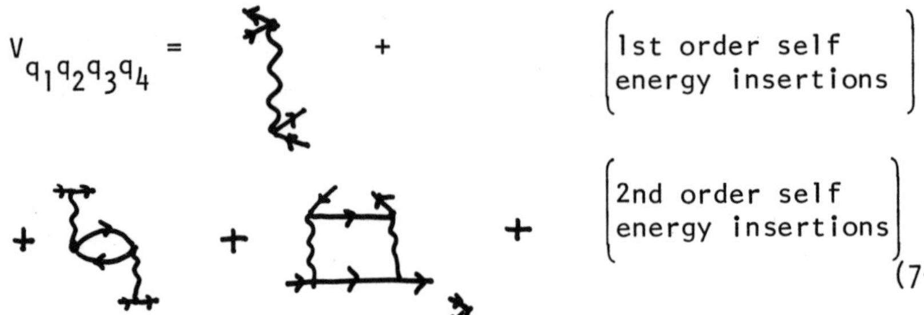

where the matrix $\left[\overleftrightarrow{1} - \chi\, \overleftrightarrow{\tilde{G}^{(2,0)} V}\right]^{-1}_{\mathbf{k}_1\mathbf{k}_2;\mathbf{q}_1\mathbf{q}_2}$ is the inverse of the matrix defined by

$$\left[\overleftrightarrow{1} - \chi\, \overleftrightarrow{G^{(2,0)} V}\right]_{\mathbf{q}_1\mathbf{q}_2;\mathbf{q}_3\mathbf{q}_4}$$

$$= \delta_{\mathbf{q}_1\mathbf{q}_3}\, \delta_{\mathbf{q}_2,\mathbf{q}_4} - \chi \sum_{\mathbf{q}_1'\mathbf{q}_2'} \tilde{G}^{(2,0)}_{\mathbf{q}_1\mathbf{q}_2\mathbf{q}_1'\mathbf{q}_2'}\, V_{\mathbf{q}_1'\mathbf{q}_2'\mathbf{q}_3\mathbf{q}_4} \tag{7-5}$$

Equations (7-1) and (7-4) are analogous to Eq. (4-2) and its solution for $G^{(1)}$ in the equation which follows (4-2). The inversion of the matrix (7-5) is non-trivial in general in this case.

Before turning to applications of (7-4), we describe a method by which all the self-energy insertions in Eqs. (7-2) and (7-3) can be included if $G^{(1)}$ is known exactly: Consider the diagramatic series for $G^{(2)}$. In any diagram such as for example the diagram

in the case that A_3 and A_4 are creation operators, we consider all the diagrams generated by putting self energy insertions into the four lines . This gives

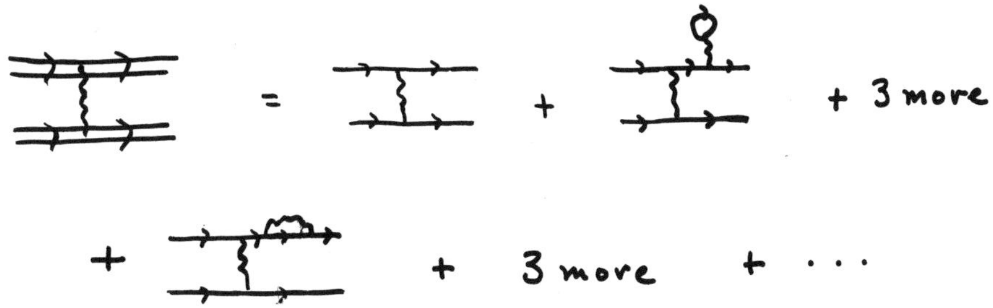

It is not hard to see that the added terms are generated in our previous formulation by the self energy insertions in (7-2). We can take account of all the self energy insertions in V by replacing bare $G^{(1,0)}$ lines  by "dressed" $G^{(1)}$ lines  in all diagrams if we appropriately define a new $V^{red}_{\underset{\sim}{q}_1\underset{\sim}{q}_2\underset{\sim}{q}_3\underset{\sim}{q}_4}$ as follows.

$V^{red}_{\underset{\sim}{q}_1\underset{\sim}{q}_2\underset{\sim}{q}_3\underset{\sim}{q}_4}$ is found by selecting all the terms in $V_{\underset{\sim}{q}_1\underset{\sim}{q}_2\underset{\sim}{q}_3\underset{\sim}{q}_4}$ which cannot be separated by drawing a horizontal line without cutting anything. From these terms, the contributions to V^{red} are found by replacing $\tilde{G}^{(1,0)}$ by $\tilde{G}^{(1)}$, in the corresponding diagram in the series for $V_{\underset{\sim}{q}_1\underset{\sim}{q}_2\underset{\sim}{q}_3\underset{\sim}{q}_4}$: we then have instead of Eqn. (7-1) the following:

$$\tilde{G}^{(2)}_{\underset{\sim}{k}_1\underset{\sim}{k}_2\underset{\sim}{k}_3\underset{\sim}{k}_4}(i\omega_\ell) =$$

$$\sum_{\underset{\sim}{q}_1\underset{\sim}{q}_2}\left[\overleftrightarrow{1} - \chi\, \tilde{G}^{(2,0)'}\, V^{red}\right]^{-1}_{\underset{\sim}{k}_1\underset{\sim}{k}_2\underset{\sim}{q}_1\underset{\sim}{q}_3} \tilde{G}^{(2,0)'}_{\underset{\sim}{q}_1\underset{\sim}{q}_2,\underset{\sim}{k}_3\underset{\sim}{k}_4} \qquad (7\text{-}6)$$

in which, diagramatically, $\tilde{G}^{(2,0)'}_{\underset{\sim}{k}_1\underset{\sim}{k}_2\underset{\sim}{k}_3\underset{\sim}{k}_4}(\tau)$ is obtained from the diagrams of (6-10) and (5-11) by the replacements of bare lines by full lines. Analytically the corresponding expressions are obtained from the third expressions in Eqs. (6-1) and (6-3) by replacing $\tilde{G}^{(1,0)}$ by $\tilde{G}^{(1)}$ everywhere. The leading terms in $V^{red}_{\underset{\sim}{k}_1\underset{\sim}{k}_2;\underset{\sim}{k}_3\underset{\sim}{k}_4}$ are given by

$V^{red}_{\underset{\sim}{k}_1\underset{\sim}{k}_2\underset{\sim}{k}_3\underset{\sim}{k}_4} =$ 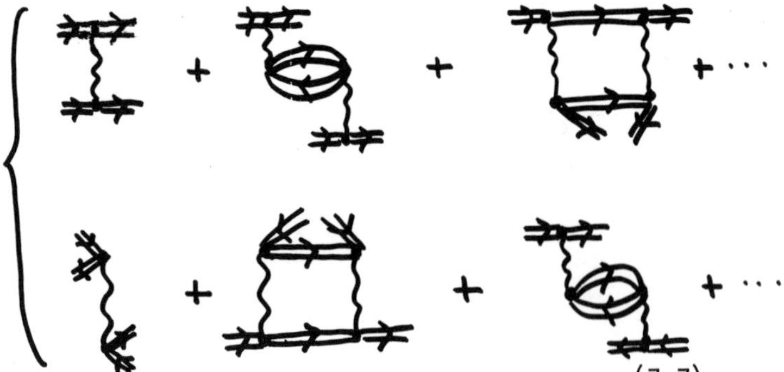

(7-7)

in the two cases of interest here.

Equation (7-6) is called the Bethe-Saltpeter equation in some books.[8]

8. Applications of First Order Approximation for $\tilde{G}^{(2)}_{\mathbf{k}_1\mathbf{k}_2\mathbf{k}_3\mathbf{k}_4}$.

We now apply the approximation resulting from taking only the first term in (7-2) or (7-7) and inserting it back into (7-1) or (7-6) respectively to get $G^{(2)}$. Reflection shows that neither of these procedures is strictly of first order in the potential in any rigorous sense. In practice, one usually proceeds by inserting (7-3) into (7-4) and then using phenomenological forms for $G^{(1)}$ whose general form is fixed by a combination of physical arguments, sum rules and general properties of the perturbation series for $G^{(1)}$. Before turning to the applications it is useful to consider the meaning of these approximations. From (7-6) with $V^{red}_{\mathbf{q}_1\mathbf{q}_2\mathbf{q}_3\mathbf{q}_4}$ replaced by its first term one sees that, in lowest order in the series for V^{red} we have

$$\tilde{G}^{(2,1)}_{\mathbf{k}_1,\mathbf{k}_2,\mathbf{k}_3,\mathbf{k}_4} = \tilde{G}^{(2,0)'}_{\mathbf{k}_1,\mathbf{k}_2,\mathbf{k}_3,\mathbf{k}_4}$$

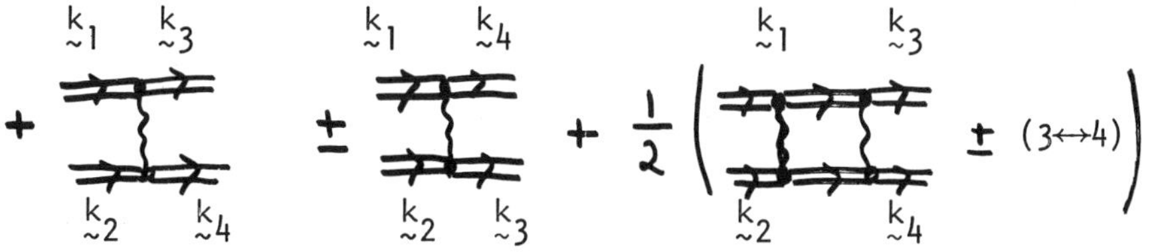

$+ \ldots$

in the case that A_3 and A_4 are both creation operators. Thus, diagramatically in the first case this approximation corresponds to summing "ladder diagrams." In the second case, when A_3 is a creation operator and A_4 a destruction operator one has

$$\tilde{G}^{(2,1)}_{\mathbf{k}_1,\mathbf{k}_2,\mathbf{k}_3,\mathbf{k}_4} = \tilde{G}^{(2,0)'}_{\mathbf{k}_1,\mathbf{k}_2,\mathbf{k}_3,\mathbf{k}_4}$$

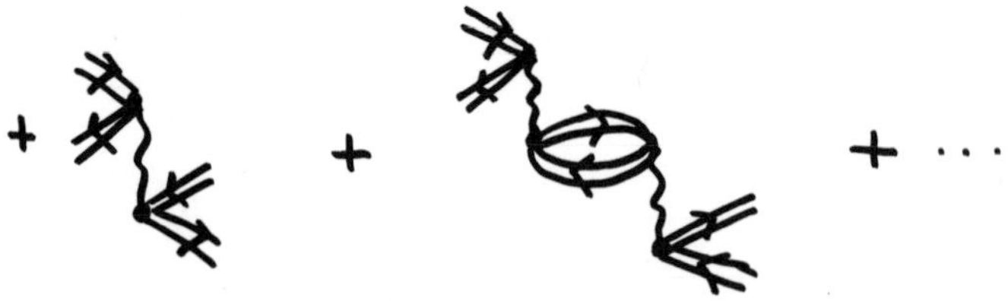

(omitting effects of the first term in (6-3), which vanishes in most cases of interst.) Thus, in this second case, $G^{(2)}$ is a sum of "ring diagrams." In any case, we can write

$$\tilde{G}^{(2)} = \tilde{G}^{(2,0)'} +$$

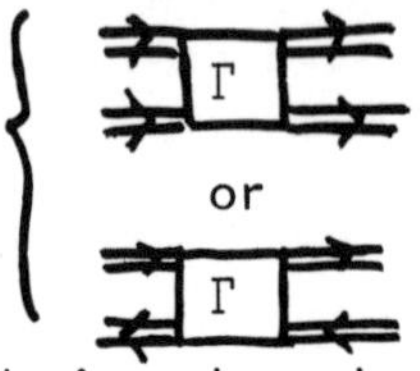

where Γ is a "vertex part" given here by

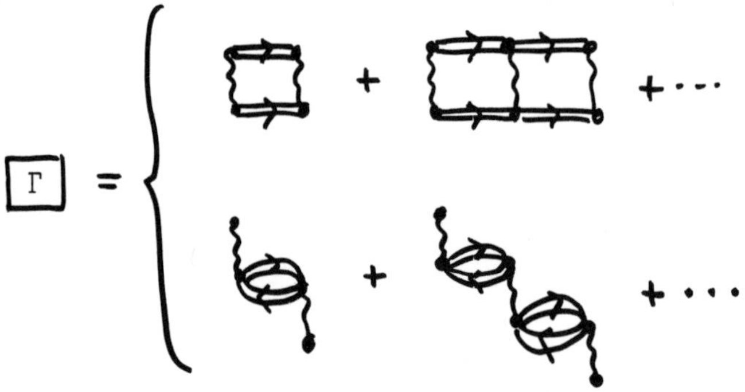

Case 1

Case 2

This formulation, however, makes systematic improvement of the approximation much less easy to formulate than does the description of the preceding section. The physical meaning of these approximations is most evident after doing some examples. Briefly, one can say this: In Case 1, one is dealing with particle-particle scattering and the approximation considered has a structure closely related to that of the exact equation for the scattering amplitude in a two-body scattering problem. In the second case, that of particle-hole scattering, the approximation is a generalization of the Thomas-Fermi screening approximation to the particle-hole interaction.

We now consider the two examples considered in zero order in Section 5 in order to illustrate these points and to understand the approximations in more detail. In each case, we find that the approximations lead to qualitatively new collective behavior in $G^{(2)}$ which was not exhibited in the zero th order approximation.

As an example of the particle-particle problem we again take the quasiparticle model of liquid helium.[17] From (1-12), (1-9), and (6-5), the Green function of interest in the Raman scattering in this problem is

$$\langle\langle \alpha_{-\underset{\sim}{q}'}\alpha_{\underset{\sim}{q}'} \, ; \, \alpha^{\dagger}_{-\underset{\sim}{q}}\alpha^{\dagger}_{\underset{\sim}{q}} \rangle\rangle_r \, (-\omega)$$

which is given by

$$\langle\langle \alpha_{-\underset{\sim}{q}'}\alpha_{\underset{\sim}{q}'} \, ; \, \alpha^{\dagger}_{-\underset{\sim}{q}}\alpha^{\dagger}_{\underset{\sim}{q}} \rangle\rangle_r \, (-\omega)$$

$$= \frac{1}{2\pi} \tilde{G}^{(2)}_{\underset{\sim}{q}',-\underset{\sim}{q}',-\underset{\sim}{q},\underset{\sim}{q}} \, (i\omega_\ell \longrightarrow \omega + i0^+)$$

Setting $\chi = 1/2$ in (7-6) and taking

$$\tilde{G}^{(2,0)'}_{\underset{\sim}{k}_1\underset{\sim}{k}_2\underset{\sim}{k}_3\underset{\sim}{k}_4} = \left[\delta_{14}\,\delta_{23} + \delta_{13}\,\delta_{24}\right]\tilde{G}_{20}(\underset{\sim}{k}_1,\underset{\sim}{k}_2) \qquad (8\text{-}1)$$

one finds

$$\tilde{G}^{(2)}_{-\underset{\sim}{q}'\,\underset{\sim}{q}',\,-\underset{\sim}{q},\,\underset{\sim}{q}}(i\omega_\ell)$$

$$= \left[\delta_{-\underset{\sim}{q}',\underset{\sim}{q}} + \delta_{\underset{\sim}{q}',\underset{\sim}{q}}\right]\tilde{G}_{20}(\underset{\sim}{q},\underset{\sim}{q}')$$

$$+ \frac{1}{2}\sum_{\underset{\sim}{k}_3,\underset{\sim}{k}_4} G_{20}(\underset{\sim}{q},\underset{\sim}{q}')\left[\overline{V}_{-\underset{\sim}{q}'\underset{\sim}{q}'\underset{\sim}{k}_3\underset{\sim}{k}_4} + V_{-\underset{\sim}{q}'\underset{\sim}{q}'\underset{\sim}{k}_3\underset{\sim}{k}_4}\right] \times \tilde{G}^{(2)}_{\underset{\sim}{k}_3\underset{\sim}{k}_4,-\underset{\sim}{q},\underset{\sim}{q}}(i\omega_\ell)$$

In general one can take

$$V_{\underset{\sim}{k}_1\underset{\sim}{k}_2\underset{\sim}{k}_3\underset{\sim}{k}_4} = g_4(\underset{\sim}{k}_1-\underset{\sim}{k}_4)\,\delta_{\underset{\sim}{k}_1-\underset{\sim}{k}_4,\underset{\sim}{k}_3-\underset{\sim}{k}_2}$$

in a translationally invariant system. Thus

$$\tilde{G}^{(2)}_{-\underset{\sim}{q}',\underset{\sim}{q}',\,-\underset{\sim}{q},\underset{\sim}{q}}(i\omega_\ell) =$$

$$\left(\delta_{-\underset{\sim}{q}',\underset{\sim}{q}} + \delta_{\underset{\sim}{q}',\underset{\sim}{q}}\right) \tilde{G}_{20}(\underset{\sim}{q},\underset{\sim}{q}')$$

$$+ \frac{\tilde{G}_{20}(\underset{\sim}{q},\underset{\sim}{q}')}{2} \sum_{\underset{\sim}{k}_3} \left[g_4(\underset{\sim}{q}' + \underset{\sim}{k}_3) + g_4(-\underset{\sim}{q}' + \underset{\sim}{k}_3)\right] \tilde{G}^{(2)}_{\underset{\sim}{k}_3,-\underset{\sim}{k}_3,-\underset{\sim}{q},\underset{\sim}{q}}(i\omega_\ell) \tag{8-2}$$

To solve this equation analytically we write

$$g_4(\underset{\sim}{q}+\underset{\sim}{k}_3) = 4\pi \sum_{\ell,m} g_4^\ell \, \eta_\ell(q)\eta_\ell(k_3) \; Y_{m\ell}(\hat{q}) \; Y_{m\ell}(\hat{k}_3) \tag{8-3}$$

It is then not hard[17] to find the solution to (8-2) in general. Here we will look only at the case $g_4^{\ell=0} \equiv g_4^o \neq 0$, $g_4^{\ell\neq 0} = 0$. We define the quantity

$$M(\underset{\sim}{q}',\underset{\sim}{q}) = \sum_{\underset{\sim}{k}_3} \left[g_4(\underset{\sim}{q}'+\underset{\sim}{k}_3) + g_4(\underset{\sim}{q}'-\underset{\sim}{k}_3)\right] \tilde{G}^{(2)}_{\underset{\sim}{k}_3,-\underset{\sim}{k}_3,\underset{\sim}{q},-\underset{\sim}{q}} \tag{8-4}$$

Multiplying (8-2) by $g_4(\underset{\sim}{k}+\underset{\sim}{q}') + g_4(\underset{\sim}{k}-\underset{\sim}{q}')$ and summing on q' one has

$$M(\underset{\sim}{k},\underset{\sim}{q}) = \sum_{\underset{\sim}{q}'} \left[g_4(\underset{\sim}{k}+\underset{\sim}{q}') + g_4(\underset{\sim}{k}-\underset{\sim}{q}')\right] (\delta_{\underset{\sim}{q}',-\underset{\sim}{q}} + \delta_{\underset{\sim}{q}',\underset{\sim}{q}}) \; \tilde{G}_{20}(\underset{\sim}{q},\underset{\sim}{q}')$$

$$+ \frac{1}{2} \sum_{\underset{\sim}{q}'} \left[g_4(\underset{\sim}{k}+\underset{\sim}{q}') + g_4(\underset{\sim}{k}-\underset{\sim}{q}')\right] G_{20}(\underset{\sim}{q},-\underset{\sim}{q}')M(\underset{\sim}{q}',\underset{\sim}{q}) \tag{8-5}$$

With (8-3), (8-2) becomes

$$M(\underset{\sim}{q},\underset{\sim}{q}') = g_4^o \left[\eta_o(\underset{\sim}{q}') + \eta_o(-\underset{\sim}{q}')\right] M(\underset{\sim}{q}) \tag{8-6}$$

in which

$$M(\underset{\sim}{q}) = \sum_{\underset{\sim}{k}_3} \eta_o(\underset{\sim}{k}_3)\ \tilde{G}^{(2)}_{\underset{\sim}{k}_3 - \underset{\sim}{k}_3,\ -\underset{\sim}{q},\underset{\sim}{q}} \tag{8-7}$$

Putting (8-6) and (8-7) in (8-5) one solves for $M(q)$:

$$M(\underset{\sim}{q}) = \frac{2\eta(q)\ \tilde{G}_{20}(\underset{\sim}{q},\ \underset{\sim}{q})}{1-g_4^o \sum_{\underset{\sim}{q}'} \tilde{G}_{20}(\underset{\sim}{q}',\ \underset{\sim}{q}')\eta^2(\underset{\sim}{q}')} \tag{8-8}$$

Then combining (8-8), (8-6), (8-4) and (8-2):

$$\tilde{G}^{(2)}_{-\underset{\sim}{q}',\underset{\sim}{q}',\ -\underset{\sim}{q},\underset{\sim}{q}}(i\omega_\ell) = (\delta_{\underset{\sim}{q}',\underset{\sim}{q}} + \delta_{-\underset{\sim}{q}',\underset{\sim}{q}})\ \tilde{G}_{20}(\underset{\sim}{q},\underset{\sim}{q})$$

$$+ \frac{2g_4{}^o\ \tilde{G}_{20}(\underset{\sim}{q},\underset{\sim}{q})\eta_o(\underset{\sim}{q})\ \tilde{G}_{20}(\underset{\sim}{q}',\underset{\sim}{q}')\eta_o(\underset{\sim}{q}')}{\left[1-g_4^o \sum_{\underset{\sim}{q}'} \tilde{G}_{20}(\underset{\sim}{q}',\underset{\sim}{q}')\eta^2(\underset{\sim}{q}')\right]} \tag{8-9}$$

Reinserting (8-9) in (1-16) one has

$$h^{(1)}(\omega) = \rho^3/(2\pi)^2\ (\omega_n/c)^4\ \frac{1}{N^3} \sum_{\underset{\sim}{q},\underset{\sim}{q}'} F(\underset{\sim}{q})\ F(\underset{\sim}{q}')$$

$$\times\ S(\underset{\sim}{q})\ S(\underset{\sim}{q}')\ 2\mathrm{Im}\left\{(\delta_{\underset{\sim}{q},\underset{\sim}{q}'} + \delta_{\underset{\sim}{q}',-\underset{\sim}{q}})\ \tilde{G}_{20}(\underset{\sim}{q},\omega)\right.$$

$$\left. + \frac{2g_4^o\ G_{20}(\underset{\sim}{q}',\omega)\eta_o(\underset{\sim}{q}')\ G_{20}(\underset{\sim}{q},\omega)\eta_o(\underset{\sim}{q})}{1-g_4^o \sum_{\underset{\sim}{q}'} \eta_o^2(\underset{\sim}{q}')\ G_{20}(\underset{\sim}{q}',\omega)}\right\} \frac{1}{(e^{-\omega\beta}-1)} \tag{8-10}$$

This is to be compared with the results of Section 6. The first term in (8-10) was obtained there. In (8-10) we have written

$$\tilde{G}_{20}(\underset{\sim}{q},\underset{\sim}{q}) \xrightarrow[i\omega_\ell \to \omega + i0^+]{} G_{20}(\underset{\sim}{q},\omega)$$

The effects of the new term are best illustrated by using the form (6-2) for $G_{20}(\underset{\sim}{q},\omega)$. At 0 temperature one has

$$G_{20}(\underset{\sim}{q},\omega) = \frac{1}{\omega - 2\varepsilon_{\underset{\sim}{q}} + i0^{+}}$$

Taking $\eta_o=1$ for illustrative purposes one has, from the 2nd term in (8-10) the following contribution

$$(g_4^o)^2 \, \mathrm{Re}\left\{G_{20}(\underset{\sim}{q}',\omega)\ G_{20}(\underset{\sim}{q},\omega)\right\} \frac{\pi\rho_2(\omega)}{\left[1 - g_4^o \int d\omega' \frac{\rho_2(\omega')}{\omega'-\omega}\right]^2 + (g_4^o)^2 \left[\pi\rho_2(\omega)\right]^2} \tag{8-11}$$

in which

$$\rho_2(\omega) = \sum_{\underset{\sim}{q}'} \delta(\omega - 2\varepsilon_{\underset{\sim}{q}'}) \tag{8-12}$$

In a region where $\rho_2(\omega)$, the two-quasiparticle density of states is small, (8-11) is approximately

$$(g_4^o)^2 \mathrm{Re}\left\{G_{20}(\underset{\sim}{q}',\omega)\ G_{20}(\underset{\sim}{q},\omega)\right\} \delta\left\{1 - g_4^o \int d\omega' \frac{\rho_2(\omega')}{\omega'-\omega}\right\} \tag{8-13}$$

leading to a sharp spike in the light scattering spectrum when

$$\frac{1}{g_4^o} = \int d\omega' \frac{\rho_2(\omega')}{(\omega'-\omega)} \tag{8-14}$$

The physical interpretation of (8-14) is that it is the condition for formation of a bound state of two quasiparticles. To illustrate its meaning we take[18]

$$\varepsilon_{\underset{\sim}{q}} = \Delta + (q-q_o)^2 / 2\mu \qquad (8\text{-}15)$$

Inserting this in (8-12) and doing the integral in (8-14) one finds

$$\frac{1}{g_4^o V} = \begin{cases} \dfrac{-1}{2\pi} \dfrac{q_o^2 \mu^{1/2}}{|\omega - 2\Delta|^{1/2}}, & \omega<2\Delta \\ 0, & \omega>2\Delta \end{cases} \qquad (8\text{-}16)$$

in this model. A bound state is thus always formed below the 2 particle continuum as long as $g_4^o<0$ in this model. This was first pointed out by Ruvalds.[18] The reason this occurs is because (8-15), which roughly approximates the liquid helium spectrum, is isotropic. This makes the singularity in the density of states $\rho_o(\omega)$ a divergence, unlike most 3 dimensional problems. The existence of the bound state in this model is thus qualitatively related to the existence of bound states in an arbitrarily weak potential in one dimension, which has been known much longer. We recall the caution of Section 1 concerning applying this model to real liquid helium.

We now turn to application of Eq. (7-6) to the particle hole problem in the electron gas. Using (1-5) and (6-7) we want the quantity

$$\tilde{G}^{(2)}_{\underset{\sim}{q},\lambda,\ \underset{\sim}{k}+\underset{\sim}{q},\ \lambda,\ \underset{\sim}{q}',\ \lambda',\ -\underset{\sim}{k}+\underset{\sim}{q}',\ \lambda'}(i\omega_\ell)$$

The potential has the property

$$V_{\underset{\sim}{k}_1\lambda_1,\ \underset{\sim}{k}_2\lambda_2,\ \underset{\sim}{k}_3\lambda_3,\ \underset{\sim}{k}_4\lambda_4}$$

$$= \frac{4\pi e^2}{|\underset{\sim}{k}_1-\underset{\sim}{k}_2|^2}\ \delta_{\underset{\sim}{k}_1-\underset{\sim}{k}_2,\ \underset{\sim}{k}_3-\underset{\sim}{k}_4} = V(|\underset{\sim}{k}_1-\underset{\sim}{k}_2|)\ \delta_{\underset{\sim}{k}_1-\underset{\sim}{k}_2,\ \underset{\sim}{k}_3-\underset{\sim}{k}_4}$$

Using this relation in (7-6) one finds

$$\tilde{G}^{(2)}_{\underset{\sim}{q},\lambda,\ \underset{\sim}{k}+\underset{\sim}{q},\ \lambda,\underset{\sim}{q}',\lambda',\ -\underset{\sim}{k}+\underset{\sim}{q}',\lambda'}(i\omega_\ell)$$

$$= \delta_{\lambda,\lambda'}\ \delta_{\underset{\sim}{k}+\underset{\sim}{q},\underset{\sim}{q}'}\ \tilde{G}^{(2,0)}_{\underset{\sim}{q},\lambda,\underset{\sim}{k}+\underset{\sim}{q},\lambda,\underset{\sim}{q}',\lambda',-\underset{\sim}{k}+\underset{\sim}{q}',\lambda'}$$

$$+ \sum_{\underset{\sim}{q}'',\lambda''} V(\underset{\sim}{k})\ \tilde{G}^{(2,0)}_{\underset{\sim}{q},\lambda,\underset{\sim}{k}+\underset{\sim}{q},\lambda',\underset{\sim}{k}+\underset{\sim}{q},\lambda,\underset{\sim}{q},\lambda}$$

$$\times\ \tilde{G}^{(2)}_{\underset{\sim}{q}'',\lambda'',\underset{\sim}{k}+\underset{\sim}{q}'',\lambda'',\underset{\sim}{q}',\lambda',-\underset{\sim}{k}+\underset{\sim}{q}',\lambda'}$$

We sum this equation on $\underset{\sim}{q},\lambda$ and solve for

$$\sum_{\underset{\sim}{q},\lambda} \tilde{G}^{(2)}_{\underset{\sim}{q},\lambda,\underset{\sim}{k}+\underset{\sim}{q},\lambda,\underset{\sim}{q}',\lambda',-\underset{\sim}{k}+\underset{\sim}{q}',\lambda'} :$$

$$\sum_{\underset{\sim}{q},\lambda} \tilde{G}^{(2)}_{\underset{\sim}{q},\lambda,\underset{\sim}{k}+\underset{\sim}{q},\lambda,\underset{\sim}{q}',\lambda',-\underset{\sim}{k}+\underset{\sim}{q}',\lambda'}$$

$$= \frac{\tilde{G}^{(2,0)}_{\underset{\sim}{q}'-\underset{\sim}{k},\lambda',\underset{\sim}{q}',\lambda',\underset{\sim}{q}',\lambda',-\underset{\sim}{k}+\underset{\sim}{q}',\lambda'}}{1- V(\underset{\sim}{k}) \sum_{\underset{\sim}{q},\lambda} \tilde{G}^{(2,0)}_{\underset{\sim}{q},\lambda,\underset{\sim}{k}+\underset{\sim}{q},\lambda,\underset{\sim}{k}+\underset{\sim}{q},\lambda,\underset{\sim}{q},\lambda}}$$

Then, using (1-5) one has

$$\langle\langle \rho_{\underset{\sim}{k}}\ ;\ \rho_{-\underset{\sim}{k}} \rangle\rangle_r\ (\omega) =$$

$$\frac{1}{2\pi}\;\frac{\sum_{\underset{\sim}{q}'\lambda'} \tilde{G}^{(2,0)}_{\underset{\sim}{q}',\underset{\sim}{k},\lambda',\underset{\sim}{q}',\lambda',\underset{\sim}{q}',\lambda',-\underset{\sim}{k}+\underset{\sim}{q}',\lambda'}(i\omega_\ell \to \omega+i0^+)}{1-V(k)\sum_{\underset{\sim}{q},\lambda} \tilde{G}^{(2,0)}_{\underset{\sim}{q},\lambda,\underset{\sim}{k}+\underset{\sim}{q},\lambda,\underset{\sim}{k}+\underset{\sim}{q},\lambda,\underset{\sim}{q},\lambda}(i\omega_\ell \to \omega+i0^+)}$$

We note that, using (6-9)

$$\sum_{\underset{\sim}{q},\lambda} \tilde{G}^{(2,0)}_{\underset{\sim}{q},\lambda,\underset{\sim}{k}+\underset{\sim}{q},\lambda,\underset{\sim}{k}+\underset{\sim}{q},\lambda,\underset{\sim}{q},\lambda}(i\omega_\ell \to \omega+i0^+)$$

$$= -\chi_0(\underset{\sim}{k},\omega)$$

Therefore

$$\langle\langle \rho_{\underset{\sim}{k}};\ \rho_{-\underset{\sim}{k}} \rangle\rangle\ (\omega) = \frac{1}{2\pi}\,\frac{-\chi_0(\underset{\sim}{k},\omega)}{1+V(\underset{\sim}{k})\chi_0(\underset{\sim}{k},\omega)}$$

Using (6-8) then gives, for the longitudinal dielectric constant

$$\varepsilon_{\shortparallel}(\underset{\sim}{k},\omega) = 1+ V(\underset{\sim}{k})\chi^0(\underset{\sim}{k},\omega)$$

This is a famous result of the random phase approximation,[19] which is equivalent to approximation of keeping the 1st term in the series for V^{red} in (7-6). The approximation is also called the ring approximation in the particle hole case because, as discussed at the beginning of this section, it involves summing ring graphs for $G^{(2)}$.

To see that one gets a collective mode in the particle-hole case, we look at imaginary part of the susceptibility $\chi''(k,\omega)$ (giving the energy absorption rate--see Chapter 1):

$$\chi''(\underset{\sim}{k},\omega) = \mathrm{Im}\,\frac{\chi_0(\underset{\sim}{k},\omega)}{1+V(\underset{\sim}{k})\chi_0(\underset{\sim}{k},\omega)}$$

$$= \frac{\chi_0''(\underset{\sim}{k},\omega)}{\left[1 + V(\underset{\sim}{k})\chi_0'(\underset{\sim}{k},\omega)\right]^2 + \left[V(\underset{\sim}{k})\chi_0''(\underset{\sim}{k},\omega)\right]^2}$$

$$\xrightarrow[\chi_0''(\underset{\sim}{k},\omega) \to 0]{} \pi\,\delta\left[1+V(\underset{\sim}{k})\chi_0'(\underset{\sim}{k},\omega)\right] \tag{8-17}$$

Thus one gets a sharp pole if the equation

$$\frac{1}{V(\underset{\sim}{k})} = -\chi_0'(\underset{\sim}{k},\omega) \tag{8-18}$$

can be solved in a region where $\chi_0''(\underset{\sim}{k},\omega)\to 0$. Expanding the real part of $\chi_0(\underset{\sim}{k},\omega)$ using (6-9) for small $\underset{\sim}{k}$ one finds, with $\varepsilon_k=\hbar^2k^2/2m$ that (8-18) is equivalent to

$$\omega^2 = 4\pi Ne^2/mV$$

so that this peak corresponds to the plasma mode. The similarity of (8-18) and (8-16) should be obvious. In the coulomb gas, the particle-particle interactions are repulsive but the particle-hole interactions are effectively attractive, leading to the possibility of a "bound state." In each case, single particle modes arising from the zero order approximation are clearly also present in the response.

9. Extensions and Other applications.

Other applications of the ideas presented here are easily made to phonon systems, for which an extensive literature on bound state problems exists.[20-23] Applications to spin problems are non-trivial because there is no simple proof of Wick's Theorem for spin operators[24]. The extension to other interactions in (1-21) has proceeded quite far: The case at $g_3 \neq 0$ has been treated for the quasiparticle model[25], for microscopic models of liquid helium[26] and for phonon problems in solids[27]. An extremely important case is that of the electron-phonon interaction in metals.[27] For many purposes the latter problem is almost isomorphic to the case $g_4 \neq 0$ considered here, the main difference being that the quantity associated with the wavy line has a frequency dependence associated physically with the time for virtual phonons to propagate between electrons.

The case of superconductivity in fermion systems requires an extension which is described in several books.[28] A large recent literature applying these techniques to superfluidity in ^{3}He exists.[29] The case of superfluidity in boson systems is also treated in text books.[30] It is closely related formally[25] to the problem with $g_3 \neq 0$ in Eq. (1-21).

Cases where the techniques described here have not proved useful exist in systems for which the interaction potential contains a hard core. In these cases, variational techniques[31] have largely supplanted efforts to deal with the problems which arise by improvements on the approximations of Section 8. Hard core systems include the extremely important cases of superfluidity in ^{3}He and ^{4}He and nuclear physics. Unfortunately, variational techniques have been less useful for dynamical properties than they are for ground state properties. For dynamical properties, the perturbation theory results often remain our best guide to the physics, despite their serious deficiencies.

References

1. C. E. Campbell and E. Feenberg, Phys. Rev. 188, 396 (1969); H. W. Jackson, Phys. Rev. A8, 1529 (1973); R. Hastings and J. W. Halley, Phys. Rev. A10, 2488 (1974).

2. A. D. B. Woods and R. A. Cowley, Rep. Prog. in Phys. 36, 1135 (1973); P. Kleban and J. W. Halley, Phys. Rev. B11, 3520 (1975).

3. R. Loudon, Adv. in Phys. 13, 423 (1964).

4. P. Kleban, Phys. Lett. A49, 19 (1974); P. Kleban and R. Hastings, Phys. Rev. B11, 1878 (1975).

5. F. Pinski and C. Campbell, to be published.

6. A. A. Abrikosov, L. P. Gorkov and I. E. Dzialoskinski, Methods of Quantum Field Theory in Statistical Physics, trans. by R. A. Silverman, Prentice-Hall, Englewood Cliffs, N.J. (1963).

7. A. Fetter and D. Walecka, Quantum Theory of Many-Particle Systems, McGraw-Hill, N.Y. (1971).

8. P. Nozieres, Theory of Interacting Fermi Systems, W. A. Benjamin, N. Y. (1964).

9. D. N. Zubarev, Usp. Fiz. Nauk. 71, 71-116 (1960) (Sov. Phys. -Uspekhi 3, 320 (1960)).

10. This is done for the weakly interacting Bose gas in section of Reference 7.

11. M. Stephen, Phys. Rev. 187, 279 (1969).

12. D. Pines, Elementary Excitations in Solids, W. A. Benjamin, N.Y. (1964).

13. J. Lindhard, Kgl. Danske Videnskab. Selskab, Mat-fys. Medd. 28, 8 (1954).

14. T. Kawasaki, J. Phys. Soc. Jap. 29, 1144 (1970).

15. See Chapter 6 of Reference 8 for a more thorough discussion of this point.

16. A. Zawadowski, J. Ruvalds, J. Solana, Phys. Rev. A5, 399 (1972) and two last references in 1.

17. P. Kleban and R. Hastings, Phys. Rev. B11, 1878 (1975).

18. See first reference in 16.

19. Reference 12.

20. M. H. Cohen and J. Ruvalds, Phys. Rev. Lett. 23, 1378 (1969); J. Ruvalds and A. Zawadowski, Phys, Rev. B2, 1172 (1970).

21. R. Tubino and J. Birman, Phys. Rev. Lett. 35, 670 (1975).

22. S. Go, H. Bilz and M. Cardona, Phys. Rev. Lett. 34, 580 (1975).

23. A. A. Maradudin, in Phonons (edited by M. Nusimovici), p. 427, Flammarion, Paris (1971).

24. M. G. Cottam and R. B. Stinchcombe, Phys. C: Sol. St. Phys. 3, 2283-304 (1970), and the article by Cottam in this volume.

25. J. W. Halley and R. Hastings, Phys. Rev. B15, 1404 (1977) and the article by Hastings in this volume.

26. Chapter 5 of Reference 6.

27. Reference 6.

28. Chapter 7 of Reference 6; Chapter 7 of Reference 8. G. M. Eliashberg, Sov. Phys. JETP 13, 333 (1971); B. I. Ivlev, S. G. Lisityn, S. G. Lisityn and G. M. Eliashberg, J. Low Temp. Phys. 10, 449 (1973).

29. D. Rainer and J. W. Serene, Phys. Rev. B13, 4745 (1976),

30. Chapter 5 of Reference 6.

31. E. Feenberg, The Theory of Quantum Fluids, Academic Press, N.Y. (1969).

RENORMALIZATION GROUP APPROACH TO DYNAMIC CRITICAL PHENOMENA

Gene F. Mazenko

The James Franck Institute and Department of Physics

The University of Chicago, Chicago, Illinois 60637

I. INTRODUCTION

In these lectures I want to discuss the recent developments in dynamic critical phenomena using renormalization group techniques. An attractive feature of this topic is that it brings together ideas from several areas of theoretical physics. We will discuss the renormalization group ideas which have their roots in quantum field theory, the statistical mechanics of phase transformations and the principles of non-equilibrium transport phenomena. I hope to show how these principles can be amalgamated into a single theory describing time dependent processes in systems near second order phase transitions. The theory I will discuss not only leads to a good description of dynamics of phase transitions but has suggested new ideas in treating the frontier problems of turbulence[1] and spinodal decomposition.[2]

These lectures are not intended for experts. I will assume the reader is unfamiliar with phase transition theory and field theoretic techniques. Clearly I cannot be complete in my discussion. I will limit myself to the simplest examples and emphasize general principles. There now exist many[3-8] excellent reviews on this topic which go into much greater detail.

These lectures will break rather naturally into four parts. In the next section I will discuss the phenomenology of second order phase transitions; their characterization and the relationship between different systems. Key points here are universality among different systems and the scaling hypothesis. In the third section I will discuss how the ideas of universality and scaling lead one to formulate the problem of static or equilibrium critical phenomena in terms of a simple quantum field theory--the Ginzburg-Landau-Wilson (GLW) model. I will also discuss Wilson's renormalization group (RG) ideas for "solving" this model. In section IV I will introduce the ideas of kinetic or transport theory relevant to critical phenomena. This will be done in the context of a Langevin equation approach. The development leads to a generalization of the GLW model to dynamics. Finally in section V I will discuss methods for treating the non-linear equations of motion appropriate for studying dynamic critical phenomena. In particular I will discuss perturbation theory methods and the dynamic renormalization group. I will, where possible, indicate how these ideas may be used in treating higher order correlation functions.

II. PHENOMENOLOGY OF SECOND ORDER PHASE TRANSITIONS

A. Examples of Second Order Phase Transitions

Let us begin with a discussion of the phenomenology of continuous "second-order" phase transitions. Examples of such transitions are the Curie transition in ferromagnets, the Néel transition in anti-ferromagnets, the liquid-gas and phase separation transitions in fluids as well as the λ-transition in helium and the superconducting transition in metals and He^3. These systems seem to have nothing in particular in common. One of the key points in the development of our modern theory of second-order phase transitions was the realization that all of the transitions listed above are essentially equivalent if one uses the appropriate language. The first step in the development of this language was Landau's mean field theory[9] of second order phase transitions. It was not, however, until the precise measurements of the 1960's, made possible to a large extent by modern scattering techniques, that it was realized that there is something very deep and fundamental relating various second-order phase transitions. Before I can discuss these fundamental relationships we need to develop a bit of language and discuss some results. For now let us specialize to the case of an isotropic ferromagnet. We will return to the general case later.

B. Characterization of Ferromagnetic Transitions

A typical ferromagnet can be thought of as a set of classical magnetic moments $\vec{M}_i$ (or spins) localized on periodic lattice sites at positions $\vec{x}_i$.

We assume for simplicity that the system can be described by a Heisenberg Hamiltonian

$$\mathcal{H} = -\sum_{i,j} J(\vec{x}_i - \vec{x}_j)\, \vec{M}_i \cdot \vec{M}_j \tag{2.1}$$

where the exchange interaction $J(\vec{x})$ is short ranged and gives the interaction between neighboring spins. For ferromagnets J is positive and tends to align the spins. In the presence of an external magnetic field $\vec{H}$ we must add a Zeeman term to obtain the total Hamiltonian

$$\mathcal{H}_T = \mathcal{H} - \vec{H} \cdot \vec{M}_T \tag{2.2}$$

where

$$\vec{M}_T = \sum_i \vec{M}_i \tag{2.3}$$

is the total magnetization. The magnetization density

$$\vec{M}(\vec{x}) = \sum_i \vec{M}_i\, \delta(\vec{x} - \vec{x}_i) \tag{2.4}$$

is related to the total magnetization by

$$\vec{M}_T = \int d^d x\, \vec{M}(\vec{x}) \tag{2.5}$$

for a d-dimensional system.

The partition function is defined as

$$Z = \mathrm{Tr}\, e^{-\beta \mathcal{H}_T} \tag{2.6}$$

where Tr means that we sum over all spin configurations and $\beta = (k_B T)^{-1}$ where k_B is Boltzmann's constant and T is temperature. The thermodynamic properties of interest are the equilibrium averaged magnetization per unit volume,

$$M^\alpha(T,H) = \langle M^\alpha(\vec{x})\rangle = \frac{\langle M_T^\alpha\rangle}{\Omega} = (\beta\Omega)^{-1}\frac{\partial}{\partial H^\alpha}\ln Z, \tag{2.7}$$

(where Ω is the volume of the system) and the magnetic susceptibility in zero external field:

$$\begin{aligned} \chi_M &= \lim_{H\to 0} \frac{1}{3\Omega}\sum_{\alpha=1}^{3}\frac{\partial\langle M_T^\alpha\rangle}{\partial H^\alpha} \\ &= \frac{1}{3}\frac{\beta}{\Omega}\langle(\delta\vec{M}_T)^2\rangle \end{aligned} \tag{2.8}$$

where

$$\delta\vec{M}_T = \vec{M}_T - \langle\vec{M}_T\rangle . \tag{2.9}$$

We will also be interested in the spatial correlations between the spins and introduce the magnetization density autocorrelation function

$$\chi(\vec{x}-\vec{x}', TH) = \frac{1}{3}\sum_{\alpha=1}^{3}\langle\delta M^\alpha(\vec{x})\,\delta M^\alpha(\vec{x}')\rangle . \tag{2.10}$$

χ depends only on $\vec{x} - \vec{x}'$ because of the translational invariance of the lattice. We also define the Fourier transform

$$\begin{aligned} \chi(q, TH) &= \int\frac{d^dx\,d^dx'}{\Omega}\, e^{-i\vec{q}\cdot(\vec{x}-\vec{x}')}\,\chi(\vec{x}-\vec{x}'; TH) \\ &= \frac{1}{3}\sum_{\alpha=1}^{3}\langle\delta M_q^\alpha\,\delta M_{-q}^\alpha\rangle \end{aligned} \tag{2.11}$$

where

$$\vec{M}_q = \int\frac{d^dx}{\sqrt{\Omega}}\, e^{-i\vec{q}\cdot\vec{x}}\,\vec{M}(\vec{x}) . \tag{2.12}$$

In this summer school we are becoming aware of the importance of higher order spin correlation functions like

$$\chi_{\alpha\beta\mu\nu}(\vec{x}_1,\vec{x}_2,\vec{x}_3,\vec{x}_4) \equiv \langle \delta M^{\alpha}(\vec{x}_1)\,\delta M^{\beta}(\vec{x}_2)\,\delta M^{\mu}(\vec{x}_3)\,\delta M^{\nu}(\vec{x}_4)\rangle . \tag{2.13}$$

Note that we can express the magnetic susceptibility as

$$\chi_M = \beta \int \frac{d^d x\, d^d x'}{\Omega}\, \chi(\vec{x},\vec{x}';TH) = \lim_{q\to 0} \beta\, \chi(q,TH) \tag{2.14}$$

We know, of course, that $\vec{M}(T, H)$ and χ can be accurately measured using various spin resonance techniques. It will be convenient, from a pedagogical point of view, to think in terms of magnetic neutron scattering[10] as our experimental probe. The elastic or Bragg part of the neutron scattering cross section is proportional to the magnetization squared. Thus, using neutrons, we can measure $\vec{M}(T, H)$. The inelastic part of the cross-section gives the correlation function $\chi(q, T, H)$. The magnetic susceptibility χ_M can be obtained experimentally as the small wave number limit of $\chi(q, T, H)$.

What do we find in our neutron scattering experiments? If we map out the magnetization as a function of T in zero external field $(M(T) = M(T, 0))$ we find the qualitative behavior in Fig. 1. For $T > T_c$ ($\equiv$the Curie temperature) one is in the paramagnetic phase and the average magnetization is zero. Below T_c a spontaneous magnetization develops starting at zero for $T = T_c$ and increasing to some finite value at $T = 0$. M(T) can take on only two particular values (corresponding to up or down orientations) for a given $T < T_c$. There is therefore a region of unstable states (as labelled in Fig. 1). If we concentrate on the region where T is just below T_c, then, since M must vanish as $T \to T_c$ we can fit our data to the form

$$M(T) \sim \varepsilon^{\beta} \tag{2.15}$$

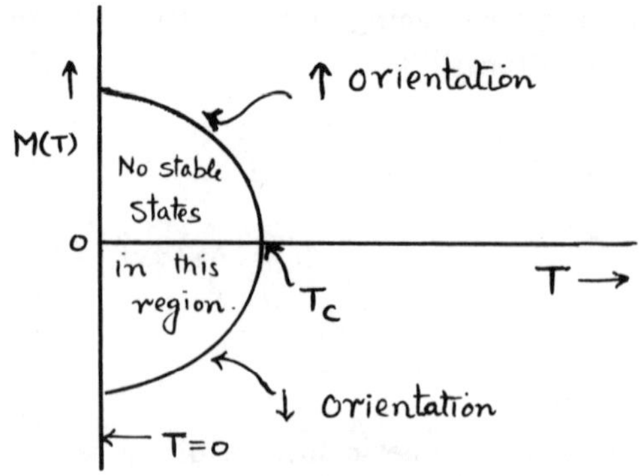

Fig. 1. Magnetization as a function of temperature in zero magnetic field for a ferromagnet.

where $\varepsilon = |(T_c - T)|/T_c$ (the reduced temperature) and β is the first of a number of exponents we will introduce.

In our measurements we can also plot M versus H at several values of temperature (See Fig. 2). If, in our experiments, we sit at $T = T_c$ and map

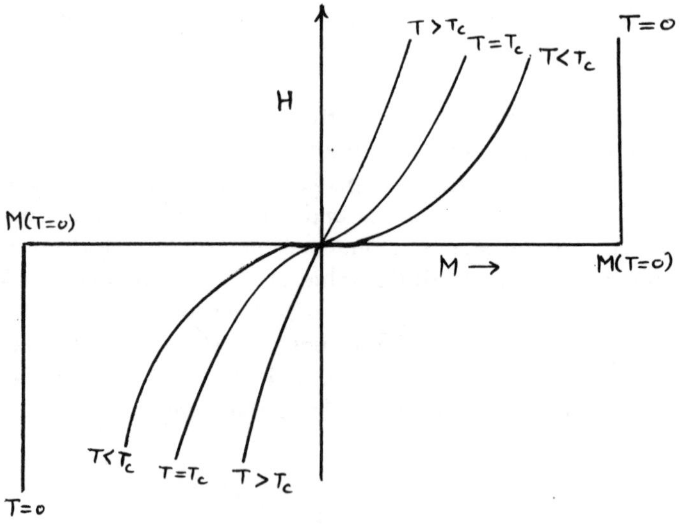

Fig. 2. Magnetization as a function of the magnetic field for various temperatures for a ferromagnet.

out $M(T_c, H)$ as a function of H, we see that $M \to 0$ as $H \to 0$ and we assume we can fit our data with the form

$$M(T_c, H) \sim H^{1/\delta} . \tag{2.16}$$

We can also see from the curves in Fig. 2 that the critical temperature is not only where the spontaneous magnetization begins to develop, it is also where

$$\left.\frac{\partial H}{\partial M}\right|_{H=0} = 0 . \tag{2.17}$$

This means the magnetic susceptibility, defined by (2.8), becomes infinite as as $T \to T_c$. As a result of this infinity the magnetic scattering near zero wave-number will be dramatically enhanced as one approaches T_c. We can fit our data if we assume the divergence is of the power law form

$$\chi(0,T,0) = \beta \chi_M \sim \varepsilon^{-\gamma} \tag{2.18}$$

for $T \approx T_c$.

If we measure the q dependence of $\chi(q)$ using neutron scattering, then it should be clear for consistency with the result at $q = 0$ that if we sit at $T = T_c$ and allow q to approach zero, then the correlation function must become infinite. The neutron data is normally fit to a form

$$\chi(q, T_c, 0) \sim q^{-2+\eta} . \tag{2.19}$$

Since the magnetic scattering is enhanced for small q we guess that very long wavelength correlations develop between the spins as we approach the critical point. In other words we expect that $\chi(\vec{x})$ decays slowly to zero for large x. The growth of these correlations can be conveniently characterized through the introduction of a correlation length

$$\xi^2 \equiv 2d \frac{\int d^dx\, \chi(x)\, x^2}{\int d^dx\, \chi(x)} , \tag{2.20}$$

or, in terms of Fourier transforms,

$$\xi^2 = \lim_{q \to 0} - \frac{\partial}{\partial q^2} \ln \chi(q^2) \tag{2.21}$$

We expect ξ to grow as we approach the critical point and it is conventional to fit data to the form

$$\xi \sim |\varepsilon|^{-\nu} \tag{2.22}$$

It is worth noting an approximate interpolation form for the correlation function for $H = 0$ that gives reasonable agreement for small q and ϵ is

$$\chi(q, T, 0) = \frac{C}{q^2 + \xi^{-2}} \tag{2.23}$$

which is called the Ornstein-Zernike form. Taking the inverse Fourier transform, we find in 3-dimensions that

$$\chi(\vec{x} - \vec{x}', T, 0) = \frac{C\, e^{-|\vec{x} - \vec{x}'|/\xi}}{|\vec{x} - \vec{x}'|} . \tag{2.24}$$

One can easily see why we have called ξ the correlation length. For finite ξ χ falls off exponentially for large $|\vec{x} - \vec{x}'|$, but as we approach the critical point $\xi \to \infty$ and the exponential approaches 1. There are then correlations that extend over very large distances. This is associated with the development of long-range order.

We have introduced a number of critical indices γ, ν, β, δ, η that characterize the behavior of the system near the critical point. Their definitions and "experimental" values are given in table 1. One of the central problems in the theory of critical phenomena has been the calculation of these critical indices from basic theoretical principles.

Quantity measured	definition of index	measured value of index
$M(T, 0)$	ϵ^{β}	$\beta = 0.34 \pm 0.02$
$M(T_c, H)$	$H^{1/\delta}$	$\delta = 4.2 \pm 0.1$
$\chi(0, T, 0) \sim \chi_M$	$\epsilon^{-\gamma}$	$\gamma = 1.33 \pm 0.02$
$\chi(q, T_c, 0)$	$q^{-2+\eta}$	$\eta = 0.07 \pm .07$
ξ	$\epsilon^{-\nu}$	$\nu = 0.65$

Table I: Relationship of measured quantities and critical indices. The measured values of the indices are taken from Ref. 7. The reduced temperature ϵ is defined as $\epsilon = | T - T_c | / T_c$.

C. Universality Among Second Order Phase Transitions

There are a number of systems that undergo second order phase transitions: antiferromagnets, fluids, superconductors, superfluids, phase-separation in binary mixtures. A rather incredible empirical fact is that the behavior near all of these physically very different transitions can be described by a set of indices β, δ, γ, ν and η similar to those for the ferromagnet if we parameterize things properly. The first parameter one must choose is the order parameter. The magnetization density is the order parameter for a ferromagnet and in general is characterized by the requirements:

(i) The order parameter goes to zero as $T \to T_c$ from below.

(ii) The Fourier transform of the order-parameter-order-parameter correlation function diverges as $q \to 0$ in zero conjugate external field.

The conjugate field is that field that couples linearly to the order parameter in the Hamiltonian. We list, in Table II, the order parameters for a number of transitions.

As an example of the usefulness of this description consider the liquid-gas case. In analogy with the ferromagnetic case we expect that at the critical point the density-density correlation function will diverge as

$$\chi_{nn}(q) \sim q^{-2+\eta_L} \tag{2.25}$$

where η_L is the index characteristic of a liquid-gas transition. $\chi_{nn}(q)$ has been accurately measured using light scattering techniques. In a similar

Critical point	Order Parameter	Example
Ferromagnetic	Magnetization	Fe
Antiferromagnetic	Sublattice magnetization	FeF_2
λ - transition	He^4 - amplitude	He^4
Liquid-gas	Density	H_2O
Superconductivity	Electron pair amplitude	Pb
Binary fluid mixture	Concentration of one fluid	CCl_4-C_7F_{14}
Binary alloy	Density of one kind of sublattice	Cu-Zn

Table II: Order parameters for various second order phase transitions.

manner we can define all of the indices γ, β, δ, η and ν for all of the systems listed above once we identify the order parameter and conjugate external field. It turns out that we cannot only characterize seemingly very different systems by analogous sets of indices, but these indices have the same order of magnitude values for all the systems. We list a few representative examples in table III.

To a first approximation all of these systems look very similar near the critical point. Are these indices the same or do they depend on different properties of different systems? All of our best information now indicates that the indices depend basically only on two properties of the system: the spatial dimensionality and the vector nature of the order parameter. This can best be understood by considering a generalization of the Heisenberg Hamiltonian:

$$\mathcal{H} = -J \sum_{\langle i,j \rangle} \sum_{\alpha=1}^{n} M_i^\alpha M_j^\alpha \tag{2.26}$$

where i and j label lattice sites in a d-dimensional lattice, J is the inter-

Critical point	Material	β	γ	δ	η
Anti-ferro.	$RbMnF_3$	.316±.008	1.397 ±.034		.067 ±.01
Liquid-gas	CO_2	.3447±.0007	1.20 ±.02	4.2	
	Xe	.344±.003	1.203 ±.002	4.4 ±0.4	
	He^3	.361±0.001	1.15 ±.03		
	He^4	.3554 ±0.0028	1.17 ±.0005		
Bin. Mix.	CCl_4-C_7F_{14}	.335±0.02	1.2		
Bin. Alloy	Co-Zn	.305 ±0.005	1.25 ±0.02		

Table III: Experimental values for critical indices for various second order phase transitions. The values quoted here are from Ref. 7.

action between nearest neighbor spins, M_i^α is the α th component of an n component spin on lattice site i. $\langle i\ j \rangle$ means we sum only over nearest neighbor spins. We assume

$$\sum_{\alpha=1}^{n} (M_i^\alpha)^2 = \text{constant}. \tag{2.27}$$

If n = 1

$$\mathcal{H} = -J \sum_{\langle i,j \rangle} M_i^z M_j^z \tag{2.28}$$

and we have the Ising model. In this case the spin can point in one of two directions and is a scalar. In the case n = 2

$$\mathcal{H} = -J \sum_{\langle i,j \rangle} (M_i^x M_j^x + M_i^y M_j^y) \tag{2.29}$$

and we have the x-y or planar model. In this case the spins are free to rotate in a plane and the order parameter is a two-vector. For n = 3 we have the Heisenberg model

$$\mathcal{H} = -J \sum_{\langle i,j \rangle} \vec{M}_i \cdot \vec{M}_j \tag{2.30}$$

where the spins are 3-vectors of unit length. It is interesting that the case $n \to \infty$ can be solved exactly.[11]

The n-vector model has been studied as a function of spatial dimension d and n using high temperature expansion techniques. Some of the results are presented in tables IV and V.

Critical index ↓	dimension →	2	3	4
ν		1	.65	.50
δ		15	5	3

Table IV: Variation of Ising model critical indices with spatial dimension, (taken from ref. 12).

n	γ	ν	η	β	δ
1	1.25	0.638	0.041	0.313	5
2	1.32	0.675	0.04	0.33	5
3	1.38	0.70	0.03	0.345	5
∞	2	1	0	1/2	5

Table V: Variation of n-vector model critical indices with n in three dimensions, (taken from ref. 12).

We see that the indices do depend on spatial dimensionality and the n-vector nature of the order parameter. The indices do not appear to depend on the particular type of lattice, the magnitude of the spin or on quantum effects. These properties are quite local properties. We have argued that critical phenomena are associated with very long wavelength correlations. It makes some sense then that critical indices do not depend on very local properties. It turns out that the indices may be weakly dependent on other properties (like long range interactions). This is discussed in some detail in the review by Fisher.[4] We will avoid these more special cases in the rest of our discussion.

All of these considerations about critical indices and the similarities between systems with seemingly very different microscopic interaction mechanisms can be summarized in the hypothesis of universality:[13]

All phase transition problems can be divided into a small number of different classes depending upon the dimensionality of the system and the symmetries of the ordered state. Within each class, all phase transitions have identical behavior in the critical region, only the names of the variables are changed.

It appears that a primary activity for people interested in critical phenomena is to classify various systems in various (n, d) groups. We first try to determine the appropriate spatial dimensionality. This will usually be 3 but in some layered compounds (d = 2) may be appropriate. We then try to identify the value of n for the order parameter in the system. We expect, for example, fluids and fluid mixtures to show "Ising-like" behavior since the density and concentration are scalars. Superfluids (helium) and super conductors have complex (n = 2) order parameters so we expect them to be "x-y like." Isotropic magnets (ferromagnets and anti-ferromagnets) have 3-vector order-parameters and are expected to be "Heisenberg-like."

D. The Scaling Hypothesis

Closely related to the idea of universality is the idea of scaling. Universality implies that the microscopic details of a system are unimportant near the critical point. The indices cannot depend on the lattice separation a in a ferromagnet or the hard-core separation distance in a fluid. If these lengths are to be removed from the problem, then there must be one length that is much longer and which dominates these microscopic lengths. This length is clearly the correlation length ξ, and it dominates all other lengths in the problem:

$$a/\xi << 1 \tag{2.31}$$

near T_c. According to the scaling hypothesis the divergence of ξ is responsible for the singular dependence on $T - T_c$ of physical quantities, and, as far as the singular dependence is concerned, ξ is the only relevant length. We will see that the scaling hypothesis is an example of how a simple assumption can have far reaching consequences.

Consider first the correlation function. We note that χ is a function, in general, of q, H, T and the temperature independent variables like, for a magnet, the lattice spacing a and J the nearest neighbor interaction. We can in principle eliminate the temperature in terms of the correlation length and write

$$\chi = \chi(q, \xi, H, a, J) \ . \tag{2.32}$$

We can rewrite this in the form

$$\chi = \chi(\xi, q\xi, H\xi^{x_1}, a/\xi, J/\xi^{x_2}) \tag{2.33}$$

where x_1 and x_2 are indices characteristic of the system. In the limit $T \to T_c$, $\xi \to \infty$, a/ξ will go to zero and J/ξ^{x_2} will either go to zero or infinity (unless for some special reason $x_2 = 0$) since a and J are fixed-numbers independent of temperature. Therefore no matter how large a is on a microscopic scale ξ will eventually become much larger. We are said to be in the scaling region when $a/\xi << 1$ and J/ξ^{x_2} is either very large or very small. In practice how close to T_c or how large ξ must be before we are in the scaling region depends on the system. Assuming we are in the scaling region we have the expansion

$$\chi = \xi^{x_3}\left(J/\xi^{x_2}\right)^{x_4}\left(a/\xi\right)^{x_5} f_1(q\xi, H\xi^{x_1}) \tag{2.34}$$

plus correction terms in powers of a/ξ and (J/ξ^{x_2}). Note that $q\xi$ and $H\xi^{x_1}$ need not go to 0 or infinity for large but finite ξ since q and H are externally adjustable. We then have the scaling result

$$\chi(q, \xi, H) = \xi^{x} f(q\xi, H\xi^{x_1}) \tag{2.35}$$

where $x \equiv x_3 - x_2 x_4 - x_5$ and

$$f(q\xi, H\xi^{x_1}) \equiv J^{x_4} a^{x_5} f_1(q\xi, H\xi^{x_1}) . \tag{2.36}$$

The indices x and x_1 are unknown. The scaling assumption has immediate consequences concerning the critical indices. We see that

$$\chi(0, \xi, 0) = \xi^{x} f(0,0) \sim \varepsilon^{-\gamma} . \tag{2.37}$$

Since

$$\xi \sim \varepsilon^{-\nu} \tag{2.38}$$

we can identify

$$-x\nu = -\gamma \tag{2.39}$$

or

$$x = \gamma/\nu . \tag{2.40}$$

Similarly

$$\chi(q, T_c, 0) = \lim_{\xi \to \infty} \xi^x f(q\xi, 0) . \tag{2.41}$$

Since χ $(q, T_c, 0)$ is finite for $q \neq 0$ it must be independent of ξ as $\xi \to \infty$ and we require

$$f(q\xi, 0) \longrightarrow (q\xi)^{-x} \tag{2.42}$$

as $q\xi \to \infty$. This means

$$\chi(q, T_c, 0) \sim q^{-x} . \tag{2.43}$$

If we compare this with (2. 19) we can identify

$$x = 2 - \eta . \tag{2.44}$$

Combining this with (2. 40) we obtain

$$\gamma = (2 - \eta)\nu . \tag{2.45}$$

The scaling hypothesis allows us to find relations between the various critical indices. They are not all independent. Analysis of the magnetic field dependence of χ allows one to derive a further scaling relation

$$\gamma = \beta(\delta - 1) . \tag{2.46}$$

It appears that only three of the 5 indices (η, β, δ, γ, ν) are independent. Notice that we have two ways of calculating γ:

$$\gamma = (2 - \eta)\nu \tag{2.47}$$

$$\gamma = \beta(\delta - 1) \tag{2.48}$$

If we use the results for the critical indices from Table V for n = 1, (2.47) gives $\gamma = 1.2498$ while (2.48) gives $\gamma = 1.252$. These values agree to within the error on the exponents.

We can go further and analyze the consequences of scaling for the free energy and derive the scaling relation ($\delta = \frac{2 + d - \eta}{d - 2 + \eta}$), reducing the number of independent exponents to two, but this would take us too far afield. The scaling hypothesis played an important role in the development of our modern theory of critical phenomena. Initially, as developed by Kadanoff[14] and Widom,[15] the scaling hypothesis served to correlate the growing amount of data for critical indices. It also established that a single length was dominating the physics for temperatures near T_c and that the local microscopics are not important in critical phenomena.

While the scaling hypothesis was a major step it was still far from a complete theory. While the scaling hypothesis reduced the number of independent indices to two, it did not tell us anything about how to calculate them. More fundamentally, one didn't know how to establish the scaling hypothesis from first principles.

III. THE GINZBURG-LANDAU-WILSON THEORY OF STATIC CRITICAL PHENOMENA

A. The Ginzburg-Landau Free Energy

We now want to develop a theoretical model that allows one to explicitly calculate the critical indices as a function of n and d. The establishment of this model relies heavily on the ideas of universality and scaling. In particular, we will use the idea that only long wavelength behavior is important near the critical point, and that the local microscopics of the system are not important. Our model theory will be one step removed from a microscopic description of a given system but one step closer to the essential physics in the problem.

We want to calculate the partition function

$$Z = \mathrm{Tr}\, e^{-\beta \mathcal{H}} \tag{3.1}$$

where $\mathcal{H}$ is a **microscopic** Hamiltonian depending on some microscopic order parameter field M(x). To obtain our model we want to perform the statistical average in two steps. We first average over the local microscopic degrees of freedom in the system. This corresponds to a coarse graining procedure. One way of looking at this is to think of averaging M(x) over some distance L which is much bigger than the "lattice spacing" a and much smaller than the correlation length ξ,

$$\xi >> L >> a \; . \tag{3.2}$$

As a result of this local averaging, which I denote Tr_L, one obtains an "effective-free energy" $F_T\{\tilde{M}\}$

$$\mathrm{Tr}_L\, e^{-\beta \mathcal{H}} \equiv e^{-F_T\{\tilde{M}\}} \; . \tag{3.3}$$

This free energy is a functional of the coarse grained or semi-macroscopic order parameter $\tilde{M}(\vec{x})$. One of several possible ways of defining this coarse grained order parameter is

$$\tilde{M}(\vec{x}) = \frac{1}{L^d} \int_{L^d} d^d y \, \langle M(\vec{y}) \rangle_{L^d} \tag{3.4}$$

where the average is over a volume L^d centered at the position $\vec{x}$. After this averaging, we have a continuous field. The discrete lattice sites and the atomic structure is then removed from the problem. We require in this averaging that the volume of averaging be much larger than microscopic dimensions (contain many lattice sites or atoms), but much less than ξ^d. To complete our averaging and obtain the physical partition function we must average over all of the possible configurations of the coarse grained order parameter $\tilde{M}(\vec{x})$. We write

$$Z = \left\langle e^{-F_T[\tilde{M}]} \right\rangle_{\tilde{M}} \tag{3.5}$$

where $\langle \ \rangle_{\tilde{M}}$ indicates the average over the various possible configurations of $\tilde{M}$. Later we will come back and say more about this averaging over $\tilde{M}$. The point I wish to make here is that if

$$P = \frac{e^{-\beta \mathcal{H}}}{Z} \tag{3.6}$$

is the probability that our microscopic system is in some microstate, then

$$\begin{aligned} P[\tilde{M}] &= \mathrm{Tr}_L\, e^{-\beta\mathcal{H}}/Z \\ &= \frac{e^{-F_T[\tilde{M}]}}{Z} \end{aligned} \tag{3.7}$$

should be the probability that our coarse grained system is in the semi-macrostate characterized by the order-parameter configuration $\tilde{M}(\vec{x})$. If we are to preserve the normalization

$$\mathrm{Tr}\, P = 1 \tag{3.8}$$

then

$$\langle P[\tilde{M}]\rangle_{\tilde{M}} = \frac{1}{Z}\langle e^{-F_T[\tilde{M}]}\rangle_{\tilde{M}} = 1 \tag{3.9}$$

or

$$Z = \langle e^{-F_T[\tilde{M}]}\rangle_{\tilde{M}} \tag{3.10}$$

Then since $P[\tilde{M}]$ gives the probability of a certain configuration in $[\tilde{M}]$ it is clear that the average order parameter can be written:

$$\begin{aligned} \langle M(\vec{x})\rangle &= \langle P[\tilde{M}]\,\tilde{M}(\vec{x})\rangle_{\tilde{M}} \\ &= \langle e^{-F_T[\tilde{M}]}\,\tilde{M}(\vec{x})\rangle/Z, \end{aligned} \tag{3.11}$$

while the correlation function can be calculated as

$$\chi(\vec{x}-\vec{x}') = \langle \delta\tilde{M}(\vec{x})\,\delta M(\vec{x}')\,e^{-F_T[\tilde{M}]}\rangle_{\tilde{M}}/Z. \tag{3.12}$$

The local averaging, as I have stated it, is not very well defined. I intend it only to be a schematic device. Since $\mathcal{H}$ and the meaning of the trace changes from system to system, this local averaging will also change from system to system. The simplest case is a Bose gas system, where the order parameter, the annihilation operator ψ , appears naturally in the Hamiltonian. In this case one simply averages overall quantum effects. In fluid and spin systems one can perform this local average using rather sophisticated functional techniques, while in superconductors there is a rather tedious field theoretical calculation. One definition for our course graining procedure, which is still rather formal but which will be useful later, is to write

$$P[\tilde{M}] = \mathrm{Tr}\, e^{-\beta\mathcal{H}}\, \delta(\tilde{M}-M)/Z \qquad (3.13)$$

where

$$\delta(\tilde{M}-M) = \prod_i \delta(\tilde{M}_i - M_i)\,, \qquad (3.14)$$

i labels all the degrees of freedom of the classical coarse gained field $\tilde{M}_i$ and the trace is over all the degrees of freedom of the microscopic field M_i.

While it is difficult in practice to carry out this coarse graining procedure, what is useful and in keeping with universality is that the coarse grained free energy F_T and the meaning of $\langle\ \rangle_{\tilde{M}}$ are more or less the same for all of these systems. The form for F_T follows, in fact, from rather general arguments due to Ginzburg and Landau.[9] We assume that $F_T\{\tilde{M}\}$ can be expressed in a functional power series expansion in the coarse grained order parameter $\tilde{M}(x)$. The most general such expansion is of the form (for simplicity we drop the tilda on M henceforth).

$$F_T[M] = \sum_{m=0}^{\infty} \sum_{\alpha_1,\alpha_2,\cdots,\alpha_{2m}} \int d^dx_1\, d^dx_2 \cdots d^dx_{2m}\, \tilde{U}_{\alpha_1,\alpha_2,\cdots\alpha_{2m}}(\vec{x}_1,\vec{x}_2\cdots\vec{x}_{2m})$$
$$\times\, M^{\alpha_1}(\vec{x}_1)\, M^{\alpha_2}(\vec{x}_2) \cdots\cdots M^{\alpha_{2m}}(\vec{x}_{2m})\,. \qquad (3.15)$$

The m = 0 term is the contribution from the short-ranged degrees of freedom that don't contribute to any divergent behavior near a phase transition. The expansion will be in even powers of M if the original microscopic Hamiltonian was invariant under the transformation $M \rightarrow -M$. We then assume that the U's are short ranged in space and can be Fourier transformed·

$$U_{\alpha_1,\alpha_2,\cdots\alpha_{2m}}(\vec{q}_1,\vec{q}_2,\cdots\vec{q}_{2m-1})\,\delta(\vec{q}_1+\vec{q}_2+\cdots\vec{q}_{2m})(2\pi)^d$$
$$= \int \prod_{i=1}^{2m} d^dx_i\, e^{i\,\vec{q}_i\cdot\vec{x}_i} \times$$

$$\times \; \tilde{U}_{\alpha_1,\alpha_2 \cdots \alpha_{2m}}(\vec{X}_1, \vec{X}_2 \cdots\cdots \vec{X}_{2m})$$

(3.16)

We have extracted the δ-function, due to translational invariance, explicitly. The effective free energy is then characterized by the functions U. We define a parameter space characterizing a general Ginzburg-Landau (GL) effective free energy as

$$\mu \equiv (u_2, u_4, u_6, \cdots\cdots) \quad . \tag{3.17}$$

Each of these entries contain more than one parameter since U_2, for example, is a function of q_1. We include in our parameter space all of the coefficients in a Taylor series expansion of U_m in powers of the wavenumbers upon which U_m depends. The parameter space characterizing a general GL free energy functional therefore is spanned by an infinite number of parameters. In order to gain some feeling for the theory let us follow the arguments of Landau which drastically reduce the number of non-zero parameters in μ.

Landau argued that we need only keep terms of $0(M^4)$ since M should be "small" near the critical point. Later we will see to what extent this argument is sensible. The coefficients $U^{(2)}$ and $U^{(4)}$ can be simplified by a few simple but reasonable assumptions. If our system is isotropic we should build this symmetry into $U^{(2)}$ and $U^{(4)}$ and write:

$$\tilde{U}^{(2)}_{\alpha_1,\alpha_2}(\vec{X}_1, \vec{X}_2) = \delta_{\alpha_1,\alpha_2} \tilde{U}^{(2)}(\vec{x}_1 - \vec{x}_2) \tag{3.18}$$

$$\tilde{U}^{(4)}_{\alpha_1,\alpha_2,\alpha_3,\alpha_4}(\vec{X}_1, \vec{X}_2, \vec{X}_3, \vec{X}_4) = \frac{1}{3}\Big[\delta_{\alpha_1,\alpha_2}\delta_{\alpha_3,\alpha_4} + \delta_{\alpha_1,\alpha_3}\delta_{\alpha_2,\alpha_4} + \delta_{\alpha_1,\alpha_4}\delta_{\alpha_2,\alpha_3}\Big] \tilde{U}^{(4)}(\vec{X}_1, \vec{X}_2, \vec{X}_3, \vec{X}_4) . \tag{3.19}$$

We also assume that the associated Fourier transforms can be expanded in powers of the wavenumber since only small wavenumbers (large wavelengths) $qa << 1$ should matter near the critical point. We write therefore

$$U^{(2)}(q_1) = r_0 + c q_1^2 + O(q_1^4) , \tag{3.20}$$

$$U^{(4)}(q_1, q_2, q_3) = u + O(q_i^2, \vec{q}_i \vec{q}_j) \tag{3.21}$$

and we assume henceforth that there is a large wavenumber cut-off Λ in all integrals. Physically Λ corresponds to an inverse lattice spacing. Since the physics near the critical point is dominated by small wavenumbers our results will be essentially independent of Λ. In coordinate space we can write

$$\tilde{U}^{(2)}(\vec{x}_1 - \vec{x}_2) = (r_0 - \nabla_{x_1}^2)\, \delta(\vec{x}_1 - \vec{x}_2) \tag{3.22}$$

$$\tilde{U}^{(4)}(\vec{x}_1, \vec{x}_2, \vec{x}_3, x_4) = u\, \delta(\vec{x}_1 - \vec{x}_2)\, \delta(\vec{x}_1 - \vec{x}_3)\, \delta(\vec{x}_1 - \vec{x}_4) \tag{3.23}$$

and the effective free-energy can then be written as

$$\begin{aligned} F[M] &\equiv F_T[M] - F_T[0] \\ &= \int d^d x \left\{ \sum_{\alpha=1}^{n} \left[r_0 (M^\alpha(x))^2 + c \sum_{j=1}^{d} (\nabla_x^j M^\alpha(x))^2 \right] \right. \\ &\quad \left. + u (\vec{M}(\vec{x}) \cdot \vec{M}(\vec{x}))^2 - \vec{H}(\vec{x}) \cdot \vec{M}(\vec{x}) \right\} \end{aligned} \tag{3.24}$$

We have included for later convenience a spatially inhomogeneous external field $\vec{H}(x)$ in F which couples to the coarse grained order parameter. F [M] above is the conventional Ginzburg-Landau free energy functional characterized (for $\vec{H} = 0$) by the parameter set

$$\mu = (r_o, c, u) \tag{3.25}$$

Clearly r_a, u, and c contain the remaining information about the original microscopic Hamiltonian. The important point is that one expects that these parameters have a regular analytic dependence on temperature. Thus, we assume they can be expanded in a power series in the temperature near T_c:

$$r_o = r_o^o + r_1 (T - T_c) + \cdots \tag{3.26a}$$

$$u = u_o + u_1 (T - T_c) + \cdots \tag{3.26b}$$

$$c = c_o + c_1 (T - T_c) + \cdots \tag{3.26c}$$

If our model is to describe second order phase transitions we must require:

$$r_o^o = 0 \quad , \quad r_1 > 0 \tag{3.27}$$

$$u_o > 0 \quad , \quad c_o > 0$$

We will see why we make these choices as we proceed.

B. The Landau or Mean Field Theory

The Ginzburg-Landau free energy F [M] is to be used in the final averaging to obtain Z, $\langle M \rangle$ and χ. We still have not specified this final averaging over M. In the Landau or mean field theories this precise averaging is not important. It is assumed that the average of exp (– F {M}) over M can be replaced by evaluating F [M] for values of $\langle M \rangle$ which minimize the free energy. Thus, we assume

$$\left\langle e^{-\beta[M]} \right\rangle_M \simeq e^{-F[\bar{M}]} \tag{3.28}$$

where $\bar{M}$ is the most probable value of the order parameter, and is the one which minimizes the free energy. Thus $\bar{M}$ is determined by the variational condition

$$\left.\frac{\delta F[M]}{\delta M(x)}\right|_{M=\bar{M}} = 0$$

$$= 2r_0\bar{M}(x) - 2c\,\nabla_x^2\,\bar{M}(x) + 4u\bar{M}^3(x) - H(x) \tag{3.29}$$

In this section, for simplicity, we assume that the order parameter is Ising-like (n = 1). The Landau theory ignores the fluctuations of the order parameter about its most probable value. We will come back to this point later. For now let us concentrate on our equation for $\bar{M}$,

$$2r_0\bar{M}(x) - 2c\,\nabla_x^2\,\bar{M}(x) + 4u\,\bar{M}^3(x) = H(x)\,. \tag{3.30}$$

If we choose H to be uniform in space, then we have translational invariance, $\bar{M}$ is independent of x and

$$\left(2r_0 + 4u\bar{M}^2\right)\bar{M} = H\,. \tag{3.31}$$

For H = 0 this equation has the solutions,

$$\bar{M} = 0 \tag{3.32}$$

and, for r_o or u negative,

$$\bar{M} = \pm \sqrt{-r_o/2u} \; . \tag{3.33}$$

We choose between these solutions for $\bar{M}$ by requiring that the physical value of $\bar{M}$ give the lowest free energy. We easily obtain

$$F = \Omega \bar{M}^2 (r_o + u\bar{M}^2)$$

$$= \begin{cases} 0 & \bar{M} = 0 \\ -\Omega\left(\frac{r_o^2}{4u}\right) & \bar{M} = \pm\sqrt{\frac{-r_o}{2u}} \end{cases} \tag{3.34}$$

The $\bar{M} \neq 0$ solution gives a lower free energy if $u > 0$. This solution is admissible only if r_o/u is negative. Thus we require $u > 0$ and that $\bar{M}$ is given by (3.33) only for $r_o < 0$.

In summary the free energy is minimized by

$$\bar{M} = 0 \qquad \text{for} \qquad r_o > 0 \tag{3.35a}$$

$$\bar{M} = \pm\sqrt{-r_o/2u} \qquad \text{for} \qquad r_o < 0 \tag{3.35b}$$

Since we want the magnetization to vanish above T_c and be non-zero below T_c, we must choose $r_o > 0$ for $T > T_c$ and $r_o < 0$ for $T < T_c$. Remembering from (3.26a) that r_o should have an analytic dependence on $T - T_c$, we see that the consistent choice for r_o is

$$r_o = r_1 (T - T_c) \tag{3.36}$$

where $r_1 > 0$. We have then, for $T < T_c$, that

$$\bar{M} = \pm\sqrt{\frac{-r_1(T-T_c)}{2u}} = \pm\sqrt{\frac{r_1}{2u}}\,(T_c-T)^{1/2}. \quad (3.37)$$

This tells us immediately that $\beta = 1/2$. Going back to (3.31) for finite H we see that at $T = T_c$ or $r_o = 0$ that

$$\bar{M}^3 = \frac{H}{4u} \quad (3.38)$$

or

$$\bar{M} = \left(\frac{H}{4u}\right)^{1/3}. \quad (3.39)$$

We can then identify $\delta = 3$.

One may now ask the question: Can the Landau theory give us information about the correlation function? It is clear that we can compute the static susceptibility $(\partial M/\partial H)_{H=0}$, but can we determine the wavenumber dependence of the correlation function within the Landau Theory? The answer to this question is not obvious and is facilitated by the introduction of functional derivatives. This point is discussed in detail in Appendix A. The final result is that the correlation function is given as the functional derivative of the average magnetization with respect to an inhomogeneous external field:

$$\chi(\vec{x},\vec{x}') = \frac{\delta\langle M(\vec{x})\rangle}{\delta H(\vec{x}')} \quad (3.40)$$

If we take the functional derivative of (3.30) with respect to $H(x')$ and use (3.40) we obtain

$$\left[2r_o - 2c\nabla_x^2 + 12u\bar{M}^2(x)\right]\chi(\vec{x},\vec{x}') = \delta(\vec{x}-\vec{x}'). \quad (3.41)$$

If we now let H be zero $\bar{M}$ will be uniform (independent of x) and $\chi(x,x') = \chi(x-x')$. If we then Fourier transform (3.41) over space we obtain

$$\chi(q,T) = [2r_0 + 2cq^2 + 12u\bar{M}^2]^{-1}. \tag{3.42}$$

Using (2.21) we can immediately express the correlation length in the Landau theory as

$$\xi^2 = C[r_0 + 6u\bar{M}^2]^{-1} \tag{3.43}$$

and write the correlation function in the Ornstein-Zernike form (2.23) with $C = 1/2c$. We note, using (3.35), that

$$\xi^2 = \begin{cases} \dfrac{c}{r_1(T-T_c)} & T > T_c \\ \dfrac{c}{2r_1(T_c-T)} & T < T_c \end{cases} \tag{3.44}$$

and we see, comparing (2.22) and (3.44), that $\nu = 1/2$ for both the ordered and disordered regions. Going back to (3.42) we see, again using (3.35), that for $q = 0$

$$\chi(0,T,0) = [2(r_0 + 6u\bar{M}^2)]^{-1}$$
$$= \begin{cases} [2r_1(T-T_c)]^{-1} & T > T_c \\ [4r_1(T_c-T)]^{-1} & T < T_c \end{cases} \tag{3.45}$$

Comparing (2.18) and (3.45) we can identify $\gamma = 1$ and the index is the same above and below the transition.

If we set $T = T_c$ in (3.44) we obtain

$$\chi(q,T_c,0) = \frac{1}{2cq^2} \tag{3.46}$$

We find, comparing this with (2.19), that $\eta = 0$.

We summarize the results of the Landau theory in Table VI and compare with the "expected" values for the indices for n = 1 and d = 3. We see that the Landau theory is not very good. It is a "30%" theory. It also turns out that the Landau theory ignores the variation of the critical indices with n and d.

It is worth noting that the Landau theory which gives "classical" values for the indices is equivalent, with respect to critical properties, to various theories for individual microscopic systems. These theories include:

The Weiss molecular field theory for ferromagnets.

The van der Waals theory for the liquid-gas transition.

The Bogoliubov theory for the λ - transition in He^4.

The Bardeen-Cooper-Schrieffer (BCS) theory of superconductivity.

Although these theories are very different in their microscopic development in the end they all lead to the same classical or mean field indices near T_c. As we go closer and closer to the critical point they all break down. The reason is simple. The Landau theory assumes that fluctuations in the order parameter are small, but we know that very near the critical point the fluctuations are very large. Thus, one must go back and discuss how one can treat fluctuations.

C. Critical Phenomena as a Field Theory

We now want to define the process of averaging over the various configurations of the coarse grained order parameter M. We said earlier that the $M(\vec{x})$ appearing in the effective free energy can be interpreted as the average of the microscopic order parameter in a volume L^d centered about the position $\vec{x}$ (see (3.4)). Suppose we break all of space up into these cells of volume L^d, $M(x_i)$ can then be interpreted as the average value of the order parameter in the i th cell. The free-energy, defined in the continuum by

$$F[M] = \int d^dx \, f[M(\vec{x})] \tag{3.47}$$

critical index	γ	η	ν	β	δ
Landau Theory	1	0	1/2	1/2	3
Experimental values	1.25	0.041	0.638	0.313	5

Table VI: Comparison of the Landau or mean field theory values for the critical indices with the "experimental values" for d = 3 and n = 1 from Table V.

can be written in the discrete case as

$$F[M] = L^d \sum_i f[M_i] \; . \tag{3.48}$$

The sum over i is over all of the cells. The partition function is then obtained by summing over all possible values of M_i, or

$$Z = \int_{-\infty}^{\infty} \prod_i \frac{dM_i}{N(L)} \exp -\left(L^d \sum_j f(M_j)\right) \tag{3.49}$$

where N(L) is a normalization factor that depends only on the size of L. Clearly any physical result can not depend on L since it is some arbitrary length satisfying $\xi >> L >> a$. Since we are interested in phenomena on a length scale comparable to ξ we should take the limit $L \to 0$. In this limit the cells reduce back to a "coarse-grained" continuum (it looks like a continuum on the scale of ξ) where

$$\lim_{L \to 0} L^d \sum_i f(M_i) = \int d^d x \; f(M(x)) \tag{3.50}$$

and

$$\lim_{L \to 0} \int \prod_i \frac{dM_i}{N(L)} \equiv \int \mathcal{D}(M) \tag{3.51}$$

The partition function can then be written as

$$Z = \int \mathcal{D}(M) \; e^{-F[M]} \; . \tag{3.52}$$

This defines our configurational average over M as a functional integral over M. We discuss a few relevant properties of functional integrals in Appendix A. The average order parameter and the correlation function are likewise given as functional integrals:

$$\langle M(x)\rangle = \int \mathcal{D}(M)\, M(x)\, e^{-F[M]}/Z \tag{3.53}$$

$$\chi(\vec{x},\vec{x}') = \int \mathcal{D}(M)\, \delta M(\vec{x})\,\delta M(\vec{x}')\, e^{-F[M]}/Z \ . \tag{3.54}$$

The Ginzburg-Landau-Wilson model is then completely specified once we identify F as the GL free energy given by (3.24).

It turns out that this model is well known in quantum field theory (with a few small variations). This model is called "φ^4" field theory. The "φ^4" corresponds to the interaction term M^4 being quartic. This is essentially the simplest non-trivial field theory one can construct.

I have spent some time on this development because the basic ideas are fundamental to the modern theory of critical phenomena. In the GLW formulation one notes at the very start that the critical fluctuations take place on a long distance scale and the short distance behavior is irrelevant and should be averaged out. Only after carrying out this average does one obtain a manageable theory where the strong critical fluctuations can be treated. Equally important is the idea that the final results should not depend on the details of the microscopic or the coarse-grained theory. We are, therefore, at liberty to choose the simplest theory that includes strong critical fluctuations. This is a conceptually difficult notion and plays a central role in developing a dynamical theory. We will try to make these ideas clearer in the next section.

D. The Renormalization Group

1. Discussion

Now that we have the GLW formulation what can we do with it? The first thing we might try is to calculate the observables in a perturbation theory expansion in the quartic coupling u. We see, using (A. 23), that for $u = 0$ we can solve the problem exactly. We find, for example, that the correlation function is given by

$$\chi(q,T,0) = \frac{1}{2(r_0 + cq^2)} \tag{3.55}$$

and the $u = 0$ theory is equivalent to the Landau theory for $T \geq T_c$ giving classical values for the exponents. In order to obtain deviations from the Landau theory we must have a non-zero u (>0). The obvious thing to try, since we cannot solve the problem exactly for non-zero u, is to develop a perturbation theory expansion for $\chi(q, T, 0)$ in powers of u. The difficulty with this program is that the perturbation theory is infrared (small q) divergent term by term as $T \to T_c$ and we don't know how to treat these divergences. An even more fundamental problem is that while any corrections to the classical values for the indices must arise from a non-zero u they must be independent of the particular non-zero value of u due to universality. This is a difficult theoretical dilemma which confounded theorists for a long time. The resolution of this difficulty was realized through the renormalization group approach due to Wilson.[16]

There are two main classes of renormalization group transformations--the so-called real space renormalization group and the GLW formulation. The real space method[17] is designed to treat spins on a discrete lattice and is very close in spirit to the original arguments of Kadanoff motivating the scaling hypothesis. The GLW formulation is a continuum field theory. Both approaches have been fruitful[18] and can be viewed as complementary. However, to date, only the continuum formulation has been successfully applied to dynamic phenomena. Since we will be concentrating on dynamics, I will emphasize the continuum description.

The renormalization group is built upon two observations:

(i) If a system is at the phase transition temperature ($T = T_c$) the correlation length ξ is infinite and the system is scale invariant. This means we can change our length scale and all observables are unchanged.

(ii) If the system is near but slightly removed from the critical point $\epsilon \equiv \frac{|T - T_c|}{T_c} << 1$ then ξ is finite and measures the degree to which the scale invariant symmetry is broken.

Let us draw an analogy between scale invariance and rotational invariance in a spin system. We all learned in quantum mechanics that rotational invariance corresponds to the invariance of a Hamiltonian under a group of transformations--the rotation group. If there is an applied field that picks out a particular direction then the invariance is broken. The field that breaks rotational invariance is analogous to the correlation length that breaks scale invariance near a phase transition. In the case of the rotation group the definition and application of the group is direct. Unfortunately the corresponding group transformation testing scale invariance near a critical point is far more complicated and requires a lot of technical machinery. I will now outline one of several formulations of the renormalization group transformation. I will emphasize the conceptual points and not worry about the many technical details. There are a number of good and complete discussions in the literature.[3-7] I will follow the treatment due to Ma[6] rather closely.

2. The Renormalization Group Transformation

We know that we want to set up a transformation that checks scale invariance. Thus we want to define some transformation under which the GLW probability distribution

$$P_0[M] = \frac{1}{Z} e^{-F_0[M]} \tag{3.56}$$

is invariant at the critical point. The first thing we might try is a straightforward scale transformation. We let

$$q \longrightarrow q' = bq \tag{3.57}$$

and rescale the fields via

$$M_q \longrightarrow M'_q = \alpha_b M_q \;. \tag{3.58}$$

The GL free energy, before the transformation, can be written in terms of the Fourier transformed fields as

$$\begin{aligned} F_0[M] &= \Omega \int^{\Lambda} \frac{d^dq}{(2\pi)^d} \left\{ (cq^2 + r_0) M_q M_{-q} \right. \\ &+ \Omega^2 u \int \frac{d^dq_1}{(2\pi)^d} \frac{d^dq_2}{(2\pi)^d} \frac{d^dq_3}{(2\pi)^d} M_{q_1} M_{q_2} M_{q_3} M_{-\vec{q}_1 - \vec{q}_2 - \vec{q}_3} \end{aligned} \tag{3.59}$$

where $\int^{\Lambda}$ means we restrict all wavenumbers to be less than Λ. After the appropriate transformations we obtain:

$$F_0[M'] = \Omega' \int^{\Lambda b} \frac{d^d q'}{(2\pi)^d} \frac{1}{b^d \alpha_b^2} \left(\frac{c q'^2}{b^2} + r_0 \right) M'_{q'} M'_{-q'}$$

$$+ \frac{u \Omega'^2}{b^{3d} \alpha_b^4} \int^{\Lambda b} \frac{d^d q'_1}{(2\pi)^d} \frac{d^d q'_2}{(2\pi)^d} \frac{d^d q'_3}{(2\pi)^d} M'_{q'_1} M'_{q'_2} M'_{q'_3} M'_{-q'_1 - q'_2 - q'_3} \,. \tag{3.60}$$

Note that this scale transformation can be viewed as a transformation on the parameters r_o, c, u and Λ. If we define the scale transformation

$$R_s \mu \equiv \mu' \tag{3.61}$$

where μ (r_o, c, u, . . .) then we easily obtain

$$r'_o = \frac{r_o}{\alpha_b^2} \tag{3.62a}$$

$$c' = \frac{c}{b^2 \alpha_b^2} \tag{3.62b}$$

$$u' = \frac{u}{b^d \alpha_b^4} \tag{3.62c}$$

and $\Lambda' = \Lambda b$. Scale invariance would imply $\mu' = \mu$ only at $T = T_c$ or $r_o = 0$ and $r_o' = 0$. Thus we can satisfy the first condition. We can satisfy $c' = c$ if we choose

$$\alpha_b = b^{-1} \quad . \tag{3.63}$$

We are left with the condition

$$u' = u b^{4-d} \tag{3.64}$$

We see in general that $u' = u$ can be satisfied only in four dimensions. Away from four dimensions the "interaction" will increase or decrease under the transformation depending on whether b is greater than or less than 1. The main point is that in three dimensions we do not have simple scale invariance. It is worth pointing out that we can already see that four dimensions will play a special role in the theory.

The difficulty with our direct scale transformation is that it attempts to rescale all the degrees of freedom on an equal footing. We expect however that the short wavelength degrees of freedom are not scale invariant. Our transformation must be a refined scale transformation that smears out the short wavelength degrees of freedom.

The probability distribution $P_o[M]$ couples, through the quartic term, all the Fourier components M_q over the range $0 \leq q \leq \Lambda$. Let us divide the M_q somewhat arbitrarily into two groups: $M_q^>$ has wavenumbers in the range $\Lambda \geq q \geq \Lambda/b$, and $M_q^<$ has wavenumber $q < \Lambda/b$ where $b > 1$. We said earlier that the small q behavior should dominate the physics near a phase transition. Thus the $M_q^<$ components must be treated carefully. The $M_q^>$ components should not be so important and represent degrees of freedom which are not scale invariant at the transition. Consequently we should eliminate the $M_q^>$ degrees of freedom so we can concentrate on the $M_q^<$. Elimination of the $M_q^>$ corresponds to integrating them out of the probability distribution . We then obtain a new distribution depending only on the low wavenumber component $M_q^<$:

$$P_1[M^<] \equiv \int \mathcal{D}(M^>)\, P_o[M^> + M^<] \tag{3.65}$$

$P_1[M^<]$ will in general have a far more complicated form than P_o. If we write

$$P_1[M^<] = \frac{1}{Z} e^{-F_1[M^<]} \tag{3.66}$$

then F_1 will be a complicated functional of all powers of $M^<$. Thus if F_o is of the simple GL form and simply characterized by the set of parameters $\mu = (r_o, c, u)$, F_1 will have the form of eq. (3.15) with M replaced by $M^<$ and is specified by the infinite set of parameters $\mu_1 = (U_2, U_4, \ldots)$ as in (3. 17). At this point our new problem appears much more difficult than the original problem. Let us nevertheless push on.

After we have carried out this coarse graining procedure, which has left us with a new effective free-energy F_1 and a cut-off Λ/b, we want to carry out the rescaling R_s defined by (3.57) and (3.58) and α_b need not satisfy (3.63). The combined operations

(i) Integrate out the $M^>$ components

(ii) Rescale wavenumbers and fields

is known as a renormalization group transformation. The result of this transformation is to take the initial probability distribution P_o into another distribution P'. Since these distributions can be characterized by the set of parameters μ we see that the transformation takes one from one set of parameters μ to another. We can write

$$\mu' = R_b \mu \tag{3.67}$$

where μ' is the new set of parameters. The cut-off is now back to its original value Λ. The set of transformations R_b, $1 \leq b \leq \infty$ is called the "renormalization group." We do not define an inverse of R_b so it is not quite a group.

Thus far the rescaling factor α_b is arbitrary. If we have two successive transformations R_b and $R_{b'}$, then they have the same effect as a single transformation $R_{bb'}$ except that $M_q \to \alpha_b \alpha_{b'} M_{bb'q}$ and not $M_q \to \alpha_{bb'} M_{bb'q}$. If we require

$$R_b R_{b'} \mu = R_{bb'} \mu \tag{3.68}$$

for any μ then we must demand

$$\alpha_b \alpha_{b'} = \alpha_{bb'} . \tag{3.69}$$

It is convenient to restrict α_b so that (3.69) is satisfied by choosing

$$\alpha_b = b^y \tag{3.70}$$

where y is a constant.

Let us consider the relationship of the RG and correlation functions. We have

$$\chi(q;\mu) = \frac{\int \mathcal{D}(M)\, e^{-F_c[M]}\, M_q^\alpha M_{-q}^\alpha}{Z} \tag{3.71}$$

If we let

$$M^\alpha = M^{\alpha,<} + M^{\alpha,>} \tag{3.72}$$

in the functional integral, then $\mathcal{D}(M) = \mathcal{D}(M^>)\mathcal{D}(M^<)$. If we assume the external wavenumber $q < \Lambda/b$, we can do the $M^>$ integration using (3.65) to obtain:

$$\chi(q;\mu) = \frac{1}{Z}\int \mathcal{D}(M^<)\, e^{-F_1[M^<]}\, M_q^{\alpha,<} M_{-q}^{\alpha,<} \tag{3.73}$$

where F_1 is given above. We then write everything in terms of the rescaled variables with the result

$$\chi(q;\mu) = \alpha_b^2 \int \mathcal{D}(M)\, M_{qb}^\alpha M_{-qb}^\alpha\, e^{-F'[M]}$$

$$= \alpha_b^2\, \chi(qb; R_b\mu)$$

(3.74)

for $q < \Lambda/b$. In the case of, for example, the four spin correlation function

$$\chi_{\alpha_1,\alpha_2,\alpha_3,\alpha_4}(\vec{q}_1,\vec{q}_2,\vec{q}_3,\vec{q}_4)\,(2\pi)^d\,\delta(\vec{q}_1+\vec{q}_2+\vec{q}_3+\vec{q}_4)$$

$$= \int d^dx_1\,d^dx_2\,d^dx_3\,d^dx_4\; e^{-\sum_{j=1}^{4}\vec{q}_j\cdot\vec{x}_j}$$

$$\times \langle \delta M^{\alpha_1}(\vec{x}_1)\,\delta M^{\alpha_2}(\vec{x}_2)\,\delta M^{\alpha_3}(\vec{x}_3)\,\delta M^{\alpha_4}(\vec{x}_4)\rangle$$

$$= \Omega^2 \langle M^{\alpha_1}_{q_1}\,M^{\alpha_2}_{q_2}\,M^{\alpha_3}_{q_3}\,M^{\alpha_4}_{q_4}\rangle \;, \qquad (3.75)$$

we see then that under the RG

$$\chi_{\alpha_1,\alpha_2,\alpha_3,\alpha_4}(q_1,q_2,q_3,q_4;\mu) = b^{d+4y}$$

$$\times\, \chi_{\alpha_1,\alpha_2,\alpha_3,\alpha_4}(bq_1,bq_2,bq_3,bq_4;\mu')$$

$$(3.76)$$

and $|\vec{q}_1|$, $|\vec{q}_2|$, $|\vec{q}_3|$, $|\vec{q}_4| < \Lambda/b$. These relations will be useful later.

3. Renormalization Group Phenomenology

In practice explicit implementation of the RG is very difficult. This is because it is difficult to carry out the initial coarse graining step explicitly. Why do we bother? The reason is that the RG formalism gives a very convenient format for understanding scaling behavior. If we are at the critical point, $T = T_c$, then the scale invariance will be reflected in the invariance of the parameters μ under the RG transformation. We find in practice, however, for an initial choice for the parameters $\mu^{(0)}$ that

$$\mu^{(1)} = R_b\,\mu^{(0)} \qquad (3.77)$$

and $\mu^{(1)} \neq \mu^{(0)}$. We do not have $\mu^{(1)} = \mu^{(0)}$ and scale invariance because the high wavenumber degrees of freedom in $\mu^{(0)}$ (those in the range $\Lambda/b < q < \Lambda$ for example) are not scale invariant. If we apply R_b to $\mu^{(1)}$ we obtain

$$\mu^{(2)} = R_b \mu^{(1)} = R_{2b} \mu^{(0)} , \tag{3.78}$$

and we expect, since $\mu^{(1)}$ has fewer degrees of freedom that are not scale invariant, that $\mu^{(2)}$ will be "closer" to $\mu^{(1)}$ than $\mu^{(1)}$ is to $\mu^{(0)}$. As we repeatedly apply R_b, we remove the non-scale invariant degrees of freedom until we finally converge as shown schematically in Fig. 3 to a scale invariant result:

$$\lim_{n \to \infty} R_{nb} \mu^{(0)} = \mu^* \tag{3.79}$$

and

$$R_b \mu^* = \mu^* \tag{3.80}$$

This scale invariant solution is known as a fixed point and its existence governs much of the physics near a critical point. Let us assume that there is such a fixed point μ^* and our initial parameter space μ is, in some sense, near μ^* :

$$\mu = \mu^* + \delta\mu . \tag{3.81}$$

Our RG transformation equation,

$$\mu' = R_b \mu , \tag{3.82}$$

can then be written as

$$\delta\mu' = R_b \delta\mu \tag{3.83}$$

if we define

$$\mu' = \mu^* + \delta\mu' \tag{3.84}$$

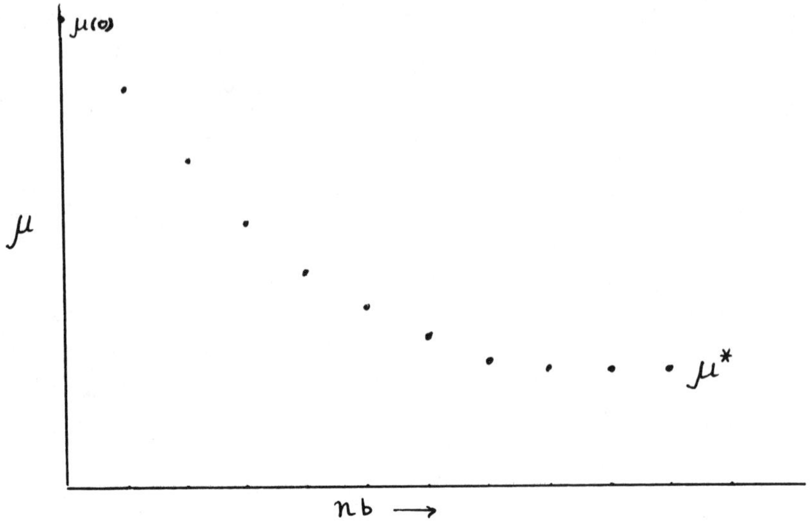

Fig. 3. Highly schematic flow of the set of parameters μ toward a fixed point μ* under repeated application of the RG.

and use (3.80). Since $\delta\mu$ is assumed to be small we can drop terms of $O(\delta\mu^2)$ and write

$$\delta\mu' = R_b^L \, \delta\mu \tag{3.85}$$

where R_b^L is a linear operator. Then, in principle, we can construct a matrix to represent R_b^L and determine the eigenvalues and eigenvectors of this matrix.

Let me now outline the situation one expects to find in the case of a phase transition. Suppose we are able to construct R_b^L explicitly and determine the eigenvalues $\lambda_i(b)$ and the corresponding eigenvectors e_i, $i = 1, 2, 3 \ldots$:

$$R_b^L e_i = \lambda_i(b) \, e_i \quad . \tag{3.86}$$

We then label the eigenvalues in the order $\lambda_1 \geq \lambda_2 \geq \lambda_3 \ldots$ Because of the semi-group property

$$R_b R_{b'} e_i = R_{bb'} e_i \ , \tag{3.87}$$

we have

$$\lambda_i(b)\,\lambda_i(b') = \lambda_i(bb') \tag{3.88}$$

or

$$\lambda_i(b) = b^{y_i} \tag{3.89}$$

where the y_i are constants and $y_1 \geq y_2 \geq y_3 \ldots$, since $b \geq 1$. We can then write $\delta\mu$ as a linear combination of the eigenvectors

$$\delta\mu = \sum_j t_j e_j \ . \tag{3.90}$$

Thus the e's give various "directions" in the function space defined by (3.15) and the t's give the projections along those directions. We find immediately using (3.86) that

$$\begin{aligned} \delta\mu' &= R_b^L \delta\mu = \sum_j t_j R_b^L e_j \\ &= \sum_j t_j b^{y_j} e_j \ . \end{aligned} \tag{3.91}$$

Ordinarily this would not be very useful, but near a critical point simplicity appears since it turns out that only $y_1 > 0$ and all other y_i's are negative. Thus if b is very large (3.91) can be written as

$$\delta\mu' = t_1 b^{y_1} e_1 + O(b^{y_2}) \tag{3.92}$$

and the first term dominates. If $t_1 = 0$ to begin with then

$$\delta\mu' = R_b^L \delta\mu \longrightarrow 0 \tag{3.93}$$

as $b^{y_2} \to 0$ and μ is pushed toward the fixed point by R_b. We see that the variable t_1 is very different from the other t's. If $t_1 \neq 0$ we do not go to the fixed point. As far as reaching the fixed point, which is identified with the critical point, t_1 is a "relevant" variable. All the other variables do not matter and are called "irrelevant." In general t_1 will be associated with the deviation of the temperature from its critical value. If $t_1 \neq 0$, $T \neq T_c$ and we clearly will not reach a fixed point or scale invariant region. It is conventional to introduce the notion of a critical surface in the parameter space. The critical surface is mapped out by $(0, t_2 e_2, t_3 e_3, \ldots)$. All points on the critical surface will "flow" to the fixed point $\mu^* = (0, t_2^* e_2, t_3^* e_3 \ldots)$ under iteration of the RG. Those points not on the critical surface will eventually flow away from μ^*.

The notions of scale invariance and universality follow from this simple picture. Suppose we start with a particular set of parameters $\mu = (r_o, c, u)$ characterizing a GL free energy. Under the RG μ will be pushed around in "parameter" space. Under repeated application of the RG μ will either flow to a fixed point or it will not. The fixed point solution corresponds to scale invariance and is only reached if the system is at $T = T_c$. If the variable t_1, which we can identify with $T - T_c$ (and indirectly the correlation length), is zero we are on the critical surface and independent of the initial values of the other parameters t_2, t_3 (or u in the original GL free energy) we will be driven to the scale invariant fixed point value of μ. In this case there is no characteristic length in the problem. If t_1 is small but non-zero the scale invariance is slightly broken and t_1 alone controls the deviation of the system from the scale invariant solution This is a restatement of the observation that near the critical point the correlation length $\xi \sim t_1^{-\nu}$ is the only important length.

We now understand, at least qualitatively, the origins of universality in second order phase transitions. If we start out with some free-energy F[M] characterized by the set of parameters μ then if we are on the critical surface determined by $t_1 = 0$ or $T = T_c$ then any such F[M] or μ will be driven to the fixed point F*[M] or μ^*. Whole classes of F[M]'s will be driven to the same F*. Thus we determine the universality class of a model by determining the appropriate fixed point. These fixed points are essentially independent of any microscopic information in F[M] or μ.

4. The Renormalization Group Near Four Dimensions

All of this is pretty abstract and in fact these ideas are quite difficult to implement in general. There are situations however where the RG can be carried out explicitly and the picture described above confirmed. Remember when we discussed the naive scaling transformation we saw that for dimensions greater than 4 the effect of u could be scaled to zero for large b and the mean field theory was correct. We might suspect that for d slightly less than 4 we could treat u as small and carry out the RG using a perturbation theory expansion in u in order to perform the coarse graining step. I will not discuss the technical details of implementing the RG; these have been extensively discussed in the literature. The point is that very near four dimensions the equation

$$\mu' = R_b \mu , \tag{3.94}$$

where $\mu = (r_o, u)$, can be written in the recursion relation form[6]

$$r_o' = b^2 r_o + \frac{(n+2)}{16\pi^2} u \left[\frac{\Lambda^2}{2}(1-b^{-2}) - r_o \ln b \right] b^2 + O(u^2) \tag{3.95a}$$

$$u' = u\left[1 - \frac{(n+8)}{16\pi^2} u \ln b\right] b^{4-d} + O(u^3) . \tag{3.95b}$$

We have noted that c, which appears as the coefficient of $(\nabla M)^2$ in the free-energy, is invariant under the RG. For convenience, and to set the length scale, we have set c = 1. While we generate other new terms in μ' they all scale rapidly to zero for small ϵ and large b. The fixed point solution of these equations is found by setting $r_o' = r_o = r_o^*$ and $u' = u = u^*$ and letting $b \to \infty$. We find then that

$$r_o^* = \lim_{b\to\infty} (1-b^2)^{-1} \frac{(n+2)}{16\pi^2} u^* b^2 \left[\frac{\Lambda^2}{2}(1-b^{-2}) - r_o^* \ln b \right] \tag{3.96a}$$

and

$$u^* = \lim_{b\to\infty} u^*\left[1 - \frac{(n+8)u^*}{16\pi^2} \ln b\right] b^{4-d} . \tag{3.96b}$$

A few words are in order here. We require that u* be small. We can then write, correct to this order in u* ,

$$1 - \frac{(n+8)u^*}{16\pi^2}\ln b + \cdots = b^{-(n+8)u^*/16\pi^2} + O(u^*)$$

(3.97)

and (3.96b) can be written as

$$1 = \lim_{b\to\infty} b^{4-d-(n+8)u^*/16\pi^2} \tag{3.98}$$

For a fixed point solution we require

$$u^* = \frac{16\pi^2}{(n+8)}(4-d) \quad . \tag{3.99}$$

For consistency u* must be small so $\epsilon = 4 - d$ must be small. Going back to (3.96a) we see that $r_o^* \sim u^*$ and we can, therefore, drop the $r_o^* \ln b$ term to this order in u* and obtain

$$r_o^* = -\frac{(n+2)u^*}{16\pi^2}\frac{\Lambda^2}{2} + O(\varepsilon^2)$$

(3.100)

$$= -\frac{(n+2)}{(n+8)}\frac{\varepsilon\Lambda^2}{2} + O(\varepsilon^2)$$

using (3.99) in the second line. We have, therefore, been able to find a fixed point near d = 4.

The next step is to look at the "stability" of this fixed point. A fixed point is stable within a model if the temperature t_1 is the only relevant variable.[19] We therefore want to find the eigenvalues λ_i defined by (3.86). First we linearize R_b by letting

$$\delta r_o = r_o - r_o^* \tag{3.101a}$$

and

$$\delta u = u - u^*$$

(3.101b)

then

$$\delta u' = \delta u\, b^{\varepsilon}\left[1 - \frac{(n+8)}{8\pi^2} u^* \ln b + \cdots\right] \tag{3.102a}$$

$$\delta r_0' = \delta r_0\, b^2 \left[1 - \frac{(n+2)}{16\pi^2} u^* \ln b\right] + \delta u A \tag{3.102b}$$

where

$$A = \frac{\Lambda^2}{32\pi^2}(n+2)(b^2-1) + O(\varepsilon)\ . \tag{3.103}$$

We can rewrite these equations, correct to $O(\varepsilon)$, as

$$\delta u' = \delta u\, b^{y_2} \tag{3.104a}$$

$$\delta r_0' = \delta r_0\, b^{y_1} + \delta u A \tag{3.104b}$$

where

$$\begin{aligned} y_1 &= 2 - \frac{(n+2)u^*}{16\pi^2} \\ &= 2 - \frac{(n+2)}{(n+8)}\varepsilon \quad > 0 \end{aligned} \tag{3.105}$$

and

$$y_2 = -\varepsilon < 0\ . \tag{3.106}$$

We easily read off that

$$R_b^L = \begin{pmatrix} b^{y_1} & A \\ 0 & b^{y_2} \end{pmatrix} . \tag{3.107}$$

The eigenvalues λ_i and the corresponding eigenvectors e_i are constructed as usual with

$$\lambda_1 = b^{y_1} \tag{3.108}$$

corresponding to

$$e_1 = (1,0) \tag{3.109}$$

and

$$\lambda_2 = b^{y_2} \tag{3.110}$$

corresponding to

$$e_2 = (a, 1) \tag{3.111}$$

where

$$a = -\frac{\Lambda^2 (n+2)}{32\pi^2} . \tag{3.112}$$

The critical surface is a line passing through the fixed point in the e_2 direction. e_1 and e_2 are not orthogonal since $R_b{}^L$ is not a symmetric matrix.

The main point is that $t_1 \sim T - T_c$ is the only relevant variable and the fixed point is stable. Consequently the picture we outlined previously holds near four dimensions.

5. Scaling and the RG

We now want to discuss how scaling follows from our RG analysis. Let us consider the situation where $\mu(T)$ is close to μ^*. Then, using (3.92), we have

$$\mu'(T) = R_b \mu(T)$$

$$= \mu^* + t_1(T)b^{y_1}e_1 + O(b^{y_2}) \tag{3.113}$$

for large b and t_1 small enough that $t_1 b^{y_1}$ is small and our linear analysis holds. Since t_1 is a smooth function of temperature (it is constructed from the local variables like r_0 and u) we can write

$$t_1(T) = A(T-T_c) + B(T-T_c)^2 + \cdots \tag{3.114}$$

and assume $A \neq 0$. Then for T very near T_c

$$\mu'(T) = \mu^* + A(T-T_c)b^{y_1}e_1 + O(b^{y_2}). \tag{3.115}$$

Recall the relation (3.74) between the transformed and untransformed correlation function:

$$\chi(q;\mu(T)) = b^{2y}\chi(qb;\mu'(T)) \tag{3.116}$$

where we have used $\alpha_b = b^y$ so that y is the index characterizing the transformation of the field in (3.58). If we use (3.115) and (3.116) we find in the region of large b and small t_1,

$$\chi(q;\mu(T)) = b^{2y}\chi(qb;\mu^* + t_1 b^{y_1}e_1 + O(b^{y_2})). \tag{3.117}$$

This is a fundamental result in the RG approach. Consider first the case $T = T_c$. Since b is arbitrary we can choose $b = \Lambda/q$ and write

$$\chi(q;\mu(T_c)) = \left(\frac{q}{\Lambda}\right)^{-2y}\left[\chi(\Lambda,\mu^*) + O\left(\left(\frac{\Lambda}{q}\right)^{y_2}\right)\right]. \tag{3.118}$$

Since $y_2 < 0$ the correction term is negligible in the asymptotic region $q/\Lambda << 1$ and we easily identify the index y from (2.19) as

$$y = 1 - \eta/2 \ . \tag{3.119}$$

Noting that y_2 controls the correction term $(\Lambda/q)^{y_2}$ which gives the approach to the asymptotic or scaling region. Consider now the case $t_1 > 0$ and $b = (t_1)^{-1/y_1} 1/\Lambda$ so that

$$\chi(q;\mu) = t_1^{-(2-\eta)/y_1} \Lambda^{-(2-\eta)} \chi\left(\frac{q}{\Lambda} t_1^{-1/y_1}; \mu^* + e_1 + \cdots\right). \tag{3.120}$$

We have then from (2.21) that the correlation length is given by

$$\begin{aligned}
\xi^2 &= \lim_{q \to 0} -\frac{\partial^2}{\partial q^2} \ln \chi(q;\mu) \\
&= \lim_{q \to 0} -\frac{\partial^2}{\partial q^2} \ln \chi(q t_1^{-1/y_1}/\Lambda, \mu^* + e_1 + \cdots) \\
&= t_1^{-2/y_1} \Lambda^{-2} \left(\lim_{x \to 0} \frac{\partial}{\partial x} \ln \chi(x^{1/2}; \mu^* + e_1 + \cdots) \right)
\end{aligned} \tag{3.121}$$

and since $\xi \sim t_1^{-\nu}$ we can identify

$$y_1 = 1/\nu \ . \tag{3.122}$$

Going back to (3.120) we can write

$$\chi(q;\mu(T)) = \xi^{(2-\eta)} \chi(q\xi; \mu^* + e_1 + \cdots) \tag{3.123}$$

where we have chosen $\xi = \Lambda^{-1} t_1^{-\nu}$. This is the scaling result we discussed earlier (see (2.35)) and from which we showed that $\gamma = (2-\eta)\nu$. We see that scaling follows if we ignore the $0(b^{y_2})$ and higher order terms in μ'.

These scaling ideas apply immediately to higher order correlation functions. Consider, for example, the 4-point correlation function we discussed earlier. If we consider only the connected part

$$\begin{aligned}\chi^{c}_{\alpha_1,\alpha_2,\alpha_3,\alpha_4}(\vec{x}_1,\vec{x}_2,\vec{x}_3,\vec{x}_4) &= \chi_{\alpha_1,\alpha_2,\alpha_3,\alpha_4}(\vec{x}_1,\vec{x}_2,\vec{x}_3,\vec{x}_4)\\ &- \chi_{\alpha_1,\alpha_2}(\vec{x}_1,\vec{x}_2)\,\chi_{\alpha_3,\alpha_4}(\vec{x}_3,\vec{x}_4)\\ &- \chi_{\alpha_1,\alpha_3}(\vec{x}_1,\vec{x}_3)\,\chi_{\alpha_2,\alpha_4}(\vec{x}_2,\vec{x}_4)\\ &- \chi_{\alpha_1,\alpha_4}(\vec{x}_1,\vec{x}_4)\,\chi_{\alpha_2,\alpha_3}(\vec{x}_2,\vec{x}_3),\end{aligned} \tag{3.124}$$

then its Fourier transform as in (3.75) satisfies (3.76) with χ_4 replaced by χ_4^c. If we are near the critical surface, ignore $0(b^{y_2})$ corrections, set all $q_i = 0$ and let $b = \xi/\Lambda$ we obtain the scaling result

$$\chi^{c}_{\alpha_1,\alpha_2,\alpha_3,\alpha_4}(\vec{0},\vec{0},\vec{0},\vec{0};\mu) = (\xi/\Lambda)^{d+4-2\eta}\,\chi^{c}_{\alpha_1,\alpha_2,\alpha_3,\alpha_4}(\vec{0},\vec{0},\vec{0},\vec{0};\mu^*+e_1). \tag{3.125}$$

It is in many cases useful to work with the one-particle irreducible vertex Γ defined by

$$\chi_{\alpha_1,\alpha_2,\alpha_3,\alpha_4}(\vec{q}_1,\vec{q}_2,\vec{q}_3,\vec{q}_4;\mu) \equiv \chi_{\alpha_1}(q_1)\;\chi_{\alpha_2}(q_2)$$

$$\times \chi_{\alpha_3}(q_3)\, \chi_{\alpha_4}(q_4)\, \Gamma_{\alpha_1,\alpha_2,\alpha_3,\alpha_4}(\vec{q}_1,\vec{q}_2,\vec{q}_3,\vec{q}_4) \tag{3.126}$$

where

$$\chi_\alpha(q) = \langle M_q^\alpha M_{-q}^\alpha \rangle . \tag{3.127}$$

Since $\chi_\alpha(0)$ scales as $\xi^{(2-\eta)}$ we easily obtain the result

$$\Gamma_{\alpha_1,\alpha_2,\alpha_3,\alpha_4}(\vec{0},\vec{0},\vec{0},\vec{0};\mu(T)) \sim \xi^{d-4+2\eta} . \tag{3.128}$$

6. Comments on the ϵ - expansion

We will discuss later the explicit implementation of the RG in the dynamical case. Before proceeding to discuss dynamics, however, it is worth saying a few words about the present state of affairs with regard to the static theory and the ϵ-expansion.

Many of you are aware that the RG method has allowed people to carry out a large number of calculations in the expansion parameters 1/n and in $\epsilon = 4 - d$. While the 1/n expansion gave us a great deal of insight into the structure of the theory it has not been seen as a practical method for calculating physical quantities. The ϵ-expansion, however, was seen from Wilson and Fisher's[20] first paper as a practical method for calculating indices. Indeed in our simple analysis we ended up calculating

$$y_1 = \frac{1}{\nu} = 2 - \frac{(n+2)}{(n+8)}\,\epsilon + \cdots\cdots \tag{3.129}$$

which makes a substantial improvement on the classical value $\nu^{-1} = 2$ for $\epsilon = 1$. Higher order terms have been worked out for all of the indices. In particular[21]

$$\nu = \frac{1}{2} + \frac{(n+2)}{4(n+8)}\varepsilon + \frac{(n+2)}{8(n+8)^3}(n^2+23n+60)\varepsilon^2$$

$$+ \frac{(n+2)}{32(n+8)^5}\Big\{ 2n^4 + 89n^3 + 1412n^2 + 5904n$$

$$+ 8640 - 192(5n+22)(n+8)T\Big\}\varepsilon^3 + O(\varepsilon^4)$$

(3.130)

and T = 0.60103. One can easily see that beyond the second order term the terms in the series are beginning to grow (for $\epsilon = 1$). The expansion is an asymptotic expansion. Because of the asymptotic nature of the expansion a number of objections have been raised concerning the whole ϵ-expansion approach. The basic line of argument has been: It is silly to carry out a perturbation series expansion in a parameter of O(1). The real world is in three dimensions while the theorists wander around near four dimensions. Put in a more mathematical form; the ϵ-expansion gives good results in three dimensions if one keeps terms of $O(\epsilon^2)$, but progressively worse results as we keep higher orders. What right do we have to keep only one $O(\epsilon^2)$ terms and throw away the higher order terms? I note that these same objections apply to quantum electrodynamics where the perturbation series is also thought to be asymptotic. Until very recently we had no real answer to these objections. All we could say was that the calculations seemed, if we only used lower order results, to give a good description of reality. Recently however there has been a breakthrough in quantum field theory which allows one to calculate the large order behavior in perturbation theory. Consider, for example, the ϵ-expansion for $1/\nu$ which we can write

$$\frac{1}{\nu} = \sum_{K=0}^{\infty} \varepsilon^K F_K \quad . \tag{3.131}$$

We know F_K from the ϵ-expansion up to K = 4. This new method due to Lipatov[22] allows one to calculate F_K for large K. Brezin et al.,[23] have found that

$$F_K = K!\, a^K K^b C\,(1 + O(1/K)) \qquad (3.132)$$

where $a = -3/(n+8)$, $b = 4 + n/2$ and C is a computable constant. The point is that (3.132) tells one the nature of the asymptotic series (3.131). Thanks to the minus sign in a the ϵ-expansion is "Borel summable." In effect one can turn the asymptotic series into a convergent series using the large K information. These new results have lead to values of the exponents which change in going from third to fourth order in ϵ by less than .1%.[24]

IV. LANGEVIN EQUATION APPROACH TO CRITICAL DYNAMICS

A. Dynamics and Time Correlation Functions

We now turn to a discussion of dynamics. We will be interested in the manner in which condensed matter systems respond to time dependent external perturbations near a critical point. Thus we take a system in equilibrium and hit it at $t = 0$ and then watch the way it decays back to equilibrium. The relationships between different types of perturbations and the measured properties of a condensed matter system are given, for moderately weak perturbations, by linear and second order response theory. These relationships have been discussed extensively in the literature[25-29] and at this summer school. For our purposes it will be sufficient to think in terms of experiments like inelastic neutron scattering. In the case of magnetic scattering the scattering cross section is proportional to the dynamic structure factor

$$C(q,\omega) = \int d^d(\vec{x}-\vec{x}')\, d(t-t')\, e^{i\vec{q}\cdot(\vec{x}-\vec{x}')}\, e^{-i\omega(t-t')} \times C(\vec{x}-\vec{x}', t-t') \tag{4.1}$$

where

$$C(\vec{x}-\vec{x}', t-t') = \langle \delta M^\alpha(\vec{x},t)\, \delta M^\alpha(\vec{x}',t') \rangle \tag{4.2}$$

is the magnetization density auto correlation function. In (4.1) $\hbar\vec{q}$ and $\hbar\omega$ are the momentum and energy exchange of the neutron with the macroscopic system. $C(q, \omega)$ is related to the static structure factor discussed earlier by

$$\chi(q) = \int \frac{d\omega}{2\pi} C(q,\omega) . \tag{4.3}$$

We might guess that in analogy with the static theory we will want to develop a theory of dynamic critical phenomena that is one step removed from a direct microscopic analysis. In this sense we must develop the dynamic extension of the coarse graining procedure discussed in section III. C. This will require us to become familiar with the Langevin equation approach to the theory of irreversible processes.

B. Langevin Equations

1. Formal Development. Let us begin with a discussion of the formal aspects of time-dependent statistical mechanics. After we have established some notation and a few results we will come back to the physics motivating our development.

We will consider a set of dynamical variables ψ_i where i labels the type of variable as well as any vector or coordinate index. In some cases, but not all, it is convenient to choose ψ_i such that it has a zero equilibrium average:

$$\langle \psi_i \rangle = 0 . \tag{4.4}$$

For a ferromagnet we might choose ψ_i to be the magnetization density $\vec{M}(\vec{x})$. In a fluid we might include in ψ_i the particle density $n(\vec{x})$, the momentum density $\vec{J}(\vec{x})$ and the energy density $\varepsilon(\vec{x})$. In general the variables ψ_i satisfy the equation of motion

$$\frac{\partial \psi_i(t)}{\partial t} = iL\psi_i(t) \tag{4.5}$$

where L is the Liouville operator. Quantum mechanically

$$L\psi_i = \frac{1}{\hbar}[H, \psi_i]_- \tag{4.6}$$

is just the commutator of the dynamical variable with the Hamiltonian. Classically L is the Poisson bracket of ψ_i with the Hamiltonian. We can formally integrate the equation of motion to obtain

$$\psi_i(t) = e^{iLt}\psi_i \tag{4.7}$$

where $\psi_i \equiv \psi_i\ (t = 0)$.

We will find it very convenient to deal with the time Laplace transform of $\psi_i(t)$ defined

$$\psi_i(z) \equiv -i\int_0^\infty dt\ e^{izt}\ \psi_i(t) \tag{4.8}$$

where z has a small positive imaginary piece. If we put (4.7) in (4.8) we easily obtain

$$\psi_i(z) = R(z)\,\psi_i \tag{4.9}$$

where the resolvant operator is defined

$$R(z) = (z+L)^{-1}. \tag{4.10}$$

The "equation of motion" for $\psi(z)$ can now be obtained by using the operator identity

$$z R(z) = 1 - L R(z). \tag{4.11}$$

We find

$$z \Psi_i(z) = \Psi_i - L \Psi_i(z). \tag{4.12}$$

In order to appreciate the content of (4.12) and to make things more concrete let us consider two examples. First let us consider a quantum mechanical Heisenberg ferromagnet described by the Hamiltonian

$$H = -\sum_{i \neq j} J_{ij} \vec{M}_i \cdot \vec{M}_j \tag{4.13}$$

where J_{ij} is the exchange interaction between spins at lattice sites i and j. We can then easily calculate using the usual spin commutation relations

$$\begin{aligned} L\vec{M}_i &= \frac{1}{\hbar}\left[H, \vec{M}_i\right]_- \\ &= -2i \sum_j J_{ij} \vec{M}_i \times \vec{M}_j \end{aligned} \tag{4.14}$$

and we see that

$$L\vec{M}_i(z) = -i\int_0^\infty dt\, e^{izt} (-2i) \sum_j J_{ij} \vec{M}_i(t) \times \vec{M}_j(t) \tag{4.15}$$

is a non-linear function of $\vec{M}_i(t)$ As a second example we consider a classical fluid with particles described by their phase-space coordinates $\vec{R}_i$ and $\vec{P}_i$. $\vec{R}_i$ is the position of the ith particle and $\vec{P}_i$ its momentum. The Hamiltonian describing N such particles is just

$$H = \sum_{i=1}^{N} \frac{p_i^2}{2m} + \sum_{i \neq j = 1}^{N} V(\vec{R}_i - \vec{R}_j) \tag{4.16}$$

where m is the particle mass and $V(\vec{R})$ is the pair potential acting between particles. An interesting dynamical variable in this case (our choice for ψ_i) is the phase space density[31]

$$f(\vec{x},\vec{P}) = \sum_{i=1}^{N} \delta(\vec{x}-\vec{R}_i)\,\delta(\vec{P}-\vec{P}_i)\,. \tag{4.17}$$

It is not difficult to show that

$$L f(\vec{x},\vec{P}) = -L_0(\vec{x},\vec{P}) f(\vec{x},\vec{P}) - \int d^d x' d^d p' \, L_I(\vec{x}-\vec{x}', p) \times f(\vec{x},\vec{P})\, f(\vec{x}',\vec{P}') \tag{4.18}$$

where

$$L_0(\vec{x},\vec{P}) = -i\,\vec{p}\cdot\nabla_x/m \tag{4.19}$$

is from the kinetic energy term, and

$$L_I(\vec{x}-\vec{x}';p) = i\,\nabla_x V(\vec{x}-\vec{x}')\cdot\nabla_p \tag{4.20}$$

is from the interaction term. Note that in the fluid case there is a piece of Lf that is proportional to f (the kinetic energy term). There is also a non-linear term as in the spin case.

It is now clear that in general $L\psi_i(z)$ is a complicated non-linear functional of $\psi_i(t)$. Despite this it is useful to assume that $L\psi_i(z)$ can be written as the sum of two pieces, one of which is linear in $\psi_i(z)$:

$$-L\psi_i(z) \equiv \sum_j K_{ij}(z)\,\psi_j(z) + i f_i(z). \tag{4.21}$$

In (4.21) $K_{ij}(z)$ is a function and $f_i(z)$ is an operator in the same sense as $\psi_i(z)$. The first piece on the right hand side of (4.21) can be thought of as the projection of $-L\psi_i(z)$ back along $\psi_i(z)$ and $f_i(z)$ is what is left over. We first note, taking the equilibrium average of (4.21), that

$$\langle f_i(z)\rangle = 0 \tag{4.22}$$

since $\langle\psi_i(z)\rangle = 0$ (from 4.4) and

$$\langle L\psi_i(z)\rangle = -i\int_0^\infty dt\, e^{izt}\, i\frac{d}{dt}\langle\psi_i(t)\rangle = 0 \tag{4.23}$$

which follows from time translational invariance of the equilibrium ensemble.

It should be clear that (4.21) is still not unique since we have one equation and two unknowns. A second equation comes from requiring that $f_i(z)$ have no projection onto the initial value of $\psi_i(t)$:[32]

$$\langle\psi_j f_i(z)\rangle = 0 \quad . \tag{4.24}$$

As we shall see (4.21) and (4.24) uniquely determine K and f. If we put (4.21) back into (4.12) we obtain

$$\sum_j [z\delta_{ij} - K_{ij}(z)]\psi_j(z) = \psi_i + i f_i(z), \tag{4.25}$$

If we take the inverse Laplace transform we find

$$\frac{\partial \psi_i}{\partial t} + \sum_j \int_0^t d\bar{t}\, K_{ij}(t-\bar{t})\,\psi_j(\bar{t}) = f_i(t) \tag{4.26}$$

This equation is known as the generalized Langevin equation. The integral kernal $K_{ij}(t-t')$ is known as the memory function and f_i is the noise. The classic work on the generalized Langevin equation is due to Mori[33] who used projection operator techniques. We will follow a somewhat different procedure.

In the Langevin equation formulation one thinks in terms of two sets of dynamical variables ψ and f_i which are initially independent. As time procedes however the ψ_i and f_i are mixed by the non-linearities in the equation of motion and $\psi_i(z)$ has components along both ψ_i and $f_i(z)$.

We will be interested in calculating the time correlation functions

$$C_{ij}(t) \equiv \langle \psi_j\, \psi_i(t) \rangle . \tag{4.27}$$

It is convenient to introduce the Fourier transform

$$C''_{ij}(\omega) = \int_{-\infty}^{\infty} dt\; e^{i\omega t}\, C_{ij}(t) \tag{4.28}$$

and the Laplace transform

$$C_{ij}(z) = -i\int_0^\infty dt\, e^{izt}\, C_{ij}(t) \tag{4.29}$$

$$= \int_{-\infty}^{\infty} \frac{d\omega}{2\pi}\, \frac{C''_{ij}(\omega)}{(z-\omega)} \tag{4.30}$$

and the last equation relates the Laplace and Fourier transforms. On comparing (4.8), (4.27) and (4.29) we can identify

$$C_{ij}(z) = \langle \psi_j\, \psi_i(z) \rangle . \tag{4.31}$$

If we multiply (4.25) by ψ_j, take the equilibrium average, and use (4.24) we obtain

$$z\, C_{ij}(z) = \chi_{ij} + \sum_{\ell} K_{i\ell}(z)\, C_{\ell j}(z) \tag{4.32}$$

where

$$\chi_{ij} = \langle \psi_j\, \psi_i \rangle \tag{4.33}$$

is the static or equal time correlation function. If we know $K_{il}(z)$ then we can solve for $C(z)$. However, to some extent, (4.32) simply defines the new function $K_{il}(z)$, known as the memory function, in terms of the correlation function. The introduction of $K_{il}(z)$ is useful only if it is easier to calculate or approximate than $C(z)$ itself. Obviously this will turn out to be the case or we would not introduce it. I show in appendix B that the memory function can be written as

$$\sum_k K_{ik} \chi_{kj} = \sum_k K_{ik}^{(s)} \chi_{kj} + \sum_k K_{ik}^{(d)}(z) \chi_{kj} \tag{4.34}$$

where the "static" or z-independent part of the memory function is given by

$$\sum_k K_{ik}^{(s)} \chi_{kj} = \langle \psi_j L \psi_i \rangle \tag{4.35}$$

and the dynamical part by

$$\sum_k K_{ik}^{(d)}(z) \chi_{kj} = -\langle (L\psi_j) R(z) L\psi_i \rangle$$

$$+ \sum_{k,\ell} \langle \psi_k R(z) L\psi_i \rangle C_{k\ell}^{-1}(z) \langle (L\psi_j) R(z) \psi_\ell \rangle. \tag{4.36}$$

The static part of K can be expressed in terms of static correlation functions. The evaluation of $K^{(d)}$, however, involves a direct confrontation with the many-body dynamics. In the case of fluids where $\psi_i \rightarrow f(\vec{x}, \vec{p})$ there has been considerable work carried out to evaluate $K^{(d)}$ and equation (4.36) is a convenient starting point for detailed microscopic calculations.[34-35] Similarly for spin systems a rather straightforward analysis of (4.36) leads to the mode coupling approximation for $K^{(d)}$ discussed by Resibois and De Leener[36] and by Kawasaki.[37] The specifics of these microscopic calculations would take us outside our main interest which is dynamics near a critical point. The main point I want to make is that we can, in principle, calculate $K_{ij}^{(d)}$ microscopically.

Now that we, again in principle, know something about K_{ij} we can view (4.21) as a defining equation for the "noise" $f_i(z)$. It is the part of $-L\psi_i(z)$ that is not "along" the vector $\psi_i(z)$. It is important to note that the auto-correlation of the noise with itself is given by the simple result:

$$\langle f_j(t') f_i(t) \rangle = \sum_\ell K_{i\ell}^{(d)}(t-t')\chi_{\ell j} \tag{4.37}$$

We prove this result in appendix C.[38] This result says that there is a fundamental relationship between the noise and the memory function. This relationship must be maintained in any approximate treatment of the generalized Langevin equation.

2. Separation of Time Scales: The "Markoff" Approximation. The basic physical picture we want to associate with our Langevin equation is that there are two time scales in our problem. A short time is associated with a set of rapidly decaying variables. A second longer time is associated with slowly decaying variables. The main idea is to include in the ψ_i the slowly decaying variables while the noise f_i represents the effects of the rapidly decaying variables. We expect then that $\langle f_j f_i(t) \rangle$ decays to zero much faster than $\langle \psi_j \psi_i(t) \rangle$. Consequently, if we are interested in long time phenomena, as we are in critical phenomena, then $\langle f_j f_i(t) \rangle$ can be taken as very sharply peaked near $t = 0$. We therefore write, to a good first approximation,

$$\langle f_j f_i(t) \rangle = 2\,\Gamma^{ij}\delta(t) \tag{4.38}$$

where

$$\Gamma^{ij} = \int_0^\infty dt \, \langle f_j f_i(t) \rangle . \tag{4.39}$$

This Markoffian approximation for the noise immediately implies, using the relationship (4.37) between the noise and the memory function, that

$$\sum_\ell K_{i\ell}^{(d)}(t-t')\chi_{\ell j} = 2\,\Gamma^{ij}\delta(t-t') \tag{4.40}$$

or[39]

$$K_{ij}^{(d)}(t-t') = 2\sum_{\ell} \Gamma^{i\ell} \chi_{\ell j}^{-1} \delta(t-t') . \tag{4.41}$$

Inserting this result in the Langevin equation (4.26) we obtain

$$\frac{\partial \psi_i(t)}{\partial t} + i\sum_j K_{ij}^{(s)} \psi_j + \sum_{\ell,j}{}' \Gamma^{i\ell} \chi_{\ell j}^{-1} \psi_j = f_i(t) . \tag{4.42}$$

Within the Markoffian approximation we completely specify the dynamics of a set of variables ψ_i by giving $K_{ij}^{(s)}$, χ_{ij} and Γ^{ij}. Specification of different forms for $K^{(s)}$, χ and Γ will characterize different types of dynamical variables. These variables are typically of three types: relaxational, hydrodynamical and oscillatory.

The simplest example of a relaxational process is the time evolution of the velocity of a tagged particle of mass m in a fluid. In this case the variable we choose is the velocity of some particle, say for definiteness, particle labelled as 1:

$$\psi_i(t) \rightarrow \vec{V}_1(t) , \tag{4.43}$$

The Langevin equation is specified completely in the Markoffian approximation by noting (i is now a vector component label)

$$\chi_{ij} = \langle v_1^i v_1^j \rangle = \delta_{ij} \frac{k_B T}{m} \tag{4.44}$$

from the equipartition theorem, from (4.35) and symmetry

$$K_{ij}^{(s)} = 0 , \tag{4.45}$$

and

$$\Gamma^{ij} = \delta_{ij} \Gamma . \tag{4.46}$$

The Langevin equation is then given by

$$\frac{\partial V_1^i(t)}{\partial t} + \frac{\Gamma m}{k_B T} V_1^i(t) = f^i(t) \tag{4.47}$$

and we have

$$\langle f_j(t') f_i(t) \rangle = 2\Gamma \delta_{ij} \delta(t-t'). \tag{4.48}$$

If we Fourier transform (4.47) over time we obtain

$$V_1^i(\omega) = \int_{-\infty}^{\infty} dt\, e^{i\omega t} V_1^i(t)$$

$$= \frac{f^i(\omega)}{\left(-i\omega + \Gamma m / k_B T\right)}. \tag{4.49}$$

We see from (4.49) that the noise "drives" the variable of interest. The average of $V_1^i(\omega)$ is then

$$\langle V_1^i(\omega) V_1^j(\omega') \rangle = \frac{\langle f^i(\omega) f^j(\omega') \rangle}{\left(-i\omega + \frac{\Gamma m}{k_B T}\right)\left(-i\omega' + \frac{\Gamma m}{k_B T}\right)}. \tag{4.50}$$

The statistical properties of V_1^i are controlled by those of the noise. Since, after Fourier transforming (4.48),

$$\langle f^i(\omega) f^j(\omega') \rangle = 2\Gamma\, \delta_{ij}\, 2\pi\, \delta(\omega+\omega') \tag{4.51}$$

we have

$$C(\omega)\, 2\pi\, \delta(\omega+\omega') = \langle V_1^i(\omega)\, V_1^j(\omega') \rangle$$

$$= \frac{2\Gamma\, \delta_{ij}\, 2\pi\, \delta(\omega+\omega')}{\omega^2 + (\Gamma m/k_B T)^2} \tag{4.52}$$

or

$$C_{ij}(\omega) = \frac{2\Gamma\, \delta_{ij}}{\omega^2 + \left(\frac{\Gamma m}{k_B T}\right)^2} \, . \tag{4.53}$$

Note that (4. 53) preserves the equal time relations (4. 3) and (4. 44). If we take the inverse Fourier transform of (4. 53) we obtain

$$C_{ij}(t) = \delta_{ij} \frac{k_B T}{m}\, e^{-\left(\frac{\Gamma m_i}{k_B T}\right)|t|} \tag{4.54}$$

and the velocity autocorrelation function relaxes exponentially to zero for long times.[41] This simple Langevin model is, of course, closely related to the Drude model[42] in solid state physics.

A slightly more complicated dynamical situation concerns a relaxational variable that is spatially dependent. An example would be the staggered-magnetization $\vec{N}(\vec{x}, t)$ in an anti-ferromagnet. In this case the index i stands for a vector index and a spatial index :

$$\Psi_i(t) \longrightarrow \vec{N}(\vec{x},t) = \sum_i \eta_i \vec{M}_i(t)\, \delta(\vec{x} - \vec{x}_i) \tag{4.55}$$

where $\vec{M}_i$ is the spin value on the lattice site $\vec{x}_i$, the sum is over all lattice sites in the magnet and $\eta_i = +1$ for spins on the "up" sublattice and $\eta_i = -1$ for spins on the "down" sublattice. We recall from our previous discussion that $\vec{N}$ is the order parameter for an anti-ferromagnet and

$$\chi_N(\vec{x}-\vec{x}')\,\delta_{\alpha\beta} = \langle N^\alpha(\vec{x})\,N^\beta(\vec{x}')\rangle \tag{4.56}$$

is the order parameter static correlation function. Again in this case the static part of the memory function $K^{(s)}$ is zero for temperatures above the Néel or transition temperature due to symmetry. The Langevin equation can be written, in the Markoffian approximation, as

$$\frac{\partial \vec{N}(\vec{x},t)}{\partial t} + \int d^dx'\,d^dx''\,\Gamma(\vec{x}-\vec{x}')\,\chi_N^{-1}(\vec{x}'-\vec{x}'')\,\vec{N}(\vec{x}'',t) = \vec{f}(\vec{x},t) \tag{4.57}$$

and

$$\langle f^\alpha(\vec{x},t)\,f^\beta(\vec{x}',t')\rangle = 2\,\Gamma(\vec{x}-\vec{x}')\,\delta_{\alpha\beta}\,\delta(t-t') \tag{4.58}$$

where Γ is space dependent kinetic coefficient. Note that we have assumed spin isotropy (which leads to the $\delta_{\alpha\beta}$ in eq. (4.58)) and spatial translational invariance.

If we Fourier transform (4.57) and (4.58) over space and time, we see that the calculation of the auto correlation function

$$C_N(\vec{x}-\vec{x}',t-t')\,\delta_{\alpha\beta} = \langle N^\alpha(\vec{x},t)\,N^\beta(\vec{x}',t')\rangle \tag{4.59}$$

follows essentially from the same analyses that led from (4.47) and (4.48) to (4.54). We have then that

$$C_N(q,t-t') = \chi_N(q)\,e^{-\Gamma(q)\chi_N^{-1}(q)|t-t'|} \quad . \tag{4.60}$$

Thus we have a relaxational decay, but the decay constant is now wavenumber dependent. If we are interested in long times and distances it will be the small q limit of $\Gamma(q)\,\chi_N^{-1}(q)$ that will control the long time decay. $\Gamma(0)$

is known as a kinetic coefficient.

Another important physical situation is where one of the "slow" variables is conserved. Consider some density (energy, particle, spin, etc.) $\psi(\vec{x})$ which is conserved. This means that the spatial integral of the quantity commutes with the Hamiltonian and is time independent.

$$\frac{d}{dt}\int d^d\vec{x}\,\psi(\vec{x},t) = \frac{i}{\hbar}\left[H\,,\int d^d\vec{x}\,\psi(\vec{x},t)\right] = 0\,. \tag{4.61}$$

This equation infers the existence of a continuity equation

$$\frac{\partial\psi(\vec{x},t)}{\partial t} + \vec{\nabla}\cdot\vec{J}(\vec{x},t) = 0 \tag{4.62}$$

where $\vec{J}(\vec{x},t)$ is the current density associated with ψ. It is characteristic of conserved variables that there exists a phenomenological constitutive relation expressing the current in terms of a gradient of the density:

$$\vec{J}(\vec{x},t) = \lambda\,\vec{\nabla}\psi(\vec{x},t)\,. \tag{4.63}$$

The proportionality constant in this case is related to a transport coefficient. An example is, of course, Fourier's law relating the heat current to the gradient of the temperature and λ is related to the thermal conductivity. If we place the constituitive relation (4.63) back into the continuity equation we obtain

$$\frac{\partial\psi(\vec{x},t)}{\partial t} = -\lambda\,\nabla^2\psi(\vec{x},t) \tag{4.64}$$

which is the usual diffusion equation. This equation is only valid for long times and large distances. This is because the constitutive equation is only valid in this region. If we multiply (4.64) by $\psi(\vec{x}',t')$ and average, we obtain

$$\frac{\partial}{\partial t}C(\vec{x}-\vec{x}',t-t') = -\lambda\,\nabla^2 C(\vec{x}-\vec{x}',t-t')\,. \tag{4.65}$$

If we Fourier transform (4.65) over space, we obtain

$$\frac{\partial}{\partial t} C(q,t-t') = \lambda q^2 C(q,t-t') \tag{4.66}$$

and we can solve this equation to obtain

$$C(q,t-t') = \chi(q)\, e^{-\lambda q^2 |t-t'|} . \tag{4.67}$$

We see in this case that the decay time is inversely proportional to q^2 and becomes very long for small q. Consequently the conserved variables can "live" for a very long time if we look at the small wavenumber behavior.

We note that eq. (4.64) is characteristic of a hydrodynamic equation[43] (like the Navier-Stokes equation in fluids) and we will in many cases refer to a conserved variable as a hydrodynamic variable.

Within our Langevin formulation a conserved variable can be characterized in the Markoffian approximation by choosing

$$\langle f(x,t)\, f(x',t')\rangle = -2 \nabla^2 \Gamma(\vec{x}-\vec{x}')\, \delta(t-t') . \tag{4.68}$$

It is easy to see that this choice leads to (4.67) with λ replaced by $\Gamma(q)\, \chi^{-1}(q)$.

Thus far we have been treating systems where $K^{(s)}$ is zero. There are cases where $K^{(s)}$ is not zero and where, characteristically, one develops damped oscillatory motion. It is easy to see from (4.35) that $K^{(s)}$ is zero from symmetry if one has only one scalar variable in the set ψ_i. Consider, however, the case of a harmonic oscillator where there are two dynamical variables, the position and momentum. If we let $\psi_i \rightarrow \{x(t), p(t)\} = \{\psi_1(t), \psi_2(t)\}$ then it is easy to show[44] that the $K^{(s)}$ term in the Langevin equation generates the usual coupled oscillator equations of motion. In a similar manner in a fluid the $K^{(s)}$ term, in the case where ψ_i includes the particle density, particle current and energy density, completely determines the oscillatory or sound wave motion in the fluid. $K^{(s)}$ and its role in determining oscillatory motion is discussed in detail by Mori.[45]

C. The Conventional or Van Hove Theory

Let us now see how this Langevin equation formulation can be used to develop a theory of dynamic critical phenomena. For simplicity I will concentrate on the case of the isotropic ferromagnet. We know from our analysis of static critical phenomena that only small wavenumbers are important near the critical point. Similarly we expect that only small frequencies are important in treating critical dynamics. Thus one expects that the "Markoffian" approximation, eq. (4.41), will be appropriate. In treating the ferromagnet the obvious choice for ψ_i is the magnetization density $\vec{M}(\vec{x}, t)$. The magnetization is a conserved quantity for an isotropic system. The simplest theory, originally due to Van Hove,[46] is equivalent to the assumption that the noise for the ferromagnet is "local":

$$\langle f^{\alpha}(\vec{x},t)\, f^{\beta}(\vec{x}',t')\rangle = -2\,\Gamma_M\,\nabla^2\delta(\vec{x}-\vec{x}')\,\delta(t-t')\,\delta_{\alpha\beta} \tag{4.69}$$

where Γ_M is the associated transport coefficient. A key assumption in van Hove's theory is that Γ_M, which arises due to the "rapid" noise variables f, will be insensitive to the long time long wavelength degrees of freedom important in critical phenomena. In other words Γ_M is essentially temperature independent near the phase transition.

We will, for convenience, restrict ourselves to the disordered region above the Curie temperature where, due to symmetry,

$$K^{(s)}_{\alpha\beta}(\vec{x}-\vec{x}') = 0 \tag{4.70}$$

We then have the Langevin equation

$$\frac{\partial M^{\alpha}(\vec{x},t)}{\partial t} - \int d^dx'\;\nabla_x^2\,\Gamma_M\,\chi_M^{-1}(\vec{x}-\vec{x}')\,M^{\alpha}(\vec{x}',t) = f^{\alpha}(\vec{x},t) \tag{4.71}$$

where $\chi_M(\vec{x}-\vec{x}')$ is the static correlation function we discussed in detail in section III. If we Fourier transform (4.71) over space and time, we easily find

$$M^{\alpha}(\vec{q},\omega) = \int d^dx \int_{-\infty}^{\infty} dt \; e^{-i\vec{q}\cdot\vec{x}} e^{i\omega t} M^{\alpha}(\vec{x},t)$$

$$= \left[-i\omega + \Gamma_M q^2 \chi_M^{-1}(q)\right]^{-1} f^{\alpha}(q,\omega) \tag{4.72}$$

where

$$\langle f^{\alpha}(\vec{q},\omega) f^{\beta}(\vec{q}\,',\omega')\rangle = 2\Gamma_M q^2 (2\pi)^{d+1} \delta(\omega+\omega')\,\delta(\vec{q}+\vec{q}\,')\,\delta_{\alpha\beta}. \tag{4.73}$$

Note how the noise serves as a source for the field. We can then easily calculate the dynamic structure factor

$$C_M(q,\omega)\,\delta_{\alpha\beta}(2\pi)^{d+1}\delta(\vec{q}+\vec{q}\,')\,\delta(\omega+\omega') = \langle M^{\alpha}(\vec{q},\omega) M^{\beta}(\vec{q}\,',\omega')\rangle$$

$$= \frac{\langle f^{\alpha}(\vec{q},\omega) f^{\beta}(\vec{q}\,',\omega')\rangle}{\left(-i\omega + \Gamma_M q^2 \chi_M^{-1}(q)\right)\left(-i\omega' + \Gamma_M q'^2 \chi_M^{-1}(q')\right)}$$

$$= \delta_{\alpha\beta}(2\pi)^{d+1}\delta(\omega+\omega')\,\delta(\vec{q}+\vec{q}\,')\,\frac{2\Gamma_M q^2}{\left[\omega^2 + \left(\Gamma_M q^2 \chi_M^{-1}(q)\right)^2\right]} \tag{4.74}$$

or

$$C_M(q,\omega) = \frac{2\Gamma_M q^2}{\omega^2 + \left(\Gamma_M q^2 \chi_M^{-1}(q)\right)^2}\,. \tag{4.75}$$

This is a key result and we will analyse it in some detail. We note first, using (4. 3), that (4. 75) preserves the equal time value of $C(\vec{q}, t)$. Next we note that at T_c, where $\chi_M^{-1}(q) \sim q^2$, that[47]

$$C_M(q,\omega) = 2\Gamma_M q^2 \left[\omega^2 + (\Gamma_M q^4)^2 \right]^{-1} . \tag{4.76}$$

Since the inelastic neutron scattering cross-section is proportional to $C_M(q, \omega)$ we see that for small q's and ω's that the scattering is greatly enhanced and sharply peaked near $\omega = 0$. If we invert the Fourier transform we obtain

$$C_M(q, t) = q^{-2} e^{-q^4 \Gamma_M t} . \tag{4.77}$$

We see that for small wavenumbers the system decays very slowly. This effect is called critical slowing down. It is characteristic of systems near a critical point that the dynamics become sluggish.

At this point it should occur to us to inquire into the scaling properties of the dynamic structure factor. In the static case we were able to scale all lengths by factors of the correlation length--the characteristic length. In the dynamical case we look for a characteristic frequency. In the van Hove theory it is easy to identify from inspection of (4. 75) the characteristic frequency

$$\omega_c \equiv q^2 \Gamma_M \chi_M^{-1}(q) . \tag{4.78}$$

If we scale the frequency by ω_c,

$$\nu \equiv \omega / \omega_c , \tag{4.79}$$

we can rewrite (4. 75) as

$$C_M(q, \omega) = \frac{\chi_M(q)}{\omega_c} \frac{2}{(\nu^2 + 1)} . \tag{4.80}$$

If we use the Ornstein-Zernike form (2. 23) for $\chi_M(q)$ then the characteristic frequency can be rewritten in the form

$$\omega_c = q^4 \Gamma_M \left(1 + (q\xi)^{-2} \right) . \tag{4.81}$$

As we will see later in our more general analysis we will be able to associate a characteristic frequency with all systems near a critical point. It is convenient, as first pointed out by Halperin and Hohenberg,[48] to write all characteristic frequencies near the critical point in the scaling form

$$\omega_c(q) = q^z f(q\xi) \tag{4.82}$$

where z is called the dynamic critical index. In our simple "conventional" theory, where we assume Γ_M is essentially temperature independent near T_c, we can easily identify $z = 4$.

It is worth noting that much of what we have said about the ferromagnet can be taken over to the anti-ferromagnet. The only difference is that the staggered magnetization is not conserved and we see comparing (4.60) and (4.77) that the characteristic frequency for the anti-ferromagnet is

$$\begin{aligned} \omega_c &= \Gamma_N \chi_N^{-1}(q) \\ &= q^2 \Gamma_N \left(1 + (q\xi)^{-2}\right) \end{aligned} \tag{4.83}$$

and we can identify $z = 2$. We note that while ferromagnets and anti-ferromagnets are in the same universality class with respect to static critical properties, they are not in the same dynamic universality class because of the conserved nature of the order parameter for a ferromagnet.

Unfortunately, the experimental results do not agree with this conventional result. It is found, from neutron scattering, that $z = 5/2$ for ferromagnets[49] and $z = 3/2$ for anti-ferromagnets.[50] This has the interesting consequence that the transport coefficient Γ_M for a ferromagnet must have a strong temperature dependence as $T \to T_c$. If we insert

$$\Gamma_M \sim (\xi\Lambda)^x \tag{4.84}$$

into (4.78) we find

$$\omega_c \sim q^{4-x} (q\xi)^x \left(1 + (q\xi)^{-2} \right) \tag{4.85}$$

and we can identify

$$z = 4 - x . \tag{4.86}$$

$z = 5/2$ requires $x = 3/2$ and

$$\Gamma_M \sim \xi^{3/2} \sim (T - T_c)^{-1} . \tag{4.87}$$

Therefore the transport coefficient <u>diverges</u> as $T \to T_c$. Divergences of transport coefficients near T_c is not restricted to ferromagnets. The thermal conductivity in a fluid diverges, for example, and this can be measured precisely using light scattering techniques.[51] We note, also, following the same arguments as for the ferromagnet, that the kinetic coefficient for the anti-ferromagnet must diverge as $\xi^{\frac{1}{2}}$ in three dimensions.

The simple van Hove theory is the dynamic analog of the Landau theory for the statics. It gives a rough picture of the behavior near a critical point but breaks down as one gets very close to the transition and fluctuations become important. There is an inconsistency in the reasoning which becomes evident in the region of strong fluctuations. In the van Hove model it is assumed that only short time and short wavelength phenomena contribute to Γ and it will therefore be insensitive to the critical fluctuations. As we pointed out above, this idea is not correct. Since Γ develops a strong temperature dependence near the transition there must be physical processes that occur on a long wavelength scale that contribute to Γ. Therefore the simple conventional theory, which assumes that all contributions to Γ come from very short wavelengths, must break down. The problems are similar to those encountered in "fixing up" the Landau theory.

D. Higher Order Correlation Functions

It is interesting to inquire into the implications of the van Hove theory concerning the behavior of higher order correlation functions. Consider, for example, the "energy-energy" correlation function

$$C_E(\vec{x}-\vec{x}',t-t') = \langle \delta(M^2(\vec{x}',t'))\,\delta(M^2(\vec{x},t))\rangle . \tag{4.88}$$

This correlation function is a limiting case of the four point correlation function

$$C_4(1234) = \langle M(1)\,M(2)\,M(3)\,M(4)\rangle \tag{4.89}$$

where 1 is a short-hand for x_1, t_1, α_1. Then

$$C_E(\vec{x}_1-\vec{x}_3, t_1-t_3) = \sum_{\alpha_1,\alpha_3} \left\{ C_4(1234) - C(12)\,C(34) \right\} \Big|_{\substack{1=2\\3=4}} \tag{4.90}$$

If we Fourier transform over space and time then

$$\begin{aligned} C_E(q,\omega)(2\pi)^{d+1}\delta(\vec{q}+\vec{q}')\,\delta(\omega+\omega') \\ = \sum_{\alpha_1,\alpha_3} \int \frac{d^d q_1}{(2\pi)^d}\frac{d\omega_1}{2\pi}\frac{d^d q_3}{(2\pi)^d}\frac{d\omega_3}{2\pi} \Big\{ C_4(q\omega\alpha_1,\, q-q_1\ \omega-\omega_1 \\ \alpha_1,\, q_3\,\omega_3\,\alpha_3\,,\, q'-q_3\ \omega'-\omega_3\ \alpha_3) - C(q_1\omega_1\alpha_1,\, q-q_1 \\ \omega-\omega_1\,\alpha_1)\; C(q_3\,\omega_3\,\alpha_3\,,\, q'-q_3,\, \omega'-\omega_3,\, \alpha_3)\Big\} \end{aligned} \tag{4.91}$$

If we then use (4.7) relating M and the noise, we obtain

$$C_4(1\,2\,3\,4) = \frac{1}{-i\omega_1 + \omega_c(1)} \; \frac{1}{-i\omega_2 + \omega_c(2)} \; \frac{1}{-i\omega_3 + \omega_c(3)} \; \frac{1}{i\omega_4 + \omega_c(4)}$$

$$\times \langle f(1)\, f(2)\, f(3) f(4) \rangle \tag{4.92}$$

where, in this case, 1 stands for $\vec{q}_1$, ω_1, α_1. We must then evaluate the average of four noise factors. In fact, we are told nothing about such averages from our Langevin equation. We must make further assumptions. The simplest assumption, which also makes good physical sense, is that the noise variables are Gaussian random variables. Mathematically this means that an average over a product of f's is to be interpreted as

$$\langle f_i(t_1) f_j(t_2) \cdots \rangle \equiv N \int \mathcal{D}[f] \left(f_i(t_1) f_j(t_2) \cdots \right)$$

$$\times \exp - \int dt \, \frac{1}{4} \sum_{\ell, m} f_\ell (\Gamma^{-1})_{\ell m} f_m \tag{4.93}$$

where

$$N^{-1} \equiv \int \mathcal{D}[f] \exp - \int_{-\infty}^{\infty} dt \; \frac{1}{4} \sum_{\ell, m} f_\ell (\Gamma^{-1})_{\ell m} f_m . \tag{4.94}$$

You can convince yourself, using our results for Gaussian functional integrals, eq. (A.23), that the definition (4.93) preserves the relation

$$\langle f_i(t_1) f_j(t_2) \rangle = 2 \Gamma^{ij} \delta(t_1 - t_2) . \tag{4.95}$$

With this assumption for the statistical properties of the noise all averages of products of f factor into products of averages of pairs of f's. The net result is that all averages of products of f's factor into their disconnected parts. Since M is proportional to f this means that all connected correlation functions, except the two point function $C(q, \omega)$, vanish within this approximation.

If we use this result to evaluate the average of four f's appearing in (4.92) we easily find

$$C_E(q,\omega) = 3\int \frac{d^d q'}{(2\pi)^d} \frac{d\omega'}{2\pi} C(\vec{q}-\vec{q}', \omega-\omega') C(\vec{q}', \omega') \tag{4.96}$$

Using (4.75) for $C(q, \omega)$, we can immediately do the frequency integral to obtain

$$C_E(\vec{q},\omega) = 6\int \frac{d^d q'}{(2\pi)^d} \frac{\chi(\vec{q}-\vec{q}')\,\chi(\vec{q}')\,[\omega_c(\vec{q}') + \omega_c(\vec{q}-\vec{q}')]}{\omega^2 + (\omega_c(\vec{q}') + \omega_c(\vec{q}-\vec{q}'))^2} \tag{4.97}$$

where ω_c is given by (4.78). We can evaluate this integral explicitly at $T = T_c$ and $q = 0$ if we use (4.81) with $\xi \to \infty$. We find

$$C_E(0,\omega) = \frac{3\pi}{2} \frac{\Gamma_M K_d}{\sin \frac{d\pi}{8}} \left(\frac{\omega}{2\Gamma_M}\right)^{-(8-d)/4} \tag{4.98}$$

where

$$K_d = \frac{2^{1-d}}{\pi^{d/2}\,\Gamma(d/2)} \tag{4.99}$$

is the surface area of a unit sphere in d-dimensional space divided by $(2\pi)^d$. We see that $C_E(0, \omega)$ is strongly peaked near $\omega = 0$.

E. Non-linear Models

We are now at a stage very similar to that encountered in treating static phenomena where we had to generalize the Landau theory to include non-linear interactions. We must extend our analysis to non-linear equations of motion. Again, as in the static case, as soon as we have a non-linear theory the problem becomes much more difficult. It is at this point that we again appeal to the ideas behind universality. In principle we want to look at classes of non-linear equations of motion and determine the relevance or irrelevance of various terms within a renormalization group analysis. In this way we can construct the dynamic universality classes by identifying the relevant terms in our non-linear equation of motion. Such a general analysis has not yet been carried out. Thus far people have been able to analyse only the simplest non-linear equations of motion.

Our first concern is the development of a method for constructing non-linear equations that are valid on a long time and distance scale. These equations must be dissipative in nature and should fit into our Langevin equation approach. Our treatment will follow the pioneering work of Green,[52] Zwanzig,[53] Mori[54] and Kawasaki.[55] One major constraint in developing these non-linear equations of motion is that they must be consistent with the statics of the Ginzburg-Landau-Wilson theory. The point is that once we start writing non-linear equations of motion, the equal time correlation functions that we calculate from these equations will depend on the various non-linear couplings. We want to preserve the property of the GLW theory that the static correlation functions are given as functional integrals over the GL probability distribution given by (3. 7). If we recall the coarse graining procedure defined by (3. 13) for obtaining the GL free energy,

$$W_\varphi = e^{-F[\varphi]} \equiv \langle \delta(\varphi - \hat{M}) \rangle, \tag{4.100}$$

where φ corresponds to the coarse grained variable and $\hat{M}$ the microscopic variable, then it might occur to us to treat the variable

$$g_\varphi(t) = \prod_{i=1} \delta(\varphi_i - \hat{M}_i(t)) \equiv \delta(\varphi - \hat{M}(t)) \tag{4.101}$$

as the basic dynamical variable in our Langevin equation. The field φ thus serves as the label i for the variable ψ_i in (4.25) and $g_\varphi(t)$ will generate all non-linear equal time couplings between the variables $\hat{M}_i$. We see, for example, that

$$\int \mathcal{D}[\varphi]\, \varphi_i(t)\, \varphi_j(t)\, g_\varphi(t) = \hat{M}_i(t)\, \hat{M}_j(t)\,. \tag{4.102}$$

The equal time correlation functions in this case are just

$$\begin{aligned}\chi_{\varphi\varphi'} &= \langle g_\varphi g_{\varphi'} \rangle \\ &= \delta(\varphi - \varphi') \langle g_\varphi \rangle \\ &= \delta(\varphi - \varphi')\, W_\varphi\end{aligned} \tag{4.103}$$

and the Langevin equation is of the form:[56]

$$z g_\varphi(z) - \int \mathcal{D}[\varphi']\, K_{\varphi\varphi'}(z)\, g_{\varphi'}(z) = g_\varphi + i R_\varphi(z) \tag{4.104}$$

where we use $R_\varphi(z)$ to represent the noise term. We can make progress in evaluating the memory function if we note the identity

$$L g_\varphi = - \sum_i \frac{\delta}{\delta \varphi_i} g_{\varphi_i} L \hat{M}_i \tag{4.105}$$

which, in the classical limit, follows from the chain-rule for differentiation. We show in appendix D that the static part of the memory function can be written as

$$K^{(s)}_{\varphi\varphi'} = - i \sum_i \frac{\delta}{\delta \varphi_i} V_i(\varphi)\, \delta(\varphi - \varphi') \tag{4.106}$$

where

$$V_i(\varphi) = - \sum_j \left[\frac{\delta}{\delta \varphi_j} Q_{ij}[\varphi] - Q_{ij}[\varphi] \frac{\delta F_\varphi}{\delta \varphi_j} \right] \tag{4.107}$$

is called the "streaming velocity" and

$$Q_{ij}(\varphi) = \beta^{-1} \langle g_\varphi \{ \hat{M}_i , \hat{M}_j \}_{P.B.} \rangle W_\varphi^{-1} \tag{4.108}$$

is anti-symmetric under the interchange of i and j and reflects the underlying Poisson bracket algebra satisfied by the fundamental fields $\hat{M}_i$.

If we use (4.105) in (4.36) we see that $K^{(d)}_{\varphi\varphi'}$ must be of the form

$$K^{(d)}_{\varphi\varphi'}(z) W_{\varphi'} = \sum_{i,j} \frac{\delta}{\delta \varphi_i} \frac{\delta}{\delta \varphi'_j} T^{ij}_{\varphi\varphi'}(z) \tag{4.109}$$

where $T^{ij}_{\varphi\varphi'}(z)$ is a complicated functional of φ and φ'.

We now have a quite general non-linear equation of motion. The contribution of $K^{(s)}$ directly reflects the underlying symmetry of the microscopic variables via the Poisson bracket relations while $K^{(d)}$ represents the effects of dissipative processes. In many cases we can evaluate $K^{(s)}$ explicitly, as, for example, in the case of an isotropic ferromagnet where, in terms of Fourier transformed variables,

$$Q_{\alpha\beta, qq'}[\Phi] = \sum_{\gamma=1}^{3} \epsilon_{\alpha\beta\gamma} \varphi^{\gamma}_{q+q'} \tag{4.110}$$

and

$$V^{\alpha}_{q}[\varphi] = \sum_{\beta,\gamma,k<\Lambda} \varepsilon_{\alpha\beta\gamma} \varphi^{\beta}_{k+q} \frac{\delta F[\varphi]}{\delta\varphi^{\gamma}_{k}} . \tag{4.111}$$

We cannot in general evaluate $K^{(d)}$ explicitly. We must approximate it in some way. In principle one would like to systematically analyse $K^{(d)}$ using (4.36). Very little has been done along these lines. Thus far people have been satisfied with making the simplest approximations for $K^{(d)}$ compatible with (4.109). The simplest approximations are as "local" as possible and of the form

$$K^{(d)}_{\varphi\varphi'} W_{\varphi'} = \sum_{i,j} \frac{\delta}{\delta\varphi_i} \frac{\delta}{\delta\varphi'_j} \Gamma^{ij}_{0} \delta(\varphi - \varphi') W_{\varphi} \tag{4.112}$$

where Γ^{ij}_{0} is independent of φ and φ'. After a few rearrangements (4.112) can be rewritten as

$$K^{(d)}_{\varphi\varphi'} = -\sum_{i,j} \frac{\delta}{\delta\varphi_i} \Gamma^{ij}_{0} \left[\frac{\delta}{\delta\varphi_j} + \frac{\delta F_{\varphi}}{\delta\varphi_j} \right] \delta(\varphi - \varphi') . \tag{4.113}$$

We could, of course, discuss more complicated approximations. We could, for example, let $\Gamma_0^{ij} \longrightarrow \Gamma_0^{ij}[\varphi]$ in (4. 112). The situation is similar to that for the statics where we replaced (3. 15) by the GLW form (3. 24). As long as the new terms do not take the model into a new universality class, then many of our results will be independent of whether we include these complicating terms or not. Of course we must determine which equation of motion fits into which universality class.

Using (4. 106) and (4. 113) in (4. 42) we obtain the equation of motion

$$\frac{\partial g_\varphi(t)}{\partial t} = D_\varphi g_\varphi(t) + R_\varphi(t) \tag{4. 114}$$

where

$$D_\varphi = -\sum_{i,j} \frac{\delta}{\delta\varphi_i}\left[V_i \delta_{ij} - \Gamma_0^{ij}\left(\frac{\partial}{\partial\varphi_j} + \frac{\delta F_\varphi}{\delta\varphi_j}\right)\right] \tag{4. 115}$$

is defined as the generalized Fokker-Planck operator. Equation (4. 114) has a number of nice features. One of the most important follows from taking the equilibrium average of both sides. Then, since we require,

$$\langle R_\varphi(t)\rangle = 0 , \tag{4. 116}$$

and $\langle g_\varphi(t)\rangle$ is independent of time, we have

$$D_\varphi \langle g_\varphi(t)\rangle = 0 \tag{4. 117}$$

If we recall from appendix D that the streaming velocity $V_i[\varphi]$ is divergenceless,

$$\sum_i \frac{\partial}{\partial\varphi_i}\left[V_i[\varphi]\, e^{-F_\varphi}\right] = 0 , \tag{4. 118}$$

we see that

$$\langle g_\varphi(t) \rangle = e^{-F_\varphi} \tag{4.119}$$

is a solution of (4.117). Thus we see that our non-linear equation will preserve the GLW statics if we require that F be of the GL form.

It is useful, in interpreting the physical significance of (4.114), to realize that the approximate equation for g_φ (4.114) generates an equation of motion for the coarse grained variable $M_i(t)$[57] by multiplying (4.114) by φ_i and integrating over φ:

$$\begin{aligned} \frac{\partial M_i(t)}{\partial t} &= \int \mathcal{D}[\varphi]\, \varphi_i\, D_\varphi\, g_\varphi(t) + \int \mathcal{D}[\varphi]\, \varphi_i\, R_\varphi(t) \\ &= V_i[M] - \Gamma_0^{ij} \frac{\delta F_M(t)}{\delta M_i(t)} + f_i(t) \end{aligned} \tag{4.120}$$

where we have defined a "moment" of the noise

$$f_i(t) = \int \mathcal{D}[\varphi]\, \varphi_i\, R_\varphi(t) \tag{4.121}$$

and, in taking the functional derivative, the time variable is held fixed. If we combine (4.37), (4.112) and (4.121) we see that

$$\begin{aligned} \langle f_i(t)\, f_j(t') \rangle &= \int \mathcal{D}[\varphi]\, \varphi_i \int \mathcal{D}[\varphi']\, \varphi'_j \langle R_\varphi(t)\, R_{\varphi'}(t') \rangle \\ &= 2 \sum_{l,m} \int \mathcal{D}[\varphi]\, \varphi_i \int \mathcal{D}[\varphi']\, \varphi'_j \frac{\delta}{\delta \varphi_l} \frac{\delta}{\delta \varphi'_m} \Gamma_0^{ij} \\ &\quad \times \delta(\varphi - \varphi')\, W_\varphi\, \delta(t-t') \end{aligned}$$

$$= 2\Gamma_0^{ij} \int \mathcal{D}[\varphi]\, W_\varphi\, \delta(t-t')$$

$$= 2\Gamma_0^{ij}\, \delta(t-t') \, . \tag{4.122}$$

Thus we preserve the noise auto-correlation function in the linear theory. It is not obvious but we can find a more direct relationship between R_φ and f. We show in appendix E that

$$R_\varphi(t) = -\sum_{i,j} \frac{\partial}{\partial\varphi_i}\left\{ f_i(t)\,\delta_{ij} + \Gamma_0^{ij}\frac{\delta}{\delta\varphi_j}\right\} g_\varphi(t) \, . \tag{4.123}$$

We completely specify our model, in the same sense discussed in section IV. D, if we demand that the noise $f_i(t)$ be Gaussianly distributed as in equation (4.93). There are a number of formal relationships we could develop between $R_\varphi(t)$, $g_\varphi(t)$ and $f_i(t)$. Some of these are discussed by Ma and Mazenko.[58] We will move on to investigate the equation satisfied by $M_i(t)$ given by (4.120). We note, however, following the analysis in appendix E, that this equation is equivalent to the equation for $g_\varphi(t)$.

Suppose we ignore the streaming velocity ($Q_{ij} = 0$) in our non-linear equation of motion and assume the field M_i is labelled by a coordinate index $\vec{x}$ and a vector index α,

$$\frac{\partial M^\alpha(\vec{x})}{\partial t} = -\Gamma_0^\alpha(\vec{x})\,\frac{\delta F[\vec{M}(t)]}{\delta M^\alpha(\vec{x},t)} + f^\alpha(\vec{x},t) \tag{4.124}$$

and we have assumed $\Gamma_0^{\alpha\beta}(\vec{x}) = \delta_{\alpha,\beta}\,\Gamma_0^\alpha(\vec{x})$. If we replace F with its Landau-Ginzburg form (3.24), we obtain

$$\frac{\partial M^\alpha(x)}{\partial t} = \int d^dx' \, \Gamma_0^\alpha(x) \left(\chi_0^{-1}(\vec{x}-\vec{x}') \, M^\alpha(\vec{x}';t) \right)$$
$$+ 4u M^\alpha(\vec{x},t) \, M^2(\vec{x},t) \Gamma_0(x) + f^\alpha(\vec{x},t) \qquad (4.125)$$

where

$$\chi_0^{-1}(\vec{x}-\vec{x}') = \left[2r_0 - 2c \nabla_x^2 \right] \delta(\vec{x}-\vec{x}') \qquad (4.126)$$

is the inverse Ornstein-Zernike correlation function. We see if we set $u = 0$ in (4. 125) that we obtain the van Hove theory. This also has the implication that van Hove theory is compatible only with a Gaussian probability distribution. The inclusion of non-linear terms in F allows one to treat more complicated "initial" probability distribution. In particular for $T < T_c$ the underlying GL free energy has a double well shape. This corresponds to r_0 being negative and the quartic term is necessary to ensure the existence of the functional integral. Another interesting physical situation where the quartic non-linearity is essential is in the case of nucleation and spinodial decomposition. [2]

It is important to note, once we include non-linear terms in the equation of motion that Γ_0 can no longer be interpreted as the measured transport coefficient. It must instead be thought of as a "bare" or local transport coefficient determined by very short range, linear interactions. The physical transport coefficient can only be extracted after solving the model.

This model, defined by (4. 125), is called the time dependent Ginzburg-Landau (TDGL) model and has been analyzed in detail recently by Halperin, Hohenberg and Ma[59] and a number of other workers. This model is applicable to certain systems undergoing structural phase transitions and anisotropic magnetic systems.[60] Historically the TDGL model has been identified with treatments of the time dependent behavior of superconductors.

It turns out that it is very important in many cases to include the streaming velocity term in our equation of motion. In the case of the isotropic ferromagnet we can write, in coordinate space,

$$\vec{V}(\vec{x},t) = \vec{M}(\vec{x},t) \times \vec{H}(\vec{x},t) \qquad (4.127)$$

where $\vec{H}(\vec{x}, t)$, defined by

$$H^{\alpha}(\vec{x},t) = \frac{\delta F[M]}{\delta M^{\alpha}(\vec{x},t)} , \tag{4.128}$$

can be interpreted as an effective internal magnetic field and $\vec{V}$ expresses the precession of the spin density about this internal field. If we insert the Ginzburg-Landau form for F we easily obtain

$$\vec{V}(\vec{x},t) = -\vec{M}(\vec{x},t) \times 2c \nabla^2 \vec{M}(\vec{x},t) \tag{4.129}$$

We see that $\vec{V}(\vec{x}, t)$ is essentially $-iLM(\vec{x}, t)$ for a microscopic Heisenberg system in the long wavelength limit.[61] Our coarse grained equation of motion can be written as

$$\frac{\partial \vec{M}(\vec{x},t)}{\partial t} = \lambda \vec{M} \times \vec{H} - \Gamma_M^0 \nabla^2 \vec{H} + \vec{f} \tag{4.130}$$

where we have chosen $\Gamma_c^{\alpha}(x) = -\Gamma_M^0 \nabla_x^2$ to reflect the conserved nature of the magnetization density and λ is a coupling parameter absorbing any constant terms in the streaming velocity. This equation has a rather direct physical interpretation. Suppose we start with the microscopic equation of motion for $\hat{M}^{\alpha}(\vec{x}\, t)$. We then break M into short ($M^>$) and long-wavelength ($M^<$) pieces and average the equation of motion over the $M^>$ components. The $\vec{M} \times \nabla^2 \vec{M}$ term results from those $M^<$ components that are not averaged over, the noise term comes from that part of the equation depending only on the $M^>$ terms and the $\Gamma_M^0 \nabla^2 \vec{H}$ term comes from the terms coupling $M^>$ and $M^<$. With this interpretation it is reasonable to assume Γ_M^0, the "bare" or background transport coefficient, which comes from the $M^>$ degrees of freedom, should be regular in temperature at the critical point. This equation of motion is the dynamical generalization for the ferromagnet of the Wilson functional expression for the static properties. Just as in the static case, solution of the resulting nonlinear problem is difficult.

Historically an important limiting case of our non-linear equation of motion corresponds to setting $u = 0$ in (4.120)

$$\frac{\partial M^{\alpha}(\vec{x},t)}{\partial t} = V^{\alpha}[M,x] - \int d^dx' \, \Gamma_0^{\alpha}(x) \chi_0^{-1}(\vec{x}-\vec{x}') M^{\alpha}(\vec{x}',t) + f^{\alpha}(\vec{x},t) . \tag{4.131}$$

This equation is compatible only with Gaussian statics and emphasizes the role of the streaming velocity term. This equation was studied extensively in the pioneering work by Kawasaki [62] on a number of different physical systems. The theory associated with this equation is known as "mode-coupling" and corresponds to the non-linear couplings generated by the streaming velocity. Kawasaki and others [63] definitively established the importance of the mode coupling terms in determining the appropriate dynamic universality class.

V. PERTURBATION THEORY AND THE DYNAMIC RENORMALIZATION GROUP

A. Perturbation Theory

We do not, in general, know how to solve our non-linear equations of motion. The best we can do, in most cases, is to construct perturbation theory solutions as a power series in the non-linear couplings. There now exist a number of sophisticated methods[64-67] for carrying out such perturbation series expansions. In these lectures I will discuss only a direct brute force approach which is sufficient for many purposes. In any case these methods can serve as an introduction to more sophisticated perturbation theory approaches.

For simplicity we will consider the n = 1 TDGL model. Extension of the method to the more general models is direct and is discussed by Ma and Mazenko[58] and Ma.[7] Let us consider the TDGL model where the bare transport coefficient $\Gamma_o(q)$ is a constant for a nonconserved order parameter and $\Gamma_o q^2$ for a conserved order parameter. The Fourier transform of the equation of motion (4.125) can be written as

$$-i\omega M(\vec{q},\omega) = -\Gamma_o(\vec{q})\,\chi_o^{-1}(q)\,M(\vec{q},\omega) + f(\vec{q},\omega)$$

$$-\Gamma_o(q)4\Omega u \int \frac{d^dq'}{(2\pi)^d}\frac{d\omega'}{2\pi}\frac{d^dq''}{(2\pi)^d}\frac{d\omega''}{2\pi} M(\vec{q}-\vec{q}\,'-\vec{q}\,'',\omega-\omega'-\omega'')$$

$$\times\; M(\vec{q}\,',\omega')\,M(\vec{q}\,'',\omega'') \;. \tag{5.1}$$

If we define

$$G_o(q,\omega) = \left[-i\omega + \Gamma_o(q)\,\chi_o^{-1}(q)\right]^{-1}\Gamma_o(q) \tag{5.2}$$

we can write (5.1) as

$$M(\vec{q},\omega) = M^0(\vec{q},\omega) - G_0(q,\omega)\, 4u\Omega \int \frac{d^dq'}{(2\pi)^d} \frac{d^dq''}{(2\pi)^d} \frac{d\omega'}{2\pi} \frac{d\omega''}{2\pi}$$

$$\times\, M(\vec{q}-\vec{q}\,'-\vec{q}\,'', \omega-\omega'-\omega'')\, M(\vec{q}\,',\omega')\, M(\vec{q}\,'',\omega'') \tag{5.3}$$

where

$$M^0(\vec{q},\omega) = \frac{G_0(q,\omega)}{\Gamma_0(q)} \frac{f(\vec{q},\omega)}{\sqrt{\Omega}} . \tag{5.4}$$

M^0 is the $u = 0$ expression for M given previously by (4.72). It is clear that we can generate corrections to $M = M^0$ by iterating (5.3). It is convenient to carry out this iteration graphically. We can represent (5.3) as

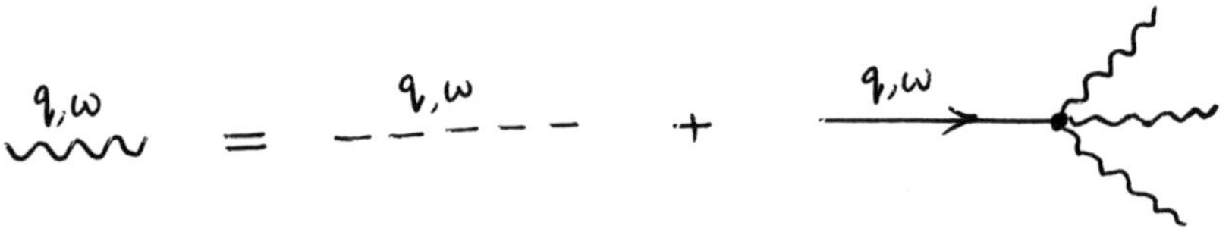

(5.5)

where ~~~~ stands for M, - - - - - stands for M^0, ——>—— stands for G_0 and a dot · represents the interaction vertex $-4u$. Note that at a vertex the wavenumber and frequency are "conserved." If we assign, for example, a frequency to each line entering a vertex, then it is easy to see that the frequency entering the vertex via ——>—— must equal the sum of the frequencies leaving via the ~~~~ lines.

A direct iteration of this equation correct to second order in u gives

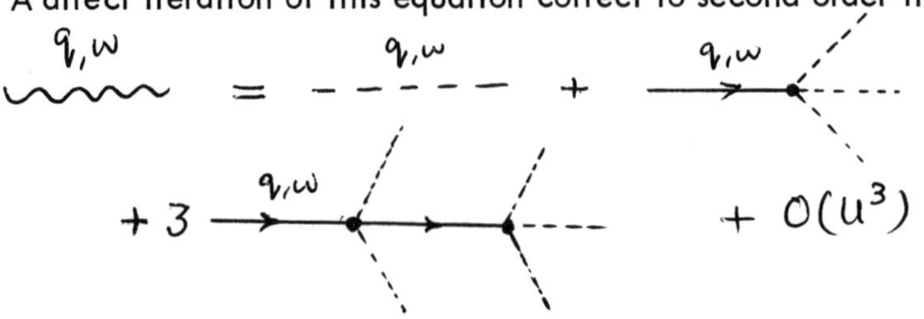

(5.6)

where we get a factor of 3 since the vertex is symmetric under interchange of the dotted lines. When we want to calculate correlation functions we will have products of M's or wavy lines that we average over. Within a perturbation theoretic approach we expand all M's in terms of - - - - - , ——>—— and • and then average. Each dotted line is essentially a factor of the noise. Consequently our average is essentially over the noise factors. We have postulated, however, that the noise is a Gaussian random variable. This means that the average of a product of noise terms is equal to the sum of all possible pairwise averages. Since M^o is proportional to the noise, the average of a product of M^o's factors into a sum of all possible pairwise averages. Algebraically we write

$$\langle M_i^o M_j^o M_k^o M_\ell^o \rangle = \langle M_i^o M_j^o \rangle \langle M_k^o M_\ell^o \rangle + \langle M_i^o M_k^o \rangle \langle M_j^o M_\ell^o \rangle + \langle M_i^o M_\ell^o \rangle \langle M_j^o M_k^o \rangle . \tag{5.7}$$

Graphically we introduce the following convention: If before averaging we have four noise lines

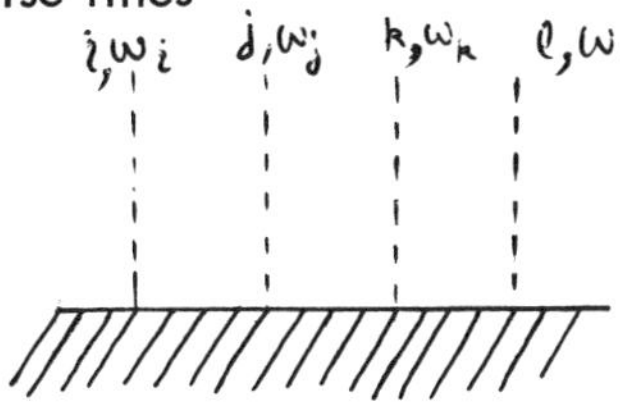

then on averaging we tie these lines together in all possible pairs

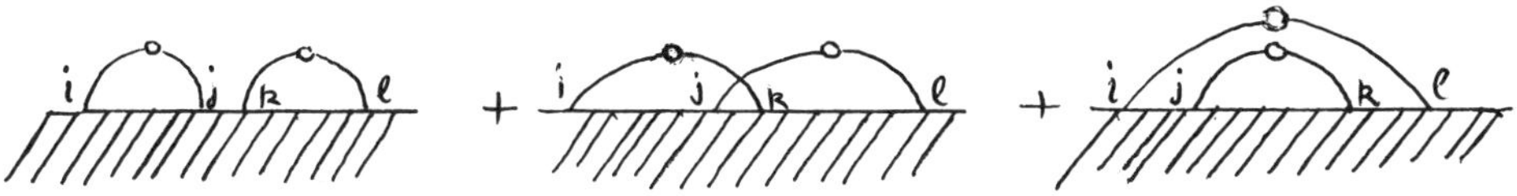

(5.8)

and

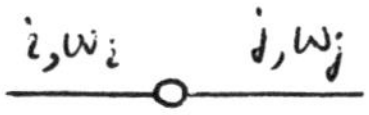

$$= \langle M_i^o(\omega_i)\, M_j^o(\omega_j) \rangle$$

$$= C^{0}_{ij}(\omega_i) \; 2\pi \; \delta(\omega_i + \omega_j) \tag{5.9}$$

is just the zeroth order correlation function. With these simple rules we can calculate any correlation function to any order in the coupling u. Let us calculate, for example, the two point correlation function to second order in u. We see that we generate the graphs:

$$C(q,\omega)\,(2\pi)^d\,\delta(\vec{q}+\vec{q}\,')\;2\pi\,\delta(\omega+\omega')$$

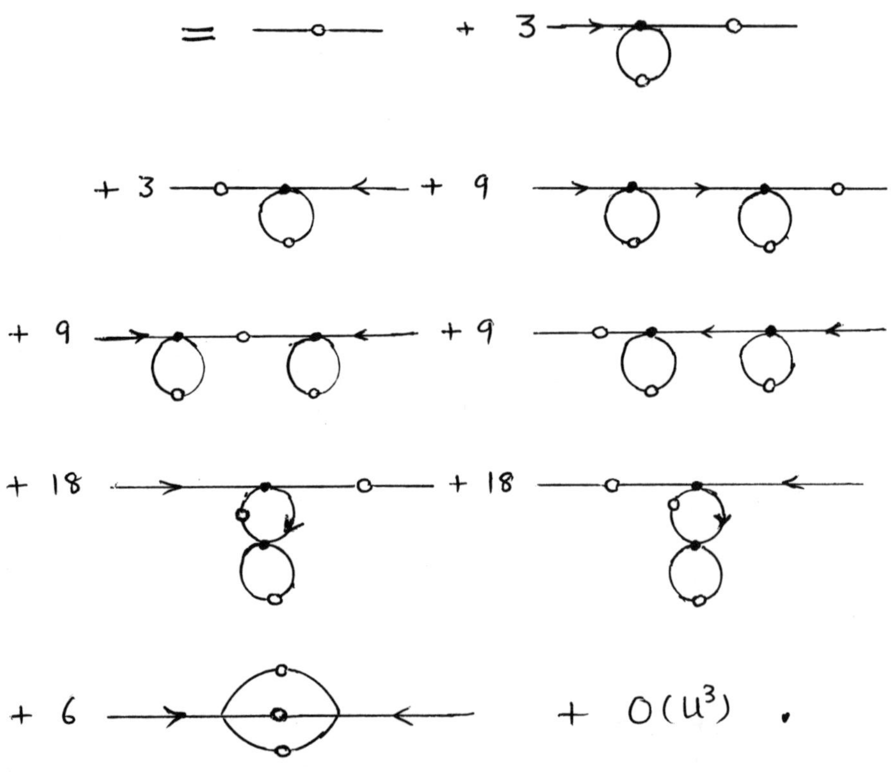

(5.10)

This method will generate all of the terms in the perturbation theory expansion. One must be careful in interpreting the consequences of this expansion however. It is always essential to have some idea of the underlying physics governing the behavior of some quantity before carrying out an expansion. These words of caution have nothing in particular to do with the fact we are dealing with dynamics and are applicable in the purely static theory. We can demonstrate how problems arise with a simple example. We believe that the static two-point correlation function at zero wavenumber is given approximately by

$$\chi(0) \sim (T-T_c)^{-1} \tag{5.11}$$

for $T \geq T_c$ (where we are not worrying about the deviations of γ from its classical value (= 1) at the moment). The zeroth order approximation for χ is given by

$$\chi_0(0) \sim (T-T_c^{\circ})^{-1} . \tag{5.12}$$

In general $T_c^{\circ} \neq T_c$. The non-linear interactions shift T_c from its zeroth order value. If we make a direct perturbation series expansion

$$\chi(0) \simeq \chi_0(0) + \text{-----},$$

then perturbation theory will tell us that $\chi(0)$ is blowing up at the "wrong" temperature. We can remedy this problem by realizing that we must "resum" an infinite number of graphs--perform essentially a mass or temperature renormalization. This amounts to requiring that we not expand χ in a perturbation series expansion but its inverse:

$$\chi^{-1}(q) \equiv \chi_0^{-1}(q) - \Sigma(q) \tag{5.13}$$

where the "self-energy" $\Sigma(q)$ depends on the non-linear interaction. Then at $q = 0$ we have (somewhat schematically) on comparing (5.11) and (5.13)

$$T-T_c = T-T_c^{\circ} - \Sigma(0,T) \tag{5.14}$$

and we obtain T_c by solving the equation

$$T_c = T_c^{\circ} + \Sigma(0,T_c) .$$

The non-linear interaction generally acts to reduce the transition temperature from its zeroth order value.

It should be clear that problems like temperature renormalization plague the expansion given by (5.10). The resolution of these difficulties involves resummation of classes of graphs as in the example above. There is, however, one complication. It turns out that the structure of the graphical expansion for $C(q, \omega)$ is not directly suited for the needed graphical resummation. The quantity which does have an expansion suited to resummation is the linear response function.

The linear response function $G(q, \omega)$ is defined as the response of the field $M(x, t)$ to an external field $h(x, t)$ introduced by adding to the GL free-energy a term $-\int d^d x\, M(x)h(x, t)$. In our simple TDGL model the introduction of the external field, as we see from (4.124), simply adds a term $+\Gamma(q)\, h(q, \omega)$ to the right hand side of (5.1), a term $G_o(q, \omega)\, h(q, \omega)$ to (5.3), and we add to (5.5) the contribution

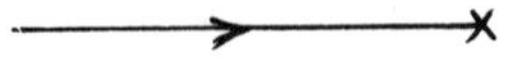

where the x indicates the external field. The linear response function is defined in general as

$$\lim_{h \to 0} \frac{\langle M(q,\omega)\rangle - \langle M(q,\omega)\rangle|_{h=0}}{h(q,\omega)} \equiv G(q,\omega) . \tag{5.15}$$

It is clear from (5.3) with the added h term that G_o is the linear response function for $u = 0$. On comparing (4.74) and (5.2) we see that the zeroth order correlation function and response function are related by

$$C^{o}(q,\omega) = \frac{2}{\omega} \operatorname{Im} G_o(q,\omega) . \tag{5.16}$$

It has been shown by Ma and Mazenko[58] that this "fluctuation-dissipation" theorem holds for the class of non-linear equations defined by (4.120) for the full correlation function and response functions:[68]

$$C(q,\omega) = \frac{2}{\omega} \operatorname{Im} G(q,\omega) . \tag{5.17}$$

Thus if we can calculate G(q, ω) we can obtain C(q, ω) via the fluctuation-dissipation theorem. The linear response function has one particularly interesting feature. If we note the spectral representation for the response function

$$G(q,\omega) = \int \frac{d\omega'}{\pi} \frac{\mathrm{Im}\, G(q,\omega')}{(\omega' - \omega - i0^+)} , \tag{5.18}$$

and use the fluctuation-dissipation theorem, we obtain

$$G(q,\omega) = \int \frac{d\omega'}{2\pi} \frac{C(q,\omega')\, \omega'}{(\omega' - \omega - i0^+)} . \tag{5.19}$$

If we set ω = 0 we find

$$G(q,0) = \int \frac{d\omega'}{2\pi} C(q,\omega') = \chi(q) . \tag{5.20}$$

Therefore we can obtain the static susceptibility from the linear response function simply by setting ω = 0.

The graphical expansion for G is relatively simple. We first iterate equation (5.5) including the - - - - - x term keeping only those terms with one x or less. We therefore obtain to second order in u

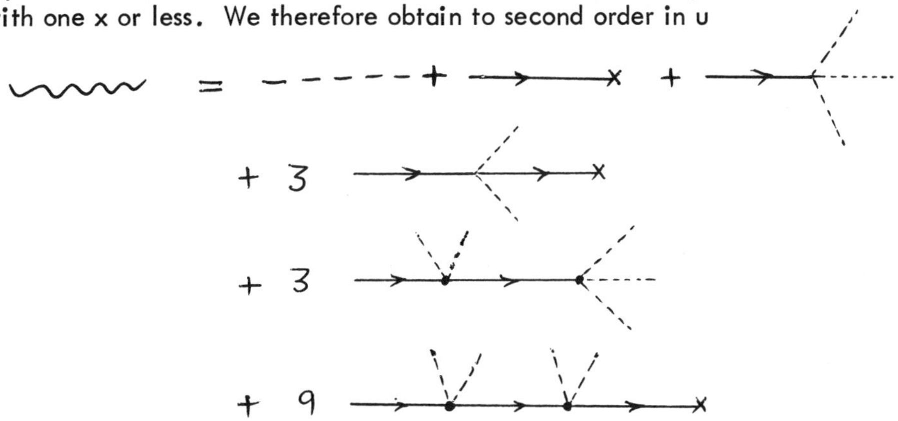

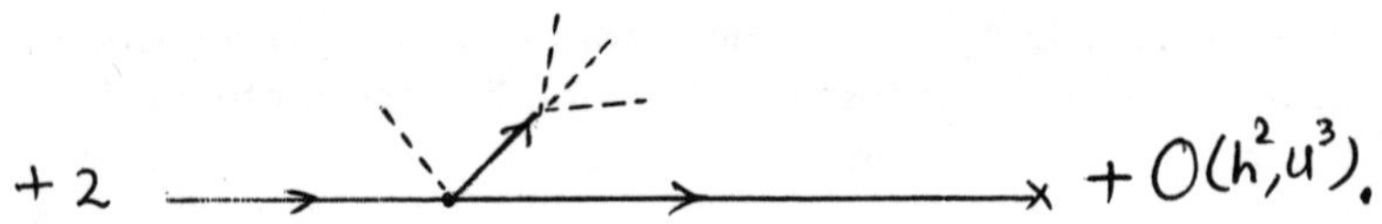

$$+ 2 \quad \text{———→} \underset{}{\bigvee} \text{———→}_{\times} \quad + O(h^2, u^3), \quad (5.21)$$

We then average over the noise. All terms independent of h vanish since they have an odd number of noise lines. We obtain

$$G(q, \omega)\, h(q, \omega) \quad = \quad \text{———→}_{\times} \quad + \; 3 \quad \text{———→}\!\!\underset{}{\bullet}\!\!\text{———→}_{\times}$$

$$+ \; 9 \quad \text{———→}\!\!\underset{}{\bullet}\!\!\text{———→}\!\!\underset{}{\bullet}\!\!\text{———→}_{\times}$$

$$+ \; 18 \quad \text{———→}\!\!\bigotimes\!\!\text{———→}_{\times}$$

$$+ \; 6 \quad \text{———→}\!\!\underset{}{\bullet}\!\!\text{———→}_{\times} \quad + \; O(h^2, u^3)\,.$$

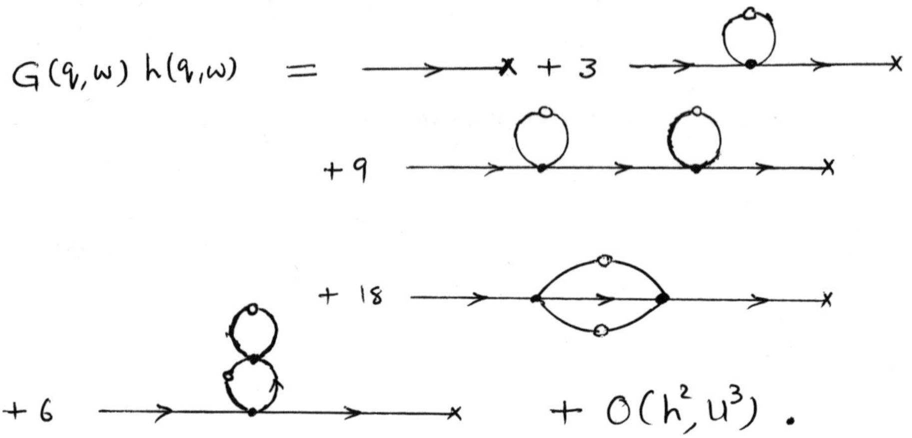

$$(5.22)$$

If we wrote down the graphs to a very high order and studied their structure we would see that they can be written in the form

$$G(q, \omega)\, h(q, \omega) \quad = \quad \text{———→}_{\times} \quad + \quad \text{———→}\!\!\langle\!\!\rangle\!\!\text{———→}_{\times}$$

$$+ \quad \text{———→}\!\!\langle\!\!\rangle\!\!\text{→}\!\!\langle\!\!\rangle\!\!\text{———→}_{\times}$$

$$+ \quad \text{———→}\!\!\langle\!\!\rangle\!\!\text{→}\!\!\langle\!\!\rangle\!\!\text{→}\!\!\langle\!\!\rangle\!\!\text{———→}_{\times} \quad + \; - - - -$$

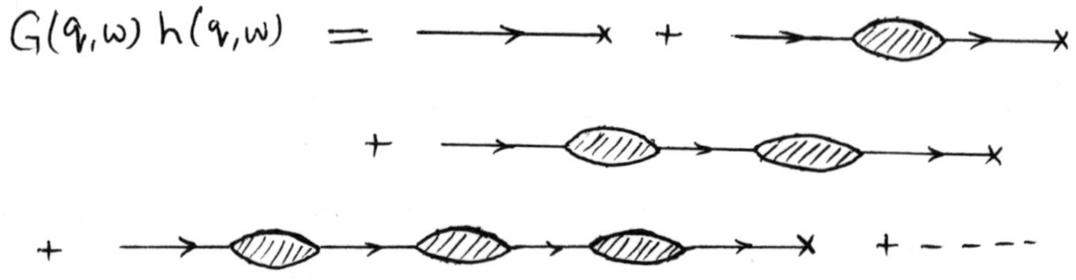

where the blob is the sum of all "one-line irreducible" graphs. These are graphs which can not be separated into two pieces by severing one line. Note that this graphical equation can be written as

$$G(q, \omega) \quad = \quad G_0(q, \omega) \quad + \quad G_0(q, \omega)\, \Sigma(q, \omega)\, G(q, \omega)$$

$$(5.23)$$

where $\Sigma(q, \omega)$ stands for  . We see then that

$$G^{-1}(q,\omega) = G_0^{-1}(q,\omega) - \Sigma(q,\omega) . \tag{5.24}$$

We see immediately from the graphs for G that the contributions to Σ to second order are

$\Sigma(q,\omega) = 3$  $+ 18$ 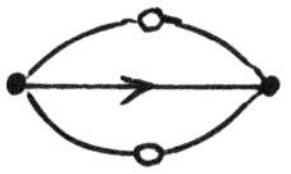

$+ 6$  $+ O(u^3)$. (5.25)

It is a good exercise to convince oneself that these graphs can be transcribed into the appropriate algebraic expressions:

$\Sigma_1(q,\omega) = 3$ 

$$= 3(-4u) \int \frac{d^d q'}{(2\pi)^d} \frac{d\omega'}{2\pi} C^0(q',\omega')$$

(5.26)

$\Sigma_2(q,\omega) = 6$ 

$$= 6(-4u) \int \frac{d^d q'}{(2\pi)^d} \frac{d\omega'}{2\pi} C^0(q',\omega') (-4u) \int \frac{d^d q''}{(2\pi)^d} \frac{d\omega''}{2\pi} C^0(q'',\omega'') G_0(q'',\omega'')$$

(5.27)

while

$\Sigma_3(q,\omega) = 18$ 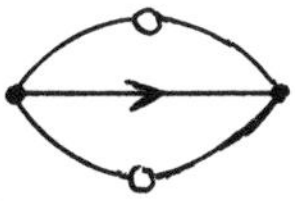

$$= 18(-4u)^2 \int \frac{d^d q'}{(2\pi)^d} \frac{d\omega'}{2\pi} \frac{d^d q''}{(2\pi)^d} \frac{d\omega''}{2\pi} G_0(q-q'-q'', \omega-\omega'-\omega'')$$

$$\times C^0(q',\omega') C^0(q'',\omega'') .$$

(5.28)

Note that the contributions from Σ_1 and Σ_2 are independent of q and ω. In particular these terms only serve to change the static q = 0 susceptibility. Suppose we keep only the 0(u) term, then

$$G^{-1}(q,\omega) = G_0^{-1}(q,\omega) - \Sigma_1$$

$$= \frac{-i\omega}{\Gamma_0(q)} + \chi^{-1}(q) \tag{5.29}$$

where

$$\chi^{-1}(q) = 2r_0 - \Sigma_1 + 2cq^2 . \tag{5.30}$$

Eq. (5. 30) can be rewritten as

$$\chi^{-1}(q) = 2r(T) + 2cq^2 \tag{5.31}$$

where

$$2r(T) = 2r_0 - \Sigma_1 . \tag{5.32}$$

The temperature where r(T) = 0 is the new transition temperature correct up to terms of $0(u^2)$. Let us go back to the more general expression for $G^{-1}(q,\omega)$.

$$G^{-1}(q,\omega) = -\frac{i\omega}{\Gamma_0(q)} + \chi_0^{-1}(q) - \Sigma(q,\omega) . \tag{5.33}$$

It is convenient to write this in the general form

$$G^{-1}(q,\omega) = \frac{-i\omega}{\Gamma(q,\omega)} + \chi^{-1}(q) \tag{5.34}$$

where $\chi(q)$ is the full static susceptibility and $\Gamma(q, \omega)$ is a frequency and wavenumber dependent kinetic coefficient. Since, in general,

$$\chi^{-1}(q) = \chi_0^{-1}(q) - \Sigma(q,0) \tag{5.35}$$

we easily find, after some algebra, that

$$\Gamma(q,\omega) = \Gamma_0(q)\left[1 + \Gamma_0(q)\left(\Sigma(q,\omega) - \Sigma(q,0)\right)/i\omega\right]^{-1}. \tag{5.36}$$

In the TDGL model only Σ_3 contributes to $\Gamma(q,\omega)$ to $O(u^3)$ and we find

$$\Gamma(q,\omega) = \Gamma_0(q)\left[1 - \frac{\Gamma_0(q)}{i\omega}\left(\Sigma_3(q,\omega) - \Sigma_3(q,0)\right) + O(u^3)\right]. \tag{5.37}$$

If we consider the case of a non-conserved order parameter, $\Gamma_0(q) = \Gamma_0$, then one finds, for $d = 4$, $T = T_c$[59], and small ω

$$\Gamma(0,\omega) = 1 + \frac{u^2\, 9 \ln(4/3) \ln \omega}{8\pi^4} + \cdots \tag{5.38}$$

This means simply, near the phase transition, perturbation theory does not work. The first "correction" is larger than the zeroth order contribution for small ω. It will be convenient, at this point, to switch the discussion from TDGL models to those with mode coupling or streaming velocity terms.

We can set up the diagrammatic rules to include the streaming velocity terms in a manner directly analogous to the TDGL case. One main point of difference is that the external field will now enter into the equation of motion in two places. It will enter as the TDGL model but it will also enter into the streaming term. This follows from (4.107) since when $F_\varphi \rightarrow F_\varphi - \sum_i h_i(t)\,\varphi_i$ we generate a term

$$- \sum_j Q_{ij}[\Phi]\, h_j(t) \tag{5.39}$$

in the equation of motion. Suppose for the moment we set u = 0 and use our perturbation theory to treat the ferromagnet with the mode coupling parameter λ (eqn. 4. 130) as a small parameter. Without going into the details (these are given by Ma and Mazenko[58]) we find that we can again write

$$G_M^{-1}(q,\omega) = \left[\frac{-i\omega}{q^2 \Gamma_M(q,\omega)} + \chi_M^{-1}(q) \right] \tag{5.40}$$

where the q^2 multiplying Γ_M is due to the conserved nature of the order parameter and,

$$\Gamma_M(0,0) = \Gamma_M^0 + \frac{4\lambda^2}{d} \int_0^\infty dt \int \frac{d^dk}{(2\pi)^d}\, k^2\, C_M(k,t)\, C_M(-k,t) + O(\lambda^4)\,. \tag{5.41}$$

This is a characteristic mode coupling correction to the bare transport coefficient to lowest order in λ. If we insert the zeroth order expressions for C(k, t) from (4.75) and do the time and angular integrations we obtain

$$\Gamma_M(0,0) = \Gamma_M^0 + \frac{2K_d\lambda^2}{d\,\Gamma_M^0} \int_0^\Lambda k^{d-1} \chi_M^3(k)\, dk \tag{5.42}$$

where K_d is defined by (4.99). If we insert the Ornstein-Zernike form for $\chi_M(k)$ we find, for $\Lambda\xi >> 1$ that

$$\Gamma_M(0,0) = \Gamma_M^0 + \frac{2K_d\lambda^2}{d\,\Gamma_M^0}\begin{cases}\frac{1}{4}\xi^{6-d}\,\Gamma(d/2)\,\Gamma\left(\frac{6-d}{2}\right) & d<6\\ \ln(\Lambda\xi)-3/4 & d=6\\ \frac{\Lambda^{d-6}}{(d-6)} + O\left((\Lambda\xi)^{-1}\right) & d>6.\end{cases}$$

(5.43)

Then, in three dimensions, for example

$$\Gamma_M(0,0) = \Gamma_M^0\left[1 + \frac{1}{48\pi}\left(\frac{\lambda}{\Gamma_M^0}\right)^2 \xi^3 + O(\lambda^4)\right]. \quad (5.44)$$

We can see immediately that the usefulness of perturbation theory is strongly coupled to the dimensionality of the system. For dimensions d > 6 perturbation theory "works." For small $f \equiv \frac{\lambda\Lambda^{\frac{d-6}{2}}}{\Gamma_M^0}$ (the dimensionless coupling constant) the expansion

$$\Gamma_M(0,0) = \Gamma_M^0\left(1 + \frac{2f^2K_d}{d(d-6)} + O(f^4)\right) \quad (5.45)$$

makes sense. In 6 dimensions the correction is logrithmically divergent and for dimensions less than 6 the perturbation theory expansion

$$\Gamma_M(0,0) = \Gamma_M^0\left(1 + \frac{2f^2K_d}{4d}\Gamma(d/2)\Gamma\left(\frac{6-d}{2}\right)(\Lambda\xi)^{6-d} + O(f^4)\right)$$

(5.46)

doesn't make sense as it stands since no matter how small f is, eventually, as $T \to T_c$, $(\Lambda\xi)^{6-d} f^2$ will become large. We see from our naive perturbation theory expansion that

(i) the mode coupling terms are very important near the critical point and

(ii) 6 dimensions is the cut-off dimension for a ferromagnet between conventional and non-conventional behavior.

We also see that our direct perturbation theory approach (where the perturbation is large compared to the zeroth order term) is dubious. The situation is formally identical to the case in treating statics where we carry out a direct perturbation expansion of χ in (3.54). The resolution of the problem--a well defined method for interpreting perturbation theory--is offered by the renormalization group.

B. The Dynamic Renormalization Group

The dynamic generalization of the renormalization group we discussed earlier is fairly straightforward. In the static case the parameter space on which the RG acts specifies a probability distribution $e^{-F[M]}$. In the dynamical case we must extend the parameter space to include all the parameters specifying the non-linear equation of motion including the streaming velocity and the bare transport coefficient. Thus in the case of the isotropic ferromagnet we would start with

$$\mu = \{\lambda, \Gamma_M^0, c, r_0, u\}. \tag{5.47}$$

The implementation of the group again consists of two steps.

(i) Eliminate the components $M_k(t)$

$$\Lambda/b < k < \Lambda$$

from the equation of motion.

(ii) In the resulting equation of motion for $M_q(t)$ with $q < \Lambda/b$, replace $M_q(t) \to b^{1-\eta/2} M_{qb}(tb^{-z})$, rescale wavenumbers, $q \to q' = bq$, times $t \to t' = tb^{-z}$ and L by bL'.

This is the same as step (ii) in the static RG except that we rescale times with a factor b^{-z}. The new exponent z plays a role in scaling times similar to that played by η in scaling distances. η and z must be adjusted so that we reach a fixed point.

The new equations of motion are then written in the old form (4.120). The new parameters are identified as entries in $\mu' = R_b\mu$. We will discuss the implementation of these steps for the ferromagnet below. First we discuss the scaling implications of the dynamic RG.

The dynamic generalization of (3.74) is

$$C(q,t;\mu) = b^{2-\eta} C(qb, tb^{-z}; \mu') . \quad (5.48)$$

If we Fourier transform over time we obviously obtain

$$C(q,\omega;\mu) = b^{2-\eta+z} C(qb, \omega b^{z}; \mu') . \quad (5.49)$$

If we use (5.15) we easily obtain for the response function

$$G(q,\omega;\mu) = b^{2-\eta} G(qb, \omega b^{z}; \mu') . \quad (5.50)$$

Suppose we can find a fixed point solution of our dynamic RG recursion relations:

$$\mu^* = R_b \mu^* . \quad (5.51)$$

Then our analysis can follow that of Section III.D.3. We expect again to obtain, as in (3.92)

$$\delta\mu' = t_1 b^{y_1} e_1 + O(b^{y_2}) \quad (5.52)$$

and $y_2 < 0$. y_2 need not be the same as in the static case. We find then if we are near the critical surface that

$$C(q,\omega;\mu(T)) = b^{2-\eta+z} C\left(qb, \omega b^{z}; \mu^* + \left(b/\xi\right)^{1/\nu} e_1 + O(b^{y_2})\right) . \quad (5.53)$$

If we set $b = \Lambda\xi$ then

$$C(q,\omega;\mu(T)) = (\Lambda\xi)^{2-\eta+z} C\left(q\xi, \omega(\xi\Lambda)^{z}; \mu^* + \Lambda^{1/\nu} e_1 + O((\Lambda\xi)^{y_2})\right) . \quad (5.54)$$

If terms of $O((\Lambda\xi)^{y_2})$ can be ignored then we have

$$C(q,\omega;\mu(T)) = \xi^{2-\eta+z} f(q\xi, \omega\xi^z) \tag{5.55}$$

which is a statement of dynamic scaling. We see that while q is naturally scaled by ξ as $T \to T_c$, ω is scaled by ξ^z. Alternatively if we let $T = T_c$ we can write

$$C(q,\omega;\mu(T_c)) = b^{2-\eta+z} C(qb, \omega b^z; \mu^* + O(b^{y_2})). \tag{5.56}$$

If we then choose $b = (q/\Lambda)^{-1}$ then

$$C(q,\omega;\mu(T_c)) = \left(q/\Lambda\right)^{2-\eta+z} C(\Lambda, \omega(q/\Lambda)^{-z}; \mu^*) \tag{5.57}$$

while for $b = (\omega/\omega_0)^{1/z}$, where ω_0 is some frequency constructed from Γ_0 and Λ,

$$C(q,\omega;\mu(T_c)) = \left(\omega/\omega_0\right)^{(2-\eta+z)/z} C(k(\omega/\omega_0)^{1/z}, \omega_0; \mu^*) \tag{5.58}$$

and

$$C(0,\omega;\mu(T_c)) \sim \omega^{(2-\eta+z)/z}. \tag{5.59}$$

It is convenient to define, as we did in the van Hove theory, a characteristic frequency. One convenient definition is

$$\omega_c \equiv \chi^{-1}(q)\left[\frac{\partial}{\partial(-i\omega)} G^{-1}(q,\omega)\right]^{-1}_{\omega=0} . \tag{5.60}$$

If we use (5.34) we see that

$$\omega_c(q) = \chi^{-1}(q)\,\Gamma(q,0) \tag{5.61}$$

and $\Gamma(q, 0)$ has the interpretation of a wavenumber dependent kinetic coefficient. Note that this definition for a characteristic frequency agrees with that used in the conventional theory (see eqs. (4.78) and (4.83)). We see from (3.116), (3.119) and (5.50) that

$$\omega_c(q;\mu(T)) = b^{-z}\omega_c(qb;\mu') . \tag{5.62}$$

Following now familiar arguments if we let $b = \Lambda\xi$ and ignore terms of $0((\Lambda\xi)^{y_2})$ we obtain the scaling result:

$$\omega_c(q;\mu(T)) = \xi^{-z} f(q\xi) . \tag{5.63}$$

Clearly one of the first orders of business in dynamic critical phenomena is to determine the dynamic index z.

C. Implementation of the Dynamic Renormalization Group for A Ferromagnet

To see howthe RG works in practice let us, for simplicity set u and r_o equal to zero[69] and investigate how the parameters $\Gamma_m{}^o$ and λ change Under the group transformation. Our Fourier transformed equation of motion in zero external field can then be written as (using (5.4) and (4. 130))

$$\vec{M}(q,\omega) = \vec{M}^o(q,\omega) + G_o(q,\omega)\int\frac{d^dq'}{(2\pi)^d}\frac{d\omega'}{2\pi}\frac{\lambda}{2\Gamma_M^o q^2}$$

$$\times\left[\vec{q}'^2-(\vec{q}-\vec{q}')^2\right]\vec{M}(\vec{q}',\omega')\,\vec{M}(\vec{q}-\vec{q}',\omega-\omega') \,. \tag{5.64}$$

Again it will be useful to use a graphical technique. If we identify,

$M(\vec{q},\omega) =$  (5.65a)

$M^0(\vec{q},\omega) =$  (5.65b)

$G_0(\vec{q},\omega) =$  (5.65c)

as before, and the three legged vertex

$\frac{\lambda}{2\Gamma_M^0 q^2}\left[\vec{q}'^2-(\vec{q}-\vec{q}')^2\right] =$ 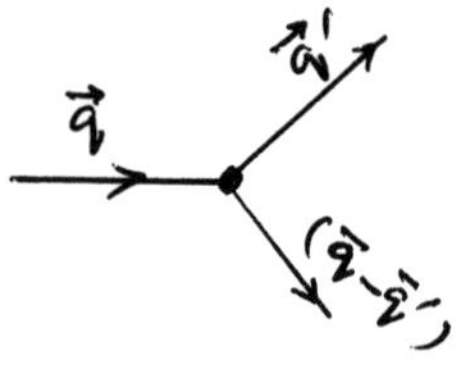

, (5.65d)

then the equation of motion can be written as

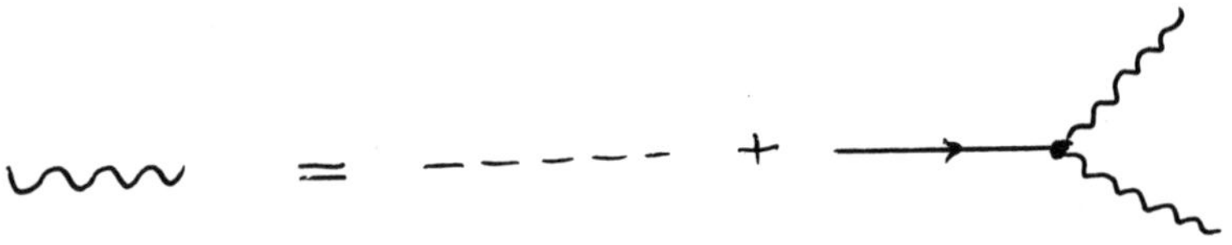
. (5.66)

In carrying out step (i) in the RG we write:

$$\begin{aligned}\vec{M}(\vec{q},\omega) &= \Theta(|q|>\Lambda)\,\vec{M}(\vec{q},\omega) + \Theta(|q|<\Lambda)\,\vec{M}(\vec{q},\omega)\\ &\equiv \vec{M}^>(\vec{q},\omega) + \vec{M}^<(\vec{q},\omega)\end{aligned} \tag{5.67}$$

We can then break up  into its high and low wavenumber components,

$$\vec{M}^{>}(q,\omega) \equiv$$  (5.68a)

$$\vec{M}^{<}(q,\omega) \equiv$$  (5.68b)

and similarly,

$$G_0^{>}(q,\omega) =$$  (5.69a)

$$G_0^{<}(q,\omega) =$$  (5.69b)

We are then led to the coupled set of equations

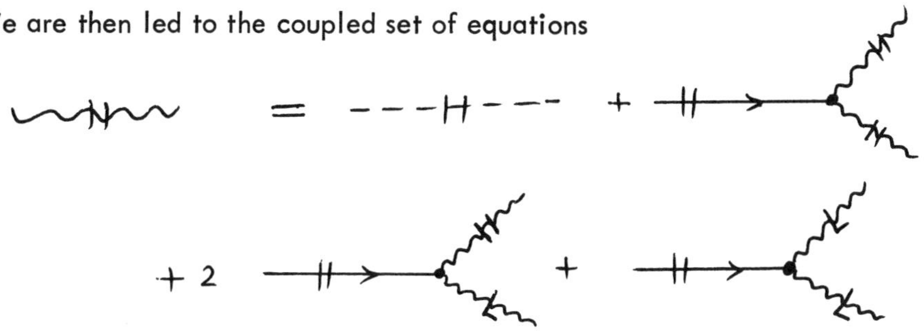

(5.70)

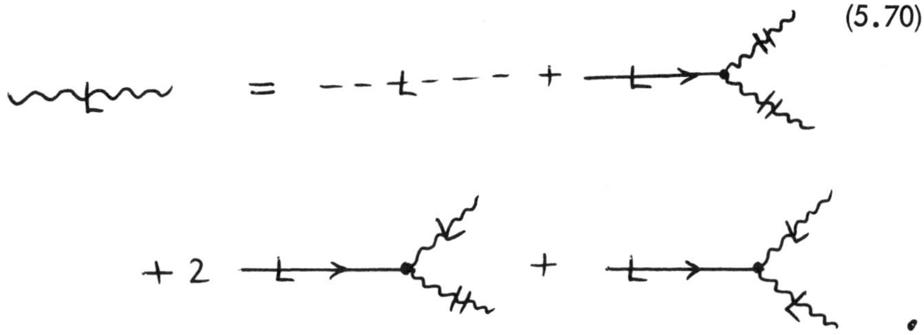

(5.71)

We need to "solve" for  in (5.70), put it into (5.71) and then average over the noise associated with high wavenumbers. We will work this out to second order in λ . Note that on iteration

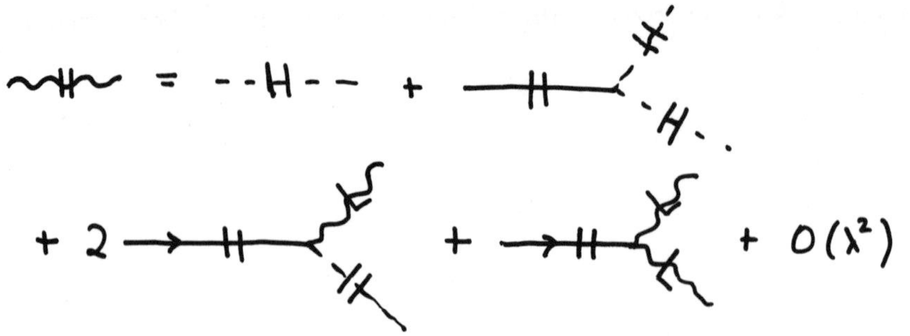

(5.72)

Putting this result for  into (5.71) we obtain

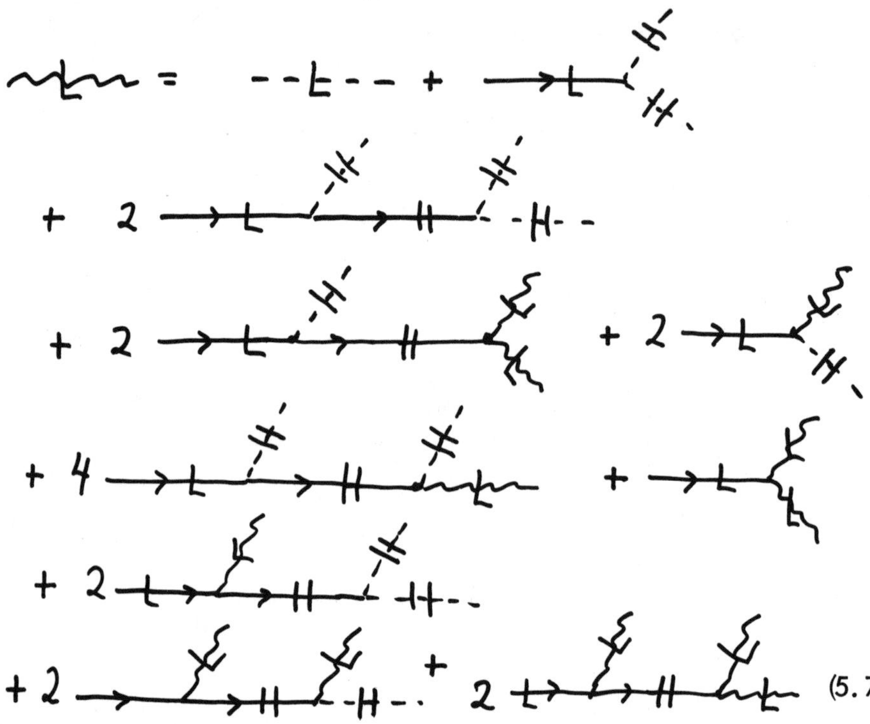

(5.73)

This looks like a mess. However we must average over - - H - - -. All of the graphs with an odd number of - - H - - - vanish. When we connect up two - - H - - - lines we obtain a factor:

 $\equiv \langle M_0^{>}(q,\omega)\, M_0^{>}(-q,-\omega)\rangle = C_0^{>}(q,\omega)$ (5.74)

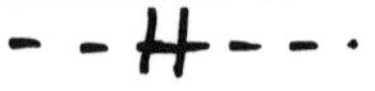

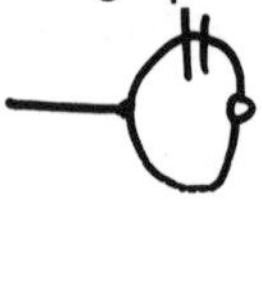

It is also useful to note that graphs with contributions like

vanish due to symmetry. We are then left with the result:

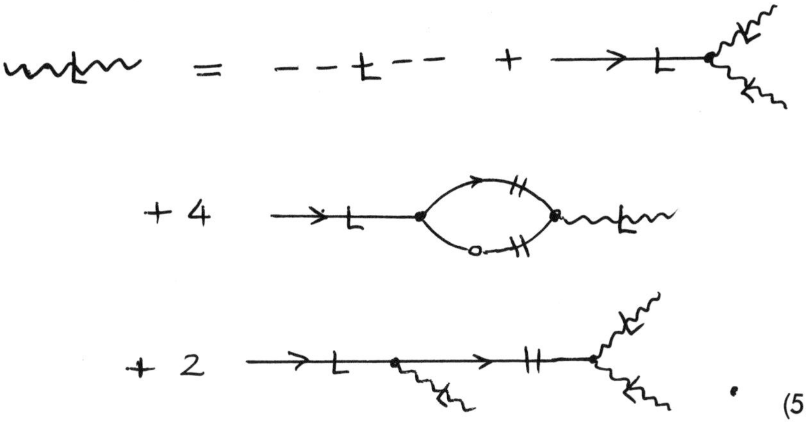

(5.75)

The first terms are those we would obtain from ignoring  altogether. The last term gives no contribution for small enough q due to a non-compatibility of step functions. We are then left with the equation,

$$\vec{M}^{<}(q,\omega) = G_0^{<}(q,\omega)\frac{1}{\Gamma_M^0 q^2}\vec{f}^{<}(q,\omega) + G_0^{<}(q,\omega)$$
$$\times \overline{\Sigma}(q,\omega)\,\vec{M}^{<}(q,\omega) + G_0^{<}(q,\omega)\int\frac{d^dk}{(2\pi)^d}\frac{d\bar{\omega}}{2\pi}\frac{\lambda}{\Gamma_M^0 q^2}$$
$$\times\left[\vec{k}^2 - (\vec{q}-\vec{k})^2\right]\vec{M}^{<}(\vec{k},\bar{\omega})\times\vec{M}^{<}(\vec{q}-\vec{k},\omega) \tag{5.76}$$

where we have defined

$\overline{\Sigma}(q,\omega) =$  . (5.77)

We can bring $G_0^{<}\overline{\Sigma}M^{<}$ term to the left hand side of (5.76) and write

$$\vec{M}^{<}(q,\omega) = \left[1 - G_0^{<}(q,\omega)\,\overline{\Sigma}(q,\omega)\right]^{-1} G_0^{<}(q,\omega)$$

$$\times f^{<}(q,\omega) + [1 - G_0^{<}(q,\omega)\,\overline{\Sigma}(q,\omega)]^{-1} G_0^{<}(q,\omega)$$

$$\times \frac{\lambda}{\Gamma_M^0 q^2} \int \frac{d^d\vec{k}}{(2\pi)^d} \frac{d\bar{\omega}}{2\pi} [(\vec{k})^2 - (\vec{q}-\vec{k})^2]$$

$$\times \vec{M}^{<}(\vec{k},\bar{\omega}) \times \vec{M}(\vec{q}-\vec{k},\omega-\bar{\omega}) \,. \tag{5.78}$$

We then note, using (5.2), that

$$[1 - G_0^{<}(q,\omega)\,\overline{\Sigma}(q,\omega)]^{-1} G_0^{<}(q,\omega) \frac{1}{\Gamma_M^0 q^2}$$

$$= \frac{1}{G_0^{<-1}(q,\omega) - \overline{\Sigma}(q,\omega)} \cdot \frac{1}{\Gamma_M^0 q^2}$$

$$= \frac{1}{-i\omega + \Gamma_M^0 q^2 (\chi^{-1}(q) - \overline{\Sigma}(q,\omega))} \,. \tag{5.79}$$

After writing out (5.77) explicitly and carrying out one internal frequency integral we find in the small q and ω limit that

$$\overline{\Sigma}(q,\omega) = -\chi_0^{-1}(q) \left(\frac{\lambda}{\Gamma_M^0}\right)^2 I_d \tag{5.80}$$

where

$$I_d \equiv \frac{2}{d}\int \frac{d^dk}{(2\pi)^d}\, \chi_0^3(k)\, \theta(k > \Lambda/b)\, \theta(\Lambda > k). \tag{5.81}$$

Our new "equation of motion" can then be written as

$$\vec{M}^1(q,\omega) = M_0^1(q,\omega) + G_0^1(q,\omega)\frac{\lambda}{\Gamma_M^1 q^2}\int \frac{d^dk}{(2\pi)^d}\frac{d\bar{\omega}}{2\pi}$$

$$\times [\vec{k}^2 - (\vec{q}-\vec{k})^2]\, \vec{M}^1(\vec{k},\bar{\omega}) \times \vec{M}^1(\vec{q}-\vec{k}, \omega-\bar{\omega}) \tag{5.82}$$

where

$$\vec{M}_0^1(q,\omega) = \frac{G_0^1(q,\omega)}{\Gamma_M^1 q^2}\, \vec{f}^<(q,\omega), \tag{5.83}$$

$$G_0^1(q,\omega) = \left[\frac{-i\omega}{\Gamma_M^1 q^2} + \chi_0^{-1}(q)\right]^{-1} \tag{5.84}$$

and

$$\Gamma_M^1 \equiv \Gamma_M^0\left[1 + (\lambda/\Gamma_M^0)^2 I_d\right]. \tag{5.85}$$

We see that the full effect of step (i) of the dynamic RG is to simply replace $\Gamma_M^0 \to \Gamma_M'$ and λ is unchanged for small q and ω. Higher order corrections in powers of q and ω are found to be irrelevant in the sense discussed earlier. If we carry out step (ii) of the RG and rescale we obtain the recursion relations

$$\Gamma_M' = b^{z-4}\,\Gamma_M^\circ\left(1 + \left(\frac{\lambda}{\Gamma_M^\circ}\right)^2 I_d + \cdots\right) \tag{5.86}$$

$$\lambda' = b^{(z-1-d/2)}\,\lambda \tag{5.87}$$

We can use these equations to write a recursion relation for the dimensionless coupling $f = (\lambda/\Gamma_M^{\circ})\,\Lambda^{(d-6)/2}$:

$$f' = b^{3-d/2}\,f\,(1 - f^2\Lambda^{6-d} I_d + O(f^4)). \tag{5.88}$$

Remembering that we have set $r_o = 0$ we can evaluate I_d in (5.81) explicitly to obtain

$$I_d = \frac{2K_d}{d}\int_{\Lambda/b}^{\Lambda} k^{d-7}\,dk$$

$$= \frac{2K_d}{d}\begin{cases}\dfrac{1}{(d-6)}\,\Lambda^{d-6}(1-b^{6-d}) & d \neq 6\\[2ex] \ln b & d = 6.\end{cases} \tag{5.89}$$

For $d > 6$ the recursion relation (5.88) can be written for large b as

$$f' = b^{-(d-6)/2}\,f\left(1 - \frac{2f^2K_d}{d(d-6)} + \cdots\right) \tag{5.90}$$

and f scales to zero for large g. The conventional theory is correct. If $d < 6$ then

$$f' = b^{(6-d)/2} f \left(1 + \frac{2K_d b^{6-d}}{d(d-6)} f^2 + \cdots\right) \tag{5.91}$$

and we see that for large b the perturbation theory expansion breaks down. If we compare the analysis here with that for the static case we see that we obtain consistency with perturbation theory and a fixed point if we assume that we are very near 6 dimensions and assume $f^{*2} \sim (6 - d) \equiv \epsilon$. Then we can evaluate I_d in 6 dimensions and write our fixed point equation as

$$f^* = b^{(6-d)/2} f^* \left(1 - \frac{f^{*2} K_6 \ln b}{3} + O(\varepsilon^2)\right). \tag{5.92}$$

Since f^{*2} is very small by assumption, we have

$$1 = b^{(6-d)/2} \, b^{-K_6 (f^*)^2/3} \tag{5.93}$$

or

$$(f^*)^2 = \frac{3}{K_6} \frac{(6-d)}{2} + O(\varepsilon^2). \tag{5.94}$$

Substituting $K_6 = (64\,\pi^3)^{-1}$ we obtain the fixed point value

$$(f^*)^2 = 96(6-d)\pi^3. \tag{5.95}$$

If we then go back to our recursion relation (5.86) for Γ_M we see that a fixed point solution requires

$$\Gamma_M^* = b^{z-4} \Gamma_M^* \left(1 + \frac{(f^*)^2 K_6 \ln b}{3} + \cdots\right) \tag{5.96}$$

or

$$z - 4 + \frac{(f^*)^2 K_6}{3} = 0 \tag{5.97}$$

Thus to first order in $\epsilon = 6 - d$

$$\begin{aligned} z &= 4 - \frac{K_6 (f^*)^2}{3} \\ &= 4 - \epsilon/2 + O(\epsilon^2) \\ &= 1 + d/2 + O(\epsilon^2) . \end{aligned} \tag{5.98}$$

This is compatible with a fixed point solution for (5.87). If we are bold and let $\epsilon = 3$, $d = 3$ we find using (5.98) that $z = 5/2$ in agreement with experiment. It has now been shown that this result for z holds to all orders in ϵ due to certain symmetries of the equation of motion. This does not mean, however, that $(f^*)^2$ is uneffected by higher order terms in ϵ.

Combining these results with those we found previously for the static parameters we see that we have found a dynamic fixed point. We could go further and show[58] that this fixed point is stable. The analysis is along the same lines as in the static case. Instead I want to point out how, once we know the structure of the fixed point, we can make our perturbation theory analysis useful. Let us go back to the expression we had for the transport coefficient resulting from a direct perturbation theory analysis given by (5.43). If we assume that our expansion is in powers of f* and we are near 6 dimension we can write (5.43) as

$$\Gamma_M(0,0) = \Gamma_M^0 \left(1 + \frac{(f^*)^2 K_6}{3} \left[\ln \Lambda \xi - 3/4 \right] + O(\epsilon^2) \right) \tag{5.99}$$

with $\epsilon = 6 - d$. Since f^{*2} is small we can rewrite this as

$$\Gamma_M(0,0) = \Gamma_M^0(\Lambda\xi)^{(f^*)^2 K_6/3}\left(1 - \frac{f^{*2} K_6}{4} + O(\varepsilon^2)\right). \tag{5.100}$$

Remembering (5.94) we can rewrite (5.100) as:

$$\Gamma_M(0,0) = \Gamma_M^0(\Lambda\xi)^{\varepsilon/2}\left(1 - \frac{3\varepsilon}{8} + O(\varepsilon^2)\right). \tag{5.101}$$

If we compare this result with (4.87) we see that we can now understand the divergence of the transport coefficient as $T \to T_c$.

We could go further and calculate $\Gamma_M(0, 0)$ to higher powers in ε or, in fact, calculate the complete correlation function[70] $C(q, \omega)$ using ε as a small parameter. This would take us far afield.

While we can groan over the fact the small parameter for ferromagnets in $\varepsilon = 6 - d$, this is not the case for essentially all other models. It appears that for the dynamics of helium,[71] planar ferromagnets,[72] isotropic anti-ferromagnets[72, 73] and fluids[74] $d = 4$ is again the cross over dimension between classical and non-classical behavior.

D. Final Remarks

In these lectures I have tried to emphasize the basic ideas behind the modern theory of dynamic critical phenomena. These are, of course, the ideas of scaling, universality and the non-linear interaction of long wavelength degrees of freedom. These ideas lead to a renormalization group analysis of semi-macroscopic non-linear equations of motion. I have made no effort to survey state of the art calculations or the wide range of dynamical models that have been analyzed. There are by now a number of elegant methods for carrying out perturbation theory calculations and Hohenberg and Halperin[8] give an excellent review of the various non-linear models that have been analyzed thus far. I should add however that the ideas we have covered in these lectures are the essential ingredients on the way to developing more sophisticated methods for handling more sophisticated models.

There is still a great deal of work to be done in the study of dynamic critical phenomena. We are far from being finished in identifying all of the dynamic universality classes: models which share the same critical surface of a fixed point. We seem to have established that dynamical fixed points are characterized by the underlying static fixed point, the conservation laws governing a system and the poisson bracket algebra satisfied by the slow variables in the system. We still have not been able to classify all of the various interrelationships between these various factors.

In these lectures I have emphasized dynamical behavior above the phase transition in the disordered state. Very interesting things happen in the ordered state, but little has been done. There are many other areas where there is work to be done. There are a number of open questions concerning dynamics near the λ-transition in helium.[11] The effects of impurities on dynamic critical behavior has been studied only using the simplest models. Presumably these ideas could be useful in treating the dynamics of superfluid He^3. Going off into other directions we expect that some of these ideas may be important in understanding the non-equilibrium statistical mechanics of metastable or unstable states, including nucleation and spinodial decomposition,[2] and turbulence.[1]

Let me finish by reminding you that the basic ideas we have developed are not intrinsically tied to the ϵ-expansion. The ϵ-expansion has been useful in implementing our ideas of scaling, fixed points and universality. In principle, however, we don't need the ϵ-expansion and if we are clever enough we should be able to calculate directly in the dimensionality of interest.

ACKNOWLEDGEMENTS

This work was supported in part by the National Science Foundation (Grant No. NSF DMR 76-21298) and The Materials Research Laboratory of the National Science Foundation. I would like to thank P. Sahni for his help in preparation of these lectures.

APPENDIX A: FUNCTIONAL DERIVATIVES AND INTEGRALS

Consider our expression (3.5) for the partition function in the presence of an external inhomogeneous field H(x):

$$Z[H] = \left\langle \exp - \left(F_0 - \int d^d x_1 \, H(x_1) M(x_1) \right) \right\rangle_{\tilde{M}} \quad (A.1)$$

where F_0 is the GL free energy function for H = 0. If we expand Z in a power series in H and keep only the first few terms we obtain

$$Z(H) = Z_0 \Big\{ 1 + \int d^d x_1 \, H(x_1) \langle M(x_1) \rangle + \frac{1}{2} \int d^d x_1 \, d^d x_2 \, H(x_1) H(x_2) \left[\chi(x_1, x_2) + \langle M(x_1) \rangle \langle M(x_2) \rangle \right] + \cdots\cdots \Big\} \quad (A.2)$$

where we have defined

$$Z_0 = \langle e^{-F_0} \rangle_{\tilde{M}} \quad (A.3)$$

$$\langle M(x) \rangle = \frac{1}{Z_0} \langle M(x) \, e^{-F_0} \rangle_{\tilde{M}} \quad (A.4)$$

and

$$\chi(x - x') = \frac{1}{Z_0} \langle \delta M(x) \, \delta M(x') \, e^{-F_0} \rangle_{\tilde{M}} \, . \quad (A.5)$$

We see that $Z(H)$ contains information about $\langle M(x) \rangle$ and $\chi(x,x')$ as coefficients of powers of the external inhomogeneous magnetic fields. How can we extract this information from Z? As a first step in answering this question let us assume that space is not continuous, but can be replaced by a discrete set of points so that integrals can be replaced by sums

$$\int d^d x \, B(x) \, A(x) \longrightarrow \sum_i A_i B_i \, ,$$

then our expression for the partition function can be written

$$Z(H) = Z_0\Big[1 + \sum_i H_i \bar{M}_i + \frac{1}{2}\sum_{i,j} H_i H_j (\chi_{ij} + \bar{M}_i \bar{M}_j) + \cdots\Big]. \tag{A.6}$$

If we take a derivative with respect to H_k we obtain

$$\frac{\partial Z[H]}{\partial H_k} = Z_0\Big[\sum_i \delta_{ik}\bar{M}_i + \frac{1}{2}\sum_{i,j}(\delta_{ik}H_j + H_i\delta_{jk}) \times (\chi_{ij} + \bar{M}_i M_j)\Big]$$

$$= Z_0\Big[\bar{M}_k + \sum_j H_j(\chi_{kj} + \bar{M}_k\bar{M}_j) + \cdots O(H^2)\Big] \tag{A.7}$$

where we have assumed χ_{ij} is symmetric in i and j. If we set H = 0 we obtain the result

$$\bar{M}_k = \frac{1}{Z_0}\frac{\partial Z}{\partial H_k}\Big|_{H=0} \tag{A.8}$$

so the average magnetization at site k is the derivative of the partition function with respect to the field at site k. If we take a second derivative we find

$$\frac{\partial^2 Z}{\partial H_k \partial H_\ell} = Z_0(\chi_{k\ell} + \bar{M}_k\bar{M}_\ell) + O(H), \tag{A.9}$$

so if we set H = 0 we obtain

$$\chi_{k\ell} = \frac{1}{Z_0}\frac{\partial^2 Z}{\partial H_k \partial H_\ell}\Big|_{H=0} - \bar{M}_k\bar{M}_\ell . \tag{A.10}$$

Consequently we can find the correlation between the sites k and l from derivatives of the partition function. What if we have a continuum of "sites," how do we generalize this result? The generalization requires a generalization of the derivative. In the case of a set of points we used the result, for some function F of the set of variables, $\varphi_1, \varphi_2, \ldots \varphi_n$, that

$$\frac{\partial F\{\varphi\}}{\partial \varphi_i} = \lim_{\epsilon \to 0} \frac{1}{\epsilon}\left[F\{\varphi_1, \varphi_2, \ldots, \varphi_i + \epsilon, \ldots \varphi_n\} - F\{\varphi_1 \ldots \varphi_i \ldots \varphi_n\}\right]$$

$$= \lim_{\epsilon \to 0} \frac{1}{\epsilon}\left[F\{\varphi_j + \epsilon \delta_{ij}\} - F\{\varphi_j\}\right]. \tag{A.11}$$

We generalize this to the continuous case $\{\varphi_1, \varphi_2, \ldots \varphi_n\} \to \varphi(x)$ by the definition

$$\frac{\delta F\{\varphi\}}{\delta \varphi(x)} = \lim_{\epsilon \to 0} \frac{1}{\epsilon}\left[F\{\varphi(y) + \epsilon\, \delta(y-x)\} - F\{\varphi(y)\}\right]. \tag{A.12}$$

Then in the discrete case

$$\frac{\partial \varphi_j}{\partial \varphi_i} = \lim_{\epsilon \to 0} \frac{1}{\epsilon}(\varphi_j + \epsilon \delta_{ij} - \varphi_j) = \delta_{ij}, \tag{A.13}$$

while in the continuous case

$$\frac{\delta \varphi(y)}{\delta \varphi(x)} = \lim_{\epsilon \to 0} \frac{1}{\epsilon}(\varphi(y) + \epsilon \delta(y-x) - \varphi(y)) = \delta(x-y). \tag{A.14}$$

The $\delta / \delta\varphi(x)$ derivative is called a functional and can be treated very much like an ordinary derivative in many ways. Using the basic definition one can easily show for example that

$$\frac{\delta \varphi^n(y)}{\delta \varphi(x)} = n\, \varphi^{n-1}(y)\, \delta(\vec{x} - \vec{y}). \tag{A.15}$$

Using this result one can then show, for example, that

$$\frac{\delta}{\delta\phi(x)} \exp - [\tfrac{1}{2}\int d^dx_1 d^dx_2\, \phi(\vec{x}_1)\, A(\vec{x}_1-\vec{x}_2)\, \phi(\vec{x}_2)]$$

$$= -\exp - [\tfrac{1}{2}\int d^dx_1\, d^dx_2\, \phi(\vec{x}_1)\, A(\vec{x}_1-\vec{x}_2)\, \phi(\vec{x}_2)]$$

$$\times \int d^dx_2\, A(\vec{x}_1-\vec{x}_2)\, \phi(\vec{x}_2) \quad . \tag{A.16}$$

It should be clear, looking at (A.2), that we can pick out $\langle M \rangle$ and χ by taking functional derivatives with respect to H(x) and then setting H = 0. Going further we see that we can generate $\langle M \rangle$ and χ in the presence of the field by differentiating ln Z[H] :

$$\langle M(x) \rangle = \frac{1}{Z(H)} \langle e^{-F} M(x) \rangle_M \tag{A.17}$$

$$= \frac{\delta}{\delta H(x)} \ln Z(H) \tag{A.18}$$

and

$$\chi(\vec{x},\vec{x}') = \frac{\delta}{\delta H(\vec{x}')} \langle M(x) \rangle$$

$$= \frac{\delta^2}{\delta H(\vec{x})\, \delta H(\vec{x}')} \ln Z(H) \quad . \tag{A.19}$$

Functional integrals can be carefully defined as the limit of a multiple integral. See, in particular, the review by Gel'fand and Yaglom [75]. There are only two simple properties of functional integrals that one usually needs in developing a perturbation theory analysis. These are

$$\int \mathcal{D}(\phi)\, \frac{\delta}{\delta\phi(x)} F\{\phi\} = 0 \tag{A.20}$$

and

$$\int \mathcal{D}(\varphi)\, F[\varphi(x)+\psi(x)] = \int \mathcal{D}(\varphi)\, F[\varphi(x)] \tag{A.21}$$

where $\lim_{\varphi \to \infty} F\{\varphi\} = 0$. The first property is the functional equivalent of integrating by parts, the second is the equivalent of a change of variable. One can prove these properties by starting with the cell formulation, where these properties clearly hold, and take the continuum limit.

One key result which follows directly from (A.16) and (A.20) **above** is that one can evaluate correlation functions for systems with a Gaussian probability of the form

$$P[M] = \exp -\tfrac{1}{2} \int d^dx\, d^dx'\, M(\vec{x})\, A(\vec{x}-\vec{x}')\, M(\vec{x}') \tag{A.22}$$

as

$$\begin{aligned} \chi(q) &= \int d^dx\, e^{iq.x}\, \chi(x-x') \\ &= \frac{1}{A(q)} \end{aligned} \tag{A.23}$$

where A(q) is the Fourier transform of A(x) and

$$\chi(\vec{x}-\vec{x}') \equiv \frac{\int \mathcal{D}[M]\, P[M]\, M(\vec{x})\, M(\vec{x}')}{\int \mathcal{D}[M]\, P[M]} . \tag{A.24}$$

APPENDIX B: CORRELATION FUNCTION EXPRESSION FOR THE MEMORY FUNCTION

If we multiply (4.12) by Ψ_i and take the equilibrium average we obtain

$$z C_{ij}(z) = \chi_{ij} - \langle \Psi_j R(z) L \Psi_i \rangle \, . \tag{B.1}$$

If we compare this equation with (4.32) we can identify

$$\sum_{\ell} K_{i\ell}(z) C_{\ell j}(z) = - \langle \Psi_j R(z) L \Psi_i \rangle \, . \tag{B.2}$$

If we multiply (B.2) by z and use (4.11) in $C_{\ell j}(z)$ and on the right hand side we obtain

$$\sum_{\ell} K_{i\ell}(z) \left[\chi_{\ell j} - \langle \Psi_j L R(z) \Psi_\ell \rangle \right]$$

$$= - \langle \Psi_j L \Psi_i \rangle + \langle \Psi_j L R(z) L \Psi_i \rangle \, . \tag{B.3}$$

After using the property of the Liouville operator

$$\langle A L B \rangle = - \langle (L A) B \rangle \tag{B.4}$$

(which follows from the time translational invariance of the equilibrium distribution function) we can write

$$K_{ij}(z) = K_{ij}^{(s)} + K_{ij}^{(d)}(z) \tag{B.5}$$

where

$$\sum_{\ell} K_{i\ell}^{(s)} \chi_{\ell j} = - \langle \Psi_j L \Psi_i \rangle \tag{B.6}$$

and

$$\sum_{\ell} K_{i\ell}^{(d)}(z)\,\chi_{\ell j} = -\langle (L\psi_j)\,R(z)\,L\psi_i\rangle - \sum_{\ell} K_{i\ell}(z)\langle (L\psi_j)\,R(z)\,\psi_\ell\rangle . \tag{B.7}$$

Note however from (B.2) that

$$K_{i\ell}(z) = -\sum_{k}\langle \psi_k\,R(z)\,L\psi_i\rangle\, C_{k\ell}^{-1}(z) \tag{B.8}$$

and we can write

$$\sum_{\ell} K_{i\ell}^{(d)}(z)\,\chi_{\ell j} = -\langle (L\psi_j)\,R(z)\,(L\psi_i)\rangle + \sum_{k,\ell}\langle \psi_k\,R(z)\,L\psi_i\rangle\, C_{k\ell}^{-1}(z)\langle (L\psi_j)\,R(z)\,\psi_\ell\rangle . \tag{B.9}$$

APPENDIX C: AUTOCORRELATION FUNCTION FOR THE NOISE

We want to evaluate

$$\langle f_i(t) f_j(t') \rangle \equiv \Phi_{ij}(t-t') \qquad \text{(C.1)}$$

using the defining equation for f_i (4.21). Here we will assume $K_{ij}(t - t')$ is known from, say, an analysis of eqs. (4.34–4.36). It is extremely useful to work with the Laplace transformed equation (4.25):

$$z\,\psi_i(z) - \sum_j K_{ij}(z)\,\psi_j(z) = \psi_i + i f_i(z) \qquad \text{(C.2)}$$

and calculate

$$\begin{aligned} \Phi_{ij}(z_1, z_2) &= -i\int_0^\infty dt\, e^{iz_1 t}(-i)\int_0^\infty dt'\, e^{iz_2 t'}\,\Phi_{ij}(t-t') \\ &= \langle f_i(z_1) f_j(z_2) \rangle . \end{aligned} \qquad \text{(C.3)}$$

If we use (C.2) to eliminate f in (C.3) we obtain

$$\begin{aligned} -\Phi_{ij}(z_1, z_2) &= \Big\langle \Big(\sum_\ell (z_1\delta_{i\ell} - K_{i\ell}(z_1))\psi_\ell(z_1) - \psi_i \Big) \\ &\quad \times \Big(\sum_k (z_2\delta_{jk} - K_{jk}(z_2))\,\psi_k(z_2) - \psi_j \Big) \Big\rangle \\ &= \sum_{\ell,k} (z_1\delta_{i\ell} - K_{i\ell}(z_1))(z_2\delta_{jk} - K_{jk}(z_2)) \\ &\quad \times \langle \psi_\ell(z_1)\psi_k(z_2) \rangle - \sum_\ell \Big[\langle \psi_i\,(z_2\delta_{j\ell} - K_{j\ell}(z_2))\psi_\ell(z_2) \rangle \end{aligned}$$

$$+\langle (z_1\delta_{i\ell} - K_{i\ell})\,\Psi_\ell(z_1)\,\Psi_j\rangle\,] + \langle \Psi_i\,\Psi_j\rangle \tag{C.4}$$

If we use the result

$$\sum_\ell (z\,\delta_{j\ell} - K_{j\ell})\,\langle \Psi_i\,\Psi_\ell(z)\rangle = \langle \Psi_i\,\Psi_\ell\rangle \tag{C.5}$$

we easily find

$$-\Phi_{ij}(z_1, z_2) = \sum_{\ell,k} (z_1\delta_{i\ell} - K_{i\ell})(z_2\delta_{jk} - K_{jk}) \times \langle \Psi_\ell(z_1)\,\Psi_k(z_2)\rangle - \chi_{ij}\,. \tag{C.6}$$

Where

$$\chi_{ij} = \langle \Psi_i \Psi_j \rangle \tag{C.7}$$

Let us look at $\langle \Psi_\ell(z_1)\,\Psi_k(z_2)\rangle$. We can write

$$\langle \Psi_\ell(z_1)\,\Psi_k(z_2)\rangle = (-i)^2 \int_0^\infty dt \int_0^\infty dt'\, e^{iz_1 t}\, e^{iz_2 t'}\, C_{\ell k}(t-t')$$

$$= (-i)^2 \int_0^\infty dt \int_0^\infty dt'\, e^{iz_1 t}\, e^{iz_2 t'} \int \frac{d\omega}{2\pi}\, C''_{\ell k}(\omega)\, e^{-i\omega(t-t')}$$

$$= \int \frac{d\omega}{2\pi}\, \frac{C''_{\ell k}(\omega)}{(z_1-\omega)(z_2+\omega)}\,. \tag{C.8}$$

If we use the identities

$$(z_1 - \omega)^{-1}(z_2 + \omega)^{-1} = (z_1 + z_2)^{-1}\left[(z_1 - \omega)^{-1} + (z_2 + \omega)^{-1}\right]$$

(C. 9)

and

$$C''_{\ell k}(\omega) = C''_{k\ell}(-\omega)$$

(C. 10)

(which follows from the time translational invariance of the equilibrium ensemble) we obtain

$$\langle \psi_\ell(z_1)\psi_k(z_2)\rangle = \frac{-1}{(z_1 + z_2)}\left[C_{\ell k}(z_1) + C_{k\ell}(z_2)\right].$$

(C. 11)

We then have, using (C. 5) twice,

$$-\Phi_{ij}(z_1, z_2) = \sum_{\ell, k} \frac{1}{(z_1 + z_2)}\left(z_1 \delta_{i\ell} - K_{i\ell}(z_1)\right)$$

$$\times \left(z_2 \delta_{jk} - K_{jk}(z_2)\right)\left[C_{\ell k}(z_1) + C_{k\ell}(z_2)\right] - \chi_{ij}$$

$$= \sum_k \frac{(z_1\delta_{jk} - K_{jk}(z_1))\chi_{ik}}{(z_1+z_2)} + \sum_l \frac{(z_2\delta_{il} - K_{il}(z_2))\chi_{jl}}{(z_1+z_2)} - \chi_{ij}$$

$$= -\frac{1}{(z_1+z_2)} \sum_k \left[K_{jk}(z_1)\chi_{ik} + K_{ik}(z_2)\chi_{jk} \right] \tag{C.12}$$

Inserting (4.34) this can be written as:

$$\Phi_{ij}(z_1,z_2) = \frac{1}{(z_1+z_2)} \sum_k \left[K_{jk}^{(s)}\chi_{ki} + K_{ik}^{(s)}\chi_{kj} + K_{jk}^{(d)}(z_1)\chi_{ki} + K_{ik}^{(d)}(z_2)\chi_{kj} \right] \tag{C.13}$$

If we then note from (4.32) that

$$\sum_k K_{ik}^{(s)}\chi_{kj} = \langle \Psi_j L \Psi_i \rangle \tag{C.14}$$

$$= -\langle \Psi_i L \Psi_j \rangle$$

$$= -\sum_k K_{jk}^{(s)}\chi_{ki} \tag{C.15}$$

so

$$\sum_k \left[K_{jk}^{(s)}\chi_{ki} + K_{ik}^{(s)}\chi_{kj} \right] = 0 \ , \tag{C.16}$$

then

$$\Phi_{ij}(z_1,z_2) = \frac{1}{(z_1+z_2)} \sum_k \left[K_{jk}^{(d)}(z_1)\chi_{ki} + K_{ik}^{(d)}(z_2)\chi_{kj} \right] \tag{C.17}$$

If we use the spectral representation

$$K_{jk}^{(d)}(z_1) = \int \frac{d\omega}{2\pi} \frac{K_{jk}^{(d)}(\omega)}{(z_1-\omega)} \tag{C.18}$$

where

$$K_{jk}^{(d)}(\omega) = \int_{-\infty}^{\infty} dt \; e^{i\omega(t-t')} K_{jk}^{(d)}(t-t') , \tag{C.19}$$

we can easily take the inverse Laplace transform to obtain

$$\langle f_i(t) f_j(t') \rangle = \sum_{\ell} K_{i\ell}^{(d)}(t-t') \chi_{\ell j} \tag{C.20}$$

Here again we have used $K_{jk}^{(d)}(\omega) = K_{kj}^{(d)}(-\omega)$.

APPENDIX D: EVALUATION OF $K^{(s)}_{\varphi\varphi'}$

The static part of the memory function can, using (4.35), be written as

$$\int \mathcal{D}[\bar{\Phi}]\, K^{(s)}_{\Phi\bar{\Phi}}\, \chi_{\bar{\Phi}\Phi'} = \langle g_{\Phi'}\, L\, g_{\Phi} \rangle .$$

Using (4.105), (4.101) and (4.6) we find

$$\begin{aligned}
\int \mathcal{D}[\bar{\Phi}]\, K^{(s)}_{\Phi\bar{\Phi}}\, \chi_{\bar{\Phi}\Phi'} &= -\sum_i \frac{\delta}{\delta\Phi_i} \langle g_{\Phi'}\, g_{\Phi}\, L \hat{M}_i \rangle \\
&= -\sum_i \frac{\delta}{\delta\Phi_i}\, \delta(\Phi-\Phi') \langle g_{\Phi}\, L \hat{M}_i \rangle \\
&= -i \sum_i \frac{\delta}{\delta\Phi_i}\, \delta(\Phi-\Phi') \langle g_{\Phi} \{H, \hat{M}_i\} \rangle
\end{aligned} \tag{D.1}$$

where $\{\ ,\ \}$ is the Poisson bracket. If we note the identity

$$\begin{aligned}
\mathrm{Tr}\, e^{-\beta H} g_{\Phi} \{H, \hat{M}_i\} &= -\beta^{-1}\, \mathrm{Tr}\, g_{\Phi} \{e^{-\beta H}, \hat{M}_i\} \\
&= \beta^{-1} Z \langle \{g_{\Phi}, \hat{M}_i\} \rangle \\
&= -\beta^{-1} Z \sum_j \frac{\delta}{\delta\Phi_j} \langle g_{\Phi} \{\hat{M}_j, \hat{M}_i\} \rangle ,
\end{aligned} \tag{D.2}$$

then

$$\int \mathcal{D}[\bar{\Phi}]\, K^{(s)}_{\Phi\bar{\Phi}}\, \chi_{\bar{\Phi}\Phi'} = i \sum_{i,j} \frac{\delta}{\delta\Phi_i}\, \delta(\Phi-\Phi') \times \frac{\delta}{\delta\Phi_j} \langle g_{\Phi} \{\hat{M}_j, \hat{M}_i\} \rangle . \tag{D.3}$$

Defining

$$Q_{ij}[\varphi] = \beta^{-1}\frac{\langle g_\varphi\{\hat{M}_i,\hat{M}_j\}\rangle}{\langle g_\varphi\rangle} \tag{D.4}$$

and using

$$\chi_{\bar{\varphi}\varphi'} = \delta(\bar{\varphi}-\varphi')\,W_{\varphi'}\,, \tag{D.5}$$

we have

$$K^{(s)}_{\varphi\varphi'} = i\,W^{-1}_{\varphi'}\sum_{i,j}\frac{\delta}{\delta\varphi_i}\,\delta(\varphi-\varphi')\,\frac{\delta}{\delta\varphi_j}\,W_\varphi\,Q_{ij}[\varphi]$$

$$= i\,W^{-1}_{\varphi'}\sum_{i,j}\frac{\delta}{\delta\varphi_i}\,\delta(\varphi-\varphi')\left[W_\varphi\frac{\delta}{\delta\varphi_j}Q_{ij}[\varphi] + Q_{ij}[\varphi]\frac{\delta W_\varphi}{\delta\varphi_j}\right]$$

$$= i\sum_{i,j}\frac{\delta}{\delta\varphi_i}\left[\frac{\delta Q_{ij}[\varphi]}{\delta\varphi_j} - Q_{ij}[\varphi]\frac{\delta F_\varphi}{\delta\varphi_j}\right]\delta(\varphi-\varphi')$$

$$\equiv -i\sum_i\frac{\delta}{\delta\varphi_i}\,V_i[\varphi]\,\delta(\varphi-\varphi') \tag{D.6}$$

where

$$V_i[\varphi] = -\sum_j\left[\frac{\delta Q_{ij}[\varphi]}{\delta\varphi_j} - Q_{ij}[\varphi]\frac{\delta F_\varphi}{\delta\varphi_j}\right]. \tag{D.7}$$

Note, by definition,

$$Q_{ij} = - Q_{ji} . \tag{D.8}$$

This anti-symmetry leads to an important property for $V_i[\varphi]$:

$$\sum_i' \frac{\partial}{\partial \varphi_i} (V_i [\varphi] W_\varphi) = 0 . \tag{D.9}$$

Since

$$\sum_i \frac{\partial}{\partial \varphi_i} (V_i W_\varphi) = \sum_{i,j}' \frac{\partial}{\partial \varphi_i} \left[\frac{\partial Q_{ij}[\varphi]}{\partial \varphi_j} - Q_{ij} \frac{\partial F_\varphi}{\partial \varphi_j} \right] W_\varphi$$

$$= \sum_{i,j}' \frac{\partial}{\partial \varphi_i} \left[\left(\frac{\partial Q_{ij}[\varphi]}{\partial \varphi_j} \right) W_\varphi + Q_{ij}[\varphi] \frac{\partial W_\varphi}{\partial \varphi_j} \right]$$

$$= \sum_{i,j}' \frac{\partial^2}{\partial \varphi_i \partial \varphi_j} (Q_{ij} [\varphi] W_\varphi)$$

$$= 0 \tag{D.10}$$

which follows from interchanging i and j in the sum.

APPENDIX E: RELATION BETWEEN R_φ AND f_i

Using the chain-rule for differentiation we can write, using (4.101),

$$\frac{\partial g_\varphi(t)}{\partial t} = -\sum_i{}' \frac{\partial g_\varphi(t)}{\partial \varphi_i}\frac{\partial M_i(t)}{\partial t}. \tag{E.1}$$

If we then use (4.120) and the fact that g_φ is a product of δ-functions we obtain

$$\frac{\partial g_\varphi(t)}{\partial t} = -\sum_i{}' \frac{\delta g_\varphi(t)}{\delta \varphi_i}\left[V_i[M] - \sum_j{}' \Gamma_0^{ij}\frac{\delta F}{\delta M_j} + f_i\right]$$

$$= -\sum_{i,j}{}' \frac{\delta}{\delta \varphi_i}\left\{\left[V_i[\varphi]\delta_{ij} - \Gamma_0^{ij}\frac{\delta F}{\delta \varphi_j} + f_i\delta_{ij}\right] g_\varphi(t)\right\}. \tag{E.2}$$

We can rewrite this in the form

$$\frac{\partial g_\varphi(t)}{\partial t} = -\sum_{i,j}{}' \frac{\delta}{\delta \varphi_i}\left\{\left[V_i[\varphi]\delta_{ij} - \Gamma_0^{ij}\left(\frac{\delta}{\delta\varphi_j} + \frac{\delta F}{\delta \varphi_j}\right)\right] g_\varphi(t)\right\}$$

$$- \sum_{i,j}{}' \frac{\delta}{\delta \varphi_i}\left\{\left(f_i\delta_{ij} + \Gamma_0^{ij}\frac{\delta}{\delta\varphi_j}\right) g_\varphi(t)\right\} \tag{E.3}$$

Comparing (E.3) with (4.114) and (4.115) we can identify

$$R_\varphi(t) = -\sum_{i,j}{}' \frac{\delta}{\delta \varphi_i}\left\{\left(f_i\delta_{ij} + \Gamma_0^{ij}\frac{\delta}{\delta\varphi_j}\right) g_\varphi(t)\right\}. \tag{E.4}$$

REFERENCES

1. D. Forster, D. R. Nelson and M. J. Stephen, Phys. Rev. Lett. 36, 867 (1976) and to be published.
E. D. Siggia, Phys. Rev. A4, 1730 (1977)

2. J. W. Cahn, Act. Met. 9, 795 (1961); 10, 179 (1962).
J. W. Cahn and J. E. Hilliard, Act. Met. 19, 151 (1971).
J. S. Langer, Ann. Phys. (N. Y.) 54, 258 (1969); 65, 53 (1971);
J. S. Langer and M. Bar-on, Ann. Phys. (N. Y.) 78, 421 (1973);
J. S. Langer and L. A. Turski, Phys. Rev. A8, 3230 (1973);
J. S. Langer, M. Bar-on and H. D. Miller, Phys. Rev. A11, 1417 (1975).
K. Kawasaki, Prog. Theo. Phys. 57, 410 (1977).
K. Kawasaki, Prog. Theo. Phys. 57, 826 (1977).

3. K. G. Wilson and J. Kogut, Phys. Repts. 12C, 75 (1974).

4. M. E. Fisher, Rev. Mod. Phys. 46, 597 (1974).

5. K. G. Wilson, Rev. Mod. Phys. 47, 773 (1975).

6. S. Ma, Rev. Mod. Phys. 45, 589 (1973).

7. S. Ma, Modern Theory of Critical Phenomena, 1976 (Benjamin, New York).

8. P. C. Hohenberg and B. I. Halperin, Rev. Mod. Phys., to be published.

9. L. P. Landau and E. M. Lifshitz, Statistical Physics, 1969 (Pergamon, Oxford).

10. See, for example, C. Kittel, Quantum Theory of Solids, 1963 (Wiley, New York) for an elementary introduction.

11. H. E. Stanley, Phys. Rev. 176, 718 (1968).

12. H. E. Stanley, Introduction to Phase Transitions and Critical Phenomena (Oxford University Press, New York, 1971).

13. L. P. Kadanoff et al., Rev. Mod. Phys. 39, 615 (1967).

14. L. P. Kadanoff, Physics (N. Y.) 2, 263 (1966).

15. B. Widom, J. Chem. Phys. 43, 3989 (1965).

16. K. G. Wilson, Phys. Rev. B4, 3174 (1971).

17. L. P. Kadanoff, Annals of Phys. (N. Y.), to be published. Th. Niemejer and J. M. J. van Leeuwen, in Phase Transitions and Critical Phenomena, edited by C. Domb and M. S. Green (Academie, New York, 1976), Vol. VI.

18. There are many examples of cross fertalization between different branches of physics via the renormalization group. One particularly interesting example is the recent introduction into quantum field theory of a lattice gauge theory. This theory of strong interactions includes guarks on lattice sites interacting through "strings" which are generalizations of exchange interactions between spins to include gauge invariance. There is hope that this model will lead to an understanding of guark confinement. See K. G. Wilson, Phys. Rev. D10, 2445 (1974). and L. P. Kadanoff, Rev. Mod. Phys. 49, 267 (1977).

19. The external field conjugate to the order parameter is also a relevant variable. We assume here for simplicity that the external field is fixed at zero.

20. K. G. Wilson and M. E. Fisher, Phys. Rev. Lett. 28, 240 (1972).

21. E. Brézin, J. C. LeGuillou, and J. Zinn-Justin, in Phase Transitions and Critical Phenomena, edited by C. Domb and M. S. Green (Academie, New York, 1976), Vol. VI.

22. L. N. Lipatov, Zh. Eksp. Teor. Fiz. Pis'ma Red 25, 116 (1977).

23. E. Brézin, J. C. Le Guillou and J. Zinn-Justin, Phys. Rev. D15, 1544, 1558 (1977).

24. E. Brézin, private communication.

25. R. Kubo, in Lectures in Theoretical Physics, Vol. I, chapter 4 (Interscience, 1959).

26. L. P. Kadanoff and P. C. Martin, Ann. Phys. (N. Y.) 24, 419 (1963).

27. P. C. Martin, in *Many Body Physics*, edited by C. De Witt and R. Balian, 1968 (Gordon and Breach, N. Y.).

28. R. M. White, *Quantum Theory of Magnetism*, (McGraw-Hill, New York, 1970).

29. D. Forster, *Hydrodynamic Fluctuations, Broken Symmetry and Correlation Functions*, 1975 (W. A. Benjamin Inc. Reading, Mass.).

30. If our first choice for $\psi_i^{(1)}$ has a non-zero average $\langle\Psi_i^{(1)}\rangle$ we then choose as our variable $\Psi_i = \Psi_i^{(1)} - \langle\Psi_i^{(1)}\rangle$ which does have zero average.

31. We can construct the particle density, momentum density and kinetic energy density from $f(\vec{x}, \vec{p})$ by multiplying by 1, $\vec{p}$ and $p^2/2m$, respectively, and integrating over $\vec{p}$.

32. To keep things simple I am assuming we have a classical theory where the Ψ's commute. This is not a necessary requirement but its elimination would necessitate additional, but essentially unilluminating, further formal development.

33. H. Mori, Progr. Theor. Phys. **33**, 423 (1965).

34. G. F. Mazenko, Phys. Rev. **A9**, 360 (1974).

35. G. F. Mazenko and S. Yip, in *Statistical Mechanics*, Pt. B, edited by B. J. Berne (Plenum Press, New York, 1977).

36. P. Resibois and M. De Leener, Phys. Rev. **152**, 305, 318 (1966) and **178**, 806, 819 (1969).

37. K. Kawasaki, Ann. Phys. (N. Y.) **61**, 1 (1970).

38. This result, of course, follows from the projection operator approach due to Mori[33]. We wish to show here that the projection operators, which can present some mathematical problems when trying to develop approximation schemes (See G. F. Mazenko, Phys. Rev. **A7**, 209 (1973) for discussion and further references), are not a necessary part of the development.

39. We define X_{ij}^{-1} by $\sum_{\ell} \chi_{i\ell}^{-1} \chi_{\ell j} = \delta_{ij}$

40. In principle this equation is valid only for $t > o$. Of course we can change our time origin to t_o and then it is valid for $t > to$. We can then let $t_o \to -\infty$. We can then Fourier transform over all times.

41. This is not the actual long time behavior of the velocity autocorrelation function. There is a strong exponential decay for times of order two to three collision times. However for longer times there is a power law decay given by $t^{-d/2}$ where d is the dimensionality for long times. This long time behavior is associated with non-linear couplings similar to those we will treat in the next section. See for example Y. Pomeau and P. Resibois, Phys. Rept. 19C, 64 (1975) and J. R. Dorfman and E. G. D. Cohen, Phys. Rev. Lett. 25, 1257 (1970) and Phys. Rev. A6, 2247 (1972).

42. There is a very nice discussion of the Drude model in N. W. Ashcroft and N. D. Mermin, Solid State Physics, 1976 (Holt, Rinehart and Winston, New York).

43. These ideas are discussed extensively by Forster in ref. 29.

44. Ma[7] has a nice discussion of this point.

45. H. Mori, Prog. Theo. Phys. 28, 763 (1962).

46. L. Van Hove, Phys. Rev. 93, 1374 (1954).

47. I will, for convenience, usually choose units such that the constant C in the Ornstein-Zernike expression for $X(q)$ (See (2. 23)) can be set equal to 1.

48. B. I. Halperin and P. C. Hohenberg, Phys. Rev. Lett. 19, 700 (1967) and Phys. Rev. 177, 952 (1969).

49. V. J. Minkewicz, M. F. Collins, R. Nathans, and G. Shirane, Phys. Rev. 182, 624 (1969).

50. For a list of experimental references see R. Freedman and G. F. Mazenko, Phys. Rev. B13, 4967 (1976).

51. J. V. Sengers, in Critical Phenomena, edited by M. S. Green (Academic, New York, 1971).

52. M. S. Green, J. Chem. Phys. 20, 1281 (1952); 22, 398 (1954).

53. R. W. Zwanzig, Phys. Rev. 124, 983 (1961).

54. H. Mori and H. Fujisaka, Prog. Theor. Phys. 49, 764 (1973).

55. K. Kawasaki, in Critical Phenomena, edited by M. S. Green (Academic, New York, 1971).

56. In this case Ψ_i does not satisfy (4.4).

57. $M_i(t)$ can be identified with with those Fourier components of $\hat{M}_i(t)$ less than the cut-off Λ.

58. S. Ma and G. F. Mazenko, Phys. Rev. Lett. 33, 1383 (1974), and Phys. Rev. B11, 4077 (1975).

59. B. I. Halperin, P. C. Hohenberg and S. Ma, Phys. Rev. Lett. 29, 1548 (1972), Phys. Rev. B10, 139 (1974) and Phys. Rev. B13, 4119 (1976).

60. See the discussion in ref. 8.

61. This follows from (4.14) by expressing J_{ij} and $M_i(t)$ in terms of their Fourier transforms, expanding in powers of the wavenumber, and transforming back to coordinate space.

62. K. Kawasaki, Phys. Rev. 150, 291 (1966), in Statistical Mechanics, edited by S. A. Rice, K. F. Freed, and J. C. Light (University of Chicago, 1972) and refs. 37 and 55.

63. M. Fixman, J. Chem. Phys. 36, 310 (1962).
L. P. Kadanoff and J. Swift, Phys. Rev. 166, 89 (1968).
R. A. Ferrell, Phys. Rev. Lett. 24, 1169 (1970).
R. Zwanzig, in Statistical Mechanics, edited by S. A. Rice, K. F. Freed, and J. C. Light, (University of Chicago Press, Chicago, 1972).

64. P. C. Martin, E. D. Siggia, and H. A. Rose, Phys. Rev. A8, 423 (1973).

65. H. K. Janssen, Z. Physik B23, 377 (1976), R. Bausch, H. J. Janssen and H. Wagner, Z. Physik B24, 113 (1976).

66. K. Kawasaki, Prog. Theor. Phys. 52, 1527 (1974).

67. C. De Dominicis, Nuovo Cimento Lett. 12, 576 (1975)
C. De Dominicis, E. Brezin and J. Zinn-Justin, Phys. Rev. B12, 4945 (1975).

68. G. F. Mazenko, M. Nolan and R. Freedman, Phys. Rev. B, to be published, discuss situations where the fluctuation - dissipation theorem may have a slightly different form.

69. We have already investigated the behavior of u and r_o under the RG in the static case. Since we generate exactly the same statics from our equation of motion we would obtain the same recursion relations.

70. M. J. Nolan and G. F. Mazenko, Phys. Rev. B15, 4471 (1977).

71. B. I. Halperin, P. C. Hohenberg, and E. D. Siggia, Phys. Rev. Lett. 32, 1289 (1974), Phys. Rev. B13, 1299 (1976).
P. C. Hohenberg, B. I. Halperin, and E. D. Siggia, Phys. Rev. B14, 2865 (1976).
E. D. Siggia, Phys. Rev. B11, 4736 (1975).

72. B. I. Halperin, P. C. Hohenberg, and E. D. Siggia, Phys. Rev. Lett. 32, 1289 (1974), Phys. Rev. B13, 1299 (1976).

73. R. Freedman and G. F. Mazenko, Phys. Rev. Lett. 34, 1575 (1975) and Phys. Rev. B13, 4967 (1976).

74. Ref. 73 and E. D. Siggia, B. I. Halperin, and P. C. Hohenberg, Phys. Rev. B13, 2110 (1976).

75. I. M. Gel'fand and A. M. Yaglom, J. Math. Phys. 1, 48 (1960).

NONLINEAR RESPONSE THEORY

Irwin Oppenheim

Chemistry Department, Massachusetts Institute of Technology, Cambridge, Mass., 02139, U.S.A.
and
Chemical Physics Department, Weizmann Institute of Science, Rehovot, Israel

In recent years there has been significant progress in the molecular derivations of transport equations describing the time dependences of macroscopic variables in non-equilibrium systems. One of the most powerful approaches has been linear response theory which has been developed and extensively applied by Kubo (1) and by others (2,3). The aims of this theory, as well as those of the other molecular approaches, are the following: to obtain the form of the macroscopic equations; to obtain explicit molecular expressions for the coefficients appearing in these equations usually in the form of correlation functions; to determine the range of validity of the equations; to extend the range of validity of the equations to more complicated situations; and to compare the transport equations with those obtained phenomenologically using hydrodynamics, thermodynamics of irreversible processes, etc.

In these lectures, we shall use nonlinear response theory to second order to obtain nonlinear transport equations. As an example we shall discuss the nonlinear hydrodynamic equations in detail.

Response theory involves the following set of assumptions and procedures: a) The system of interest is assumed to be at equilibrium in the infinite past; b) Real or ficticious external forces, $\underset{\sim}{F}(\underset{\sim}{r},t)$, are turned on adiabatically starting at $t=-\infty$ and are turned off sharply at $t=0$; c) The relaxation of an <u>appropriate</u> set, $\underset{\sim}{a}(\underset{\sim}{r},t)$, of macroscopic variables from their initial ($t=0$) nonequilibrium values to their final ($t=+\infty$) equilibrium values is studied.

The equations describing this relaxation are the transport equations of interest. We note that the external forces are used only to create a nonequilibrium state at t=0 in which the macroscopic variables have the initial values $\underset{\sim}{a}(\underset{\sim}{r},0)$.

The detailed steps in c) are the following: 1) We obtain an expression for $\underset{\sim}{a}(\underset{\sim}{r},t)$, t>0, as a power series in $\underset{\sim}{F}$. 2) We use the fact that $\underset{\sim}{F}(\underset{\sim}{r},\tau)$ has the form

$$\begin{aligned} \underset{\sim}{F}(\underset{\sim}{r},\tau) &= e^{\varepsilon\tau}\ \underset{\sim}{F}(\underset{\sim}{r}) && \tau<0,\ \varepsilon\to 0+ \\ &= 0 && \tau\geqslant 0 \end{aligned}$$

to express $\underset{\sim}{a}(\underset{\sim}{r},t)$ as a power series in $\underset{\sim}{F}(t=0)$ with time dependent coefficients. 3) We differentiate this expression with respect to t to obtain $\dot{\underset{\sim}{a}}(\underset{\sim}{r},t)$ as a power series in $\underset{\sim}{F}(t=0)$ with time dependent coefficients. 4) We invert 2) to obtain $\underset{\sim}{F}(\underset{\sim}{r})$ as a power series in $\underset{\sim}{a}(t)$ and substitute the result in 3) to obtain $\dot{a}$ as a power series in $\underset{\sim}{a}$ with time dependent coefficients. 5) We use the fact that the set of a's are the only pertinent slowly varying quantities (in space and time) to obtain $\dot{\underset{\sim}{a}}(\underset{\sim}{r},t)$ as a power series in $\underset{\sim}{a}(\underset{\sim}{r},t)$ with time independent coefficients. This is the transport equation of interest. 6) As an alternative procedure which greatly simplifies the final results we write $\underset{\sim}{a}(\underset{\sim}{r},t)$ in a power series in the conjugate forces $\Phi(\underset{\sim}{r},t)$ with time independent coefficients and substitute the result in 5) to obtain $\dot{\underset{\sim}{a}}(\underset{\sim}{r},t)$ as a power series in $\underset{\sim}{\Phi}(t)$ with time independent coefficients.

In order for the response technique to be useful we must be able to demonstrate that: i) The appropriate expansion parameter for $\underset{\sim}{a}(\underset{\sim}{r},t)$ is $\underset{\sim}{F}$ and not $N\underset{\sim}{F}$ where N is the number of particles in the system; ii) the time dependent correlation functions which appear in the coefficients in 1)-3) decay properly as $t\to\pm\infty$; iii) the time dependent correlation functions which appear in 4) decay to zero rapidly compared to the time scales of interest for the time dependence of $\underset{\sim}{a}(\underset{\sim}{r},t)$; and iv) the preparation of the ensemble prescribed by response theory (5a,b) is appropriate for the description of the time dependence of $\underset{\sim}{a}(\underset{\sim}{r},t)$. We shall discuss these points in detail below but mention here that their demonstration depends on the proper choice of the set of variables $\underset{\sim}{a}(\underset{\sim}{r},t)$. Objections (4) to the use of response theory depend on the fact that the expansion parameter for the nonequilibrium distribution function, f(t), is $N\underset{\sim}{F}$ and that even very weak perturbations change the time dependence of f(t) dramatically. We do not believe that these objections apply to the use of response theory for the description of the properties of $\underset{\sim}{a}(\underset{\sim}{r},t)$.

Many of the topics that will be discussed here have appeared in the literature (5) and frequent reference to these publications will be made.

NONLINEAR RESPONSE IN ISOLATED SYSTEMS

We consider a classical system of N particles confined by walls or other external static fields to some region of space V. The state of the system at any time t is given by the phase point

$$X^N(t) \equiv \{q^N(t), p^N(t)\},$$

whose time dependence is governed by the hamiltonian $H_T(X^N,t)$. This total hamiltonian may be written as

$$\begin{aligned} H_T(X^N,t) &\equiv H(X^N)+H_1(X^N,t) \ , \\ H_1(X^N,t) &\equiv -\underset{\sim}{A}(X^N,\underset{\sim}{r})*\underset{\sim}{F}(\underset{\sim}{r},t) \ , \end{aligned} \tag{1}$$

where $H(X^N)$ is the hamiltonian of the unperturbed system, $\underset{\sim}{A}(X^N,\underset{\sim}{r})$ the set of dynamical variables whose macroscopic response is desired,

$$\underset{\sim}{A}(X^N,\underset{\sim}{r})*\underset{\sim}{F}(\underset{\sim}{r},t) \equiv \int d^3\underset{\sim}{r}\ \underset{\sim}{A}(X^N,\underset{\sim}{r})\cdot\underset{\sim}{F}(\underset{\sim}{r},t) \tag{2}$$

and where the time dependence of the forces is

$$\underset{\sim}{F}(r,t) \equiv \left\{ \begin{matrix} 0 & t \geqslant 0 \\ \underset{\sim}{F}(r)e^{\varepsilon t} & t<0;\ \varepsilon \to 0^+ \end{matrix} \right\} , \tag{3}$$

as required by the switching-on procedure described above. The set of variables $\underset{\sim}{A}(X^N,\underset{\sim}{r})$ should include all pertinent slowly varying variables.

The distribution function $f(X^N,t)$ satisfies Liouville's equation

$$\frac{\partial f(X^N,t)}{\partial t} = -iL_T(t)f(X^N,t) \ , \tag{4}$$

where the Liouville operator $L_T(t)$ is defined as

$$\begin{aligned} L_T(t) &\equiv L+L_1(t) \ , \\ L &\equiv i[H,] \ , \\ L_1(t) &\equiv i[H_1(t),] \ , \end{aligned} \tag{5}$$

the symbol [,] indicating a Poisson bracket. Eq. (4) has the formal solution

$$f(X^N,t)=T_+\exp[-\int_{-\infty}^{t} dt_1 iL_T(t_1)]f(X^N,-\infty), \tag{6}$$

where T_+ is a time ordering operator. We assume that the $t=-\infty$ system was at equilibrium which is equivalent to

$$iLf(X^N,-\infty)=0 \ . \tag{7}$$

Expanding (6) to second order in the forces and using (7) we find that

$$f(X^N,t)=f(X^N,-\infty)+\int_{-\infty}^{t} dt_1 e^{-iL(t-t_1)} \tag{8}$$

$$\times[f(X^N,-\infty),\underset{\sim}{A}(X^N,\underset{\sim}{r}_1)]*\underset{\sim}{F}(\underset{\sim}{r}_1,t_1)+\int_{-\infty}^{t} dt_1\int_{-\infty}^{t_1} dt_2 e^{-iL(t-t_1)}[\{e^{-iL(t_1-t_2)}$$

$$\times[f(X^N,-\infty),\underset{\sim}{A}(X^N,\underset{\sim}{r}_2)]\},\underset{\sim}{A}(X^N,\underset{\sim}{r}_1)]\underset{*}{*}\underset{\sim}{F}(\underset{\sim}{r}_1,t_1)\underset{\sim}{F}(\underset{\sim}{r}_2,t_2)+o((NF)^3) \ .$$

Eq. (8) is not a suitable expression for the full N body distribution function, as the expansion parameter is NF not F, and this is consistent with objections raised by others (4). However when (8) is used to compute the average macroscopic response of the dynamical variables, an expansion in powers of F is obtained.

All experiments are performed on either isolated systems or on those in contact with heat or particle reservoirs. The former requires that $f(X,-\infty)$ be a microcanonical distribution function, while the latter necessitates the inclusion of reservoir degrees of freedom. In this paper only isolated systems are considered. Thus we take

$$f_0(X^N)\equiv f(X^N,t=-\infty)\equiv\delta(E-H(X^N))/\Omega(E), \tag{9}$$

where $\delta(X)$ is the Dirac δ function and

$$\Omega(E)\equiv\int dX^N\delta(E-H(X^N)) \ . \tag{10}$$

In the next section the question concerning the equivalence of the relaxation in any isolated system to the average relaxation for some canonical ensemble of systems shall be examined.

Noting that in force-free systems

$$\dot{\underset{\sim}{A}}(\underset{\sim}{r},t) \equiv \frac{d}{dt}\underset{\sim}{A}(X^N(t),\underset{\sim}{r})=iL\underset{\sim}{A}(\underset{\sim}{r},t) \ , \tag{11}$$

(8) and (9) imply (using the definition of [,] and the equations of motion)

$$f(X^N,t)=f_0(X^N)+\Omega^{-1}(E)\{\frac{\partial}{\partial E}(\Omega(E)f_0(X^N))$$

$$\times\int_{-\infty}^{t} dt_1\dot{\underset{\sim}{A}}(\underset{\sim}{r}_1,t_1-t)*\underset{\sim}{F}(\underset{\sim}{r}_1,t_1) \tag{12}$$

$$+\int_{-\infty}^{t} dt_1 \int_{-\infty}^{t_1} dt_2 \, [\frac{\partial^2(\Omega(E)f_0(X^N))}{\partial E^2} \dot{\underset{\sim}{A}}(\underset{\sim}{r}_2,t_2-t)$$

$$\times \dot{\underset{\sim}{A}}(\underset{\sim}{r}_1,t_1-t) \overset{*}{*} \underset{\sim}{F}(\underset{\sim}{r}_1,t_1)\underset{\sim}{F}(\underset{\sim}{r}_2,t_2) + \frac{\partial(\Omega(E)f_0(X^N))}{\partial E} e^{-iL(t-t_1)}$$

$$\times [\dot{\underset{\sim}{A}}(\underset{\sim}{r}_1,t_2-t_1),\underset{\sim}{A}(\underset{\sim}{r}_2,0)] \overset{*}{*} \underset{\sim}{F}(\underset{\sim}{r}_2,t_2)\underset{\sim}{F}(\underset{\sim}{r}_1,t_1)]\}+O((NF)^3) \quad . \quad (12)$$

We want to calculate the average deviation from equilibrium of the dynamical variables $\underset{\sim}{A}(X^N,\underset{\sim}{r})$. That is of

$$\underset{\sim}{a}(\underset{\sim}{r},t) \equiv \int dX^N f(X^N,t)\underset{\sim}{A}(X^N,\underset{\sim}{r}) - \langle\underset{\sim}{A}(\underset{\sim}{r})\rangle, \quad (13)$$

where $\langle\ \rangle$ denotes an equilibrium microcanonical average. Using (12) for $f(X^N,t)$, (13) becomes

$$\underset{\sim}{a}(\underset{\sim}{r},t) \sim \Omega^{-1}(E)[\int_{-\infty}^{t} dt_1 \frac{\partial}{\partial E}\Omega(E)\langle\underset{\sim}{A}(\underset{\sim}{r},0)\dot{\underset{\sim}{A}}(\underset{\sim}{r}_1,t_1-t)\rangle$$

$$* \underset{\sim}{F}(\underset{\sim}{r}_1,t_1) + \int_{-\infty}^{t} dt_1 \int_{-\infty}^{t} dt_2 \{\frac{\partial^2}{\partial E^2}\Omega(E)\langle\underset{\sim}{A}(\underset{\sim}{r},0)\dot{\underset{\sim}{A}}(\underset{\sim}{r}_1,t_1-t)\dot{\underset{\sim}{A}}(\underset{\sim}{r}_2,t_2-t)\rangle$$

$$\overset{*}{*} \underset{\sim}{F}(\underset{\sim}{r}_2,t_2)\underset{\sim}{F}(\underset{\sim}{r}_1,t_1) + \frac{\partial}{\partial E}\Omega(E)\langle\underset{\sim}{A}(\underset{\sim}{r},t-t_1)[\dot{\underset{\sim}{A}}(\underset{\sim}{r}_1,t_2-t_1),\underset{\sim}{A}(\underset{\sim}{r}_2,0)]\rangle$$

$$\overset{*}{*} \underset{\sim}{F}(\underset{\sim}{r}_2,t_2)\underset{\sim}{F}(\underset{\sim}{r}_1,t_1)\}] \quad . \quad (14)$$

We now assume that the system is mixing. That is, for variables $\underset{\sim}{A}(\underset{\sim}{r},t)$:

$$\lim_{t\to\pm\infty}\langle A_\alpha(\underset{\sim}{r},t)A_{\alpha'}(\underset{\sim}{r}')\rangle = \langle A_\alpha(\underset{\sim}{r})\rangle\langle A_{\alpha'}(\underset{\sim}{r}')\rangle \quad . \quad (15)$$

As was discussed in refs. 5b and 6, we can expect Eq. (15) to hold only if the ensemble is constructed such that all normal constants of the motion do not fluctuate. Thus, should the system have other conserved quantities (for example total linear momentum) the correct infinite past distribution function must include additional delta functions pertaining to them.

By making use of the definition (3) for the forces in relaxation processes, Eq. (15), and integrating (14) by parts it is easily shown that for $t \geq 0$

$$\underset{\sim}{a}(r,t) \sim \Omega^{-1}(E)\Big[\frac{\partial}{\partial E}\,\Omega(E)\langle\hat{\underset{\sim}{A}}(\underset{\sim}{r},t)\hat{\underset{\sim}{A}}(\underset{\sim}{r}_1)\rangle * \underset{\sim}{F}(\underset{\sim}{r}_1)$$

$$+\frac{1}{2}\frac{\partial^2}{\partial E^2}\,\Omega(E)\langle\hat{\underset{\sim}{A}}(\underset{\sim}{r},t)\hat{\underset{\sim}{A}}(\underset{\sim}{r}_1)\hat{\underset{\sim}{A}}(\underset{\sim}{r}_2)\rangle \overset{*}{*} \underset{\sim}{F}(\underset{\sim}{r}_2)\underset{\sim}{F}(\underset{\sim}{r}_1) \tag{16}$$

$$-\frac{\partial}{\partial E}\int_{-\infty}^{0} dt_1 \lim_{t_2\to-\infty} \Omega(E)\langle\underset{\sim}{A}(\underset{\sim}{r},t-t_1)[\underset{\sim}{A}(\underset{\sim}{r}_2,t_2),\underset{\sim}{A}(\underset{\sim}{r}_1)]\rangle \overset{*}{*} \underset{\sim}{F}(\underset{\sim}{r}_1)\underset{\sim}{F}(\underset{\sim}{r}_2)]+0(\varepsilon),$$

where we have written $\underset{\sim}{A}(\underset{\sim}{r})$ for $\underset{\sim}{A}(\underset{\sim}{r},t=0)$ and

$$\hat{\underset{\sim}{A}}(\underset{\sim}{r},t) \equiv \underset{\sim}{A}(\underset{\sim}{r},t)-\langle\underset{\sim}{A}(\underset{\sim}{r})\rangle \;. \tag{17}$$

The microcanonical formulation has the advantage that the mixing assumption (Eq. 15) is well studied within the context of ergodic theory (7). In fact Sinai (8) has shown that a system consisting of a finite number of hard spheres in a box is mixing. Thus we can expect Eq. (15) to hold before the thermodynamic limit is taken, a crucial requirement if finite systems are to be studied. Performing an integration by parts in the phase averages in the last form in Eq. (16) and using Eqs. (10) and (15), we find that

$$\Omega^{-1}(E)\,\frac{\partial}{\partial E}\,\Omega(E)\int_{-\infty}^{0} dt_1 \lim_{t_2\to-\infty}\langle\underset{\sim}{A}(\underset{\sim}{r},t-t_1)[\underset{\sim}{A}(\underset{\sim}{r}_2,t_2),\underset{\sim}{A}(\underset{\sim}{r}_1)]\rangle \overset{*}{*} \underset{\sim}{F}(\underset{\sim}{r}_1)\underset{\sim}{F}(\underset{\sim}{r}_2)$$

$$= -\Omega^{-1}(E)\,\frac{\partial}{\partial E}\,\Omega(E)\int_{-\infty}^{0} dt_1[\langle[\underset{\sim}{A}(\underset{\sim}{r},t-t_1),\underset{\sim}{A}(\underset{\sim}{r}_1)]\rangle\langle\underset{\sim}{A}(\underset{\sim}{r}_2)\rangle$$

$$+\,\Omega^{-1}(E)\,\frac{\partial}{\partial E}\,\Omega(E)\langle\underset{\sim}{A}(\underset{\sim}{r},t-t_1)\underset{\sim}{A}(\underset{\sim}{r}_1)\rangle\langle\underset{\sim}{A}(\underset{\sim}{r}_2)\rangle] \overset{*}{*} \underset{\sim}{F}(\underset{\sim}{r}_2)\underset{\sim}{F}(\underset{\sim}{r}_1) \tag{18}$$

$$= -\Omega^{-1}(E)\,\frac{\partial}{\partial E}[\Omega(E)\langle\hat{\underset{\sim}{A}}(\underset{\sim}{r},t)\hat{\underset{\sim}{A}}(\underset{\sim}{r}_1)\rangle * \underset{\sim}{F}(r_1)\,\frac{\partial}{\partial E}\,\langle\underset{\sim}{A}(\underset{\sim}{r}_2)\rangle * \underset{\sim}{F}(\underset{\sim}{r}_2)] \;.$$

Combining (16) and (18) gives

$$\underset{\sim}{a}(\underset{\sim}{r},t) \sim \Omega^{-1}(E)\,\frac{\partial}{\partial E}[\Omega(E)\langle\hat{\underset{\sim}{A}}(\underset{\sim}{r},t)\hat{\underset{\sim}{A}}(\underset{\sim}{r}_1)\rangle * \underset{\sim}{F}(\underset{\sim}{r}_1)$$

$$\times\,[\,1+\frac{\partial}{\partial E}\,\langle\underset{\sim}{A}(\underset{\sim}{r}_2)\rangle * \underset{\sim}{F}(\underset{\sim}{r}_2)] \tag{19}$$

$$+\frac{1}{2}\frac{\partial^2}{\partial E^2}\Omega(E)\langle\hat{\underset{\sim}{A}}(\underset{\sim}{r},t)\hat{\underset{\sim}{A}}(\underset{\sim}{r}_1)\hat{\underset{\sim}{A}}(\underset{\sim}{r}_2)\rangle \overset{*}{*} \underset{\sim}{F}(\underset{\sim}{r}_2)\underset{\sim}{F}(\underset{\sim}{r}_1)] \;.$$

The derivation of Eq. (19) depends on the mixing assumption (Eq. 15), and on the neglect of the $O(F^3)$ terms (which should be valid, providing each of the variables $\underset{\sim}{A}$ may be expressed as a sum whose terms depend only on a small number of single particle variables). Neither of these restrictions need depend on the size of N or E and so Eq. (19) should hold for small systems.

For what will follow we assume that the system is large, that is, $E \sim O(N)$ with $N \gg 1$. Noting that the temperature, $T(N,E,V)$, of a microcanonical ensemble is defined as

$$\beta(N,E,V) \equiv (K_B T(N,E,V))^{-1} = \left(\frac{\partial \ln(\Omega(E))}{\partial E}\right)_{N.V} , \tag{20}$$

where K_B is Boltzmann's constant, Eq. (19) becomes, to lowest order in N^{-1},

$$\begin{aligned} \underset{\sim}{a}(\underset{\sim}{r},t) \sim \beta <\hat{\underset{\sim}{A}}(\underset{\sim}{r},t)\hat{\underset{\sim}{A}}(\underset{\sim}{r}_1)> * \underset{\sim}{F}(\underset{\sim}{r}_1)[1 + \frac{\partial <\hat{\underset{\sim}{A}}(\underset{\sim}{r}_2)>}{\partial E} * \underset{\sim}{F}(\underset{\sim}{r}_2)] \\ + \frac{1}{2} \beta^2 <\hat{\underset{\sim}{A}}(\underset{\sim}{r},t)\hat{\underset{\sim}{A}}(\underset{\sim}{r}_1)\hat{\underset{\sim}{A}}(\underset{\sim}{r}_2)> \overset{*}{*} \underset{\sim}{F}(\underset{\sim}{r}_2)\underset{\sim}{F}(\underset{\sim}{r}_1) + O(N^{-1}) . \end{aligned} \tag{21}$$

In analysing the order of magnitude of the various terms in Eq. (19), we make use of the fact that the correlation functions therein decay to $O(N^{-1})$ whenever any of their arguments are separated beyond some microscopic correlation length. Should the variable $A_\alpha(\underset{\sim}{r},t)$ be conserved, Eq. (21) implies that

$$\int_V d\underset{\sim}{r}\ a_\alpha(\underset{\sim}{r},t) = 0 . \tag{22}$$

In the next section we show how Eq. (21) may be rewritten in terms of canonical or grand canonical correlation functions and averages.

TRANSFORMATION TO OTHER ENSEMBLES

For many applications and for the sake of comparison we would like to express Eq. (21) in terms of ensemble averages other than the microcanonical. To do this we extend the procedure for finding the asymptotic forms of reduced canonical distribution functions of Lebowitz and Percus (9) to arbitrary correlation functions.

We consider two ensembles Γ and Γ' which differ in the fact that the set of variables $\underset{\sim}{C}(X^N)$ fluctuate in the latter but not in the former. In addition it is required that the variables $\underset{\sim}{C}$ are $O(N)$ in the Γ ensemble, and that $<\underset{\sim}{C}>_\Gamma = <\underset{\sim}{C}>_{\Gamma'}$, where the symbols $< >_\Gamma$ and $<>_{\Gamma'}$ represent averages in the Γ and Γ' ensembles, respectively. Averages in the two ensembles are related by

$$<B>_{\Gamma'} = \int d\underset{\sim}{C} <B>_\Gamma P(\underset{\sim}{C},\underset{\sim}{\gamma}) , \tag{23}$$

$$P(\underset{\sim}{C},\underset{\sim}{\gamma})\equiv[g(\underset{\sim}{C})/\Lambda(\underset{\sim}{\gamma})]\exp[-\underset{\sim}{\gamma}\cdot\underset{\sim}{C}] \ . \tag{24}$$

The γ_i are intensive variables conjugate to the C_i,$g(\underset{\sim}{C})$ is a degeneracy factor, and $\Lambda(\underset{\sim}{\gamma})$ is a normalization constant.

For large systems at finite temperature and away from critical points, the function $P(\underset{\sim}{C},\underset{\sim}{\gamma})$ should be very sharply peaked near $\underset{\sim}{C}=\langle\underset{\sim}{C}\rangle_{\Gamma'}$. This fact allows us to expand $\langle B\rangle_\Gamma$ in Eq. (23) in a Taylor series about $\underset{\sim}{C}=\langle\underset{\sim}{C}\rangle_{\Gamma'}$. Thus the two ensemble averages may be related from Eq. (23) by

$$\langle B\rangle_{\Gamma'}=\langle B\rangle_\Gamma+\frac{1}{2}\langle\underset{\sim}{\hat{C}}\underset{\sim}{\hat{C}}\rangle_{\Gamma'}:\left.\frac{\partial^2\langle B\rangle_\Gamma}{\partial\underset{\sim}{C}\partial\underset{\sim}{C}}\right|_{\underset{\sim}{C}=\langle\underset{\sim}{C}\rangle_{\Gamma'}}$$
$$+\frac{1}{3!}\langle\underset{\sim}{\hat{C}}\underset{\sim}{\hat{C}}\underset{\sim}{\hat{C}}\rangle_{\Gamma'}\vdots\left.\frac{\partial^3\langle B\rangle_\Gamma}{\partial\underset{\sim}{C}\partial\underset{\sim}{C}\partial\underset{\sim}{C}}\right|_{\underset{\sim}{C}=\langle\underset{\sim}{C}\rangle_{\Gamma'}}+O(N^{-3}), \tag{25}$$

where we have retained terms to $O(N^{-2})$. Rearranging Eq. (25) and iterating, we obtain

$$\langle B\rangle_\Gamma=\langle B\rangle_{\Gamma'}-\frac{1}{2}\langle\underset{\sim}{\hat{C}}\underset{\sim}{\hat{C}}\rangle_{\Gamma'}:\frac{\partial^2\langle B\rangle_{\Gamma'}}{\partial\langle\underset{\sim}{C}\rangle_{\Gamma'}\partial\langle\underset{\sim}{C}\rangle_{\Gamma'}}$$
$$-\frac{1}{3!}\langle\underset{\sim}{\hat{C}}\underset{\sim}{\hat{C}}\underset{\sim}{\hat{C}}\rangle_{\Gamma'}\vdots\frac{\partial^3\langle B\rangle_{\Gamma'}}{\partial\langle\underset{\sim}{C}\rangle_{\Gamma'}\partial\langle\underset{\sim}{C}\rangle_{\Gamma'}\partial\langle\underset{\sim}{C}\rangle_{\Gamma'}} \tag{26}$$
$$+\frac{1}{4}\langle\underset{\sim}{\hat{C}}\underset{\sim}{\hat{C}}\rangle_{\Gamma'}:\frac{\partial^2}{\partial\langle\underset{\sim}{\hat{C}}\rangle_{\Gamma'}\partial\langle\underset{\sim}{\hat{C}}\rangle_{\Gamma'}}\left(\langle\underset{\sim}{\hat{C}}\underset{\sim}{\hat{C}}\rangle_{\Gamma'}:\frac{\partial^2\langle B\rangle_{\Gamma'}}{\partial\langle\underset{\sim}{\hat{C}}\rangle_{\Gamma'}\partial\langle\underset{\sim}{\hat{C}}\rangle_{\Gamma'}}\right)+O(N^{-3}).$$

From Eqs. (23) and (24) it is easily shown that

$$\frac{\partial\langle C_i\rangle_{\Gamma'}}{\partial\gamma_j}=\langle\hat{C}_i\hat{C}_j\rangle_{\Gamma'} \ , \tag{27}$$

and

$$\frac{\partial\langle\hat{C}_i\hat{C}_j\rangle_{\Gamma'}}{\partial\gamma_k}=\langle\hat{C}_i\hat{C}_j\hat{C}_k\rangle_{\Gamma'} \ . \tag{27'}$$

We first consider some binary correlation function of the form

$$\langle\hat{B}\hat{D}\rangle_\Gamma=\langle BD\rangle_\Gamma-\langle B\rangle_\Gamma\langle D\rangle_\Gamma \ , \tag{28}$$

where B and D are any dynamical variables. Employing Eq. (26) to $O(N^{-1})$ it follows that

$$\langle\hat{B}\hat{D}\rangle_\Gamma = \langle\hat{B}\hat{D}\rangle_{\Gamma'} - \frac{1}{2}\langle\hat{\underset{\sim}{C}}\hat{\underset{\sim}{C}}\rangle_{\Gamma'} : \frac{\partial^2\langle\hat{B}\hat{D}\rangle_{\Gamma'}}{\partial\langle\underset{\sim}{C}\rangle_{\Gamma'}\partial\langle\underset{\sim}{C}\rangle_{\Gamma'}}$$

$$- \langle\hat{\underset{\sim}{C}}\hat{\underset{\sim}{C}}\rangle_{\Gamma'} : \frac{\partial\langle B\rangle_{\Gamma'}\partial\langle D\rangle_{\Gamma'}}{\partial\langle\underset{\sim}{C}\rangle_{\Gamma'}\partial\langle\underset{\sim}{C}\rangle_{\Gamma'}} + O(N^{-2}) , \tag{29}$$

which by Eq. (27) may be written as

$$\langle\hat{B}\hat{D}\rangle_\Gamma = \langle\hat{B}(1-P)\hat{D}\rangle_{\Gamma'}$$

$$- \frac{1}{2}\langle\hat{\underset{\sim}{C}}\hat{\underset{\sim}{C}}\rangle_{\Gamma'} : \frac{\partial^2\langle\hat{B}\hat{D}\rangle_{\Gamma'}}{\partial\langle\underset{\sim}{C}\rangle_{\Gamma'}\partial\langle\underset{\sim}{C}\rangle_{\Gamma'}} + O(N^{-2}) . \tag{29'}$$

In this last expression the operator P is defined as

$$PB \equiv \langle B\hat{\underset{\sim}{C}}\rangle_{\Gamma'}\cdot\langle\hat{\underset{\sim}{C}}\hat{\underset{\sim}{C}}\rangle_{\Gamma'}^{-1}\cdot\hat{\underset{\sim}{C}} \tag{30}$$

and is simply the projection operator onto the fluctuating parts of the variables $\underset{\sim}{C}$ in the Γ' ensemble.

We also need to know to $O(N^{-2})$ how triple correlation functions of the form $\langle\hat{B}\hat{D}\hat{E}\rangle_\Gamma$ may be expressed in terms of Γ' quantities. This is accomplished by using Eq. (26) in a rather tedious calculation, and the result is given in the appendix of Ref. 5c.

Taking the Γ' ensemble to be microcanonical and using Eqs. (26) and (29'), Eq. (21) becomes

$$\underset{\sim}{a}(\underset{\sim}{r},t) = \beta\langle\hat{\underset{\sim}{A}}(\underset{\sim}{r},t)(1-P)\hat{\underset{\sim}{A}}(\underset{\sim}{r}_1)\rangle_{\Gamma'} * \underset{\sim}{F}(\underset{\sim}{r}_1)$$

$$\times[1 + \frac{\partial\langle\underset{\sim}{A}(\underset{\sim}{r}_2)\rangle_{\Gamma'}}{\partial\langle E\rangle_{\Gamma'}} * \underset{\sim}{F}(\underset{\sim}{r}_2)] + \frac{1}{2}\beta^2[\langle[(1-P)\hat{\underset{\sim}{A}}(\underset{\sim}{r},t)]\hat{\underset{\sim}{A}}(\underset{\sim}{r}_1)\hat{\underset{\sim}{A}}(\underset{\sim}{r}_2)\rangle_{\Gamma'}$$

$$-\langle\hat{\underset{\sim}{A}}(\underset{\sim}{r},t)[P\hat{\underset{\sim}{A}}(\underset{\sim}{r}_1)]\hat{\underset{\sim}{A}}(\underset{\sim}{r}_2)\rangle_{\Gamma'} - \langle\hat{\underset{\sim}{A}}(\underset{\sim}{r},t)\hat{\underset{\sim}{A}}(\underset{\sim}{r}_1)[P\underset{\sim}{A}(\underset{\sim}{r}_2)]\rangle_{\Gamma'}$$

$$+ 2\langle\hat{C}_i\hat{C}_j\hat{C}_k\rangle_{\Gamma'} \frac{\partial\langle\underset{\sim}{A}(\underset{\sim}{r})\rangle_{\Gamma'}\partial\langle\underset{\sim}{A}(\underset{\sim}{r}_1)\rangle_{\Gamma'}\partial\langle\underset{\sim}{A}(\underset{\sim}{r}_2)\rangle_{\Gamma'}}{\partial\langle C_i\rangle_{\Gamma'}\quad\partial\langle C_j\rangle_{\Gamma'}\quad\partial\langle C_k\rangle_{\Gamma'}} \tag{31}$$

$$+ \frac{1}{2}\langle\hat{C}_i\hat{C}_j\rangle_{\Gamma'}\langle\hat{C}_k\hat{C}_l\rangle_{\Gamma'}\left[\frac{\partial^2\langle\underset{\sim}{A}(\underset{\sim}{r})\rangle_{\Gamma'}}{\partial\langle C_i\rangle_{\Gamma'}\partial\langle C_j\rangle_{\Gamma'}}\right.$$

$$\times \frac{\partial\langle\underset{\sim}{A}(\underset{\sim}{r}_1)\rangle_{\Gamma'}}{\partial\langle C_k\rangle_{\Gamma'}}\frac{\partial\langle\underset{\sim}{A}(\underset{\sim}{r}_2)\rangle_{\Gamma'}}{\partial\langle C_l\rangle_{\Gamma'}} + \frac{4\partial^2\langle\underset{\sim}{A}(\underset{\sim}{r})\rangle_{\Gamma'}}{\partial\langle C_i\rangle_{\Gamma'}\partial\langle C_k\rangle_{\Gamma'}}\frac{\partial\langle\underset{\sim}{A}(\underset{\sim}{r}_1)\rangle_{\Gamma'}\partial\langle\underset{\sim}{A}(\underset{\sim}{r}_2)\rangle_{\Gamma'}}{\partial\langle C_j\rangle_{\Gamma'}\quad\partial\langle C_l\rangle_{\Gamma'}}$$

$$+ \begin{matrix} \text{2 other permutations of the} \\ \text{position of the double derivatives} \end{matrix}]] \underset{\sim}{\overset{*}{*}} F(\underset{\sim}{r}_2) \underset{\sim}{F}(\underset{\sim}{r}_1) + O(N^{-1}), \quad (31)$$

where repeated indices are to be summed. If the set of variables C includes the energy and we impose the condition

$$\frac{\partial \langle \underset{\sim}{A}(\underset{\sim}{r}) \rangle_{\Gamma'}}{\partial \langle C_i \rangle_{\Gamma'}} * \underset{\sim}{F}(\underset{\sim}{r}) = 0 \quad (32)$$

for all i, Eq. (31) simplifies to

$$\begin{aligned} \underset{\sim}{a}(\underset{\sim}{r},t) &= \beta \langle \hat{\underset{\sim}{A}}(\underset{\sim}{r},t) \hat{\underset{\sim}{A}}(\underset{\sim}{r}_1) \rangle_{\Gamma'} * \underset{\sim}{F}(\underset{\sim}{r}_1) \\ &+ \frac{1}{2} \beta^2 \langle [(1-P) \hat{\underset{\sim}{A}}(\underset{\sim}{r},t)] \hat{\underset{\sim}{A}}(\underset{\sim}{r}_1) \hat{\underset{\sim}{A}}(\underset{\sim}{r}_2) \rangle_{\Gamma'} \overset{*}{*} \underset{\sim}{F}(\underset{\sim}{r}_2) \underset{\sim}{F}(\underset{\sim}{r}_1). \end{aligned} \quad (33)$$

It should be noted that Eq. (32) is equivalent to

$$\langle \hat{\underset{\sim}{C}} \hat{\underset{\sim}{A}}(\underset{\sim}{r}) \rangle_{\Gamma'} * \underset{\sim}{F}(\underset{\sim}{r}) = 0, \quad (34)$$

which is a consequence of Eq. (27). When Γ' is a canonical ensemble (C=E) in the homogeneous limit, the response equations (33) and constraint equation (32) are equivalent to those obtained in Ref. (5b) through correcting for projections onto conserved variables.

Let $\underset{\sim}{C}(\underset{\sim}{r},t)$ be the average displacement from equilibrium of the densities of the variables $\underset{\sim}{C}$. According to Eqs. (33) and (34)

$$\int_V d^3\underset{\sim}{r}\ \underset{\sim}{C}(\underset{\sim}{r},t) = \beta \langle \hat{\underset{\sim}{C}} \hat{\underset{\sim}{A}}(\underset{\sim}{r}) \rangle_{\Gamma'} * \underset{\sim}{F}(\underset{\sim}{r}) = 0. \quad (35)$$

Since the actual relaxation occurs in an isolated system the integrals of the densities of the nonfluctuating quantities must never change from their equilibrium values, independent of the choice of ensemble used to express the correlation functions. This is shown by Eq. (35).

TRANSPORT EQUATIONS

The first step in obtaining the transport equations is to express the fictitious forces in terms of the variables $\underset{\sim}{a}(\underset{\sim}{r},t)$. This is most easily accomplished when Γ' is a grand canonical ensemble. Here the correlation functions can be expected to decay exactly to zero beyond some microscopic correlation length and the linear transformation

$$\underset{\sim}{a}^{(1)}(\underset{\sim}{r},t) \equiv \beta <\hat{A}(\underset{\sim}{r},t)\hat{A}(\underset{\sim}{r}_1)>_{G.C.} * \underset{\sim}{F}(\underset{\sim}{r}_1) \equiv \beta K(\underset{\sim}{r}|\underset{\sim}{r}_1,t) * \underset{\sim}{F}(\underset{\sim}{r}_1) \qquad (36)$$

can be inverted. In this section the symbol $<>_{G.C.}$ denotes a grand canonical average. Defining the inverse of the kernel in Eq. (36) through

$$\underset{\sim}{K}^{-1}(\underset{\sim}{r}|\underset{\sim}{r}_1;t) * <\hat{A}(\underset{\sim}{r}_1,t)\hat{A}(\underset{\sim}{r}')>_{G.C.} = \delta(\underset{\sim}{r}-\underset{\sim}{r}')\underset{\sim}{1} \qquad (37)$$

and using Eq. (33), it is easily shown that

$$\beta \underset{\sim}{F}(\underset{\sim}{r}) = \underset{\sim}{K}^{-1}(\underset{\sim}{r}|\underset{\sim}{r}_1;t) * [\underset{\sim}{a}(\underset{\sim}{r}_1,t) - \frac{1}{2} <[(1-P)\hat{A}(\underset{\sim}{r}_1,t)]\hat{A}(\underset{\sim}{r}_2)\hat{A}(\underset{\sim}{r}_3)>_{G.C.}$$

$$* (\underset{\sim}{K}^{-1}(\underset{\sim}{r}_3|\underset{\sim}{r}_4;t) * \underset{\sim}{a}(\underset{\sim}{r}_4,t))(\underset{\sim}{K}^{-1}(\underset{\sim}{r}_2|\underset{\sim}{r}_5;t) * \underset{\sim}{a}(\underset{\sim}{r}_5,t))] + O(a^3). \qquad (38)$$

Since for a grand canonical ensemble the new fluctuating variables $\underset{\sim}{C}$ are the total number and energy, both constants of the motion, Eq. (33) implies

$$\dot{\underset{\sim}{a}}(\underset{\sim}{r},t) = \beta <\dot{\hat{A}}(\underset{\sim}{r},t)\hat{A}(\underset{\sim}{r}_1)>_{G.C.} * \underset{\sim}{F}(\underset{\sim}{r}_1)$$

$$+ \frac{1}{2} <\dot{\hat{A}}(\underset{\sim}{r},t)\hat{A}(\underset{\sim}{r}_1)\hat{A}(\underset{\sim}{r}_2)>_{G.C.} \overset{*}{*} \underset{\sim}{F}(\underset{\sim}{r}_2)\underset{\sim}{F}(\underset{\sim}{r}_1), \qquad (39)$$

which by Eq. (38) becomes to $O(a^2)$

$$\dot{\underset{\sim}{a}}(\underset{\sim}{r},t) = \underset{\sim}{M}(\underset{\sim}{r}|\underset{\sim}{r}_1;t) * \underset{\sim}{a}(\underset{\sim}{r}_1,t)$$

$$+ \underset{\blacksquare}{N}(\underset{\sim}{r}|\underset{\sim}{r}_1|\underset{\sim}{r}_2;t) \overset{*}{*} \underset{\sim}{a}(\underset{\sim}{r}_2,t)\underset{\sim}{a}(\underset{\sim}{r}_1,t). \qquad (40)$$

In the last equation

$$\underset{\sim}{M}(\underset{\sim}{r}|\underset{\sim}{r}_1;t) \equiv <\dot{\hat{A}}(\underset{\sim}{r},t)\hat{A}(\underset{\sim}{r}_2)>_{G.C.} * \underset{\sim}{K}^{-1}(\underset{\sim}{r}_2|\underset{\sim}{r}_1;t) \qquad (41)$$

and

$$\underset{\equiv}{N}(\underset{\sim}{r}|\underset{\sim}{r}_1|\underset{\sim}{r}_2;t) \equiv <\dot{\hat{A}}(\underset{\sim}{r},t)\hat{A}(\underset{\sim}{r}_3)\hat{A}(\underset{\sim}{r}_4)>_{G.C.} \overset{*}{*} \underset{\sim}{K}^{-1}(\underset{\sim}{r}_4|\underset{\sim}{r}_1;t)\underset{\sim}{K}^{-1}(\underset{\sim}{r}_3|\underset{\sim}{r}_2,t)$$

$$- \frac{1}{2} <\dot{\hat{A}}(\underset{\sim}{r},t)\hat{A}(\underset{\sim}{r}_3)>_{G.C.} * \underset{\sim}{K}^{-1}(\underset{\sim}{r}_3|\underset{\sim}{r}_4;t) \qquad (42)$$

$$* <[(1-P)\hat{A}(\underset{\sim}{r}_4,t)]\hat{A}(\underset{\sim}{r}_5)\hat{A}(\underset{\sim}{r}_6)>_{G.C.} \overset{*}{*} \underset{\sim}{K}^{-1}(\underset{\sim}{r}_5|\underset{\sim}{r}_1;t)\underset{\sim}{K}^{-1}(\underset{\sim}{r}_6|\underset{\sim}{r}_2;t).$$

The projection operator in Eq. (42) may be omitted by using the fact that

$$\dot{\underline{\underline{K}}}*\underline{\underline{K}}^{-1}+\underline{\underline{K}}*\dot{\underline{\underline{K}}}^{-1}=0 \tag{43}$$

which follows from Eq. (37). Thus

$$\begin{aligned}&\dot{\underline{\underline{K}}}(\underset{\sim}{r}|\underset{\sim}{r}_3;t)*\underline{\underline{K}}^{-1}(\underset{\sim}{r}_3|\underset{\sim}{r}_4;t)*\langle\underset{\sim}{A}(\underset{\sim}{r}_4,t)\hat{\underset{\sim}{C}}\rangle\\ &= \underline{\underline{K}}(\underset{\sim}{r}|\underset{\sim}{r}_3;t)*\underline{\underline{K}}^{-1}(\underset{\sim}{r}_3|\underset{\sim}{r}_4;t)*\langle\dot{A}(\underset{\sim}{r}_4,t)\hat{C}\rangle\\ &=0 \ .\end{aligned} \tag{44}$$

The set of variables $\underset{\sim}{A}(\underset{\sim}{r},t)$ was assumed to be slowly varying. Therefore, it is useful to rewrite Eq. (40) making as many time derivatives as possible explicit. We write

$$\underline{\underline{M}}(\underset{\sim}{r}|\underset{\sim}{r}_1;t)=\underline{\underline{M}}(\underset{\sim}{r}|\underset{\sim}{r}_1;0)+\int_0^t \dot{\underline{\underline{M}}}(\underset{\sim}{r}|\underset{\sim}{r}_1;t)dt_1 \tag{45}$$

and a similar equation for $\underline{\underline{N}}$. Substitution of Eq. (45) into Eq. (40) yields

$$\begin{aligned}\dot{\underset{\sim}{a}}(\underset{\sim}{r},t)=&\underline{\underline{M}}(\underset{\sim}{r}|\underset{\sim}{r}_1;0)*\underset{\sim}{a}(\underset{\sim}{r}_1,t)+\underline{\underline{N}}(\underset{\sim}{r}|\underset{\sim}{r}_1|\underset{\sim}{r}_2;0)\overset{*}{*}\underset{\sim}{a}(\underset{\sim}{r}_2,t)\underset{\sim}{a}(\underset{\sim}{r}_1,t)\\ &-\int_0^t dt_1\langle\underset{\sim}{I}'(\underset{\sim}{r},t_1)\underset{\sim}{I}'(\underset{\sim}{r}_1)\rangle_{G.C.}*\underline{\underline{K}}^{-1}(\underset{\sim}{r}_1|\underset{\sim}{r}_2;0)*\underset{\sim}{a}(\underset{\sim}{r}_2,t)\\ &+\int_0^t dt_1[\tfrac{1}{2}\langle\underset{\sim}{I}'(\underset{\sim}{r},t_1)\underset{\sim}{I}'(\underset{\sim}{r}_1)\rangle_{G.C.}*\underline{\underline{K}}^{-1}(\underset{\sim}{r}_1|\underset{\sim}{r}_2;0)\\ &*\langle\hat{\underset{\sim}{A}}(\underset{\sim}{r}_2)\hat{\underset{\sim}{A}}(\underset{\sim}{r}_3)\hat{\underset{\sim}{A}}(\underset{\sim}{r}_4)\rangle_{G.C.}-\tfrac{1}{2}\langle\underset{\sim}{I}'(\underset{\sim}{r},t_1)\underset{\sim}{I}'(\underset{\sim}{r}_3)\hat{\underset{\sim}{A}}(\underset{\sim}{r}_4)\rangle_{G.C.}\\ &-\tfrac{1}{2}\langle\underset{\sim}{I}'(\underset{\sim}{r},t_1)\hat{\underset{\sim}{A}}(\underset{\sim}{r}_3)\underset{\sim}{I}'(\underset{\sim}{r}_4)\rangle_{G.C.}]\overset{*}{*}(\underline{\underline{K}}^{-1}(\underset{\sim}{r}_4|\underset{\sim}{r}_5;0)*\underset{\sim}{a}(\underset{\sim}{r}_5,t))(\underline{\underline{K}}^{-1}\\ &(\underset{\sim}{r}_3|\underset{\sim}{r}_6;0)*\underset{\sim}{a}(\underset{\sim}{r}_6,t)),\end{aligned} \tag{46}$$

to second order in the smallness parameter characterizing the time rate of change of the slow variables. The generalized dissipative current, $\underset{\sim}{I}'$, is defined as

$$\underset{\sim}{I}'(\underset{\sim}{r},t)\equiv\dot{\underset{\sim}{A}}(\underset{\sim}{r},t)-\underline{\underline{M}}(\underset{\sim}{r}|\underset{\sim}{r}_1;0)*\hat{\underset{\sim}{A}}(X^N,\underset{\sim}{r}_1) \ . \tag{47}$$

These dissipative currents have the property of being orthogonal to the slowly varying quantities. That is,

$$<\underset{\sim}{J}'(\underset{\sim}{r},t)\hat{\underset{\sim}{A}}(\underset{\sim}{r}')>_{G.C.}=0 \tag{48}$$

to the first order in the smallness parameter characterizing the magnitude of the time derivative of $\underset{\sim}{A}$. If the set of variables $\underset{\sim}{A}$ contains all of the pertinent linear slow variables in the system, the time dependent correlation functions in Eq. (46) decay sufficiently rapidly to zero so that the upper limit of integration can be extended to infinity. This result is correct to second order in the smallness parameter characterizing $\dot{\underset{\sim}{A}}$ (10).

Eq. (46) is the desired transport equation for the set of variables $\underset{\sim}{a}$. We find that $\dot{\underset{\sim}{a}}$ can be expressed as a power series in $\underset{\sim}{a}$ with time independent coefficients. Eq. (46) is general since it applies to systems which are inhomogeneous in equilibrium. It is correct to second order in deviations from equilibrium $O(a^2)$ and to second order in the smallness parameter characterizing $\dot{\underset{\sim}{A}}$.

Note the important fact that the form of $\underset{\sim}{a}(\underset{\sim}{r},t)$, Eq. (33), does not arise from a local equilibrium distribution function at t=0 because of the presence of the projection operator. However, since the projection operator disappears in Eq. (42), it will be possible to write the transport equation using local equilibrium averages.

In order to accomplish this we introduce the force-like quantities $\underset{\sim}{\Phi}(\underset{\sim}{r},t)$ defined by:

$$\begin{aligned}\underset{\sim}{a}(\underset{\sim}{r},t)\equiv&\beta<\hat{\underset{\sim}{A}}(\underset{\sim}{r})\hat{\underset{\sim}{A}}(\underset{\sim}{r}_1)>_{G.C.}*\underset{\sim}{\Phi}(\underset{\sim}{r}_1,t)\\&+\frac{1}{2}\beta^2<\hat{\underset{\sim}{A}}(\underset{\sim}{r})\hat{\underset{\sim}{A}}(\underset{\sim}{r}_1)\hat{\underset{\sim}{A}}(\underset{\sim}{r}_2)>_{G.C.}\overset{*}{*}\ \underset{\sim}{\Phi}(\underset{\sim}{r}_2,t)\underset{\sim}{\Phi}(\underset{\sim}{r}_1,t).\end{aligned} \tag{49}$$

Using Eq. (37), we find

$$\begin{aligned}\beta\underset{\sim}{\Phi}(\underset{\sim}{r},t)=&\underset{\approx}{K}^{-1}(\underset{\sim}{r}|\underset{\sim}{r}_1;0)*[\underset{\sim}{a}(\underset{\sim}{r}_1,t)\\&-\frac{1}{2}<\hat{\underset{\sim}{A}}(\underset{\sim}{r}_1)\hat{\underset{\sim}{A}}(\underset{\sim}{r}_2)\hat{\underset{\sim}{A}}(\underset{\sim}{r}_3)>_{G.C.}\overset{*}{*}(\underset{\approx}{K}^{-1}(\underset{\sim}{r}_3|\underset{\sim}{r}_4;0)*\underset{\sim}{a}(\underset{\sim}{r}_4,t))\\&\times(\underset{\approx}{K}^{-1}(\underset{\sim}{r}_2|\underset{\sim}{r}_5;0)*\underset{\sim}{a}(\underset{\sim}{r}_5,t))]+O(a^3)\ .\end{aligned} \tag{50}$$

Note that $\underset{\sim}{\Phi}(\underset{\sim}{r},o)$ is not equal to $\underset{\sim}{F}(\underset{\sim}{r})$. Eq. (46) in terms of the Φ's becomes

$$\dot{\underset{\sim}{a}}(\underset{\sim}{r},t)=\beta<\dot{\hat{\underset{\sim}{A}}}(\underset{\sim}{r})\hat{\underset{\sim}{A}}(\underset{\sim}{r}_1)>_{G.C.}*\underset{\sim}{\Phi}(\underset{\sim}{r}_1,t)$$

$$+ \frac{1}{2} \beta^2 <\hat{\dot{A}}(\underset{\sim}{r})\hat{A}(\underset{\sim}{r}_1)\hat{A}(\underset{\sim}{r}_2)>_{G.C.} *\overset{*}{\underset{\sim}{\Phi}}(\underset{\sim}{r}_2,t)\underset{\sim}{\Phi}(\underset{\sim}{r}_1,t)$$

$$- \int_o^\infty dt_1 [\beta<\underset{\sim}{l}'(\underset{\sim}{r},t_1)\underset{\sim}{l}'(\underset{\sim}{r}_1)>_{G.C.} *\underset{\sim}{\Phi}(\underset{\sim}{r}_1,t) \tag{51}$$

$$+ \frac{1}{2} \beta^2 [<\underset{\sim}{l}'(\underset{\sim}{r},t)\underset{\sim}{l}'(\underset{\sim}{r}_1)\hat{A}(\underset{\sim}{r}_2)>_{G.C.} +<\underset{\sim}{l}'(\underset{\sim}{r},t)\hat{A}(\underset{\sim}{r}_1)\underset{\sim}{l}'(\underset{\sim}{r}_2)>_{G.C.}]$$

$$\overset{*}{\underset{\sim}{\Phi}}(\underset{\sim}{r}_2,t)\underset{\sim}{\Phi}(\underset{\sim}{r}_1,t)].$$

The force-like quantities may be used to construct a local equilibrium distribution function, $f_L(X^N,t)$:

$$f_L(X^N,t) \equiv \frac{f_{G.C.}(X^N)\exp[\beta A(X^N,\underset{\sim}{r})*\underset{\sim}{\Phi}(\underset{\sim}{r},t)]}{\sum_{N=0}^\infty \int dX^N f_{G.C.}(X^N)\exp[\beta A(X^N,\underset{\sim}{r})*\underset{\sim}{\Phi}(\underset{\sim}{r},t)]} \tag{52}$$

where $f_{G.C.}(X^N)$ is the grand canonical distribution function. From Eq. (49) it follows that

$$\underset{\sim}{a}(\underset{\sim}{r},t)=<\hat{A}(\underset{\sim}{r})>_L(t)+0(\Phi^3) \tag{53}$$

and from Eq. (51)

$$\underset{\sim}{\dot{a}}(\underset{\sim}{r},t)=<\hat{\dot{A}}(\underset{\sim}{r})>_L(t)-\beta \int_o^\infty dt_1 <\underset{\sim}{l}'(\underset{\sim}{r},t_1)\underset{\sim}{l}'(\underset{\sim}{r}_1)>_L(t)*\underset{\sim}{\Phi}(\underset{\sim}{r}_1,t)+0(\Phi^3), \tag{54}$$

where $<>_L(t)$ represents an average using the distribution function given by Eq. (52). The quantities $\underset{\sim}{\Phi}(\underset{\sim}{r},t)$ are now seen to be the thermodynamic forces conjugate to the macroscopic variables $\underset{\sim}{a}(\underset{\sim}{r},t)$.

HYDRODYNAMIC EQUATIONS

We shall now specialize the results of the last section to derive the hydrodynamic equations for a system which is homogeneous in equilibrium. All of the total equilibrium averages below are grand canonical and we drop the subscript G.C.

The pertinent slow dynamical variables, $\underset{\sim}{A}$, are the number density

$$N(\underset{\sim}{r},t)= \sum_{j=1}^N \delta(\underset{\sim}{r}-\underset{\sim}{r}_j(t)) \ , \tag{55}$$

the momentum density

$$\underset{\sim}{P}(\underset{\sim}{r},t)= \sum_{j=1}^{N} \underset{\sim}{p}_j(t)\delta(\underset{\sim}{r}-\underset{\sim}{r}_j(t)), \tag{56}$$

and the energy density

$$E(\underset{\sim}{r},t)= \sum_{j=1}^{N} e_j(t)\delta(\underset{\sim}{r}-\underset{\sim}{r}_j(t)) \tag{57}$$

where $r_j(t)$ is the position of particle j at time t, $p_j(t)$ is the momentum of particle j at time t , and the energy of particle j is

$$e_j = \frac{p_j^2}{2m} + \frac{1}{2} \sum_{\ell \neq j} u_{\ell j} \tag{58}$$

where $u_{\ell j}$ $(r_{\ell j})$ is the short-range interaction potential between particles ℓ and j.

The time derivatives of these variables are easily obtained. They are

$$\dot{N}(\underset{\sim}{r},t)= - \nabla_r \cdot \underset{\sim}{P}(\underset{\sim}{r},t)/m \tag{59a}$$

$$\dot{\underset{\sim}{P}}(\underset{\sim}{r},t)= - \nabla_r \cdot \underset{=}{\tau}(\underset{\sim}{r},t) \tag{59b}$$

$$\dot{E}(\underset{\sim}{r},t)= - \nabla_r \cdot \underset{\sim}{J}_E(\underset{\sim}{r},t) \tag{59c}$$

where the stress tensor $\underset{=}{\tau}$ is given by

$$\underset{=}{\tau}(\underset{\sim}{r},t)= \sum_j \{\frac{\underset{\sim}{p}_j\underset{\sim}{p}_j}{m} - \frac{1}{2} \sum_{\ell \neq j} \underset{\sim}{r}_{\ell j} \nabla_j u_{\ell j}\}\delta(\underset{\sim}{r}-\underset{\sim}{r}_j) \tag{60}$$

and the energy current by

$$\underset{\sim}{J}_E(\underset{\sim}{r},t)= \sum_j \{\frac{e_j\underset{\sim}{p}_j}{m} - \frac{1}{2m} \sum_{\ell \neq j} \underset{\sim}{r}_{\ell j} \underset{\sim}{p}_j \cdot \nabla_j u_{\ell j}\}\delta(\underset{\sim}{r}-\underset{\sim}{r}_j). \tag{61}$$

Note that the smallness parameters governing the time dependence of the $\underset{\sim}{A}(\underset{\sim}{r},t)$ are the magnitudes of the spatial gradients in the system. Eqs. (59) can be summarized by

$$\dot{\underset{\sim}{A}}(\underset{\sim}{r},t)=-\nabla \cdot \underset{\sim}{J}(\underset{\sim}{r},t) \tag{62}$$

The dissipative currents $\underset{\sim}{I}'(r,t)$ are easily obtained from Eqs. (59) and (47) as:

$$\underset{\sim}{I}'(\underset{\sim}{r},t)=-\nabla_r \cdot \underset{\sim}{I}(\underset{\sim}{r},t) \tag{63}$$

where

$$\underset{\sim}{I}_N(\underset{\sim}{r},t)=0, \quad (64a)$$

$$\underset{=}{I}_P(\underset{\sim}{r},t)=\underset{=}{\tau}(\underset{\sim}{r},t)-[(\frac{\partial p_h}{\partial n})^o_e N(\underset{\sim}{r},t)+(\frac{\partial p_h}{\partial e})^o_n E(\underset{\sim}{r},t)]\underset{=}{1}, \quad (64b)$$

$$\underset{=}{I}_E(\underset{\sim}{r},t)=\underset{\sim}{J}_e(\underset{\sim}{r},t)-\frac{e^o+p^o_h}{mn^o}\underset{\sim}{P}(r,t), \quad (64c)$$

p^o_h is the equilibrium hydrostatic presure and the superscript o implies the value of the quantity at total equilibrium. In order to obtain Eqs. (64) we have used the facts that $\langle A(\underset{\sim}{r})B(\underset{\sim}{r}')\rangle$ decays to $\langle A(\underset{\sim}{r}))\rangle\langle B|\underset{\sim}{r}')\rangle$ for $|\underset{\sim}{r}-\underset{\sim}{r}'|>\lambda$, the correlation length in the fluid, and that in homogeneous systems $\langle A(\underset{\sim}{r})B(\underset{\sim}{r}')\rangle$ is a function of $|\underset{\sim}{r}-\underset{\sim}{r}'|$ only.

Next, we turn to a specification of the force-like quantities $\Phi(\underset{\sim}{r},t)$ introduced in the last section. We do this by comparing Eq. (52) with the phenomenological local equilibrium distribution function

$$f_L'(X^N,t)=\frac{1}{\Xi_L(t)N!h^{3N}}\exp\{-\int d\underset{\sim}{r}\ \beta(\underset{\sim}{r},t)[E(\underset{\sim}{r})-\underset{\sim}{\upsilon}(\underset{\sim}{r},t)\cdot\underset{\sim}{P}(r)- -\phi(\underset{\sim}{r},t)N(\underset{\sim}{r})]\} \quad (65)$$

where $\Xi_L(t)$ is a normalization factor,

$$\beta(\underset{\sim}{r},t)=[k_BT(\underset{\sim}{r},t)]^{-1} \quad (66)$$

where $T(\underset{\sim}{r},t)$ is the local temperature, $\underset{\sim}{\upsilon}(\underset{\sim}{r},t)$ is the local velocity, and

$$\phi(\underset{\sim}{r},t)=\mu(\underset{\sim}{r},t)-\frac{1}{2}m\upsilon^2(\underset{\sim}{r},t) \quad (67)$$

is the local chemical potential, $\mu(\underset{\sim}{r},t)$, minus the local kinetic energy. Comparison of Eqs. (65) and (52) yields:

$$\beta\ \Phi_n(\underset{\sim}{r},t)=\beta(\underset{\sim}{r},t)\phi(\underset{\sim}{r},t)-\beta\mu^o \quad (68a)$$

$$\beta\ \underset{\sim}{\Phi}_p(\underset{\sim}{r},t)=\beta(\underset{\sim}{r},t)\underline{\upsilon}(\underset{\sim}{r},t) \quad (68b)$$

and

$$\beta\ \Phi_e(\underset{\sim}{r},t)=\beta-\beta(\underset{\sim}{r},t) \quad (68c)$$

where μ^o is the equilibrium chemical potential and $k_B\beta$ is the inverse equilibrium temperature. We shall demonstrate that the Φ's are conjugate forces below when we introduce the entropy function $S(t)$. The local velocity $\underset{\sim}{\upsilon}(\underset{\sim}{r},t)$ is related to the local momentum density by

$$<\underset{\sim}{P}(\underset{\sim}{r})>_L(t)=m\,\underset{\sim}{\upsilon}(\underset{\sim}{r},t)n(\underset{\sim}{r},t). \tag{69}$$

The general transport equation, Eq. (54), can now be rewritten

$$\dot{\underset{\sim}{a}}(\underset{\sim}{r},t)=-\nabla\cdot<\underset{\sim}{J}(\underset{\sim}{r})>_L(t)+\nabla\cdot\underset{\equiv}{\chi}\,(\underset{\sim}{r},t)\cdot\nabla[\beta\underset{\sim}{\Phi}(\underset{\sim}{r},t)] \tag{70}$$

where we have used Eqs. (62) and (63) and

$$\underset{\equiv}{\chi}(\underset{\sim}{r},t)=\int_o^\infty d\tau\int d\underset{\sim}{r}_1<\underset{\sim}{I}(\underset{\sim}{r},\tau)\underset{\sim}{I}(\underset{\sim}{r}_1)>_L(t)\ . \tag{71}$$

In order to obtain (71), we have also used the fact that $<\underset{\sim}{I}(\underset{\sim}{r},\tau)\underset{\sim}{I}(\underset{\sim}{r}_1)>\to 0$ for $|\underset{\sim}{r}-\underset{\sim}{r}_1|>\lambda$. In Eq. (70) we have neglected terms of order $\lambda/\xi<<1$ where ξ characterizes the spatial gradients of the macroscopic quantities.

The equations for the hydrodynamic variables are easy to obtain from Eq. (70). Since $I_N=0$, the number density equation becomes

$$\dot{n}(\underset{\sim}{r},t)=-\nabla\cdot\underset{\sim}{p}(\underset{\sim}{r},t)/m=-\nabla\cdot[n(\underset{\sim}{r},t)\underset{\sim}{\upsilon}(\underset{\sim}{r},t)] \tag{72}$$

where we have used Eqs. (59a) and (69).

The calculation of the Euler terms, $-\nabla\cdot<\underset{\sim}{J}(\underset{\sim}{r})>_L(t)$, for the momentum and energy densities are facilitated by introducing the relative momenta

$$\underset{\sim}{p}_j^+=\underset{\sim}{p}_j-m\underset{\sim}{\upsilon}(\underset{\sim}{r},t) \tag{73}$$

and the definition of the pressure tensor in a local equilibrium system

$$p_h(\underset{\sim}{r},t)\underset{\equiv}{1}=<\sum_j\{\frac{\underset{\sim}{p}_j^+\underset{\sim}{p}_j^+}{m}+\frac{1}{2}\sum_{\ell\neq j}\underset{\sim}{r}_{\ell j}\ \nabla_j\ u_{\ell j}\}\delta(\underset{\sim}{r}-\underset{\sim}{r}_j)>_L(t) \tag{74}$$

The equation for the momentum density is:

$$\dot{\underset{\sim}{p}}(\underset{\sim}{r},t)=-\nabla\cdot\{p_h(\underset{\sim}{r},t)\underset{\equiv}{1}+\underset{\sim}{\upsilon}(\underset{\sim}{r},t)\underset{\sim}{p}(\underset{\sim}{r},t)\}$$

$$-\nabla\cdot\underset{\equiv}{\chi}_{pe}(\underset{\sim}{r},t)\cdot\nabla\beta(\underset{\sim}{r},t)+\nabla\cdot\underset{\equiv}{\chi}_{pp}(\underset{\sim}{r},t):\nabla[\beta(\underset{\sim}{r},t)\underset{\sim}{\upsilon}(\underset{\sim}{r},t)] \tag{75}$$

since $\chi_{pn}=0$.

The dissipative current in Eq. (75) is

$$\underline{\underline{\Pi}}(\underset{\sim}{r},t)\equiv\underline{\underline{\chi}}_{pe}(\underset{\sim}{r},t)\cdot\nabla\beta(\underset{\sim}{r},t)-\underline{\underline{\chi}}_{pp}(\underset{\sim}{r},t):\nabla[\beta(\underset{\sim}{r},t)\underset{\sim}{v}(\underset{\sim}{r},t)] \ . \tag{76}$$

The equation for the energy density is

$$\begin{aligned}\dot{e}(\underset{\sim}{r},t)&=-\nabla\cdot[\underset{\sim}{v}(\underset{\sim}{r},t)e(\underset{\sim}{r},t)+\underset{\sim}{v}(\underset{\sim}{r},t)p_h(\underset{\sim}{r},t)]\\ &\quad+\nabla\cdot\underline{\underline{\chi}}_{ep}(\underset{\sim}{r},t):\nabla[\beta(\underset{\sim}{r},t)\underset{\sim}{v}(\underset{\sim}{r},t)]\\ &\quad-\nabla\cdot\underline{\underline{\chi}}_{ee}(\underset{\sim}{r},t)\cdot\nabla\beta(\underset{\sim}{r},t).\end{aligned} \tag{77}$$

The dissipative energy current is

$$\begin{aligned} j_e^d&\equiv j_q+\underline{\underline{\Pi}}(\underset{\sim}{r},t)\cdot\underset{\sim}{v}(\underset{\sim}{r},t)\\ &\equiv-\underline{\underline{\chi}}_{ep}(\underset{\sim}{r},t):\nabla[\beta(\underset{\sim}{r},t)\underset{\sim}{v}(\underset{\sim}{r},t)]+\underline{\underline{\chi}}_{ee}(\underset{\sim}{r},t)\cdot\nabla\beta(\underset{\sim}{r},t).\end{aligned} \tag{78}$$

Equations (72), (75) and (77) are the nonlinear hydrodynamic equations. In the next section we shall study the dissipative coefficients, $\underline{\underline{\chi}}$, in detail.

THE DISSIPATIVE COEFFICIENTS

The explicit forms of the dissipative coefficients can be found from Eqs. (71), (64) and (52). In addition we use the factorization property of the total equilibrium correlation functions when the spatial variables occuring in them are separated by distances greater than λ and the symmetry property that the total equilibrium average is zero unless the products of the various order tensors in the averages are contractable into a scalar or a unit tensor.

We find that:

$$\chi_{\alpha n}=\chi_{n\alpha}=0 \tag{79}$$

for all α;

$$\begin{aligned}\underline{\underline{\chi}}_{ee}(\underset{\sim}{r},t)=\int_0^\infty d\tau\int & d\underset{\sim}{r}_1\{<\underset{\sim}{I}_E(\underset{\sim}{r},\tau)\underset{\sim}{I}_E(\underset{\sim}{r}_1)>\\ &+\int d\underset{\sim}{r}_2<\underset{\sim}{I}_E(\underset{\sim}{r},\tau)\underset{\sim}{I}_E(\underset{\sim}{r}_1)\hat{A}_\alpha(\underset{\sim}{r}_2)>\beta\Phi_\alpha(\underset{\sim}{r},t)\}\end{aligned}$$

$$\equiv \underline{\underline{\chi}}_{ee}^{(0)} + \underline{\underline{\chi}}_{ee}^{(1)}(\underset{\sim}{r},t) \tag{80}$$

where the repeated Greek index α is summed over N and E;

$$\underline{\underline{\chi}}_{pp}(\underset{\sim}{r},t) = \int_0^\infty d\tau \int d\underset{\sim}{r}_1 \{ \langle \underline{\underline{I}}_p(\underset{\sim}{r},\tau)\underline{\underline{I}}_p(\underset{\sim}{r}_1)\rangle$$

$$+ \int d\underset{\sim}{r}_2 \langle \underline{\underline{I}}_p(\underset{\sim}{r},\tau)\underline{\underline{I}}_p(\underset{\sim}{r}_1)\hat{A}_\alpha(\underset{\sim}{r}_2)\rangle \beta\Phi_\alpha(\underset{\sim}{r},t)\} \tag{81}$$

$$\equiv \underline{\underline{\chi}}_{pp}^{(0)} + \underline{\underline{\chi}}_{pp}^{(1)}(\underset{\sim}{r},t)$$

where again α is summed over N and E;

$$\underline{\underline{\chi}}_{pe}(\underset{\sim}{r},t) = \int_0^\infty d\tau \int d\underset{\sim}{r}_1 \int d\underset{\sim}{r}_2 \langle \underline{\underline{I}}_p(\underset{\sim}{r},\tau)\underline{I}_E(\underset{\sim}{r}_1)\underline{P}(\underset{\sim}{r}_2)\rangle \cdot \beta(\underset{\sim}{r},t)\underset{\sim}{\upsilon}(\underset{\sim}{r},t)$$

$$\equiv \underline{\underline{\chi}}_{pe}^{(1)}(\underset{\sim}{r},t); \tag{82}$$

and

$$\underline{\underline{\chi}}_{ep}(\underset{\sim}{r},t) = \int_0^\infty d\tau \int d\underset{\sim}{r}_1 \int d\underset{\sim}{r}_2 \langle \underline{I}_E(\underset{\sim}{r},\tau)\underline{\underline{I}}_p(\underset{\sim}{r}_1)\underline{P}(\underset{\sim}{r}_2)\rangle \cdot \beta(\underset{\sim}{r},t)\underset{\sim}{\upsilon}(\underset{\sim}{r},t)$$

$$\equiv \underline{\underline{\chi}}_{ep}^{(1)}(\underset{\sim}{r},t). \tag{83}$$

Thus χ_{pp} and χ_{ee} depend on $\underset{\sim}{r},t$ through the variables $\beta(\underset{\sim}{r},t)$ and $\beta(\underset{\sim}{r},t)\phi(\underset{\sim}{r},t)$ while χ_{ep} and χ_{pe} depend on $\underset{\sim}{r},t$ through $\beta(\underset{\sim}{r},t)\underset{\sim}{\upsilon}(\underset{\sim}{r},t)$. In addition χ_{ep} and χ_{pe} are zero in the linear theory.

We next turn to the relation of the linear χ's, i.e. $\chi_{ee}^{(0)}$ and $\chi_{pp}^{(0)}$, to the usual transport coefficients, i.e. the thermal conductivity and the bulk and shear viscosities. The thermal conductivity, κ, is given by

$$\kappa = \frac{1}{k_B T^2} \int_0^\infty d\tau \int d\underset{\sim}{r} \langle I_{Ex}(\underset{\sim}{r},\tau) I_{Ex}(0,0)\rangle. \tag{84}$$

Thus

$$\underline{\underline{\chi}}_{ee}^{(0)} = \underline{\underline{1}}\, k_B T^2 \kappa\ . \tag{85}$$

The expressions for the shear and bulk viscosities are

$$4/3\eta + \zeta = \frac{1}{k_B T} \int_0^\infty d\tau \int d\underset{\sim}{r} \langle I_{pxx}(\underset{\sim}{r},\tau) I_{pxx}(0,0)\rangle \tag{86a}$$

and

$$\eta = \frac{1}{k_B T} \int_0^\infty d\tau \int d\underset{\sim}{r} \langle I_{pxy}(\underset{\sim}{r},\tau) I_{pxy}(0,0)\rangle . \tag{86b}$$

The components of $\underline{\underline{\chi}}^{(0)}_{pp}$ can be written

$$\chi^{(0)}_{ijk\ell} = F_1 \, \delta_{ij}\delta_{k\ell} + F_2(\delta_{ik}\delta_{j\ell} + \delta_{i\ell}\delta_{jk}) \tag{87}$$

where

$$F_2 = \int_0^\infty d\tau \int d\underset{\sim}{r}_1 \langle I_{pxy}(\underset{\sim}{r},\tau) I_{pxy}(\underset{\sim}{r}_1)\rangle = k_B T \eta \tag{88a}$$

and

$$F_1 = \int_0^\infty d\tau \int d\underset{\sim}{r}_1 \langle I_{pxx}(\underset{\sim}{r},\tau) I_{pyy}(\underset{\sim}{r}_1)\rangle = k_B T\{\zeta - 2/3\eta\} . \tag{88b}$$

Thus

$$\chi^{(0)}_{ijk\ell} = k_B T\{(\zeta - 2/3\eta)\delta_{ij}\delta_{k\ell} + \eta(\delta_{ik}\delta_{j\ell} + \delta_{i\ell}\delta_{jk})\}. \tag{89}$$

The first order corrections to χ_{pp} and χ_{ee} are easily obtained either by using general thermodynamic theory or by direct dynamical calculation of the correlation function. If we use the canonical ensemble to compute $\langle \underline{I}(\tau)\underline{I}\rangle$ the results are:

$$\underline{\underline{\chi}}^{(1)}_{pp}(\underset{\sim}{r},t) = [\beta\Phi_e(\underset{\sim}{r},t)\, \frac{\partial}{\partial\beta})_n + \Phi_n(\underset{\sim}{r},t)(\frac{\partial n}{\partial\mu})^{o}_{\beta}\, \frac{\partial}{\partial n})_\beta]\underline{\underline{\chi}}^{(0)}_{pp} \tag{90a}$$

and

$$\underline{\underline{\chi}}^{(1)}_{ee}(\underset{\sim}{r},t) = [\beta\Phi_e(\underset{\sim}{r},t)\, \frac{\partial}{\partial\beta})_n + \Phi_n(\underset{\sim}{r},t)(\frac{\partial n}{\partial\mu})_{\beta}\, \frac{\partial}{\partial n})_\beta]\underline{\underline{\chi}}^{(0)}_{ee} . \tag{90b}$$

Thus χ_{pp} and χ_{ee} are properly defined in the sense that the correlation functions which appear in them decay to zero on molecular time scales and the extension of the upper limit of time integration to infinity is appropriate.

There is a symmetry relation between $\chi^{(1)}_{pe}$ and $\chi^{(1)}_{ep}$ which is easily obtained using the fact that $\underline{\underline{I}}_p$ is a symmetric tensor and that $\int d\underset{\sim}{r}_2 \underset{\sim}{P}(\underset{\sim}{r}_2)$ is a constant of the motion. We write the correlation function in χ_{pe} as

$$\eta_{ijk\ell} \equiv \int d\underset{\sim}{r}_2 \langle \underline{\underline{I}}_p(\underset{\sim}{r},\tau)\underline{I}_E(\underset{\sim}{r}_1)\underline{P}(\underset{\sim}{r}_2)\rangle_{ijk\ell} = G_1\delta_{ij}\delta_{k\ell} + G_2(\delta_{ik}\delta_{j\ell} + \delta_{i\ell}\delta_{jk})$$

where (91)

$$G_1 = \eta_{xxyy} \tag{92}$$

and

$$G_2=\eta_{xyxy}=\eta_{xyyx} \ .$$

The correlation function in χ_{ep} is

$$\hat{\eta}_{ijk\ell}\equiv \int d\underset{\sim}{r}_2 \langle \underline{I}_E(\underset{\sim}{r},\tau)\underline{\underline{I}}_p(\underset{\sim}{r}_1)\underline{P}(\underset{\sim}{r}_2)\rangle_{ijk\ell}=H_1\delta_{ij}\delta_{k\ell}+H_2(\delta_{ik}\delta_{j\ell}+\delta_{i\ell}\delta_{jk})$$

$$= \int d\underset{\sim}{r}_2 \langle \underline{I}_E(\underset{\sim}{r})\underline{\underline{I}}_p(\underset{\sim}{r}_1,\tau)\underline{P}(\underset{\sim}{r}_2)\rangle_{ijk\ell} \tag{93}$$

$$=\eta_{jki\ell} .$$

We have used time reversal invariance to obtain the third equality and the fact that η is a function of $|\underset{\sim}{r}-\underset{\sim}{r}_1|$ to obtain the fourth equality. Thus $G_1=H_2$, $G_2=H_1$ and

$$(\chi^{(1)}_{ep})_{ijk} = (\chi^{(1)}_{pe})_{jki} \ . \tag{94}$$

Next, we show that the triple correlation function in $\chi^{(1)}_{ep}$ can be written in terms of a double correlation function. It can be shown by dynamical calculation that

$$\int_0^\infty d\tau \int d\underset{\sim}{r}_1 \int d\underset{\sim}{r}_2 \langle \underline{\underline{I}}_p(\underset{\sim}{r},\tau)\underline{I}_E(\underset{\sim}{r}_1)\underline{P}(\underset{\sim}{r}_2)\rangle_{ijk\ell}$$
$$= [\frac{\partial}{\partial\beta\upsilon_\ell}\int_0^\infty d\tau \int d\underset{\sim}{r}_1 \langle \underline{\underline{I}}_p(\underset{\sim}{r},\tau)\underline{\underline{I}}_E(\underset{\sim}{r}_1)\rangle^*_{ijk}]_{\underset{\sim}{\upsilon}=0} \tag{95}$$

where the symbol $\langle\rangle^*$ denotes an equilibrium average over a distribution function in which the momenta $\underline{p}_j$ are replaced by $p_j-m\underline{\upsilon}$. We change the variables of integration from $\underline{p}_j$ to $\underline{p}_j-m\underline{\upsilon}=\underline{P}_j$ and find that

$$(95) = \frac{1}{\beta}\int_0^\infty d\tau \int d\underset{\sim}{r}_1 \{\langle \frac{\partial}{\partial\upsilon_\ell}\underline{\underline{I}}^*_p(\underset{\sim}{r})e^{-iL\tau}\underline{I}^*_E(\underset{\sim}{r}_1)\rangle_{ijk}$$
$$+ \langle \underline{\underline{I}}^*_p(\underset{\sim}{r})e^{-iL\tau}\frac{\partial}{\partial\upsilon_\ell}\underline{I}^*_E(\underset{\sim}{r}_1)\rangle_{ijk}\}_{\underline{\upsilon}=0} \tag{96}$$

where the notation I^* implies the appropriate dissipative current, Eqs. (64), with $\underset{\sim}{p}_j$ replaced by $\underline{P}_j+m\underline{\upsilon}$. The Liouville operator is not affected by this transformation because the system is uniform and the I's depend on relative coordinates only. It is easy to see from Eqs. (64b), (60) and (58) that

$$\left(\frac{\partial I^*_{pij}(\underset{\sim}{r})}{\partial\upsilon_\ell}\right)_{\underline{\upsilon}=0} =\delta_{i\ell}\underline{P}_j(\underset{\sim}{r})+\delta_{j\ell}\underline{P}_i(\underset{\sim}{r})-\left(\frac{\partial p_h}{\partial e}\right)^o_n \delta_{ij}\underline{P}_\ell(\underset{\sim}{r}) \tag{97}$$

and from Eqs. (64c), (61) and (58) that

$$\left(\frac{\partial I^*_{E\,k}(\underset{\sim}{r}_1)}{\partial \upsilon_\ell}\right)_{\underline{\upsilon}=0} = \tau_{k\ell}(\underset{\sim}{r}_1) - \delta_{k\ell}E(\underset{\sim}{r}_1) - \frac{e^o+p^o_h}{n^o}\,\delta_{k\ell}N(\underset{\sim}{r}_1). \tag{98}$$

Substitution of Eqs. (97) and (98) into (96) and use of the orthogonality of $\underset{\sim}{I}$ to the conserved variable densities, Eq. (48), we find

$$(95) = \frac{1}{\beta}\int_o^\infty d\tau \int d\underset{\sim}{r}_1 \langle \underline{\underline{I}}_p(\underset{\sim}{r},\tau)\underline{\underline{I}}_p(\underset{\sim}{r}_1)\rangle_{ijk\ell}. \tag{99}$$

Thus,

$$\underline{\underline{\chi}}^{(1)}_{pe}(\underset{\sim}{r},t) = \frac{1}{\beta}\,\underline{\underline{\chi}}^{(0)}_{pp}\cdot\beta(\underset{\sim}{r},t)\underset{\sim}{\upsilon}(\underset{\sim}{r},t) \tag{100}$$

where we have used Eqs. (81) and (82). This coefficient is also well defined.

Substitution of Eq. (100) into the dissipative momentum current, Eq. (76), yields

$$\begin{aligned}\underline{\underline{\Pi}}(\underset{\sim}{r},t) &= \frac{1}{\beta}\,\underline{\underline{\chi}}^{(0)}_{pp}:\beta(\underset{\sim}{r},t)\underset{\sim}{\upsilon}(\underset{\sim}{r},t)\nabla\beta(\underset{\sim}{r},t)\\ &\quad - \underline{\underline{\chi}}_{pp}:\nabla[\beta(\underset{\sim}{r},t)\underset{\sim}{\upsilon}(\underset{\sim}{r},t)]\end{aligned} \tag{101}$$

which to second order in deviations from equilibrium becomes

$$\underline{\underline{\Pi}}(\underset{\sim}{r},t) = -[\beta(\underset{\sim}{r},t)\underline{\underline{\chi}}^{(0)}_{pp} + \beta\underline{\underline{\chi}}^{(1)}_{pp}]:\nabla\underset{\sim}{\upsilon}(\underset{\sim}{r},t)\;. \tag{102}$$

Eq. (102) agrees with the classical ansatz of the thermodynamics of irreversible processes (11)

$$\underline{\underline{\Pi}}(\underset{\sim}{r},t) = \underline{\underline{L}}_{pp}(\underset{\sim}{r},t):\nabla\underset{\sim}{\upsilon}(\underset{\sim}{r},t) \tag{103}$$

where L_{pp} depends on $\underset{\sim}{r},t$ via the thermodynamic variables.

Finally substitution of Eq. (100) into the dissipative energy current, Eq. (78), yields

$$\begin{aligned}\underline{j}^d_e(\underset{\sim}{r},t) &= -\frac{1}{\beta}\,[\underline{\underline{\chi}}^{(0)}_{pp}:\nabla\beta(\underset{\sim}{r},t)\underset{\sim}{\upsilon}(\underset{\sim}{r},t)]\cdot\underset{\sim}{\upsilon}(\underset{\sim}{r},t)\beta(\underset{\sim}{r},t)\\ &\quad + \underline{\underline{\chi}}_{ee}(\underset{\sim}{r},t)\cdot\nabla\beta(\underset{\sim}{r},t)\end{aligned} \tag{104}$$

which to second order in deviations from equilibrium becomes

$$\underline{j}_e^d(\underset{\sim}{r},t) = -[\beta \underline{\underline{\chi}}_{pp}^{(0)} : \nabla \underset{\sim}{\upsilon}(\underset{\sim}{r},t)] \cdot \underset{\sim}{\upsilon}(\underset{\sim}{r},t) + \underline{\underline{\chi}}_{ee}(\underset{\sim}{r},t) \cdot \nabla\beta(\underset{\sim}{r},t) \quad (105)$$

which again agrees with the classical ansatz

$$\underline{j}_e^d(\underset{\sim}{r},t) = \underline{\underline{L}}_{ee}(\underset{\sim}{r},t) \cdot \nabla k_B \beta(\underset{\sim}{r},t) + (L_{pp} : \nabla \underset{\sim}{\upsilon}(\underset{\sim}{r},t)] \cdot \underset{\sim}{\upsilon}(\underset{\sim}{r},t) \ . \quad (106)$$

We note that all the new nonlinear coefficients are non-zero as the density approaches zero except for those involving density derivatives of χ_{pp} and χ_{ee} which tend to zero.

ENTROPY PRODUCTION

We define the entropy function, S(t), and the local entropy density, $s(\underset{\sim}{r},t)$, by the equations

$$\begin{aligned} S(t) &= -k_B \langle \log f_L \rangle_L (t) \\ &= \int s(\underset{\sim}{r},t) d\underset{\sim}{r} \end{aligned} \quad (107)$$

where f_L is given by Eq. (52). It follows then that

$$\dot{S}(t) = -\frac{1}{T} \int \dot{\underset{\sim}{a}}(\underset{\sim}{r},t) \cdot \underset{\sim}{\Phi}(\underset{\sim}{r},t) d\underset{\sim}{r} \quad (108)$$

and

$$\dot{s}(\underset{\sim}{r},t) = -\frac{1}{T} \dot{\underset{\sim}{a}}(\underset{\sim}{r},t) \cdot \underset{\sim}{\Phi}(\underset{\sim}{r},t) \ . \quad (109)$$

It is clear from these equations that the Φ's are the conjugate forces to the fluxes $\dot{\underset{\sim}{a}}$.

The entropy production follows immediately from Eqs. (72), (75) and (77) as:

$$\begin{aligned} \dot{S}(t) = k_B \int d\underset{\sim}{r}\ \beta(\underset{\sim}{r},t) \{ &\nabla \cdot \underline{\underline{\chi}}_{ep} : \nabla(\beta \underset{\sim}{\upsilon}) - \nabla \cdot \underline{\underline{\chi}}_{ee} \cdot \nabla\beta \\ &+ \underset{\sim}{\upsilon} \cdot \nabla \cdot \underline{\underline{\chi}}_{pe} \cdot \nabla\beta + \underline{\upsilon} \cdot \nabla \cdot \underline{\underline{\chi}}_{pp} : \nabla\beta\underset{\sim}{\upsilon} \} \end{aligned} \quad (110)$$

which is positive semi definite and is of the same form as the classical ansatz.

SUMMARY

We have shown that nonlinear response theory is appropriate for the description of systems which are not too far from equilibrium. It is essential in the development that the set of variables $\underset{\sim}{a}(\underset{\sim}{r},t)$ be slowly varying compared to all of the other variables in the system. Otherwise the specification of the initial state by response theory will be improper.

The application of nonlinear response theory to hydrodynamics yields transport equations which are consistent with the classical irreversible thermodynamic treatment.

REFERENCES

1. R. Kubo, J.Phys.Soc.Japan 12 (1957) 570;
   R. Kubo, M. Yokota and S. Nakajima, J.Phys.Soc.Japan 12 (1957) 1203.
2. L.P. Kadanoff and P.C. Martin, Ann.Physics 24 (1963) 419.
3. P.A. Selwyn and I. Oppenheim, Physica 54 (1971) 161.
4. N.G. Van Kampen, Physica Norwegica 5 (1971) 10.

5a. J.H. Weare and I. Oppenheim, Physica 72 (1974) 1,20.

5b. I. Oppenheim, in Topics in Statistical Mechanics and Biophysics: A Memorial to J.L. Jackson, R.A. Piccirelli, ed. (A.I.P., New York, 1976) p. 111.

5c. D. Ronis and I. Oppenheim, Physics 86A (1977) 475.

6. For a discussion of ergodic theory and the choice of ensemble, see: A.I. Khinchin, Mathematical Foundations of Statistical Mechanics, chapter II (Dover Publ., New York, 1949).
7. J.L. Lebowitz, Hamiltonian Flows and Rigorous Results in Non-Equilibrium Statistical Mechanics, in the Proc.I.U.P.A.P. Conference on Statistical Mechanics, S.A. Rice, K.F. Freed and J.C. Light, eds. (Univ. of Chicago Press, 1972).
8. Ja.G. Sinai, Ergodicity of Boltzmann's Gas Model, in Proc. I.U.P.A.P. Meeting, Copenhagen, 1966, T.A. Bak, ed. (Benjamin, New York, 1967) p. 559.
9. J.L. Lebowitz and J.K. Percus, Phys.Rev., 122 (1961) 1675.
10. T. Keyes and I. Oppenheim, Phys.Rev., A7 (1973) 522;
    I.A. Michaels and I. Oppenheim, Physica 81A (1975) 522.
11. S.R. De Groot and P. Mazur, Non-Equilibrium Thermodynamics (North-Holland Publ.Comp., Amsterdam 1962).

Part II
Magnetic Systems

MAGNETIC EXCITATIONS

M. F. Thorpe

Physics Department, Michigan State University

East Lansing, Michigan 48824

1. INTRODUCTION

Spin waves (or magnons) are the elementary excitations in spin systems. In these lectures we will develop the basic theory of these excitations using both spin operators and Bose operators. During the last decade, infra-red absorption and light scattering, involving the creation of two magnons, have been studied in many materials. These experiments have a number of interesting features that are now understood in rather considerable detail. This is in contrast to other areas that you will hear about at this summer school where the second order scattering is only understood in general phenomenological terms.

The general layout of these lectures is as follows: in the next section we develop Spin Wave Theory for ferromagnets and antiferromagnets. This is mainly intended as a review and will serve to define the notation and give us some familiarity with handling spin operators and the associated Bose operators. In section 3, Localized Magnons, we will examine the impurity modes associated with an isolated defect in a magnetic system. This will also serve to introduce the Green functions that will be needed later. In the next section we discuss the Interaction of Light with Magnetic Systems concentrating mainly on the symmetry aspects, as this is the area where our detailed understanding of the interactions is least complete. Fortunately, as we shall see, it is not necessary to know too much about this interaction in order to get a good description of the experiments. In section 5, we look at Two Magnon Excitations in ferromagnets briefly (particularly the linear chain) and more extensively in antiferromagnets. The main objective here is to get a correct description of the interaction between the two magnons. We concentrate on the physics and refer you to the literature for the details of the calculations. In this section we also discuss Two Magnon Excitations in Magnetic Alloys of which we regard the case of a single impurity as being the low concentration limit. This leads to some interesting new effects that are surprisingly well understood considering the complexity of the system.

These lectures are designed to be introductory and various topics that are only touched on or ignored completely will be covered in more detail by others; finite temperature effects (M. Cottam and V. Balucani), two magnon spectra of ferromagnets (P. D. Loly), and experiments on two magnon spectra in antiferromagnets (P. Fleury) and neutron scattering (R. A. Cowley).

2. SPIN WAVE THEORY

a) General

Spin waves are the elementary excitations in spin systems and are very similar to phonons in many ways. At low temperatures, it is possible to define an ordered crystal structure (apart possibly from zero point motion) which has a certain unit cell containing a rather small number of atoms usually, that when repeated defines the crystal. This structure can be found by performing x-ray or neutron diffraction experiments. Phonons are collective excitations in which atoms perform small oscillations about their equilibrium positions under the influence of harmonic forces.

In magnetic crystals, some atoms in the crystal have magnetic moments, which at sufficiently low temperatures form a regular pattern with a characteristic unit cell. This magnetic cell is at least as large as the chemical cell and may be larger (e.g. the simple linear chain antiferromagnet shown below). When the magnetic unit is larger than the chemical unit cell, this shows up in an elastic neutron scattering experiment as additional Bragg peaks.

Spins waves are the low lying excitations from this static spin configuration. The classical picture is that the spin precesses around its equilibrium position at a small angle θ as shown in Figure 2.

At first sight spin waves would appear to be simpler objects than phonons. In the phonon problem one has to diagonalize a (3s x 3s) matrix eventually where s is the number of atoms in the unit cell, whereas in the magnon problem one only has to diagonalize an r x r matrix where r is the number of magnetic ions in the unit cell. However, this simplification is minor when compared to the

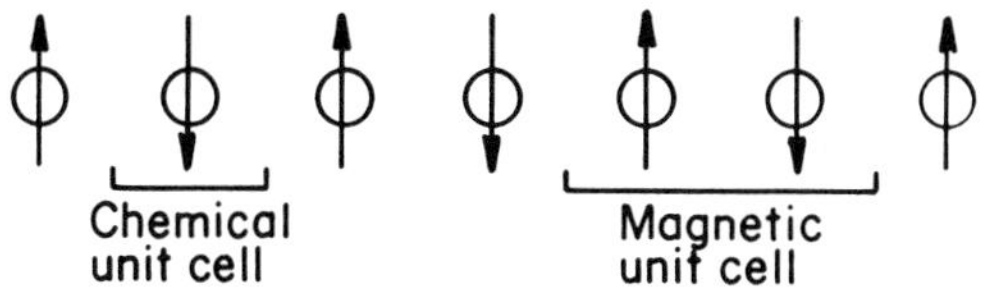

Fig. 1. A simple linear chain antiferromagnet where the magnetic unit cell is twice as large as the chemical unit cell.

complications caused by the fact that one cannot in general make the harmonic approximation in magnetic systems (one important exception is in ferromagnets at low temperatures) as we will see later, whereas in most phonon calculations one can - the anharmonic effects being $\sim \langle u^2 \rangle / a^2 \approx 10^{-2}$ where $\langle u^2 \rangle$ is the mean square displacement and a is an interatomic separation. Thus in experiments involving two phonons (see lectures by H. Bilz) one can almost always forget about interactions between the phonons. It may however be possible to see the very weak repulsion between two optic phonons in diamond (see papers by J. Ruvalds referenced in the Bilz lectures).

b) Spin Waves in Ferromagnets

The simplest example of spin waves are those in a ferromagnet[1]. We will assume a simple isotropic Heisenberg interaction between nearest neighbor spins and an external field h .

$$H = -\frac{1}{2}\sum_{ij} J_{ij}\, \underline{S}_i \cdot \underline{S}_j - h \sum_i S_i^z \qquad (1)$$

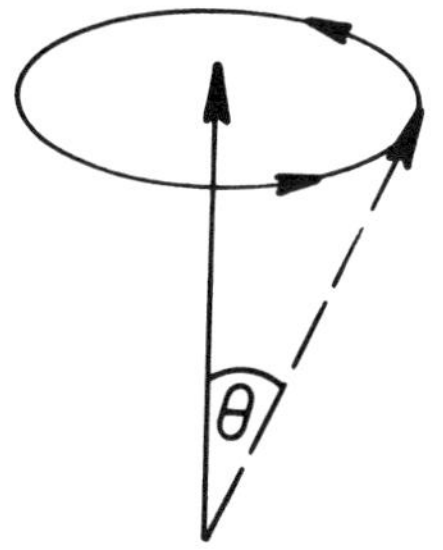

Fig. 2. A spin precessing about its static equilibrium position (shown as solid arrow). The motions of these individual spins are coupled together by the interactions between the spins.

The external field is included to stabilize one direction for the magnetization; however it can be arbitrarily small. Isotropic exchange can arise when the spin permutation operator $2\,\underline{s}_i \cdot \underline{s}_j + 1/2$, where $\underline{s}_i, \underline{s}_j$ are individual spins, is summed over all the electrons in a half-filled shell in a magnetic insulator (e.g. Mn^{++} ions in MnF_2). More complicated exchange interactions can arise when the orbital angular momentum is not quenched[2].

The ground state of (1) is clearly the state in which all the spins are parallel and have their maximum value S along the z axis. The spin operators obey the following commutation rules

$$[S_i^+, S_j^-] = 2\, S_i^z \delta_{ij} \qquad (2)$$

$$[S_i^z, S_j^\pm] = \pm\, S_i^\pm \delta_{ij}$$

where $S^\pm = S^x \pm iS^y$ are raising and lowering operators with the property that (dropping the site labels)

$$S^+|S^z{=}m\rangle = \sqrt{(S-m)(S+m-1)}\,|S^z{=}m+1\rangle \qquad (3)$$

$$S^-|S^z{=}m\rangle = \sqrt{(S+m)(S-m+1)}\,|S^z{=}m-1\rangle$$

The spin waves are excitations where the z component of the magnetization is changed by one unit. The equation of motion for the spin lowering operator is given by (putting $\hbar = 1$).

$$-i\,\frac{\partial S_i^-}{\partial t} = [H, S_i^-] = \sum_j J_{ij}(S_j^z S_i^- - S_i^z S_j^-) \qquad (4)$$

If we denote the fully aligned ground state by $|0\rangle$ and define running waves

$$S_{\underline{k}}^{-} = \frac{1}{\sqrt{N}} \sum_{i} e^{-i\underline{k}\cdot\underline{i}}\, S_{i}^{-} \tag{5}$$

then it is easy to see that

$$[S_{\underline{k}}^{-},H]\;|0\rangle = \omega_{\underline{k}}|0\rangle \tag{6}$$

where the spin wave energies $\omega_{\underline{k}}$ are given by

$$\omega_{\underline{k}} = S[J(0) - J(\underline{k})] + h$$

and (7)

$$J(\underline{k}) = \sum_{\underline{\delta}} J\, e^{i\underline{k}\cdot\underline{\delta}}$$

the summation over $\underline{\delta}$ goes over the z nearest neighbors.

These spin waves are exact eigenstates of the Hamiltonian with $\Delta M = 1$. They are not necessarily the lowest lying excitations. Indeed we will see that the two magnon bound states in a ferromagnet do have lower frequencies. However in systems that have a spontaneous magnetization at low temperatures they are the most important low frequency excitations.

These spin waves are thermally excited at low temperatures and we need to know what statistics they obey (i.e. what happens when many are excited). In fact they behave as Bosons and it is useful to develop the Holstein-Primakoff transformation[1] in order to show this

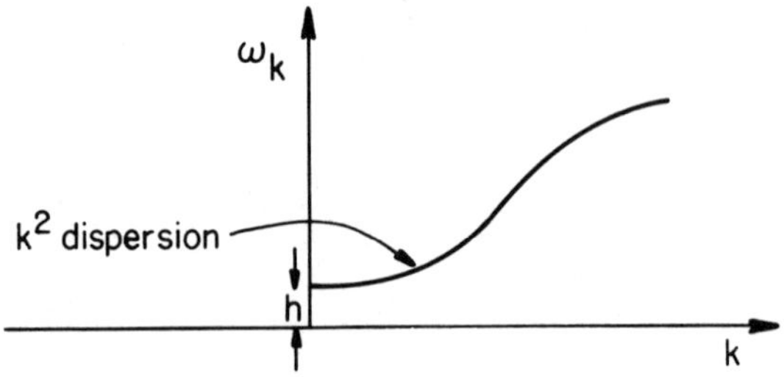

Fig. 3. The spin wave dispersion (7) for a simple ferromagnet described by the Hamiltonian (1).

Defining the number of deviations n = S-m, we can rewrite (3)

$$S^+|n\rangle = \sqrt{n(2S-n+1)}\ |n-1\rangle$$
$$S^-|n\rangle = \sqrt{(2S-n)(n+1)}\ |n+1\rangle \qquad (3a)$$

These relations (3a) are "similar" to the relations for Bose operators $a, a^\dagger$

$$a|n\rangle = \sqrt{n}\ |n-1\rangle$$
$$a^\dagger|n\rangle = \sqrt{n+1}\ |n+1\rangle \qquad (8)$$

with the commutation rule $[a, a^\dagger] = 1$.

if we write

$$S^+ = (2S-a^\dagger a)^{\frac{1}{2}} a$$

$$S^- = a^\dagger (2S-a^\dagger a)^{\frac{1}{2}} \tag{9}$$

then the relations (3a) are maintained. To get a representation of S^z in terms of the Bose operators we write

$$\begin{aligned}
2S^z &= [S^+,S^-] \\
&= (2S-a^\dagger a)^{\frac{1}{2}} a a^\dagger (2S-a^\dagger a)^{\frac{1}{2}} - a^\dagger (2S-a^\dagger a)^{\frac{1}{2}} (2S-a^\dagger a)^{\frac{1}{2}} a \\
&= (2S-a^\dagger a)^{\frac{1}{2}} (1+a^\dagger a)(2S-a^\dagger a)^{\frac{1}{2}} - a^\dagger (2S-a^\dagger a) a \\
&= (2S-a^\dagger a)(1+a^\dagger a) - a^\dagger (2S-a^\dagger a) a \\
&= 2S-a^\dagger a+2Sa^\dagger a-a^\dagger a a^\dagger a-2Sa^\dagger a+a^\dagger a^\dagger aa \\
&= 2S-a^\dagger a-a^\dagger (1+a^\dagger a)a+a^\dagger a^\dagger aa \\
&= 2S-2a^\dagger a.
\end{aligned}$$

$$\therefore\ S^z = S-a^\dagger a \tag{10}$$

A result that might have been anticipated. In order to perform useful calculations with the Bose operators, it is necessary to expand the square roots. This immediately introduces a kinematic error because there should not be more than 2S deviations on any particular site whereas the Bose operators have no such restriction. It has been shown that these kinematical errors lead to factors $\sim\exp(-J/k_BT)$ in the low temperature thermodynamics and so are not important. This means that the unphysical states with $n \gtrsim 2S$ lie above the physical states in energy so that when the lattice only contains a few spin waves at low temperatures they are not important as they are not appreciably populated.[3] Therefore keeping only the lowest terms from (9)

$$S^+ = \sqrt{2S}\, a$$
$$S^- = \sqrt{2S}\, a^\dagger \tag{9a}$$

and using (10) we find that the Hamiltonian (1) becomes

$$\begin{aligned} H = & -\frac{1}{2}\sum_{ij} J_{ij}\ \{(S-a_i^\dagger a_i)(S-a_j^\dagger a_j) + \frac{1}{2}(2S)(a_i^\dagger a_j + a_i a_j^\dagger)\} \\ & - h\sum_i (S-a_i^\dagger a_i) \\ = & -\frac{1}{2}\sum_{ij} J_{ij}\ S^2 - h\sum_i S + \sum_{ij} J_{ij}\ S(a_i^\dagger a_i - a_i^\dagger a_j) \\ & + h\sum_i a_i^\dagger a_i \end{aligned} \tag{10}$$

Again we form running waves

$$a_i = \frac{1}{\sqrt{N}}\sum_{\underline{k}} e^{i\underline{k}\cdot\underline{i}}\ a_{\underline{k}}$$
$$a_i^\dagger = \frac{1}{\sqrt{N}}\sum_{\underline{k}} e^{-i\underline{k}\cdot\underline{i}}\ a_{\underline{k}}^\dagger \tag{11}$$

then

$$H = E_o + \sum_{\underline{k}} \omega_{\underline{k}}\ a_{\underline{k}}^\dagger\ a_{\underline{k}} \tag{12}$$

where $E_o = -\frac{1}{2}NJS^2z - hNS$ is the ground state energy and $\omega_{\underline{k}}$ are the spin wave energies (7).

The low temperature thermodynamics is dominated by the small k spin waves that are quadratic in k. For a simple cubic lattice $\omega_k \sim JS(ka)^2$ if we put h=0 and a is the near neighbor distance. The magnetization is given by

$$M = S - \frac{1}{N}\sum_{\underline{k}} n_{\underline{k}}$$

$$= S - \frac{1}{N}\sum_{\underline{k}} \frac{1}{e^{\beta\omega_{\underline{k}}} - 1}$$

$$\approx S - \left(\frac{a}{2\pi}\right)^3 \int_{B.Z} \frac{4\pi k^2 dk}{e^{JS\beta(ka)^2} - 1}$$

$$\approx S - \frac{1}{2\pi^2}\left(\frac{k_BT}{JS}\right)^{3/2} \int_0^\infty \frac{x^2 dx}{e^{x^2} - 1} \qquad (13)$$

where $\beta = 1/k_BT$ and we have used the substitution $x^2 = JS\beta(ka)^2$ and extended to integral to ∞. Thus the magnetization deviates from saturation as $T^{3/2}$. This is known as the Bloch $T^{3/2}$ law and is found to be obeyed in many ferromagnets.

It is interesting to note that the integrals in (13) would diverge in 1 and 2 dimensions where k^2dk is replaced by $k^{d-1}dk$. This tells us that the deviations from the (assumed) ordered state are infinite and so the original assumption of ordering was incorrect and in fact M=0 at all temperatures (A rigourous proof of this is given by Mermin and Wagner[4]). As we noted earlier this also ties in with the fact that the two magnon bound states are very low lying in 1 and 2 dimensions.

c) Antiferromagnetic Magnons

Changing the sign of J in (1) takes us from a ferromagnet to an antiferromagnet. This has the effect of defining two sublattices in simple crystal structures (e.g. $RbMnF_3$, K_2MnF_4, MnF_2, etc), one with up spins and the other with down spins, such that an up spin only has down spins for nearest neighbors and vice versa. Because of the two sublattices we have to solve a 2x2 equation to get the spin wave spectra. The ground state of the antiferromagnet is not known exactly although very tight bounds can be put on the ground state energy[5] that can be written

$$E_0 = -\frac{JNS^2z}{2}\left(1 + \frac{\gamma}{Sz}\right) \tag{14}$$

where $0<\gamma<1$. Thus for MnF_2 where S = 5/2, z = 8 and 1/Sz = 1/20, the Néel state where the spins just alternate up and down is a very good approximation to the ground state. However for a $S = \frac{1}{2}$ linear chain where 1/Sz = 1 it is much less good. We will assume that the ground state is not too different from the Néel state so that if i is the up sublattice and j the down sublattice, we have

$$\begin{aligned} S_i^z|0\rangle &\approx S|0\rangle \\ S_j^z|0\rangle &\approx -S|0\rangle \end{aligned} \tag{15}$$

We will use an approximate equation of motion approach using the spin operators to find the spin waves. The same result can be obtained using Bosons[1]. Rewriting (1) with a staggered field h_A to stabilize the sublattice magnetization and an external field h.

$$H = \sum_{ij} J_{ij}\underline{S}_i \cdot \underline{S}_j - (h+h_A)\sum_i S_i^z - (h-h_A)\sum_j S_j^z \tag{16}$$

The equations of motion for the operators S_i^-, S_j^- are

$$i\,\frac{\partial S_i^-}{\partial t} = [S_i^-, H]$$

$$= \sum_j J_{ij}\,(S_i^- S_j^z - S_j^- S_i^z) - (h+h_A) S_i^-$$

$$i\,\frac{\partial S_j^-}{\partial t} = \sum_i J_{ij}\,(S_j^- S_i^z - S_i^- S_j^z) - (h-h_A) S_j^- \qquad (17)$$

We allow both sides of (17) to operate on the ground state and use (15)

$$i\,\frac{\partial S_i^-}{\partial t} = -\sum_j J_{ij}\,S(S_i^- + S_j^-) - (h+h_A) S_i^-$$

$$i\,\frac{\partial S_j^-}{\partial t} = \sum_i J_{ij}\,S(S_i^- + S_j^-) - (h-h_A) S_j^- \qquad (18)$$

Defining (N is the number of atoms in each sublattice)

$$S_{\underline{k}}^{1-} = \frac{1}{\sqrt{N}} \sum_i e^{i\underline{k}\cdot\underline{i}}\, S_i^-$$

$$S_{\underline{k}}^{2-} = \frac{1}{\sqrt{N}} \sum_j e^{i\underline{k}\cdot\underline{j}}\, S_j^- \qquad (19)$$

for the two sublattices and Fourier transforming with respect to time, we obtain from (18)

$$
\begin{aligned}
-\omega S^{1-}_{\underline{k}} &= -[J(0)S + h + h_A]S^{1-}_{\underline{k}} - J(\underline{k})S\, S^{2-}_{\underline{k}} \\
-\omega S^{2-}_{\underline{k}} &= [J(0)S - h + h_A]S^{2-}_{\underline{k}} + J(\underline{k})S\, S^{1-}_{\underline{k}}
\end{aligned}
\qquad (20)
$$

These equations can be diagonalized with the Bogoliubov transformation

$$
\begin{aligned}
\alpha^{\dagger}_{\underline{k}} &= \cosh\theta_{\underline{k}}\, S^{1-}_{\underline{k}} + \sinh\theta_{\underline{k}}\, S^{2-}_{\underline{k}} \\
\beta_{\underline{k}} &= \sinh\theta_{\underline{k}}\, S^{1-}_{\underline{k}} + \cosh\theta_{\underline{k}}\, S^{2-}_{\underline{k}}
\end{aligned}
\qquad (21)
$$

where $\tanh 2\theta_{\underline{k}} = -SJ(\underline{k})/[SJ(0) + h_A]$ and $\alpha^{\dagger}_{\underline{k}}$ creates a mode with energy $\omega^{\alpha}_{\underline{k}}$ and $\beta_{\underline{k}}$ destroys a mode with energy $\omega^{\beta}_{\underline{k}}$ where

$$\omega^{\alpha}_{\underline{k}} = \sqrt{(SJ(0) + h_A)^2 - (SJ(\underline{k}))^2} + h \qquad (22a)$$

$$\omega^{\beta}_{\underline{k}} = \sqrt{(SJ(0) + h_A)^2 - (SJ(\underline{k}))^2} - h \qquad (22b)$$

These normal modes are excitations that involve both sublattices except at points where $J(\underline{k}) = 0$ when α mode is entirely on the up sublattice and the β mode is entirely on the down sublattice. This condition is achieved at the zone boundary and corresponds to the maximum spin wave frequency. At this frequency only the Ising terms S^zS^z in the Hamiltonian contribute and the effect of the S^+S^- terms cancel when summed over the neighors. Notice that the α,β modes correspond to $\Delta M = -1, +1$ transitions respectively.

Finally we note that the $\alpha_{\underline{k}},\beta_{\underline{k}}$ operators obey Boson commutation rules approximately

$$[\alpha_{\underline{k}},\alpha_{\underline{g}}^\dagger] = [\beta_{\underline{k}},\beta_{\underline{g}}^\dagger] = \cosh\theta_{\underline{k}} \cosh\theta_{\underline{g}} \frac{2}{N}\sum_{i} e^{i(\underline{g}-\underline{k})\cdot\underline{i}} S_i^z$$

$$+ \sinh\theta_{\underline{k}} \sinh\theta_{\underline{g}} \frac{2}{N}\sum_{j} e^{i(\underline{g}-\underline{k})\cdot\underline{j}} S_j^z$$

$$\approx 2S\,\delta_{\underline{k},\underline{g}} \tag{23}$$

where we have taken the expectation value of the right side in the Néel state. Thus if we divide α,β by $\sqrt{2S}$, they behave as Bosons.

In the absence of anisotropy and an external field, the two branches are degenerate and $\omega_{\underline{k}} \sim k$ for long wavelengths. Thus long wavelength antiferromagnet magnons are very much like acoustic phonons and both give a T^3 contribution to the specific heat at low temperatures in 3 dimensions. Of particular interest in light scattering experiments are the $\underline{k} = 0$ modes given by

$$\omega_{\underline{k}=0} = \sqrt{h_A(h_A+2SJ(0))} \pm h \tag{24}$$

As the external field h is increased from zero, the degenerate modes split until one of the modes has zero energy at the critical field

$$h_c = \sqrt{h_A(h_A+2SJ(0))} \tag{25}$$

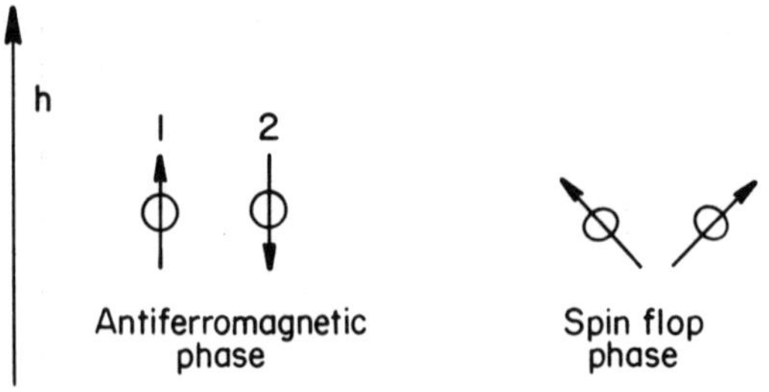

Fig. 4. Showing the spin directions of the two sublattices as h is increased to induce a spin flop transition.

For $h>h_c$, the spins try to align along the external field although the antiferromagnetic coupling prevents them from doing so perfectly (see Figure 4).

3. LOCALIZED MAGNONS

The single impurity problem has been studied in some detail in the last 30 years. The most extensive work has been on the electron problem[6] and the phonon problem[7]. The magnon problem is formally equivalent to the phonon problem as we have seen that magnons behave as Bosons at low temperatures. A great deal of the work on the phonon problem has been devoted to the "mass defect" case where the force constants do not change. The formalism is particularly simple in this case as the defect matrix contains only a single element. Unfortunately there is no equivalent to this in the magnon case. We will follow the treatment of Wolfram and Callaway[8] who treat the simple cubic ferromagnet. However, we will use Bose operators rather than spin operators. We have seen that the Hamiltonian for the pure crystal may be written as in Eq.(12)

$$H_o = \sum_{\underline{k}} \omega_{\underline{k}} a^{\dagger}_{\underline{k}} a_{\underline{k}} \tag{12a}$$

where $\omega_{\underline{k}} = S[J(0)-J(\underline{k})]$ and we omit the ground state energy E_0 for convenience. If we add a defect to the lattice at the origin with spin S' coupled ferromagnetically to its neighors at Δ with exhange J', then

$$V = -J' \sum_{\Delta} \underline{S}'_o \cdot \underline{S}_\Delta + J \sum_{\Delta} \underline{S}_o \cdot \underline{S}_\Delta$$

$$= (J'-J)Sza_o^\dagger a_o + \sum_{\Delta} (J'S'-JS)a_\Delta^\dagger a_\Delta$$

$$+ (JS-J'\sqrt{SS'}) \sum_{\Delta} (a_o^\dagger a_\Delta + a_o a_\Delta^\dagger) + \text{constant} \qquad (26)$$

where we keep only the quadratic Bose terms. This is exact when only a single excitation is present. The complete Hamiltonian for the defect system is

$$H = H_o + V \qquad (27)$$

It is convenient to discuss impurity modes using Green functions in the energy representation. We introduce a Green function operator P for the pure crystal

$$P = (\omega - H_o)^{-1} \qquad (28)$$

which is defined by its matrix elements between states

$$\left.\begin{aligned} |\underline{k}\rangle &= \frac{1}{\sqrt{N}} \sum_{i} e^{i\underline{k}\cdot\underline{i}}|i\rangle \\ |i\rangle &= a_i^{\dagger}|0\rangle \end{aligned}\right\} \qquad (29)$$

so that

$$\langle \underline{k}|P|\underline{q}\rangle = \frac{1}{\omega-\omega_{\underline{k}}}\,\delta_{\underline{k},\underline{q}}$$

and (30)

$$\langle i|P|j\rangle = \frac{1}{N}\sum_{\underline{k}} \frac{e^{i\underline{k}\cdot(\underline{i}-\underline{j})}}{\omega-\omega_{\underline{k}}}$$

The poles of the Green function give the eigenfrequencies of the system and provide a convenient way to find the eigenfrequencies of the defect system. The problem can also be solved by straightforward linear algebra[9] without the explicit use of Green functions. We also introduce a Green function for the complete system.

$$G = (\omega-H)^{-1} \qquad (31)$$

We see that

$$(\omega-H)^{-1} = (\omega-H_o)^{-1} + (\omega-H_o)^{-1}\,V(\omega-H)^{-1} \qquad (32)$$

i.e.

$$G = P + PVG \qquad (33)$$

using 27, 28 and 31. This is an operator equation that allows us to calculate G in terms of P. It is frequently referred to as a Dyson equation and occurs often in physics problems. Usually one has to resort to approximations to get a solution but in the present case we can obtain an exact solution because the matrix V only has a few non-zero elements in real space.

$$\langle 0|V|0\rangle = (J'-J)Sz = \alpha$$

$$\langle \Delta|V|\Delta\rangle = (J'S'-JS) = \beta \tag{34}$$

$$\langle 0|V|\Delta\rangle = (JS-J'\sqrt{SS'}) = \gamma$$

(The zero in $|0\rangle$ will refer to the defect site rather than the ground state from now on in this section). The solution of (33) may be written formally as

$$G = P(1-VP)^{-1} \tag{35}$$

The poles of G occur at the poles of P which give the eigenfrequencies of the pure crystal and also some extra eigenfrequencies associated with the defect when

$$\text{Det }|1-VP| = 0 \tag{36}$$

where V is a (z+1)x(z+1) matrix whose elements are given by (34) and P has matrix elements given by (30) with i,j restricted to the origin and its nearest neighbors Δ. Equation (36) has 7 roots in the case of a simple cubic crystal and can be diagonalized rather easily by using the point group symmetry of the defect as shown in Fig. 5

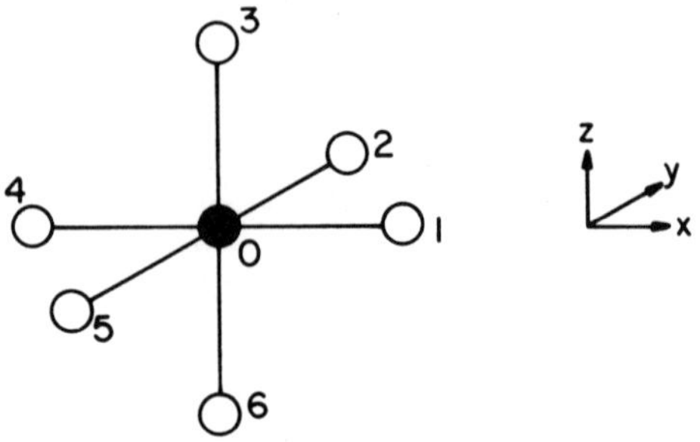

Fig. 5. Showing the defect labelled 0 and its six nearest neighbors in a simple cubic lattice.

By standard group theory techniques[10] we see that scalar quantities on the seven sites transform as $2\Gamma_1^+ + \Gamma_3^+ + \Gamma_4^-$. If we label in order 0 through 6 then from (34)

$$V = \begin{bmatrix} \alpha & \gamma & \gamma & \gamma & \gamma & \gamma & \gamma \\ \gamma & \beta & 0 & 0 & 0 & 0 & 0 \\ \gamma & 0 & \beta & 0 & 0 & 0 & 0 \\ \gamma & 0 & 0 & \beta & 0 & 0 & 0 \\ \gamma & 0 & 0 & 0 & \beta & 0 & 0 \\ \gamma & 0 & 0 & 0 & 0 & \beta & 0 \\ \gamma & 0 & 0 & 0 & 0 & 0 & \beta \end{bmatrix} \tag{37}$$

Applying the unitary transformation U to the matrices in (36), where

$$U = \begin{bmatrix} 1 & 0 & 0 & 0 & 0 & 0 & 0 \\ 0 & a & b & 0 & 0 & 0 & d \\ 0 & a & 0 & b & 0 & c & e \\ 0 & a & 0 & 0 & b & -c & e \\ 0 & a & -b & 0 & 0 & 0 & d \\ 0 & a & 0 & -b & 0 & c & e \\ 0 & a & 0 & 0 & -b & -c & e \end{bmatrix} \tag{38}$$

and $a = 1/\sqrt{6}$, $b = 1/\sqrt{2}$, $c = 1/2$, $d = 1/\sqrt{3}$, $e = 1/\sqrt{12}$. The determinant factorizes to give

$$1 - \beta(\langle 1|P|1\rangle - \langle 1|P|4\rangle) = 0 \tag{39}$$

for the three Γ_4^- states and

$$1 - \beta(\langle 1|P|1\rangle + \langle 1|P|4\rangle) - 2 \langle 1|P|2\rangle) = 0 \tag{40}$$

for the two Γ_3^+ states. We are left with a 2x2 matrix for the two Γ_1^+ states

$$\mathrm{Det}\left[\underset{\approx}{1} - \begin{bmatrix} \langle 0|P|0\rangle & \sqrt{6}\langle 0|P|1\rangle \\ \sqrt{6}\langle 0|P|1\rangle & \sum\limits_{\Delta}\langle 1|P|\Delta\rangle \end{bmatrix} \begin{bmatrix} \alpha & \sqrt{6}\gamma \\ \sqrt{6}\gamma & \beta \end{bmatrix} \right] = 0 \tag{41}$$

Notice that we have used the symmetry in putting $\langle 0|P|1\rangle = \langle 2|P|0\rangle$; $\langle 1|P|2\rangle = \langle 6|P|2\rangle$ etc.

When the frequency ω is inside the magnon band (i.e $0<\omega<12JS$), the $\langle i|P|j\rangle$ are complex and so even if the real parts of (39)-(41) vanish, the imaginary parts will not and we will have a resonance rather than a bound state.

In order to get a physical feel for the bound states, we may neglect the S^+S^- terms in the Hamiltonian and just consider the Ising terms. The spin wave band becomes flat at the average energy ω_{Is}

$$\omega_{Is} = 6JS \tag{42}$$

This corresponds to turning a single spin over costing an energy JS for each of the six neighbors. The local mode on the impurity $\Gamma_1^+(s_0)$ has an energy

$$\omega_{s_0} = 6J'S \tag{43}$$

while the energy of the other $\Gamma_1^+(s_1)$ and the $\Gamma_3^+(d)$ and $\Gamma_4^-(p)$ modes are given by

$$\omega_{Is} + (J'S'-JS) \tag{44}$$

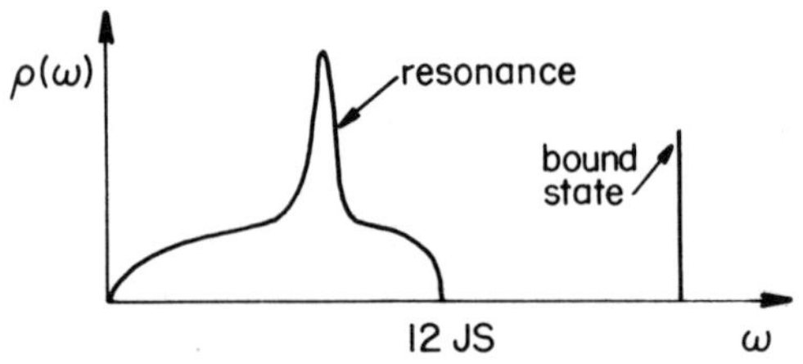

Fig. 6. Showing schematically a resonance and a bound state due to an impurity.

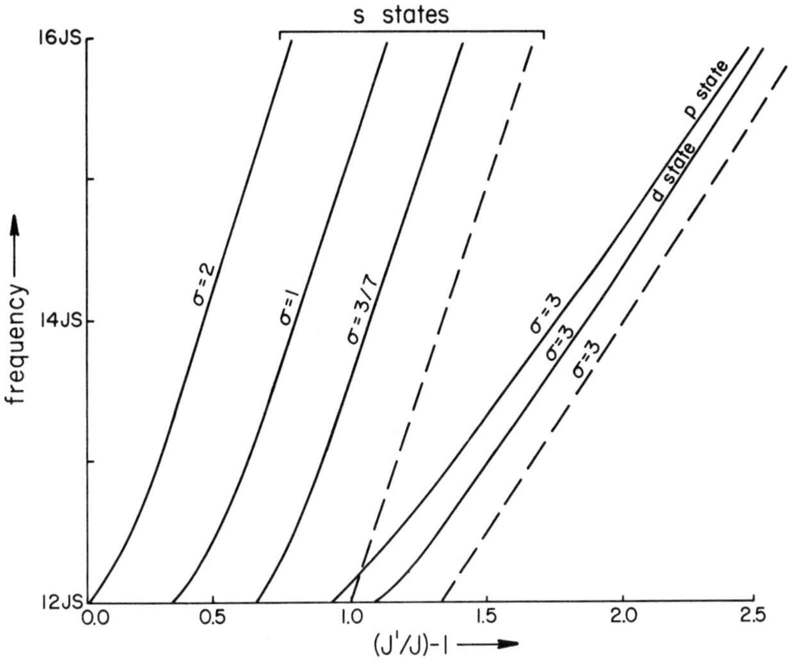

Fig. 7. The solutions of equations 39-41 for the bound states of a simple cubic ferromagnet with an impurity having spin $S'=\sigma S$ and exchange coupling J' to its near neighbors. The molecular field or Ising energies are shown as dashed lines.

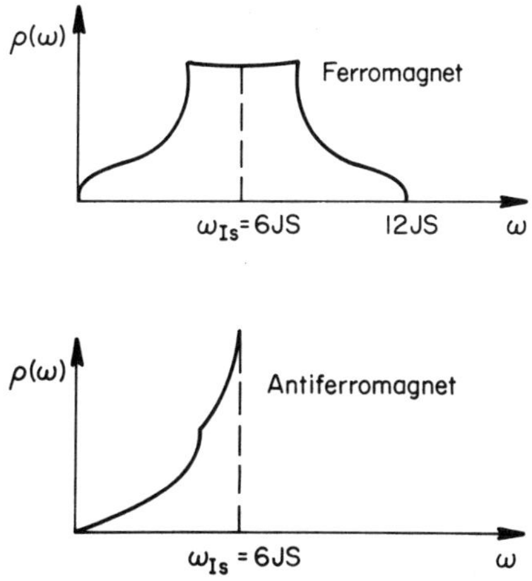

Fig. 8. Sketches of the density of spin wave states for a pure simple cubic ferromagnet and antiferromagnet.

These energies are shown in Fig. 7 together with the exact solutions from equations 39-41.

In antiferromagnets, molecular field theory of the impurity modes is much more satisfactory as the density of states shows a strong peak at the Ising energy. The theory of local modes in an antiferromagnet has been worked out by Lovesey[11].

In Fig. 9, we show the impurity modes in the antiferromagnetic system Ni:$RbMnF_3$($S=5/2$, $S'=1$, $J=4.7\ cm^{-1}$ and $J'=16.7\ cm^{-1}$).

It is instructive to compare the frequencies of the local modes shown in Fig. 9 (from the proper Green function treatment[11]) with those from the Ising model (42-44). You will find that they are very close.

4. INTERACTION OF RADIATION WITH THE MAGNETIC SYSTEM

We have discussed the various kinds of single spin-wave excitations that can exist in a pure crystal and around a defect, but we have not said how we can couple to these excitations with the radiation field in either an infra-red absorption experiment (one photon involved) or Raman scattering experiment (two photons involved). The general expression for the transition probability from a state $|i\rangle$ to a state $|j\rangle$ where these are product states of the crystal and the radiation field is given by[13]

$$W = 2\pi \sum_j \left| \langle i|V|j\rangle + \sum_m \frac{\langle i|V|m\rangle\langle m|V|j\rangle}{E_i - E_m} + \text{etc} \right|^2 \delta(E_i - E_j) \quad (45)$$

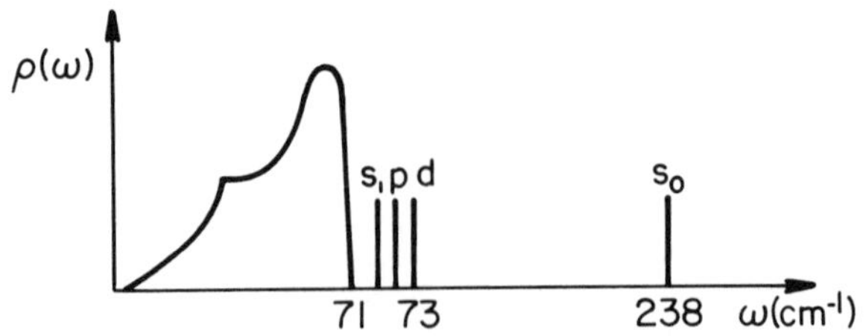

Fig. 9. Sketch of the local density of states in Ni:$RbMnF_3$ using parameters given in text. The s_0 mode has been observed by Misetich and Dietz[12] in a fluorescence experiment.

where V is the interaction between the crystal and the radiation field. The first term will be recognized as just the Fermi golden rule. We will treat the radiation field semi-classically and as $\underline{k} \cdot \underline{a}$ is very small (where $\underline{k}$ is the wavevector of the radiation and $\underline{a}$ an lattice vector); we can make the dipole approximation. The interaction between the crystal and the radiation may therefore proceed via either the electric dipole term

$$V_{E.D.} = -e\underline{E} \cdot \sum_i \underline{r}_i \tag{46}$$

or the magnetic dipole term

$$V_{M.D.} = -g\mu\beta\underline{H} \cdot \sum_i \underline{s}_i \tag{47}$$

where the summations are over all sites and electrons in the system.

In an absorption experiment, a single interaction with the radiation is involved. For example the interaction (47) is directly responsible for ferromagnetic and antiferromagnetic resonance involving a single magnon that is created by the $\sum_i S_i^-$ operator in (47) in the case of a ferromagnet. Raman scattering, on the other hand, involves two interactions with the radiation field, one at the incident frequency and one at the scattered frequency. Clearly the most direct way would be to use (47) for both of these. This was proposed by Bass and Kaganov[14] for one magnon processes. They proposed that one of the radiation fields scatters a magnon via $H^z \sum_i S_i^z \approx H^z NS$ at low temperatures in a ferromagnet and then the other radiation field causes a magnon to be created via $H^+ \sum_i S_i^- \approx H^+\sqrt{2S}\, a_0^\dagger$. Light scattering experiments are difficult in ferromagnets because the resonance frequency is usually low. However a similar mechanism is clearly operative in antiferromagnets and one should be able to see the k=0 spin waves. This is indeed the case[15] as shown in Fig. 10 for FeF_2.

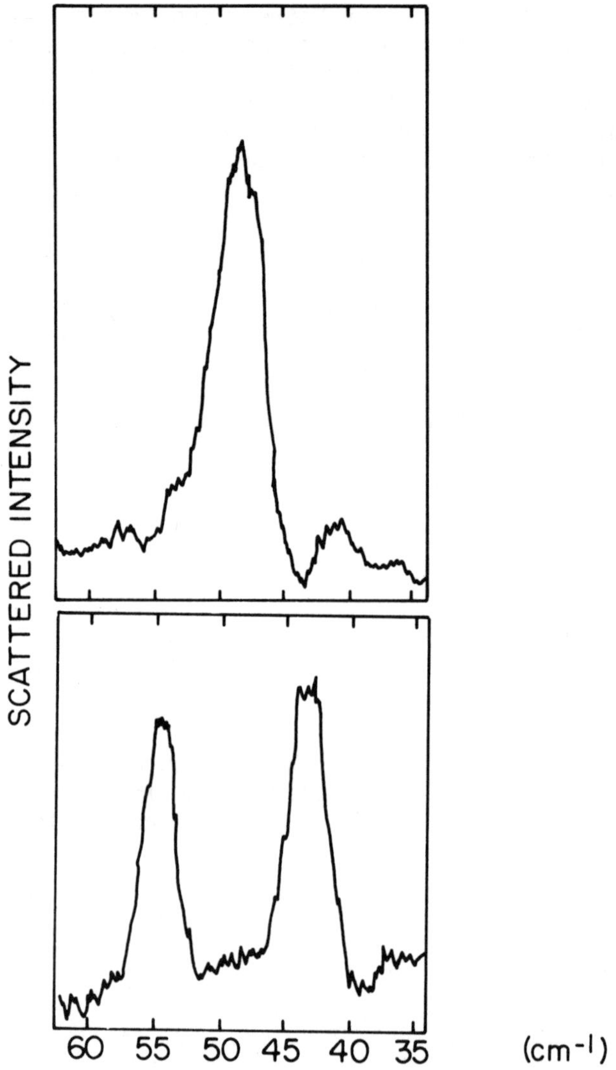

Fig. 10. The upper part of the figure shows the k=0 modes in FeF_2 in zero external field. The lower part shows the line split into two by the application of an external field as required by Eq.(24). Figure from Fleury and Loudon[15].

However the intensity of this mode is much larger than predicted by the Bass and Kaganov mechanism. Also the dependence on the polarization of the incident and scattered radiation fields is wrong. We therefore have to look to the only other mechanism capable of giving a larger cross section; the electric dipole interaction (46) which is larger by ~137, and can potentially give an additional factor $(137)^4 \approx 10^8$ in the cross section. However the electric field does not couple directly with the spin system and so we must invoke some indirect coupling to the spins.

For simplicity, let us consider the case of a half filled shell (e.g. Mn^{++}, $S = 5/2$) where the orbital angular momentum is quenched[2] in the ground state which is ^6S. We have $L = 0$ in the ground state. It was suggested by Elliott and Loudon[16] that the spin orbit coupling in an excited state could lead to a spin dependent interaction. We use the third order term in the expansion (45) [not explicitly written out] in which the operator $V_{E.D.}$ is used to connect the ground state of the atom with an excited state of opposite parity (i.e. a different configuration), the spin orbit coupling is allowed to act in this excited state and then $V_{E.D.}$ is used again to connect with the ground state. If we leave the spin as a free parameter in the ground state and take only orbital matrix elements, we get an effective interaction between the states $|i\rangle$ and $|j\rangle$ in (45)

$$H_{eff} = V + \sum_m \frac{V|m\rangle\langle m|V}{E_o - E_m} + \text{etc} \tag{45a}$$

and $E_o = E_i = E_j$ if we neglect the exchange energy as being very small compared to the atomic energies involved in (45a). Thus H_{eff} is essentially a single atom interaction and by symmetry must have the form

$$H_{eff} = A(\underline{E}_1 x \underline{E}_2 \cdot \sum_i \underline{S}_i) \tag{48}$$

where the sum goes over all sites and $\underline{E}_1$ and $\underline{E}_2$ are the electric fields of the incident and scattered radiation. The coefficient $A \sim e^2\langle r^2\rangle \lambda/\Delta E$ where $\lambda/\Delta E$ is the ratio of the spin orbit coupling energy λ to the configuration splitting ΔE and is typically ~10^{-3} . Thus this mechanism is potentially 100 times more intense than the Bass-Kaganov coupling via the magnetic dipole. The situation in FeF_2 is slightly more complicated as instead of a half filled shell, we have a singlet crystal field level in which $\langle \underline{L} \rangle = 0$. This leads to an interaction of the form (48) except that the z-z term in the dot product can have a different numerical coefficient from the x-x and y-y terms. Experiments by Fleury and Loudon[15] on FeF_2 (see Fig. 10) show that the magnitude predicted by (46) is the right order and more importantly, the polarization effects are predicted correctly.

Of more concern to us here are experiments in which two magnons are created via either an absorption process (involving 1 photon)[7] or a scattering process (involving two photons)[15]. The surprising feature of these experiments is that the second order scattering is as intense or even more intense than the first order scattering -- suggesting strongly that a different mechanism is involved.

One possible mechanism is to use the phonon modulation of the exchange integral[18] to give terms like

$$\left(\frac{\partial J_{ij}}{\partial r_{ij}}\right)_0 (\underline{u}_i - \underline{u}_j)\ \underline{S}_i \cdot \underline{S}_j$$

A virtual phonon created in a first order process will decay into two magnons in an antiferromagnet via the $S_i^+ S_j^-$ term above. Such a process should be sensitive to changes in the density of phonon states caused by say an electric field[19]. There is no evidence for this to date.

The most widely accepted mechanism for the creation of two magnons in an antiferromagnet is to use the exchange operator directly[20,15]. In an absorption experiment[20], we use $V_{E.D.}$ to couple to an excited state in one ion and then the exchange operator to get an effective coupling in the ground state of the form

$$H_{eff} = A \sum_{i,\delta} \underline{E} \cdot [(\underline{S}_i \times \underline{S}_{i+\delta}) \times \underline{\pi}_\delta] \qquad (49)$$

for $RbMnF_3$ where a pair of nearest neighbor Mn^{++} ions has D_{4h} symmetry[10]. The important feature is that the interaction is linear in $\underline{E}$ and involves terms like $S_i^- S^+_{i+\delta}$ where δ is a nearest neighbor vector. $\underline{\pi}_\delta$ is a pseudo-vector in the direction of $\underline{\delta}$. No two magnon absorption has been found in $RbMnF_3$ but it has been seen in MnF_2 where the interaction is similar to (49) but involving more constants because of the lowered symmetry.

The light scattering proceeds via a similar mechanism but involving an additional $V_{E.D.}$ for the scattered radiation. A nearest neighbor pair of manganese ions in $RbMnF_3$ has a symmetry D_{4h} and the combination of $\underline{E}_1$ and $\underline{E}_2$ that transforms as Γ_1^+ under this group is clearly

$$(B_1 - \tfrac{1}{3}B_3)(\underline{E}_1 \cdot \underline{E}_2) + B_3(\underline{E}_1 \cdot \hat{\underline{\delta}})(\underline{E}_2 \cdot \hat{\underline{\delta}})$$

where $\hat{\underline{\delta}}$ is a unit vector joining the two ions and B_1 and B_3 are constants. The complete interaction may then be formed by summing over all nearest neighbor pairs to give an effective interaction

$$H_{eff} = \sum_{i,\delta} \{(B_1 - \tfrac{1}{3}B_3)(\underline{E}_1 \cdot \underline{E}_2) + B_3(\underline{E}_1 \cdot \hat{\underline{\delta}})(\underline{E}_2 \cdot \hat{\underline{\delta}})\} \, \underline{S}_i \cdot \underline{S}_{i+\delta} \tag{50}$$

In the cubic group of $RbMnF_3$, this interaction leads to a Γ_1^+ mode (proportional to B_1^2) and a Γ_3^+ mode (proportional to B_3^2).

All the mechanisms that we have discussed in this section will also be operative in defect systems, except that the coefficients (e.g. B_1 and B_3 in (50)) will depend on position. However this may be regarded as a detail as it has a rather small effect on the shape of the observed Raman spectra. Notice that all the interactions discussed in this section have a symmetry appropriate to the paramagnetic phase rather than the ordered phase. We expect the interaction to be largely temperature independent because of the large energies of the intermediate states (this may not be true for the phonon assisted mechanism) and so it must have the form appropriate to the high symmetry phase.

The two major mechanisms discussed in this section lead to $\Delta M = 0$ processes; that is magnons with $\pm k$ from different branches are created. This is verified experimentally as the spectra are insensitive to magnetic fields. In ferromagnets there are no two magnon $\Delta M = 0$ processes. However it has been suggested[21] that a $\Delta M = 2$ process may be possible by using the spin orbit operator twice to give terms like $S_i^- S_i^-$. Such an effect has not been observed experimentally.

5. TWO MAGNON EXCITATIONS

a) Ferromagnetic Linear Chain

We will begin by looking at the two magnon states in ferromagnets. These can be claculated exactly as there is no ground state problem as in the antiferromagnet. We note that the Hamiltonian (1) has the property

$$[\sum_i S_i^z, H] = 0 \tag{51}$$

so that the eigenstates can be classified according to the value of the magnetization $\sum_i S_i^z = \mathcal{S}$. The ground state is a singlet and has $\mathcal{S} = NS$. The spin waves states have $\mathcal{S} = NS - 1$ and constitute a single band in the first Brillouin zone containing precisely N states. The two magnon states have $\mathcal{S} = NS - 2$ and there are $\frac{1}{2}N(N-1)$ states if $S = \frac{1}{2}$, when there can't be two spin deviations on the same site, and $\frac{1}{2}N(N+1)$ states if $S \geq 1$.

The energy of these states was first found for the $S = \frac{1}{2}$ linear chain by Bethe[22] and for the more general case of arbitrary spin in any dimension by Wortis[23]. We will derive Bethe's result by a slightly different route. A general wave function within the two magnon subspace may be written

$$\psi = \sum_{\ell \geq m} a_{\ell m} S_\ell^- S_m^- |0\rangle \tag{52}$$

where $a_{\ell m} = a_{m\ell}$. By carefully examing Schrödinger's equation using the Hamiltonian (1), and the relations (2) and (3), we find that the coefficients $a_{\ell m}$ obey the following equation

$$[\omega - E_0 - 2h - 2SJ(0)]a_{\ell m} + S\sum_j (J_{\ell j} a_{jm} + J_{mj} a_{j\ell})$$

$$= J_{\ell m}(a_{\ell\ell} + a_{mm} - a_{m\ell} - a_{\ell m}) \qquad (53)$$

The terms on the left hand side represent non-interacting spin waves and the term on the right hand side is the interaction that is only effective if ℓ and m are nearest neighbors. Neglecting the interactions, we can find solutions

$$a_{\ell m} = e^{i\underline{k}_1 \cdot \underline{\ell}} e^{i\underline{k}_2 \cdot \underline{m}} f(\underline{k}_1, \underline{k}_2) \qquad (54)$$

leading to

$$\omega = E_0 + \omega_{\underline{k}_1} + \omega_{\underline{k}_2} \qquad (55)$$

where $\omega_{\underline{k}}$ is just the spin wave energy (7). A more careful analysis shows that the interactions shift the frequency of these modes by $O(1/N)$. It is convenient to go to center of mass and relative momenta.

$$\underline{K} = \underline{k}_1 + \underline{k}_2$$

$$\qquad (56)$$

$$\underline{q} = \underline{k}_1 - \underline{k}_2$$

so that in one dimension for $S = 1/2$

$$\omega = E_0 + 2h + 2J[1 - \cos(\frac{K\delta}{2})\cos(\frac{q\delta}{2})] \tag{57}$$

for fixed K and $h = 0$, the two spin wave band is defined by

$$2J[1 + |\cos(\frac{K\delta}{2})|] \geq \omega - E_0 \geq 2J[1 - |\cos(\frac{K\delta}{2})|] \tag{58}$$

as is shown in Fig. 11. Notice that the spin wave lies entirely within the two spin wave band. It has been suggested that it might be possible to observe two magnon effects at finite temperatures

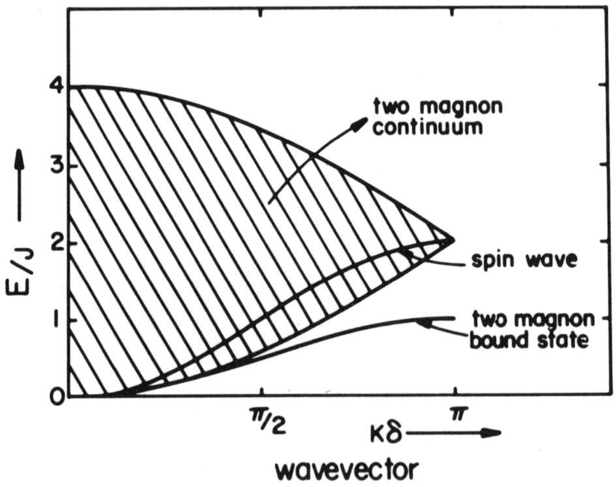

Fig. 11. Showing the spin wave, two magnon continuum and two magnon bound state for a linear chain with $S = \frac{1}{2}$.

when the two spin wave states will contribute to the self energy of the one spin wave states[24]. To see if we can find any bound state solutions of (53), in one dimension we will take out the center of mass motion by putting

$$a_{\ell m} = e^{iK(\ell+m)/2}\, F(\ell-m) \tag{59}$$

where $F(\ell-m)$ depends only on $|\ell-m|$. We find that

$$\begin{aligned}[\omega-E_0-2h-2SJ(0)]F(\ell-m) &+ 2S\sum_\delta J(\delta)\cos(\tfrac{K\delta}{2})F(\ell+\delta-m) \\ &= J(\ell-m)[\cos(\tfrac{K\delta}{2})F(0) - F(\ell-m)]\end{aligned} \tag{60}$$

where $J(\ell-m) = J_{\ell m}$

These equations look very similar to those for a one dimensional system with a single impurity in the vicinity of the origin[25] Writing these equations out in more detail with $\alpha = \omega - E_0 - 2h - 2SJ(0)$ and $c = \cos(\frac{K\delta}{2})$, we have

$$\alpha F(0) + 4JSc\, F(1) = 0 \tag{60a}$$

$$\alpha F(1) + 2JSc\, F(0) + 2JSc\, F(2) = Jc\, F(0) - J\, F(1) \tag{60b}$$

$$\alpha F(n) + 2JSc\, [F(n+1) + F(n-1)] = 0 \quad \text{for } n \geq 2 \tag{60c}$$

These equations are particularly simple for the case where

$S = \frac{1}{2}$, when $F(0) = 0$ because we can not have two spin deviations on the same site. Thus equation 60a is inoperative and F(0) is absent from in 60b. Therefore we put

$$F(n)/F(n+1) = e^{-\beta} \text{ for } n \geq 1 \tag{61}$$

and find that

$$\alpha + Jc\, e^{-\beta} = -J \tag{60b'}$$

and

$$\alpha + Jc(e^{-\beta} + e^{\beta}) = 0 \tag{60c'}$$

giving

$$e^{-\beta} = c$$

and

$$\alpha = -J - Jc^2$$

so that

$$\omega = E_0 + J/2[1 - \cos(K\delta)] \tag{62}$$

and

$$e^{-\beta} = \cos(\frac{K\delta}{2}) \tag{63}$$

The energy of the two magnon bound state is just half that of the spin wave state with the same total wave vector when h = 0. This is a surprise as we might have expected that the spin waves were the lowest lying excitations at small wave vectors. In three dimensions, this is the case[23], but in one and two dimensions the bound states have very low energies at small wave vectors and dominate the low temperature thermodynamics to such an extent that the fluctuations are large enough to prevent the system from having a spontaneous magnetization at any finite temperature[4].

It is instructive to examine the nature of the bound state in more detail. The wavefunction is particularly simple at the zone boundary $(K\delta = \pi)$ where all the $F(n) = 0$ except $F(1)$. This corresponds to the two spin deviations being on neighboring sites. The energy of this state is given by the Ising terms (i.e. those involving $S_i^z S_j^z$) in the Hamiltonian and is J. This is less than the Ising energy if the two spin deviations were further apart when it would be $2J$. At small wavevectors the bound state lies below the continuum by an energy $J(K\delta)^4/64$. In this case the wavefunction decays exponentially as the separation of the two spin deviations increases.

The equations (60) are more difficult to solve for $S \geq 1$ but again are simple at $K\delta = \pi$ when only $F(1)$ is non zero and

$$\omega = E_0 + 2h + J(4S-1)$$

Again this is just the Ising energy associated with spin deviations on nearest neighbor sites.

We see that the analogy with spin impurities is very useful in gaining insight into the mathematics and understanding the results. In higher dimensions; the modes can be classified using the appropriate point group as in the case of a real spin impurity, the only difference is that because $F(R) = F(-R)$ only even parity representations are present.

We see that spin wave interactions are attractive. This is because the attractive Ising effects are more important than the hard core repulsion that is greatest for $S = \frac{1}{2}$ when two spin deviations cannot occupy the same site.

Although these effects cannot be seen directly in ferromagnets, they are rather easy to see in antiferromagnets optically. Of course only the $K = 0$ part of the spectrum can be probed this way (i.e. along the ordinate in fig. 11) as the wavelength of light is very long compared to atomic separations. The two magnon states in ferromagnets correspond to $\Delta M = 2$ transitions whereas we will be interested in $\Delta M = 0$ transitions in antiferromagnets (i.e. one magnon from each branch (22a) and (22b) with wavevectors $\pm \underline{k}$). This is confirmed experimentally as the spectra are insensitive to an external magnetic field. A small dependence on the external field has been seen in the two magnon infra-red absorption spectrum of MnF_2 as it goes through the spin flop transition[26], due mainly to the magnetic dipole interactions coming in differently in the two phases.

b) Antiferromagnets

We will first calculate the two-magnon cross section in a simple-minded way assuming that there is no interaction between the two magnons. This fails to give agreement with the experimental results and so we will recalculate including the interaction effects.

We notice first that the Γ_1^+ part of the interaction (50) commutes with the Hamiltonian (16) in zero field apart from the h_A term. This is negligibly small in $RbMnF_3$ which is to an excellent approximation a nearest neighbor Heisenberg antiferromagnet. To calculate the Γ_3^+ cross section, we convert (50) into $\underline{k}$ space

$$H_{eff} = \frac{1}{2}B_3 \sum_{\delta} [(\underline{E}_1\cdot\hat{\underline{\delta}})(\underline{E}_2\cdot\hat{\underline{\delta}}) - \frac{1}{3}(\underline{E}_1\cdot\underline{E}_2)]\, e^{i\underline{k}\cdot\underline{\delta}}(S_{\underline{k}}^{1+}S_{\underline{k}}^{2-}+S_{\underline{k}}^{1-}S_{\underline{k}}^{2+}) \tag{64}$$

where we have neglected the longitudinal part of the interaction. (There is a small contribution to the two magnon cross section from these terms because of the presence of spin deviations in the ground state. This has been used[27] to see the two magnon cross section with neutrons in CoF_2). We transform from the S operators to the α,β operators using (21) and so (64) contains a piece

$$(\cosh^2\theta_{\underline{k}} + \sinh^2\theta_{\underline{k}})\, \alpha_{\underline{k}}^{\dagger}\, \beta_{\underline{k}}^{\dagger} \tag{65}$$

which creates an α magnon with wavevector $\underline{k}$ and a β magnon with wavevector $-\underline{k}$. This ensures that the composite excitation has $\underline{K} = 0$ and is insensitive to an external field. Henceforth we will put $h = 0$ Using the Fermi gold rule for the transition rate

$$W = 2\pi \sum_{\underline{k}} \left| B_3 \sum_{\delta} \{ (\underline{E}_1 \cdot \hat{\underline{\delta}})(\underline{E}_2 \cdot \hat{\underline{\delta}}) - \frac{1}{3}(\underline{E}_1 \cdot \underline{E}_2) \} e^{i\underline{k}\cdot\underline{\delta}} \right.$$

$$\left. \frac{1}{2}(\cosh^2\theta_{\underline{k}} + \sinh^2\theta_{\underline{k}}) 2S \right|^2 \delta(\omega - 2\omega_{\underline{k}})$$

$$= 2\pi B_3^2 [(E_1{}^x E_2{}^x)^2 + (E_1{}^y E_2{}^y)^2 + (E_1{}^z E_2{}^z)^2 - \frac{1}{3}(\underline{E}_1 \cdot \underline{E}_2)^2] \,.$$

$$\cdot\, J(0) S \sum_{\underline{k}} \frac{(\cos k_x a - \cos k_y a)^2}{\omega_{\underline{k}}} \delta(\omega - 2\omega_{\underline{k}}) \qquad (66)$$

The summation over the Brillouin zone can be done easily numerically and the result is shown as the dashed line in Fig. 12

The agreement between theory and experiment is poor. The reason for the disagreement is that we have not included the interaction between the two magnons that are created as wave packets in close proximity in real space where they feel a strong attractive interaction (similar to the one found in the ferromagnetic linear chain). As always, one has to resort to approximations to do the calculation in the antiferromagnet. It will be noticed that the calculated cross section in Fig. 12 looks very much like the density of states for the antiferromagnet sketched in Fig. 8 and peaks strongly at the maximum energy that occurs at the zone boundary in $\underline{k}$ space. This is because of the phase space weighting factor in the summation. The modes right at the maximum energy are described by the Ising part of the Hamiltonian and so we expect it to be useful in predicting the effect of magnon interactions.

If a single spin deviation is created, the excitation energy is JSz. If there are two spin deviations in the lattice the energy will be 2JSz, unless they are on neighboring sites when their energy will be lowered to 2JSz - J. These spin deviations can be thought of as a bound pair. We would expect a delta function response a fraction 1/2Sz (i.e. 1/30 for $RbMnF_3$) below the top of the two magnon band. When the transverse terms are added to the Hamiltonian the band extends from 0 up to 2JSz but the resonance is still about a fraction 1/2Sz below the top of the band. It actually shifts rather further in the present case because the critical point (due to magnons at $k\delta = (\pi/2, 0, 0)$ points in the Brillouin zone) suppresses the high frequency side

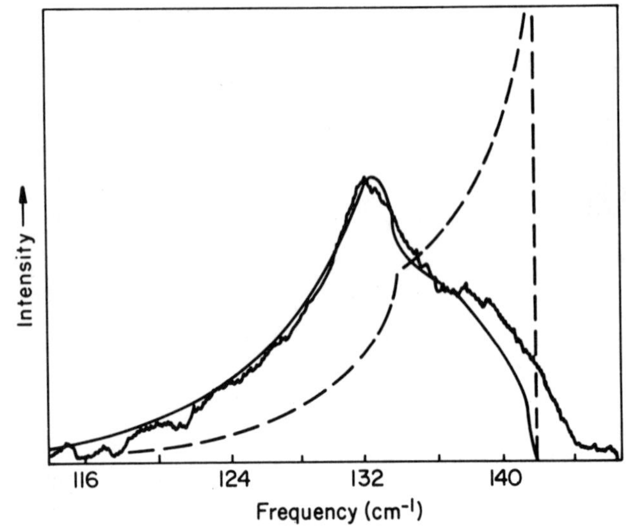

Fig. 12. Showing the experimental[28] results from Raman scattering on $RbMnF_3$. Also shown are the non-interacting spin wave theory (dashed line) and the inclusion of the interactions (solid line)[29,30].

of the resonance. We will describe the calculation of this spectra later in the section.

This experiment was very satisfying because it represented the first time the interactions between magnons had been clearly seen and there could be no argument. Previous work on MnF_2 had been rather inconclusive[15] because of the larger z (8 rather than 6) and because the existence of more distant exchange interactions made it difficult to say unequivocally that magnon interactions were being seen. These effects are greatest for small S and z and so are particularly large[31] in K_2NiF_4, a two dimensional nearest neighbor Heisenberg antiferromagnet.

In order to calculate the cross section including the interactions between the magnons it is convenient to use Green functions. We define a Green function at zero temperature

$$G(A/B) = \sum_i \left[\frac{\langle 0|A|i\rangle \langle i|B|0\rangle}{\omega - E_i + E_o} - \frac{\langle 0|B|i\rangle \langle i|A|0\rangle}{\omega + E_i - E_o}\right] \qquad (67)$$

If A, B are Bose operators, this reduces to the definition of a Green function used previously in the impurity problem (equations 28-31). Note that the zero refers to the ground state. Using a little algebra one can easily show that G(A/B) obeys the equation of motion

$$\omega G(A/B) = \langle 0|[A,B]|0\rangle + G([A,H]/B) \qquad (68)$$

The only other property of the Green function that we need is that the cross section is simply related to the imaginary part of (67) with A=B

$$2\mathrm{Im}\ G(A/A) = 2\pi \sum_i |\langle 0|A|i\rangle|^2 \delta(\omega - E_i + E_o) \qquad (69)$$

where we have used the relation

$$\mathrm{Im}\ \left(\frac{1}{x}\right) = \frac{1}{2i} \mathop{\mathrm{Lt}}_{\varepsilon \to 0^+} \left[\frac{1}{x+i\varepsilon} - \frac{1}{x-i\varepsilon}\right] = -\pi\delta(x) \qquad (70)$$

A thorough account of the properties of Green function can be found in Zubarev[32]. For the present case we take

$$A = B = \frac{1}{2}\sum_{\underline{i}} (S_i^+ S_{i+\delta}^- + S_i^- S_{i+\delta}^+) \qquad (71)$$

where again i labels the up sublattice and $\underset{\sim}{\delta}$ is a nearest neighbor vector. We will only sketch the calculation below as it is rather lengthy and not very instructive to go through in detail. Full details are given in ref. 30. Using (68) the commutator is evaluated approximately using the Néel state in place of the unknown ground state. The commutator [A,H] leads to terms of the form

$$S_\ell^z S_m^+ S_n^- \tag{72}$$

where either all three site labels are different or any pair are equal but never all three. If we just replace S_ℓ^z by $\pm S$ for the appropriate sublattice we get a closed set of equations for the Green function and find that the cross section is proportional to Im G_o where

$$G_o = \frac{1}{N}\sum_{\underline{k}} \frac{(\cos k_x a - \cos k_y a)^2}{\omega^2 - 4\omega_{\underline{k}}^2} \tag{73}$$

which is equivalent to the expression obtained previously (66) if we note that from (70)

$$\mathrm{Im}\left(\frac{1}{\omega^2 - 4\omega_{\underline{k}}^2}\right) = \frac{-\pi}{4\omega_{\underline{k}}}\ \delta(\omega - 2\omega_{\underline{k}}) \tag{74}$$

for positive frequencies. Clearly a more careful treatment of the terms (72) is required if we are to obtain an improved expression. The improved decoupling is as follows. If ℓ is not equal to m or n, we replace S_ℓ^z by $\pm S$ for the appropriate sublattice.

If $\ell = m$ and is in the up sublattice we put

$$S_\ell^z S_\ell^+ \to S S_\ell^+$$
$$S_\ell^+ S_\ell^z \to (S-1) S_\ell^+ \qquad (75)$$

Notice that these expressions maintain the commutation rule (2) and become exact for $S = \frac{1}{2}$. Similar relations are used for the down sublattice. They can be interpreted physically by noting that with two spin deviations present in the lattice, a spin can see its neighboring spin with its full value $\pm S$ if no spin deviation is present on it, or with the value $\pm(S-1)$ if a deviation is present. Using (75) we again obtain a closed system of equations for the Green function which has the form of a Dyson equation (33). This is not surprising because the first spin deviation appears as an impurity to the second. The equivalent of P (i.e. G_0) in the Dyson equation is obtained by neglecting the -1 in (75) and the interaction part V occurs when we include the -1. The Dyson equation can be solved exactly, as in the impurity case, because of the short range of the interaction in real space. We find

$$G = \frac{(1 - \frac{1}{2Sz})G_0}{1 + J(4Sz-1)G_0} \qquad (76)$$

and the cross section is proportional to Im G.

This modified form for the cross section is also shown in Fig. 12 and agrees very well with the experimental result. We note that the dependence on the polarization $\underline{E}_1$, $\underline{E}_2$ is unaffected by the interaction and is given by (66).

This kind of effect has been seen in very many antiferromagnets during the past decade. The theory sketched above is adequate to describe all these at very low temperatures but problems remain in extending it to finite temperatures. For a good recent review see U. Balucani and V. Tognetti[33]. Instead of using spin operators and a decoupling scheme, the whole problem can be reworked, perhaps in a more systematic way, using Bose operators. This gives very similar but not identical answers because the fluctuations in the ground state are treated differently.

c) Antiferromagnetic Alloys

Insulating antiferromagnetic alloys like $Rb_2Mn_xNi_{1-x}F_4$ have been grown and studied extensively in the past few years.

Clearly these systems are best understood in the limit of extreme dilution when one can think of isolated impurities. In this case the theory of section 3 is applicable and is in accord with experiment. For example the s_0 mode in Ni:$RbMnF_3$ has been seen[12] at 238 cm^{-1}. The two magnon spectra in the dilute limit offer numerous possibilities for combinations of two modes - however most of these would be expected to have a very low intensity. The Raman scattering[34] from the $s_0 + d$ combination in Ni:$RbMnF_3$ has been observed. This is excited primarily by creating the initial pair of spin deviations on the Ni and its nearest neighbor Mn with an interaction of the form (50). All the parameters for this system are known (the Ni-Mn exchange J' may be calculated from the observed frequency of the s_0 mode) and so we may calculate the pair mode, assuming that the s_0 and d mode do not interact. As shown in Fig. 13, the frequency obtained is too high and moreover is a delta function whereas the observed line has considerable width. A simple Ising model estimate shows that the interaction should lower the frequency of the pair mode by $J' = 16.7$ cm^{-1} and this agrees rather well with experiment. However the width is clearly an important property of this pair mode which cannot be obtained from the Ising part of the interaction.

We are faced with a three body problem, two magnons in the vicinity of an impurity, and so it is necessary to resort to a further approximation[35]. The creation of the $s_0 + d$ mode can be thought of as a two stage process. The s_0 mode is primarily (~ 90%) on the Ni ion and so the Ni behaves as if its spin were reduced from 1 to 0 (i.e. it becomes a vacancy). The d mode is then created around an effective vacancy where it is no longer bound. In other words the presence of the s_0 mode allows the d mode to escape. These ideas can be put together mathematically to give the line shape in Fig. 13 which is in very satisfactory agreement with experiment. The situation as of 1972 regarding impurity modes in magnetic systems in summarized by Cowley and Buyers[36].

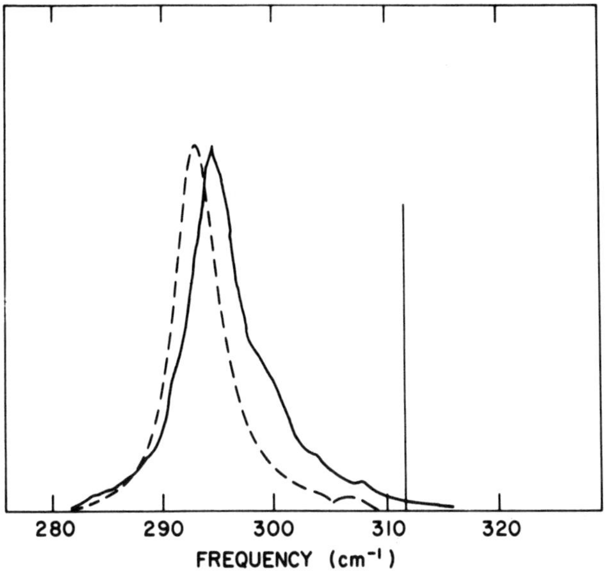

Fig. 13. Showing the $s_0 + d$ two magnon Raman scattering[34] in Ni:$RbMnF_3$. The delta function is the sum of the energies of the noninteracting s_0 and d modes both of which are localized. The interaction effectively delocalizes the composite excitation.

The alloys, away from the dilute limit, represent a formidable problem. However as the interactions are so simple, they represent perhaps the best systems in which to study single excitations. The one magnon response function $S(k,\omega)$ has been measured[37] in the two dimensional antiferromagnet $\bar{R}b_2Mn_{1/2}Ni_{1/2}F_4$. Two spin wave branches are seen as shown in Fig. 14.

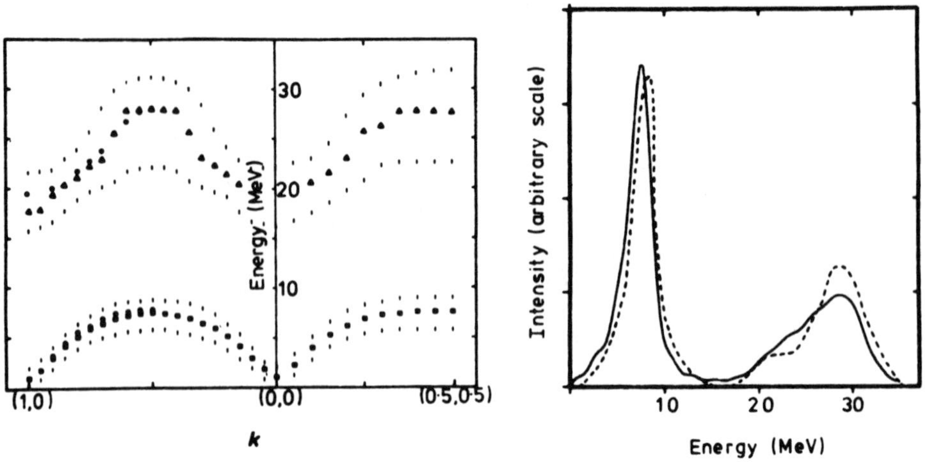

Fig. 14. Showing the one spin wave neutron scattering law $S(\underline{k},\omega)$ in $Rb_2Mn_{1/2}Ni_{1/2}F_4$. On the left the computed dispersion curves are shown. The squares and triangles represent the centers of the peaks and the dashes the width at half the peak height. The circles are the experimental results. On the right land side[38] (solid line) is compared to experiment[37] (dashed line) at the zone boundary. The horizontal scale is energy in meV and the vertical scale is the intensity in arbitrary units.

The lower branch is predominantly on the Mn ions and the upper branch on the Ni although the widths are an indication of the mixing. The theoretical curves were obtained from a computer simulation of the spin dynamics using linearized equations of motion. This is a straightforward procedure and gives excellent agreement with experiment. This is in contrast to quite elaborate C.P.A. theories[39] which even including quite large clusters are not very satisfactory[40]. The upper peak shows some structure at higher resolution in the computer simulations which the experiments were not able to pick up in $Rb_2Mn_{1/2}Ni_{1/2}F_4$. However they have recently been seen in the dilute antiferromagnet[40] $Rb_2Mn_cMg_{1-c}F_4$. The structure is due to the Ising terms in the Hamiltonian and different kinds of cluster. Thus the single spin wave excitations are understood in as much as they can be simulated very accurately on a computer.

The two magnon spectra in these systems have also been observed optically. They are easy to understand in general overal behavior although perhaps not in detail. In $Rb_2Mn_{1/2}Ni_{1/2}F_4$, two broad peaks are seen[41] corresponding to creating pairs of spin deviations on Ni-Ni and Ni-Mn nearest neighbor pairs. These peaks occur about in the positions that are predicted by the Ising model. The most surprising feature of this experiment is the relative intensities in these two peaks and the absence of a Mn-Mn peak. If we accept the excited state exchange coupling mechanism described in section 4 and if the exchange in the excited state scales with exchange in the ground state, we would expect the intensity in the central peak to be roughly the geometric mean of the other two. This is not the case[41] and suggests that we may not understand the details of the interaction with the radiation field. Presumably we have the correct form in the pure system (50) even if we do not properly understand the origin of B_3. The experiments in the dilute system[42] $Mn_{1-c}Zn_cF_2$ show that increasing the concentration of Zn acts rather like increasing the temperature in the pure system in that the excitations move to lower energies and broaden out.

It is questionable if it is worth putting too much effort into studying two magnon spectra in alloy systems as we are not going to learn much about alloys and we already understand two magnon processes in pure systems. It would be worthwhile to study the intensity and position in the various peaks in alloys like $Rb_2Mn_cNi_{1-c}F_4$ as a function of concentration to try and learn more about the optical interaction.

G. Conclusions

In these lectures we have covered quite a large territory briefly. The main impression I want to leave with you is that there are no great mysteries here. Because of the simple Hamiltonian there is now a rather sophisticated understanding of spin waves in pure systems, systems with isolated impurities and alloys. The two magnon spectra are well understood in pure systems although not quite so well in systems with isolated defects and alloys. The finite temperature behavior of these systems is more difficult to understand because of the large anharmonicity that ultimately of course leads to the phase transition.

These notes are based on unpublished lectures given at the 1970 summer school on Light Scattering in Solids held at Northwestern University. They were updated in the summer of 1977 during a visit to U.K.A.E.A., Harwell, England; whom I would like to thank for their hospitality.

REFERENCES

1. For a general reference see C. Kittel "Quantum Theory of Solids" (John Wiley and Sons, Inc., New York, 1963), Ch. 4.
2. R. J. Elliott and M. F. Thorpe, J. Appl. Phys. 39, 802 (1968).
3. F. J. Dyson, Phys. Rev. 102, 1217 (1956).
   J. Zittartz, Zeit. für Physik, 184, 506 (1965).
4. N. D. Mermin and H. Wagner, Phys. Rev. Lett. 17, 1133 (1966).
5. P. W. Anderson, Phys. Rev. 83, 1260 (1951).
6. G. F. Koster and J. C. Slater, Phys. Rev. 96, 1208 (1954).
7. P. G. Dawber and R. J. Elliott, Proc. Phys. Soc. 273, 222 (1963).
8. T. Wolfram and J. Callaway, Phys. Rev. 130, 2207 (1963).
9. Lord Rayleigh, "Theory of Sound" Vol. 1 sec. 90 (Macmillan and Co., London 1877)
   E. W. Montroll and G. H. Weiss, Rev. Mod. Phys. 30, 1751 (1958) Appendix A.
10. G. F. Koster, T. O. Dimmock, R. G. Wheeler, and H. Statz, "Properties of the 32 Point Groups" (M.I.T. Press, Cambridge, MA, 1963).
11. S.W. Lovesey, Proc. Phys. Soc. C1, 102 (1968).
12. A. Misetich and R. E. Dietz, Phys. Rev. Letters 17, 392 (1966).
13. L. D. Landau and E. M. Lifshitz, "Quantum Mechanics", (Addison-Wesley Pub. Co., Reading, Mass. 1965), Chapter 6.
14. F. G. Bass and M. J. Kaganov, J.E.T.P. 10, 986 (1960).
15. P. A. Fleury and R. Loudon, Phys. Rev. 166, 514 (1967).
16. R. J. Elliott and R. Loudon, Phys. Lett. 3, 189 (1963).
17. R. Loudon, Advances in Physics 17, 243 (1968).
18. J. W. Halley, Phys. Rev. 149, 423 (1966).
19. J. W. Halley in "Light Scattering Spectra of Solids", Ed. G. B. Wright, (Springer-Verlag, New York, 1969), p. 207.
20. Y. Tanabe, T. Moriya and S. Sugano, Phys. Rev. Lett. 15, 1023 (1965).
21. M. F. Thorpe, Phys. Rev. B4, 1608 (1977).
22. H. A. Bethe, Z. Physik 71, 205 (1931).
23. M. Wortis, Phys. Rev. 132, 85 (1963).
24. R. Silberglitt and A. B. Harris, Phys. Rev. Lett. 19, 30 (1967).
25. See for example C. Kittel "Introduction to Solid State Physics" (John Wiley and Sons, Inc., New York, 1967), p. 156.
26. T. Bernstein, A. Misetich and B. Lax, Phys. Rev. B6, 979 (1972).
27. R. A. Cowley, W. J. L. Buyers, P. Martel and R. W. H. Stevenson, Phys. Rev. Lett. 23, 86 (1969).
28. P. A. Fleury, Phys. Rev. Lett. 21, 151 (1968).
29. R. J. Elliott, M. F. Thorpe, G. F. Imbusch, R. London and J. Parkinson, Phys. Rev. Lett. 24, 147 (1968).
30. R. J. Elliott and M. F. Thorpe, J. Phys. C2, 1630 (1969).

31. S. R. Chinn, H. J. Zeiger and J. R. O'Connor, Phys. Rev. B3, 1709 (1971).
32. D. N. Zubarev, Sov. Phys. Uspekhi 3, 320 (1960), see also the lectrues by Sinchcombe in this volume.
33. U. Balucani and V. Togentti, La Rivista del Nuovo Cimento 6, 39 (1976)
34. A. Oseroff, P. S. Pershan and M. Kestigian, Phys. Rev. 188, 1046 (1969).
35. M. F. Thorpe, Phys. Rev. Lett. 23, 472 (1969).
36. R. A. Cowley and W. J. L. Buyers, Rev. Mod. Phys. 44, 506 (1972).
37. J. Als-Nielson, R. J. Birgeneau and G. Shirane, Phys. Rev. B12, 4963 (1975).
38. M. F. Thorpe and R. Alben, J. Phys. C9, 2555 (1976).
39. R. J. Elliott, J. N. Krumhansl and P. L. Leath, Rev. Mod. Phys. 46, 465 (1974).
40. R. A. Cowley, G. Shirane, R. J. Birgeneau and H. J. Guggenheim, Phys. Rev. B15, 4292 (1977).
41. P. A. Fleury and H. J. Guggenheim, Phys. Rev. B12, 985 (1975).
42. M. Buchanan, W. J. L. Buyers, R. J. Elliott, R. T. Harley, W. Hayes, A. M. Perry and I. D. Saville, J. Phys. C5, 2011 (1972).

A GREEN FUNCTION APPROACH TO TWO-MAGNON LIGHT SCATTERING IN ANTIFERROMAGNETS AT $T < T_N$

M.G. Cottam

Physics Department, University of Essex

Colchester CO4 3SQ, England

1. INTRODUCTION

In this paper we describe some calculations for the temperature dependence of two-magnon light scattering in antiferromagnets. This involves using a Green function diagrammatic perturbation expansion to generalise the zero temperature results of Elliott and Thorpe[1] to $T \neq 0$. The theory and results are described here only in outline, whilst details are given in refs.2,3 and 4. Other calculations for two-magnon light scattering at $T \neq 0$ have recently been reviewed by Balucani and Tognetti[5]. In section 2 we describe the Green function perturbation formalism and its application to evaluating the two-magnon scattering cross-section. In section 3 the theory is applied to several antiferromagnets for $T < T_N$ and comparison is made with experimental data.

2. THEORY

In two-magnon scattering the incident photon interacts with the magnetic material to create two magnons with wavevectors k and -k. As in ref.1 we assume that the interaction giving rise to this process is proportional to an isotropic combination $S_r.S_{r+\rho}$ of spin operators, where ρ is a vector linking nearest neighbour sites only. The Raman cross-section is then proportional to

$$(1 - e^{-\beta\omega})^{-1} \text{ Im } G(\rho,\rho',\omega+iO^+) \quad (1)$$

where $\beta = 1/k_BT$, $\omega = \omega_i - \omega_s$ (ω_i and ω_s are the incident and scattered photon frequencies respectively), and G is a Green function:

$$G(\rho,\rho',\omega) = \langle\langle \sum_r S_r . S_{r+\rho} \; ; \; \sum_{r'} S_{r'} . S_{r'+\rho'} \rangle\rangle \tag{2}$$

It is also convenient here to define $G_1(\rho,\rho',\omega) = \langle\langle P(\rho);P(\rho')\rangle\rangle$, $G_2(\rho,\rho',\omega) = \langle\langle P(\rho);Q(\rho')\rangle\rangle$ and $G_3(\rho,\rho',\omega) = \langle\langle Q(\rho);Q(\rho')\rangle\rangle$, where

$$P(\rho) = \tfrac{1}{2}\sum_r (S_r^+ S_{r+\rho}^- + S_r^- S_{r+\rho}^+), \quad Q(\rho) = \sum_r S_r^z S_{r+\rho}^z \tag{3}$$

By separately evaluating G_1, G_2 and G_3, we may deduce $G(\rho,\rho',\omega)$.

To do this we use a diagrammatic perturbation theory, which may be carried out in terms of either the drone-fermion method[6,7] (for spin $S = \frac{1}{2}$) or the Vaks, Larkin and Pikin method[8] (for general S), together with an expansion in powers of $1/z$, where z is the number of spins interacting with any spin. In lowest order $(1/z)^0$ this leads to molecular field theory, whilst spin fluctuation effects are obtained from higher orders. Magnon-magnon interactions are important in two-magnon scattering[1], and this necessitates carrying out calculations up to order $(1/z)^2$. To establish the diagram technique we define

$$T_{\ell,\ell'}(k,\omega) = \sum_r \exp\{ik.(r-r')\}\langle\langle S_r^+ ; S_{r'}^- \rangle\rangle$$
$$L_{\ell,\ell'}(k,\omega) = \sum_r \exp\{ik.(r-r')\}\langle\langle (S_r^z - \langle S_r^z\rangle) ; (S_{r'}^z - \langle S_{r'}^z\rangle)\rangle\rangle \tag{4}$$

where ℓ and ℓ' are sublattice labels for sites r and r' (with $\ell = 1$ and $\ell = 2$ referring to sublattices with spins up and down respectively). These Green functions may first be evaluated in molecular field theory $(1/z)^0$ and then renormalised to order $(1/z)^1$, corresponding to non-interacting magnons. Details are given in ref.7, but the calculation may be regarded as formally analogous to a diagrammatic RPA evaluation of the electron-hole propagator for an interacting fermi gas[9] (where the non-interacting pair propagator and the fermion interaction energy play a similar role to the molecular field Green functions and the exchange interaction respectively). For zero applied magnetic field the results are[7]

$$T(k,\omega) = \frac{2R}{(\omega^2-\omega_k^2)} \begin{pmatrix} -(\omega+\gamma) & -RJ(k) \\ -RJ(k) & (\omega-\gamma) \end{pmatrix} \tag{5}$$

$$L(k,\omega) = \delta_{\omega,0} \frac{R'}{\{1-(R'J(k))^2\}} \begin{pmatrix} 1 & R'J(k) \\ R'J(k) & 1 \end{pmatrix} \tag{6}$$

$$\omega_k = \{\gamma^2 - R^2J^2(k)\}^{\frac{1}{2}}, \quad \gamma = H_A + RJ(0) \tag{7}$$

where $R = |\langle S^z\rangle|$ is a sublattice magnetisation factor evaluated in the molecular field approximation, $J(k)$ is the Fourier transform of the exchange interaction, H_A is the anisotropy field, and $R' = \partial R/\partial\gamma$. At low temperatures, where $R \simeq S$, ω_k reduces to the usual expression for the spin wave energy.

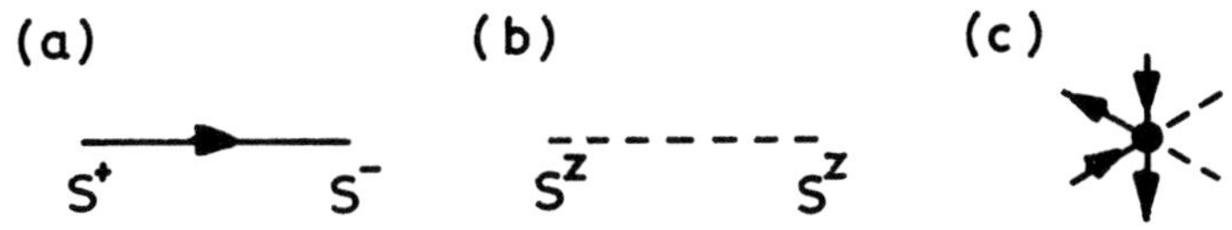

FIG.1. The diagrammatic representation

Diagrammatically we may represent $T(k,\omega)$ and $L(k,\omega)$ in order $(1/z)^1$ by a solid directed line and a broken line as in Fig.1(a) and (b) respectively. Contributions to any other quantity can be represented as diagrams containing these lines together with various interaction vertices[8] drawn as shown in Fig.1(c). Generally these vertices involve 2m solid lines and m' broken lines, where m and m' are zero or positive integers and $n = (2m+m') \gtrsim 3$. They are related to molecular field averages of the form $\langle S^{\alpha_1} S^{\alpha_2} \ldots S^{\alpha_n} \rangle$ (where α_i = +,−,z). These complicated vertices occur because there is no direct analogue of Wick's theorem for spins. The $1/z$ dependence of any diagram can be obtained from the simple rule[7] that each independent momentum label in the diagram gives a $1/z$ factor. In the present context the theory is valid for $T \lesssim \frac{3}{4} T_N$.

We first consider $G_1(\rho,\rho',\omega)$, which in fact gives the dominant contribution to the intensity in most cases. The lowest order $(1/z)^1$ diagram is shown in Fig.2(a), and it corresponds to the magnon pair propagating without interaction. The result is[3]

$$G_1^o(\rho,\rho',\omega) = -4R^2\gamma^2 \sum_k \frac{\coth(\frac{1}{2}\beta\omega_k)}{\omega_k(\omega^2-4\omega_k^2)} \exp\{-ik.(\rho-\rho')\} \tag{8}$$

By going to next order $(1/z)^2$ we include the dominant effect of magnon-magnon interactions, and the appropriate diagrams are shown in Fig.2(b). We also take into account repeated scattering diagrams (such as in Fig.2(c)) by means of a Dyson equation. For simple cubic antiferromagnets the light scattering mode of interest corresponds to Γ_3^+ symmetry, for which we obtain[3]

$$G_1(\Gamma_3^+) = -\frac{2R^2\gamma(2\gamma-J)G_1^o(\Gamma_3^+)}{\{1 + 4J\gamma^2 G_1^o(\Gamma_3^+)} \tag{9}$$

$$G_1^o(\Gamma_3^+) = \frac{1}{N}\sum_k \frac{\coth(\frac{1}{2}\beta\omega_k)}{\omega_k(\omega^2-4\omega_k^2)} (\cos k_x a - \cos k_y a)^2 \tag{10}$$

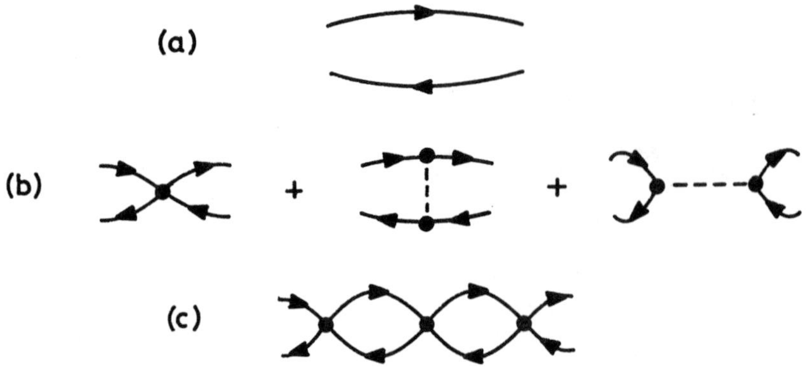

FIG.2. Some contributions to $G_1(\rho,\rho',\omega)$.

Analogous results are obtained for rutile structure antiferromagnets[4]. At low temperatures eqs.(9) and (10) become similar to results obtained by Elliott and Thorpe[1] for T = 0. Green functions G_2 and G_3 may also be evaluated using the diagram formalism. Details are given in ref.3, where we conclude that their dominant effect is to modify the terms in eq.(9) by a fraction 1/(2Rz), which is small in the region of validity of the theory.

3. APPLICATIONS AND DISCUSSION

The theory has been applied to obtain numerical results for simple cubic antiferromagnets $RbMnF_3$ and NiO (see refs. 2 and 3) and for rutile structure antiferromagnets MnF_2 and FeF_2 (see ref.4) for temperatures up to about $\frac{3}{4}T_N$. For example, we show in Fig.3 calculations for NiO (S=1 and T_N=523 K) for Γ_3^+ symmetry at (a) 1.4 K and (b) 297 K. The assumed values for J and H_A are quoted in ref.3. The predicted two-magnon spectra (solid curves) are in reasonable agreement with experiment[10] (dotted curves).

Generally it is concluded that the theory is fairly satisfactory in accounting for the principal features of the two-magnon spectra at $T \lesssim \frac{3}{4}T_N$. The variation with temperature of the resonance peak is particularly well described. However, at elevated temperatures the high frequency cut-off to the spectrum and the width of the two-magnon resonance are predicted to have lower values than found experimentally. This is probably due to our having neglected higher order terms in the perturbation expansion which lead to one-magnon energy renormalisation (to a value Ω_k) and damping Γ_k. When these effects are included (see ref.3), we conclude that eq.(9) is

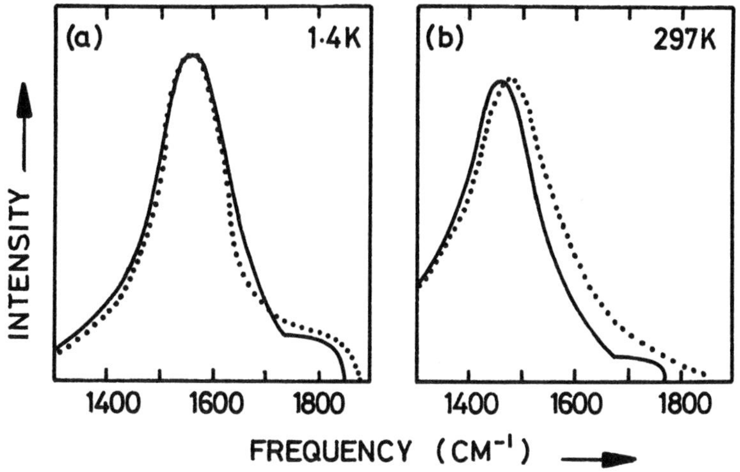

FIG. 3. Theoretical (full curves) and experimental (dotted curves) two-magnon spectra of NiO (Γ_3^+ symmetry) at (a) 1.4 K and (b) 297 K.

still applicable but with $G_1^0(\Gamma_3^+)$ redefined as

$$G_1^0(\Gamma_3^+) = \frac{1}{N}\sum_k \frac{\coth(\frac{1}{2}\beta\Omega_k)}{\Omega_k(\omega-2\Omega_k+2i\Gamma_k)(\omega+2\Omega_k+2i\Gamma_k)}(\cos k_x a-\cos k_y a)^2 \qquad (11)$$

This modification, with damping included, gives improved agreement with experiment, as discussed by other authors[5,11].

REFERENCES

1. R.J. Elliott and M.F. Thorpe, J. Phys. C 2, 1630 (1969), and preceding section of this volume.
2. M.G. Cottam, Solid St. Comm. 10, 99 (1972)
3. M.G. Cottam, J. Phys. C 5, 1461 (1972)
4. M.G. Cottam, Solid St. Comm. 11, 889 (1972)
5. U. Balucani and V. Tognetti, Riv. Nuovo Cimento 6, 39 (1976)
6. H.J. Spencer, Phys. Rev. 167, 434 (1968)
7. M.G. Cottam and R.B. Stinchcombe, J. Phys. C 3, 2305 (1970)
8. V.G. Vaks, A.I. Larkin and S.A. Pikin, JETP 26, 188 (1968)
9. See for example D. Pines and P. Nozieres, "The Theory of Quantum Liquids, 1" (Benjamin, New York, 1966), section 5.5
10. R.E. Dietz, G.I. Parisot and A.E. Meixner, Phys. Rev. B4, 2302 (1971)
11. C.R. Natoli and J. Ranninger, J. Phys. C 6, 345 (1973)

TWO-SPIN LIGHT SCATTERING IN HEISENBERG ANTIFERROMAGNETS

U.Balucani and V.Tognetti

Laboratorio di Elettronica Quantistica del C.N.R.

Via Panciatichi 56/30, Florence, Italy

Since its first experimental observations [1] two-spin Raman scattering in antiferromangets has proved to be a very useful tool to study the high-wavevector magnetic excitations in these systems. The scattering cross section turns out [2] to be proportional to the Fourier transform of $<M(0)M(t)>$ where the Raman transition operator is given by

$$M = \sum_{\vec{k}} \Phi_k (\vec{S}_{\vec{k}} \cdot \vec{S}_{-\vec{k}}) \qquad (1)$$

Here $\vec{S}_k$ is the space Fourier transform of the site spin operators and Φ_k is a weighting factor determined by the field polarizations and by crystal symmetry. In the following we shall explicitly refer to three-dimensional cubic antiferromagnets like $RbMnF_3$ and $KNiF_3$, which are well described by a Heisenberg Hamiltonian with exchange J limited to the z nearest neighbours of a given spin.

The scattering process very strongly weighs the Brillouin zone-boundary excitations. Moreover, the four-spin correlations involved in $<M(0)M(t)>$ lead to two-magnon interaction effects present even in the T=0 limit [2]. In order to fully understand the short-range spin dynamics a theoretical study of the temperature dependence of the spectra proves to be necessary. In the most part of the ordered region a description of the magnetic system in terms of interacting bosons works quite well:for our problem it is not necessary to worry much about the kinematical restrictions introduced by the non-Bose character of the spin operators [3]. An approach based on two-particle Green functions, when pursued to the 2nd order [4] has been found able to explain the main observed spectral features, i.e. a downward shift of the inelastic peak frequency and a marked broadening of the spectra at increasing temperatures. Many-body diagrammatic techniques [5] give essentially the same formal results. In both methods all the relevant T-dependent physical effects like single-magnon self-energies and two-magnon interaction terms arise consistently at the first two stages of the Green function hierarchy. The main results of the approach are the following. The downward shift can be basically interpreted with a Hartree-Fock renormalization of high $\vec{k}$ one-magnon energies: here two-magnon effects play a comparatively smaller role. Even the broadening of the experimental spectra turns out to be dominated by one-magnon properties. The zone-boundary magnon damping Γ has been evaluated in Ref.4 taking into account the sharp peaking of the density of states in this energy region. Γ increases rapidly with temperature and accounts for the increasing widths of the two-magnon spectra. Fig.1 shows the good agreement between the theoretical and observed [5,6] peak frequency and width up to $T \sim 0.8\ T_N$ in $KNiF_3$. Similar agreement has been also found for $RbMnF_3$ [4], $KMnF_3$ [7], K_2NiF_4 and K_2MnF_4 [8]. Summing up, in the most of the ordered region two-spin light scattering can be used as a powerful and fully understood technique for the investigation of zone-boundary magnons, in alternative to inelastic neutron scattering.

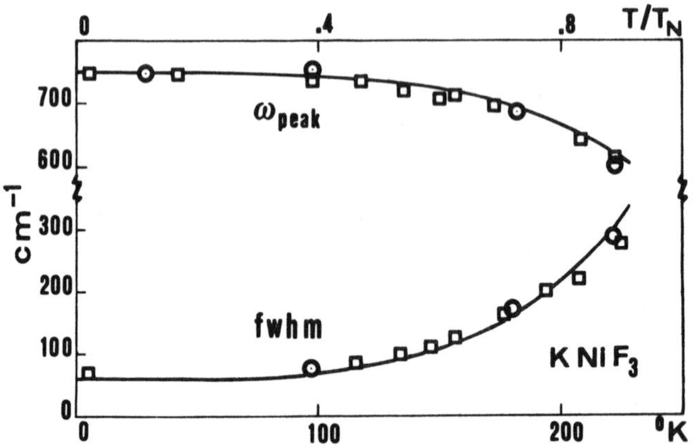

Fig.1-Stokes peak frequency and full width at half maximum of the two-magnon spectra of $KNiF_3$ in the ordered phase. Full lines: 2nd order Green function theory result. Experimental data: from Ref.5 (□) and from Ref.6 (○).

In the paramagnetic phase all experimental spectra show the persistence of a broad inelastic peak up to $T \sim 1.4\ T_N$. Only at $T >> T_N$ the spectra have a structureless shape centered around $\omega=0$. Physically, one can explain this persistence recalling that the two-spin Raman scattering process privileges high wavevectors, i.e. samples only the behaviour of clusters of neighboring spins, thus giving a measure of "short range magnetic order" which is present in the system at all finite temperatures. However, for a quantitative interpretation of the observed spectra at $T>T_N$ conventional many-body methods (all based on the concept of well defined elementary excitations) are of little use because of the overdamped character of the "excitation". The problem can be instead approached by other more general theoretical methods, such as Mori's continued fraction representation of the relevant dynamic correlation [9]. In our case, one introduces the canonical correlation function:

$$f_o(t) = \int_0^\beta d\lambda\ \langle e^{\lambda H} M(0)\ e^{-\lambda H} M(t)\rangle \cdot \langle M(0)M(0)\rangle^{-1} \qquad (2)$$

where H is the magnetic Hamiltonian and $\beta = (k_B T)^{-1}$. Then its Laplace transform $\hat{f}_o(z)$ has the following continued fraction representation:

$$\hat{f}_o(z) = \left[z + \Delta_1 \hat{f}_1(z)\right]^{-1} \quad , \quad \hat{f}_n(z) = \left[z + \Delta_{n+1} \hat{f}_{n+1}(z)\right]^{-1} \qquad (3)$$

Here, the quantities Δ_n are related to the frequency moments $\langle \Omega^{2n} \rangle$ of the even function $\hat{f}_o(z=i\omega)$. The scattering cross section turns out to be proportional to $Re\ \hat{f}_o(z=i\omega)$ times a detailed balance factor $\omega[1 - \exp(-\beta\omega)]^{-1}$. The representation (3) is formally exact, but in practice the continued fraction must be terminated at a certain (hopefully low) stage by means of physically plausible statistical approximations. The quantities controlling the latter are i) the time scales (or the frequency range) of the processes under consideration and ii) the frequency moments, the first ones of which are often amenable to a first-principle calculation.

In our case, the quantities Δ_1 and Δ_2 can be approximately evaluated in all the paramagnetic region by means of a decoupling procedure [10]. They turn out to be dominated by the high $\vec{k}$ contribution: e.g. the ratio G between Δ_1 and its pure zone-boundary value Δ_1^{ZB} is nearly 1 through all the paramagnetic phase. The structure of experimental spectra near T_N can be accounted for with a termination of the

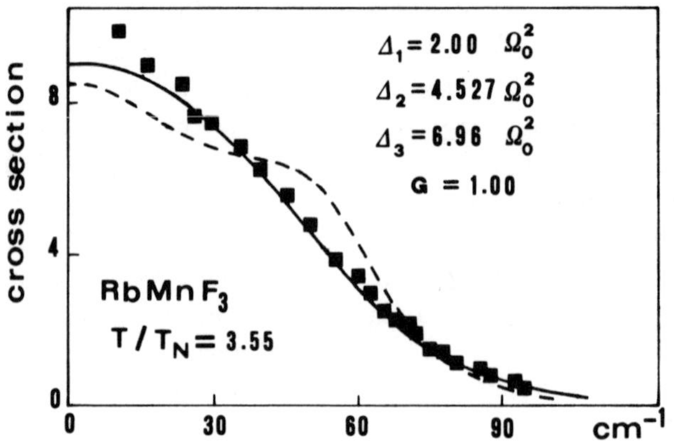

Fig.2-Two-spin Stokes spectrum in $RbMnF_3$ at room temperature. Full line: modified long-time theory; dashed line: purely Markoffian theory. $\Omega_0^2 = (2/3)J^2zS(S+1)$. Experimental data (■) from Ref.9.

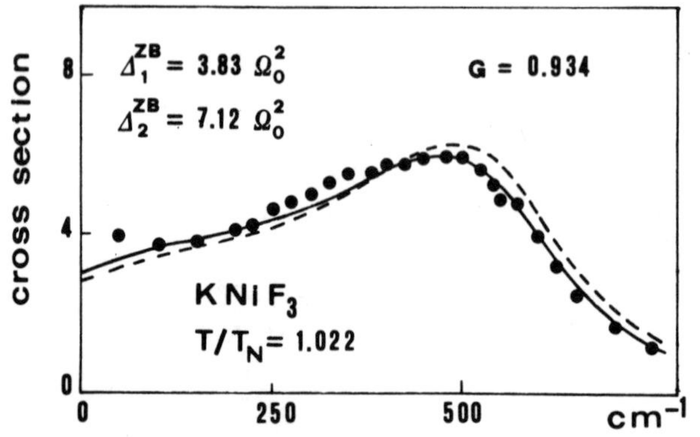

Fig.3-Two-spin Stokes spectrum in $KNiF_3$ at $T \sim 1.02\ T_N$. Full and dashed lines: theoretical shapes with G=0.934 and G=1 respectively (see text). Experimental data (●) from Ref.6.

continued fraction at the 3rd stage; moreover, since the most important spectral features lie in a rather limited frequency region around $\omega=0$, it is convenient to terminate the continued fraction with a long-time approximation for the third order "memory function" $f_3(t)$. The simplest one is the purely Markoffian termination which implies $f_3(t)\propto\delta(t)$ or $\hat{f}_3(z) \simeq \hat{f}_3(z=0)$. The parameter $\Delta_3\hat{f}_3(z=0)$ can be approximately evaluated stepping down the continued fraction to previous stages and relating $\hat{f}_3(z=0)$ with the corresponding lower order quantities $\hat{f}_j(z=0)$ with $j<3$. This procedure involves integrals like $\int_0^\infty f_j(t)dt$ which can be evaluated using simple guesses for $f_j(t)$ which, however, must satisfy the moment sum rules i.e. have the correct Δ_1 and Δ_2. In fact, the value of $\Delta_3\hat{f}_3(z=0)$ turns out to be only slightly dependent on the stage of the stepping down and on the particular form chosen for $f_j(t)$. Possible deviations from the Markoffian behaviour can be taken into account writing $\hat{f}_3(z)\simeq\hat{f}_3(z=0)+z\hat{f}'_3(z=0)$. However, the evaluation of $\hat{f}'_3(z=0)$ is generally more difficult because it involves integrals $\int_0^\infty tf_j(t)dt$ and a more detailed knowledge of $f_j(t)$ is required. Some of the theoretical shapes so calculated [10] are plotted in Figs.2,3. The first one refers to $RbMnF_3$ at room temperature where any inelastic feature has disappeared (in this $T \to \infty$ limit, it is also possible to evaluate approximately the quantity Δ_3). A satisfactory agreement with the experimental data [11] is obtained only taking into account the deviations from a purely Markoffian behaviour. Fig.3 shows the comparison between theory and experiment in $KNiF_3$ at 256^{0}K where a broad inelastic peak is present. For $T\sim T_N$ it is difficult to estimate the deviations from the Markoffian behaviour, but for the simpler and closely related one-spin neutron scattering problem at high $\vec{k}$ they are found to be very small[10]. The overall agreement is very good: in particular, the position of the inelastic peak is well reproduced. A similar agreement has been also found for $RbMnF_3$ near the transition temperature.

[1] P.A.FLEURY, these Proceedings (and references therein).

[2] M.F.THORPE, these Proceedings.

[3] See, however, the contribution by M.G.COTTAM in these Proceedings (where spin operators are used).

[4] U.BALUCANI and V.TOGNETTI, Phys.Rev. B8, 4247 (1973).

[5] R.W.DAVIES, S.R.CHINN and H.J.ZEIGER, Phys.Rev. B4, 992 (1971).

[6] P.A.FLEURY, W.HAYES and H.J.GUGGENHEIM, J.Phys.C 8, 2183 (1975).

[7] D.J.LOCKWOOD and G.J.COOMBS, J.Phys.C 8, 4062 (1975).

[8] A.VAN DER POL *et al.*, Solid State Com. 19, 177 (1976); W.LEHMANN and R.WEBER, J.Phys.C 10, 97 (1977).

[9] H.MORI, Progr.Theor.Phys. 33, 423 (1965); 34, 399 (1965).

[10] U.BALUCANI and V.TOGNETTI, Phys.Rev. B16, 271 (1977).

[11] P.M.RICHARDS and J.W.BRYA, Phys.Rev. B9, 3044 (1974).

TWO-MAGNON SPECTRA OF FERROMAGNETS

P.D. Loly

Department of Physics, University of Manitoba

Winnipeg, Manitoba, Canada, R3T 2N2

Historically the foundations of the contemporary study of two-magnon states were laid by Wortis[1] in 1963 when he located the two-magnon bound states of the n.n. isotropic hypercubic Heisenberg ferromagnet, established the connection with bound states of the corresponding Ising problem and gave a spin Green function formalism that exactly described the 2-magnon problem in ferromagnets at T=0°K. The conclusion that 2-magnon bound states in the n.n. s.c. case only occurred at large values of the total pair wavevector($\vec{K}=\vec{k}_1+\vec{k}_2$) and with an energy too large to affect the low temperature thermodynamics did not offer much hope for observing these effects. The scene shifted to broad peaks detected in optical studies of transparent antiferromagnets[2] in 1966 (infrared absorption, Raman scattering) which stimulated $\vec{K}$=0 calculations that eventually attained good agreement with experiment when magnon-magnon interactions were properly included by Elliott and Thorpe[3]. Even at T=0°K the anti-ferromagnetic formulation is subject to decoupling approximations in contrast to the ferromagnet. In recent years, partly due to the stimulus of Thorpe[4] in drawing attention to the possibility of a 2-magnon Raman effect in EuO, the ferromagnetic problem has been probed again by several groups[5-11] with the result that a better understanding of the origin of the resonance structure due to 2-magnon interactions in the continuum of 2-magnon states is beginning to emerge. A good understanding of the dynamics of bound states for the ferromagnet should help interpret the effects in antiferromagnets and magnetic alloys and guide the study of pairing effects in other systems.

We shall discuss 2-magnon spectra of the Heisenberg ferromagnet in a general sense by including their dependence on $\vec{K}$, for both single-ion and exchange excitation processes, and by allowing the possibility of additional variables, e.g. Ising and uniaxial anisotropy as well as higher order (biquadratic) exchange:

$$H = -\tfrac{1}{2}\sum_{i\neq j} J_{ij}\{\sigma_{ij} S_i^+ S_j^- + S_i^z S_j^z\} - D\sum_i (S_i^z)^2$$

It should be realised that at the present time we only have a mechanism for the single-ion Raman process[4] (K=0) and that the others are motivated as relevant analogs of the exchange process of antiferromagnets and of neutron scattering (K≠0) which are soluble exactly for the T=0°K ferromagnet.

The dispersion function, ω_k, for the n.n. s.c. Heisenberg ferromagnet has a tight-binding form:

$$\omega_k = 2SJ\{3 - \sigma\sum_{\nu}^{x,y,z} \cos k_\nu a\} + (2S-1)D$$

so that the one-magnon density of states is symmetric about mid-band and exhibits saddle point singularities of the general van Hove cusp type at 1/3 and 2/3 of the bandwidth. The density of states of two magnons which are not interacting can be expressed as a function of the total wavevector $\vec{K}=\vec{k}_1 + \vec{k}_2$:

$$S(k,K) = \omega_{k_1} + \omega_{k_2}$$
$$= 4SJ\{3 - \sigma\sum_{\nu} \alpha_\nu \cos k_\nu a\}$$

where $\alpha_\nu = \cos(K_\nu a/2)$ and $2\vec{k} = \vec{k}_1 - \vec{k}_2$.
This has the same structure as the one-magnon density of states but with a doubled energy scale and has an Ising-like narrowing as K increases from the zone centre (Γ) to (111) corner where it has zero width. In addition the existence of Ising anisotropy ($\sigma<1$) narrows the spectrum about its mid-point thereby providing a gap of $6SJ(1-\sigma)$ at $\vec{K}=0$ while for $S>\tfrac{1}{2}$ the presence of uniaxial anisotropy elevates the whole spectrum by a gap of (2S-1)D. At this point let us contrast the special case of the n.n. s.c. with the general situation where the continuum NEVER narrows to a point as exemplified by the n.n.n. s.c. case in Fig. 1a.

Let us now examine the energy of states in the Ising model which corresponds to the (111) zone boundary K point. To see this note that the α_ν in $S(K)$ create an effective Ising anisotropy. It is appropriate to examine the energies of several arrangements of spin pairs in an otherwise aligned (saturated) matrix: two spin deviations on a single site E(⇑) and two on n.n., E(↑↑), n.n.n., E(↑↓↑) compared to pairs at a greater separation than the range of the exchange interaction, 2E(↑), and which corresponds to the continuum point. As a result the binding energies relative to free pairs are

$$2E(\uparrow) - E(\Uparrow) = 2D \ , \ \text{if } S > \tfrac{1}{2}$$
$$2E(\uparrow) - E(\uparrow\uparrow) = 2J_{n.n.}S$$
$$2E(\uparrow) - E(\uparrow\downarrow\uparrow) = 2J_{n.n.n.}S$$

The latter pair are $\tfrac{1}{2}Z_{nn}$, $\tfrac{1}{2}Z_{nnn}$ degenerate respectively.

These binding energies express the effects of interactions and are the limiting case of dispersive 2-magnon states of the Heisenberg ferromagnet at K_{111} zone boundary.

A spin Green[7] function formulation of the 2-magnon spectrum can be set up exactly for the ferromagnet at T=0 K. Defining

$$G_{\delta,\delta'} = \frac{J_{n.n.}}{4S} \langle\langle A_\delta^+ ; A_{\delta'} \rangle\rangle$$

where

$$A_\delta = \sum_i S_i^- S_{i+\delta}^-$$

One can show that

$$G_{\delta,\delta'} = \left(1 - \frac{\delta_{\delta;0}}{2S}\right)\frac{1}{2}\left(g_{\delta,\delta'} + g_{\delta,-\delta'}\right) + \frac{1}{2S}\sum_x \eta_x\left[\sigma_x g_\delta - \frac{1}{2}\left(g_{\delta,x} + g_{\delta,-x}\right)\right] G_{x,\delta'} - \frac{D}{SJ_{nn}} g_\delta G_{0,\delta'}$$

where the g_δ are lattice Green functions,

$$g_\delta = 2SJ_{nn}\frac{1}{N}\sum_k \frac{e^{ik\cdot\delta}}{\omega - 2\omega_k}$$

From this formalism the cross-section for the Raman scattering of two magnons according to the spin-orbit process (Thorpe) taken to finite K is

$$\chi_{0,0}(\vec{K}) = (-)\, I_m\, G_{0,0}(\sigma') \quad , \quad \sigma' = \sigma \cos\left(\tfrac{1}{2} Ka\right)$$

while for an exchange process one is interested in

$$\chi_{1,1}(\vec{K}) = (-)\, I_m\, G_{1,1}(\sigma')$$

The bound states are located from the poles of the $G_{\delta,\delta'}$ within this formalism. The system is still soluble if biquadratic exchange, $K(\vec{S}i\cdot\vec{S}j)^2$, is added to the Hamiltonian.[9,11]

As $\vec{K}$ is reduced the bound states at the zone boundary disperse until they join the continuum, the s-wave exchange bound state grazes the continuum and fades away at lower K while the d-wave enters the continuum and as a broadened resonance is a major feature of the K=0 spectrum of the exchange probe. For D=0 the single-ion bound state (present only for $S>\frac{1}{2}$) is entirely within the continuum but does contribute a broad resonance at $\vec{K}$=0, indeed in the single-ion process it is the only resonance since the d-wave is not excited. These conclusions are drawn from a number of studies[6-9,12].

In Fig. 1b we show a close-up view of the dynamics of the d-wave resonance for a range of spin values. This shows unambiguously that it is unable to pass through the lower van Hove saddle point singularity. Interestingly the single-ion bound state is not so inhibited and for small values of D is able to enter the middle portion of the spectrum after entering the continuum.[7]

We refer to reference 7 for some representative samples of the two types of spectra for S=1 at K=0 (since the single-ion effects don't show up for $S=\frac{1}{2}$) and compare them with the density of states of non-interacting magnons. As $\vec{K}$ increases the spectra gradually

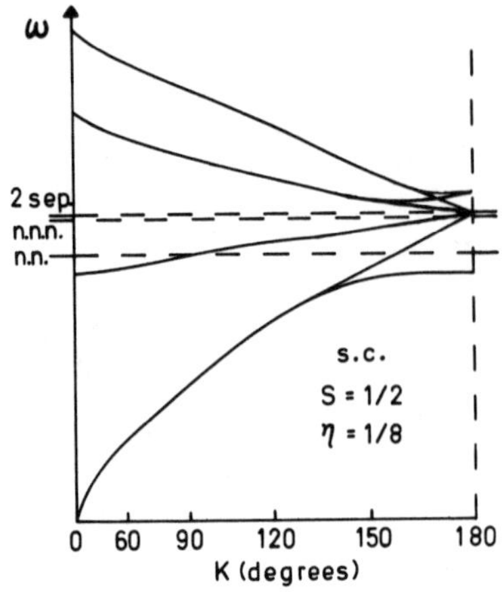

Fig. 1a. K-dependence of 2-magnon continuum and internal singularities for n.n.n. s.c. case with $J_2/J_1=1/8$.

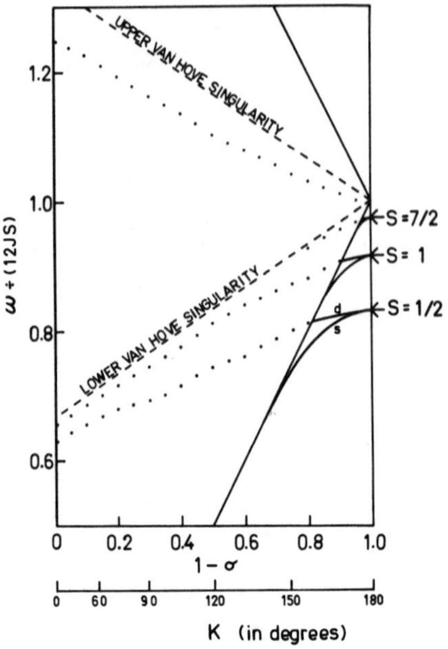

Fig. 1b. K-dependence of d- and s- wave bound state/resonances including D=0 single ion resonance for n.n. s.c. case.

change, first building up a resonance on the lower edge which exits as the s-wave bound state and then the d-wave (for G_{11} only). If D≠0 then the single-ion bound state will also exit at some point.

Having outlined the behaviour of bound state resonances in the n.n. s.c. case we turn to the more general situation where the Ising bound states are eclipsed by the continuum for all K. Such is the case for the n.n. f.c.c. ferromagnet and n.n. antiferromagnets which have mainly been studied at $\vec{K}=0$. In our own calculations[5] the Raman spectrum for EuO (S=7/2,f.c.c. with significant n.n.n. effects) is dominated by a strong peak which lies close to the top of the band. This structure is due partly to a resonance and also to a high density of states of the unperturbed magnons. It is also sensitive in position and linewidth to the ratio of exchange constants, $J_2/J_1 = \eta$. Observations of this spectrum are difficult because of the opacity of the material in the visible region but it is hoped that a study in the infrared will be forthcoming.

There remains much to be done in understanding the relationship of the Ising calculations to the position of the Raman resonances when the Ising states are eclipsed by the continuum for all $\vec{K}$ as is the case for the n.n. f.c.c. ferromagnet and the antiferromagnets.

Our studies suggest that the combined effects of anisotropy and $\vec{K}$-dependence which allow transition to discrete bound states will offer a systemmatic approach to this question.

It is hoped that some of the information obtained from the exactly soluble ferromagnetic problem will be of help in understanding pairing resonances in other systems e.g. antiferromagnetic magnons, phonons in liquid helium and diamond, librons in the solid hydrogens....Particularly intriguing is the problem of the interaction of spin waves and the Stoner continuum in itinerant magnets for which interesting data has been obtained via neutron inelastic scattering in the study of MnSi.

Acknowledgement

The author thanks Dr. A. Bahurmuz for many discussions about 2-magnon problems.

References

1. M. Wortis, Phys. Rev. 132, 85 (1963).
2. See the review by P.A. Fleury and S.P.S. Porto, J. Appl. Phys. 39, 1035 (1968).
3. R.J. Elliott and M.F. Thorpe, J. Phys. C. 2, 1630 (1969).
4. M.F. Thorpe, Phys. Rev. B 4, 1608 (1971).
5. P.D. Loly, B.J. Choudhury and W.R. Fehlner, Phys. Rev. B 11, 1996 (1975).
6. B.J. Choudhury and P.D. Loly, A.I.P. Conf. Proc. 24, 180 (1975).
7. P.D. Loly and B.J. Choudhury, Phys. Rev. B 13, 4019 (1976).
8. P.D. Loly p. 278 Proc. of Third International Conference on Light Scattering from Solids, Campinas, Brazil, published by Flammarion, 1976.
9. S-T Chiu-Tsao, P.M. Levy and C. Paulson, Phys. Rev. B 12, 1819 (1975).
10. A.M. Bonnot and J. Hanus, Phys. Rev. B 7, 2207 (1973).
11. D.A. Pink and P. Tremblay, Can. J. Phys. 50, 1728 (1972) and D.A. Pink and R. Ballard, Can. J. Phys. 52, 33 (1974).
12. R. Silberglitt and J.B. Torrance, Phys. Rev B 2, 772 (1970).

LIGHT SCATTERING DETERMINATIONS OF DYNAMIC FOUR POINT CORRELATION FUNCTIONS

P. A. Fleury

Bell Laboratories

Murray Hill, New Jersey 07974

I. INTRODUCTION

For many purposes the dynamics of a many body system are adequately described by the familiar autocorrelation function $S(\vec{q},\omega)$. $S(\vec{q},\omega)$ is the Fourier transform of $\langle u(\vec{r},t)u(\vec{r}'t')\rangle$, where $u(\vec{r},t)$ represents some characteristic dynamic variable such as spin, density, etc. Scattering experiments using light, neutrons, electrons or x-rays, are routinely directed toward the measurement of $S(\vec{q},\omega)$. However in recent years it has become increasingly appreciated that such scattering probes may also couple strongly to higher powers of the dynamic variable and may thus explore more complex dynamic correlation functions than $S(\vec{q},\omega)$. Indeed the operation of certain selection rules, particularly for light scattering, often permits rather direct and unambiguous measurement of the four point correlation function $\langle u(\vec{r}_1t_1)u(\vec{r}_2t_2)u(\vec{r}_3t_3)u(\vec{r}_4t_4)\rangle$. This type of function contains information on the interactions among those elementary excitations, which are described to lowest order by $S(\vec{q},\omega)$. In addition, various limits of this function determine such quantities as specific heat, acoustic attenuation, and other transport properties. In this paper the experimental information regarding four point correlation functions that may be obtained by the technique of inelastic light scattering is reviewed for both solids and liquids.

We begin by reciting some general remarks on the light scattering process, including both kinematics and selection rule restrictions. First, the distinction between first- and second-order light scattering is defined. (Sec. II) Experimental results are then reviewed, starting in Sec. III with magnetic systems. The two sublattice antiferromagnets at zero temperature in two- and three-dimensions are discussed first. The complications of disorder are considered next. The thermal disorder associated with increasing temperature is explored through the temperature dependence of the magnon-pair spectral lineshape, particularly as the system is taken above T_N into the paramagnetic phase. Next, the effects of configurational disorder are considered beginning with the case of random occupation by spins of ordered lattice sites.

Fully disordered systems such as the simple rare gas liquids are treated in Sec. IV. Experiments on liquid helium, in which roton pairs represent well defined excitations, are reviewed. A view of the intermolecular light scattering in classical fluids as a generalization of two phonon Raman scattering familiar in crystalline solids is proposed and supported by experiments on Ar, Kr, Ne and Xe. And some remarks are made on liquid hydrogen, which through varying orth-para mixtures admits of separate studies of translational and rotational degrees of freedom in the liquid state. Finally, in Sec. V we consider briefly the important role of higher order correlation functions the dynamics of structural phase transitions. Although some recent experiments have probed their behavior near T_c, an adequate theoretical treatment of critical dynamics for higher order correlation functions has not yet been formulated.

II. LIGHT SCATTERING PROCESSES OF FIRST- AND SECOND-ORDER

The process of inelastic light scattering within the dipole approximation may be conveniently described by the interaction Hamiltonian H_I:[1]

$$H_I = \sum_{\alpha,\beta} \sum_{\vec{r}} E_1^{\alpha}(\vec{r},t) E_2^{\beta}(\vec{r},t) P^{\alpha\beta}(\vec{r},t)$$

where $\vec{E}_1$ and $\vec{E}_2$ are the incident and scattered electric fields and $P_{\alpha\beta}$ is a dynamic polarizability tensor operator whose structure depends upon the details of the material system. The spontaneous scattering cross section is expressed in terms of matrix elements of the operator H_I between states of the overall radiation matter system; $\langle i|H_I|f\rangle$, where $\langle i| = \langle n_1=1, n_2=0, g|$ describes the state corresponding to one photon in the incident field, none in the scattered field and the material in its ground state. The state $\langle f|$ is denoted $\langle 0,1,e|$ in an obvious manner. The scattering rate is obtained via application of the Golden Rule by an average over initial and sum over final states of the squares of $\langle i|H_I|f\rangle$ matrix elements. It is more convenient for our purposes to express this rate in terms of the generalized Green's function as follows[2]

$$\frac{d^2\sigma}{d\Omega d\omega_2} = \frac{\omega_1\omega_2^3 n_2}{2\pi c^4 V n_1} \frac{1}{(1-e^{-\beta\hbar\omega})} \sum_{\alpha\beta r\delta} \varepsilon_1^\alpha \varepsilon_2^\beta \varepsilon_1^\gamma \varepsilon_2^\delta \mathrm{Im} G^{\alpha\beta\gamma\delta} \quad (2)$$

where ω_i, $\vec{k}_i$, n_i and $\vec{\varepsilon}_i$ are the frequency, wave vector, refractive index and direction of polarization, respectively, associated with the field, $\vec{E}_i$. The Green's function is then[3]

$$G^{\alpha\beta\gamma\delta}(\omega,\vec{q}) = \mathcal{F.T.}\ i\theta(t)\left\langle[P^{\alpha\beta}(\vec{r},t),P^{\gamma\delta}(0,0)]\right\rangle \quad (3)$$

where the <> are statistical averages, [,] represents the commutator, FT the Fourier transform on both $\vec{r}$ and t variables and $\theta(t)$ a time ordering operator. The values $\omega = \omega_2-\omega_1$ and $\vec{q} = \vec{k}_2-\vec{k}_1$ result from carrying out the Fourier transforms, and express the physically reasonable requirements of energy and momentum conservation in the overall scattering process. The commutator in (3) is simply related to the auto-correlation function $\langle P^{\alpha\beta}(\vec{r},t)P^{\gamma\delta}(0,0)\rangle$ which can then be viewed as the fundamental dynamic quantity governing the light scattering process.

It is the dependence of $P^{\alpha\beta}$ upon the dynamic variables of real liquids and solids which shall concern us from now on, particularly the contribution P receives which is quadratic in these variables. For purposes of simplicity and illustration let us consider the dynamic variable $u(\vec{r},t)$ to represent the spin operator $S^{\pm}(r,t)$. The P operator can then be expanded in powers of the spin variable to determine the light scattering from spin dynamics.[3]

$$P^{\alpha\beta} = a_1^{\alpha\beta}S^+ + a_2^{\alpha\beta}S^+S^- + \text{---} + \text{C.C.} \quad (4)$$

where the coefficients $a_i^{\alpha\beta}$ express both the polarizability induced by an ith order spin excitation and the requirements of crystal symmetry. The first term in Eq. 4 describes a scattering event in which a <u>single</u> elementary excitation (magnon) is created or destroyed, and is proportional to the familiar $\langle S^+S^-\rangle = S(\vec{q},\omega)$ correlation function. This process will be called "first-order scattering" and will not concern us further except insofar as we shall contrast it with higher order processes. The kinematic restrictions cited above limit our examination of the magnon dispersion curve through first-order scattering to the very long wavelength (small $\vec{q}$) regime; e.g. $q_{max} = 2k_1 \underset{\sim}{\sim} 10^6\text{cm}^{-1} \ll q_{ZB} = \frac{\Pi}{a} \underset{\sim}{\sim} 10^8\text{cm}^{-1}$. This of course follows from the facts that $q \simeq 2k_1 \sin\frac{\theta}{2}$ and that for visible light $k_1 = \frac{2\pi n_1}{\lambda_1} \sim 10^5\text{-}10^6\text{cm}^{-1}$.

Consider now the process represented in (4) by the second term. Here the correlation function F.T. $\langle S_i^+(\vec{r},t)S_j^-(\vec{r},t)S_k^+(\vec{r}',t')S_\ell^-(\vec{r}',t')\rangle$ describes the simultaneous creation or destruction of pairs of magnons whose individual momenta may be anywhere in the Brillouin zone as long as they are paired so that $\vec{q} = \vec{K}_+ + \vec{K}_- = \vec{k}_2 - \vec{k}_1$ is satisfied. This type of process is called <u>second-order scattering</u>. Fig. 1 contrasts the kinematic requirements for light scattering as manifested in the first- and second-order processes. For first-order scattering the observed scattered spectrum will exhibit a peak at $\omega_2 = \omega_1 \pm \omega_q$ whose lineshape is determined by the lifetime of a single excitation at $\vec{q} = \vec{k}_2 - \vec{k}_1$.

For the second-order process illustrated in Fig. 1 the spectral lineshape will reflect the joint density of states summed over reciprocal space.[4] Thus Brillouin zone boundary excitations will dominate this process. See Fig. 2. Complications of this simple description arising from: a) interactions between members of the excitation pairs; b) symmetry dictated preferential weighting of specific regions of the Brillouin zone; c) pair excitations whose members lie on different branches, etc. pose many of the unanswered theoretical questions in this field and will be raised below in the appropriate specific instances.

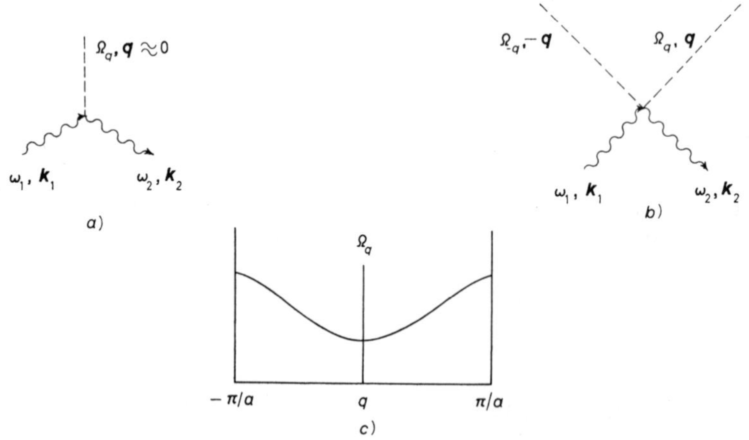

Fig. 1 Kinematics of first- and second-order light scattering. a) First-order ($\omega_2 = \omega_1 \pm \Omega_q$), b) second-order ($\omega_2 =_1 \pm [\Omega_q + \Omega_{-q}]$), c) simple dispersion curve for a single-branch excitation. In second-order scattering, excitations from opposite halves of the zone are paired.

III. EXPERIMENTAL RESULTS ON FOUR SPIN CORRELATION FUNCTIONS

A. Pure Magnetic Systems at "Zero" Temperature

Since the initial observation of light scattering from spin waves,[5] where the second-order process was found to be surprisingly strong, most of the effort and interest in this field has centered uponthe second-order or magnon pair pair scattering, primarily for the unique information it provides about the dynamic four spin correlation function. The rich and varied family of transition metal (Fe, Mn, Ni, Co) fluoride compounds has been the most extensively studied.[6-12] Practical reasons for the favoritism of these materials center upon their transparency to visible light, their convenient magnetic transition temperatures, their relative freedom from interfering phonon scattering processes (particularly when compared to their oxide relatives) and their availability as single crystals of good optical quality. Mixed crystals are also prepared with relative ease. Fortunately from a theoretical point of view, the variety of spins and crystal structures available to this family maps well onto the set of theoretically most interesting and tractable spin Hamiltonians. All of these antiferromagnetic systems can be described by the spin Hamiltonian[2]

$$H_s = \sum_{ij} J_{ij}\vec{S}_i\cdot\vec{S}_j + \sum_j \{D(S_j^z)^2 - E\left[(S_j^x)^2-(S_j^y)^2\right]\} - \beta\vec{S}_j\cdot\overleftrightarrow{g}\cdot\vec{H}\} + \sum\{E\rightarrow -E\}. \qquad (5)$$

where J_{ij} represents the exchange interaction between magnetic ions i and j, $\overleftrightarrow{g}$ is the gyromagnetic tenor, $\vec{H}$ the applied magnetic field, D the anisotropy energy determining the direction of sublattice magnetization, and E the anisotropy energy tending to cant the sublattice spins to 90° relative orientation. In cases where all of these exchange and anisotropy parameters are nonzero (e.g. NiF_2 of space group D_{4h}^{14}) the magnon dispersion curves obtained from this spin Hamiltonian are shown schematically in Fig. 2. The explicit relations between the magnon frequencies and the parameters in Eq. 5 are: $\omega_o^- = 2E$, $\omega_o^+ = (16DJ_2+E^2)^{1/2}$, $2\omega_m = D + 16(J_2-J_3)$,

$2\omega_R = D + 16J_2 - 8(J_1+J_3)$, where J_i denotes ith nearest neighbor exchange. Observations of both one- and two-magnon scattering have been made in the perovskite structure $NaNiF_3$[11] and the rutile structure NiF_2,[12] from which the spin Hamiltonian parameters have been obtained.

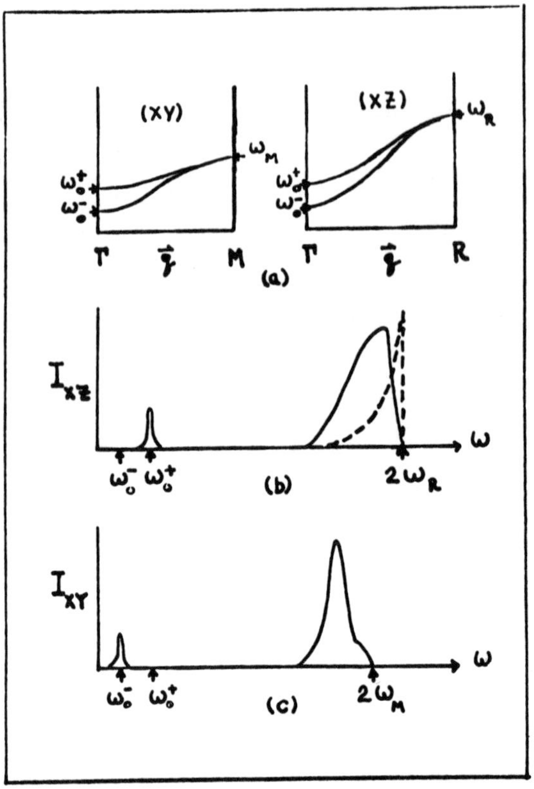

Fig. 2 Schematic of dispersion relations (a) and magnetic light scattering spectra for system with spin Hamiltonian in Eq. 5. (b) and (c) illustrate the effect of polarization selection rules on magnon features observed.

Of primary interest here are magnon pair spectra such as illustrated in Fig. 3, which reveal that even at zero temperature, magnon-magnon interactions are not negligible. That is, the four-spin correlation function is not trivially related to the two-spin correlation function. To appreciate this, consider the lowest order expression for the two magnon Green's function, G^o_{II}, in which there is no interaction between the members of the magnon pair:[13]

$$G^o_{II}(\omega,q\approx 0) = \frac{\text{const}}{N}\sum_k \frac{\Phi^2_{\alpha\beta}(k)}{\omega^2-4\omega_k{}^2};\ (T=0). \qquad (6)$$

This is essentially the two magnon density of states, weighted by the trigonometric factors $\Phi^2_{\alpha\beta}$ which result from crystal symmetry requirements. The resultant lineshape is indicated in Fig. 3 by the dashed curve and fails to agree quantitatively with the observed low temperature spectrum. It was first pointed out by Elliott and Thorpe[13] that magnon interaction effects upon G_{II} can be accounted for so as to produce essentially perfect agreement with experiment at low temperature. The resulting expression is:

$$G_{II} = \frac{G^0_{II}}{1 + \alpha G_{II}{}^0} \tag{7}$$

where $\alpha = 4J_2(D+2\langle S\rangle(4J_2-J_1))$ and $\langle S\rangle$ is the sublattice magnetization. The resultant predicted magnon-pair lineshape is illustrated by the dotted curve in Fig. 3. This advance permits magnon-pair scattering to be used for quantitative determination of spin Hamiltonian parameters. Thorpe[14] has computed magnon interaction corrections for MnF_2 where the magnon-pair lineshape dependence on polarization is appreciable.

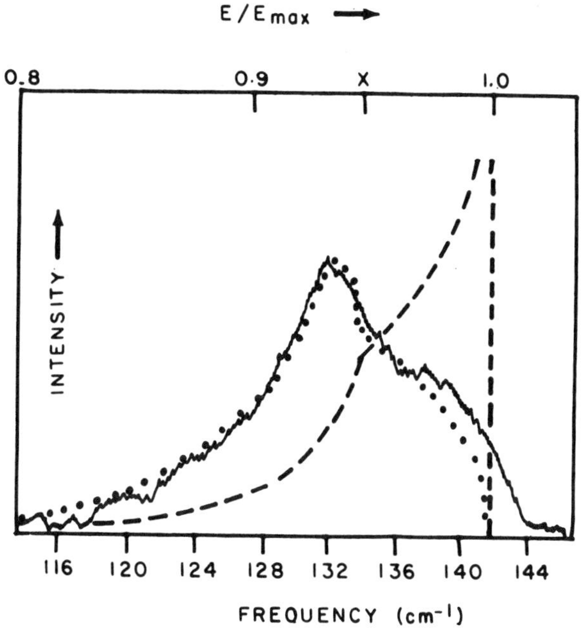

Fig. 3 Magnon pair light scattering in $RbMnF_3$ at ∿10K. Solid line is experimental observation. Dashed curve is theory without magnon-magnon interactions. Dotted curve is theory including interactions. After Ref. 8.

The crystal structure for $KNiF_3$ (cubic perovskite) is identical to that for $RbMnF_3$ and the low temperature effects of magnon interactions are formally the same in the two materials.[6] Interestingly, a closely related set of crystals typified by K_2NiF_4 exhibit two dimensional-rather than three dimensional spin excitation spectra. In the two dimensional case the Green's functions analogous to Eqs. 6 and 7 can be expressed in closed analytical form. Parkinson's calculated[15] magnon pair lineshapes including magnon interactions are essentially identical to the experimental observations.[16] No adjustable parameters are required.

The present situation regarding four-spin correlation functions in pure systems at zero temperature may therefore be regarded as quantitatively satisfactory. In crystals of lower symmetry, having more complicated spin Hamiltonians, the magnon dispersion curves (and consequently the one- and two-magnon spectra) will be correspondingly more complex. However from an experimental point of view it is possible to sort out much of this complexity. Specifically, through the $a_2^{\alpha\beta}$ (or $\Phi^{\alpha\beta}$) terms in Eqs. 4 and 6, magnon pairs from different regions of the Brillouin zone are emphasized in spectra with different polarization selection rules. Table I illustrates this effect for the rutile structure[4] antiferromagnets (Mn-,Fe-,Co-,NiF_2).

Table I. Polarizations of incident (1) and scattered (2) electric fields with corresponding weighting factors and emphasized critical points (right).

$E_1^xE_2^x;E_1^yE_2^y;E_1^zE_2^z$	...$\cos(\frac{1}{2}\alpha k_x)\ \cos(\frac{1}{2}\alpha k_y)\ \cos(\frac{1}{2}ck_z)$...	Γ
$E_1^xE_2^y;E_1^yE_2^x$	...$\sin(\frac{1}{2}\alpha k_x)\ \sin(\frac{1}{2}\alpha k_y)\ \cos(\frac{1}{2}ck_z)$...	M
$E_1^xE_2^z;E_1^zE_2^x$	...$\sin(\frac{1}{2}\alpha k_x)\ \cos(\frac{1}{2}\alpha k_y)\ \sin(\frac{1}{2}ck_z)$...	R
$E_1^yE_2^z;E_1^zE_2^y$	...$\cos(\frac{1}{2}\alpha k_x)\ \sin(\frac{1}{2}\alpha k_y)\ \sin(\frac{1}{2}ck_z)$...	R

The effects of magnon interactions on these various spectral components may be handled in the same way as the simple case illustrated in Fig. 3.

B. Pure Magnetic Systems at Finite Temperature

The effects of increasing temperature upon collective magnetic excitations are of obvious interest, since collective behavior will completely disappear upon total destruction of magnetic order. As $T \overset{\rightarrow}{<} T_N$, zone center magnons decrease in both frequency and lifetime, disappearing above T_N with the loss of long range order. Only a few measurements of the temperature dependence of one magnon scattering have been reported,[11,17] and in at least some respects - particularly in finite field[17] - do not agree with theoretical predictions.

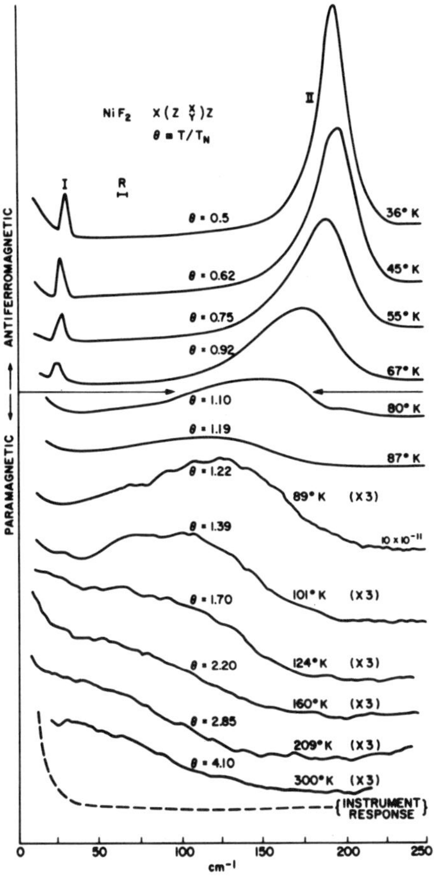

Fig. 4 Temperature evolution of one magnon (I) and magnon pair (II) spectra in NiF_2. Note gain change for $T \geq 89K$. After Ref. 12.

The temperature dependence of magnon-pair spectra is more interesting and more challenging theoretically. Fig. 4 shows the temperature evolution of both one- and two-magnon spectra in NiF_2.[12] It is typical[8] in that the magnon-pair frequency decreases at a less rapid rate with increasing T than does ω_0, and in that the magnon-pair mode remains well defined far into the paramagnetic regime. This behavior indicates that (1) the renormalization of magnon frequency with temperature is strongly dependent on wave vector and (2) the persistence of short range magnetic order above T_N sustains collective excitations whose wavelength is small compared to the range of that order. Although it was recognized very early that the magnon pair lineshape was dependent upon the k dependent temperature normalization of individual magnon frequencies, temperature dependent magnon lifetimes, and magnon-magonn interactions, it has not been possible until relatively recently to extend the successful zero temperature theory to finite temperatures - much less into the paramagnetic phase. Within the last couple of

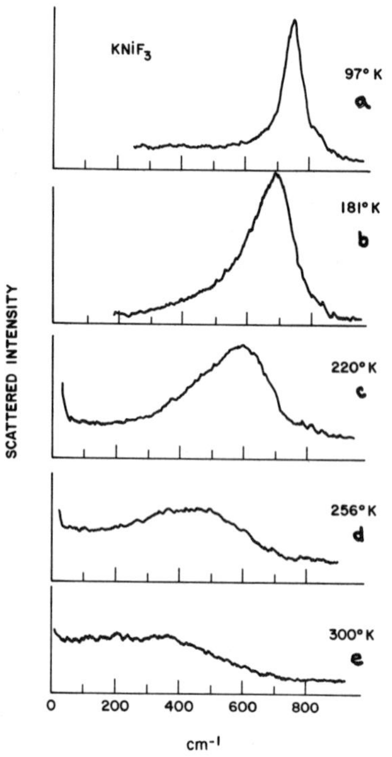

Fig. 5 Magnon pair spectra of $KNiF_3$ at (a) 0.39 T_N, (b) 0.73 T_N, (c) 0.88 T_N, (d) 1.02 T_N (e) 1.2 T_N. (T_N=250K). After Ref. 25.

years a variety of attempts at this extension were made. Those which can be regarded as replacing G^o_{II} in Eq. (6) by the form:

$$G^o_{II}(T,\omega,q\approx 0) = \frac{\text{const}}{N} \sum_k \frac{R(T)\Phi^2_{\alpha\beta}(k)\coth(\beta h\omega_k/2)}{\omega^2-4\omega_k^2} \quad (8)$$

where $\omega_k = \omega_k(T)$, were reviewed and compared in the 1973 paper by Natoli and Ranninger.[18] They noted there that addition of $4i\omega\Gamma_k(T)$ to the denominator in Eq. (8) would produce satisfactory agreement with experiment. Unfortunately calculated values of the temperature and wave vector dependent magnon damping, $\Gamma_k(T)$, available at that time were considerably smaller than required to fit the observed spectra.

More recently several groups[19,20] have made impressive progress on this problem. Calculations of $\Gamma_k(T)$ carried to higher order have produced good agreement with experiment for both the magnon pair frequency and lineshape for temperature up to about $0.8T_N$. The computed lineshapes for $KNiF_3$ and corresponding experimental curves are shown in Figs. 5 and 6. The theory of Balucani and Tognetti[19] has been applied to the two dimensional case of K_2NiF_4 by van der Pol et al.[21] to achieve equally good agreement with experiment. Fig. 7 shows the measured magnon pair frequency

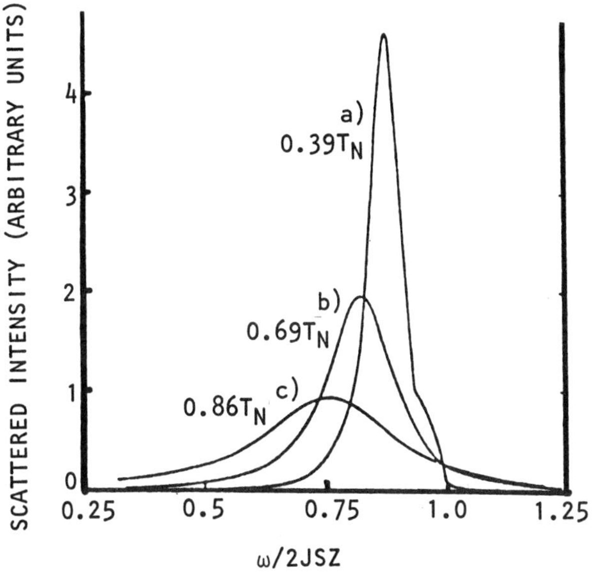

Fig. 6 Calculated magnon pair lineshapes for $KNiF_3$ at finite temperatures in the antiferromagnetic phase. After Ref. 19.

and linewidth as a function of T for the two-dimensional K_2NiF_4[16] and the closely related three-dimensional $KNiF_3$.[2,6] The van der Pol calculations for the former produces good agreement with experiment. Fig. 7 also illustrates a generally observed behavior that the renormalization of both frequency and lifetime with temperature is much weaker for two-dimensional antiferromagnets than for three-dimensional ones.

Still more recently Balucani and Tognetti[22] have used a modified Markoffian approximation for the Mori continued fraction representation to obtain the four-spin correlation function line-shape. Fig. 8 compares their calculations with experiments on $KNiF_3$ in the paramagnetic phase. The recent theoretical progress in calculations of the four spin correlations at temperatures throughout the ordered phase and into the paramagnetic phase, establishes light scattering as a quantitative probe of zone boundary magnon frequencies, lifetimes and interactions.

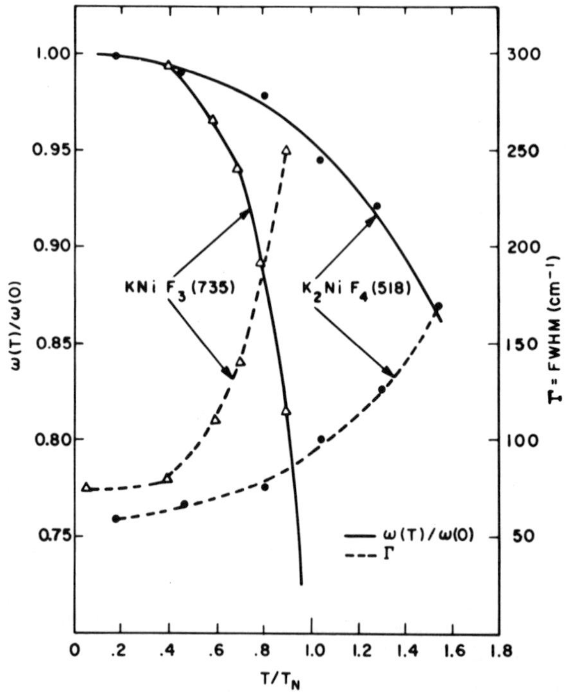

Fig. 7 Comparison of temperature dependences of magnon pair frequency ($\omega(T)$) and linewidth (Γ) for S=1 antiferromagnets which are two-(K_2NiF_4) and three-($KNiF_3$) dimensional.

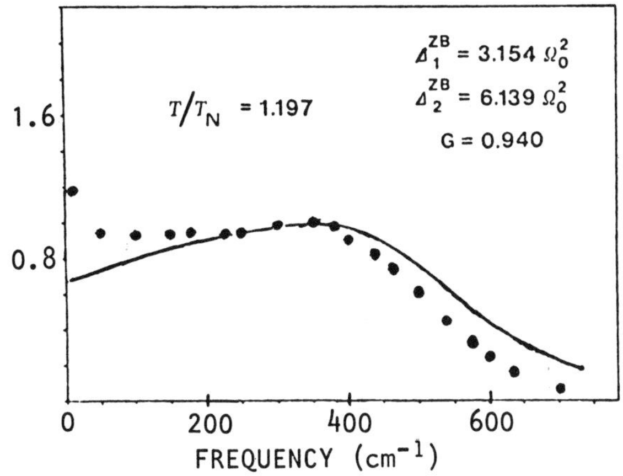

Fig. 8 Comparison of experiment (●) and theory (solid curve) for $KNiF_3$ pair mode scattering in the paramagnetic phase. After Ref. 22

The successful utilization of the continued fraction representation in the paramagnetic phase is suggestive that a similarly quantitative description of second order spectra in simple liquids may not be far off.[23]

III. C. Dilute and Mixed Antiferromagnets

The departure from perfect spin ordering has been induced by composition variation as well as by finite temperature. The first systematic experimental study of magnon pairs in this type of system was carried out by Buchanan et al.[24] in the dilute antiferromagnet $(Mn_{1-c}Zn_c)F_2$ for c between 0 and 1. At constant low temperature (5K) they observed the pair mode to soften and broaden with increasing dilution. Their results could be qualitively understood in terms of an Ising cluster model wherein the lineshape is a histogram of two-spin energies weighted in intensity according to the probability of occurrence for a cluster with that energy.

In hopes of obtaining more detailed information in the neighborhood of the critical concentration (c_o) (percolation limit) Fleury et al.[25] studied the $K(Ni_{1-c}Mg_c)F_3$ system as a function of <u>both</u> temperature and concentration. The hoped-for dramatic behavior near c_o did not emerge. However some interesting correlations between the effects on magnon pairs of finite temperature and finite dilution were obtained. One example is shown in Fig. 9. At present no theory exists for the finite temperature pair modes in a dilute or mixed spin system.

As a final topic in spin systems we consider the disordered two dimensional antiferromagnet: $Rb_2Mn_{.5}Ni_{.5}F_4$. Inelastic neutron experiments[26] on this system revealed two relatively well

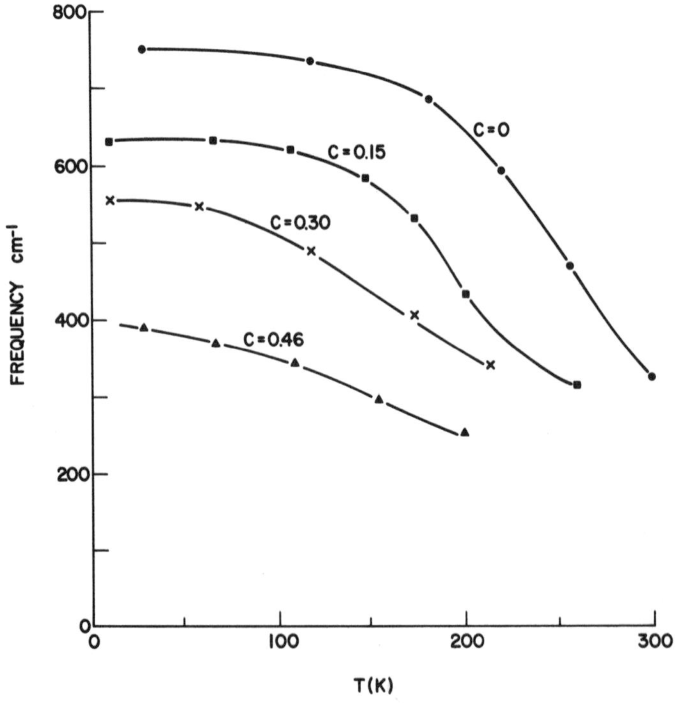

Fig. 9 Dependences of magnon pair mode frequencies upon temperature and concentration (c) of substituted nonmagnetic species for the $KNi_{(1-c)}Mg_cF_3$ system. After Ref. 25.

defined magnon dispersion branches which could approximately be associated with Ni-like and Mn-like spin waves. The magnon pair spectra observed in light scattering(27) as a function of temperature in $Rb_2Mn_{.5}Ni_{.5}F_4$ appear in Fig. 10. Magnon pairs associated with the Ni-Ni and the Ni-Mn excitations occur at 370cm^{-1} and 240cm^{-1}, respectively. These are shifted down from the sum of the relevant zone boundary frequencies by precisely the amount expected from theory of magnon-magnon interactions. The histogram in Fig. 10a was calculated for the simple Ising cluster model discussed in Ref. 24 and generalized from the dilute to the mixed antiferromagnet in Ref. 25.

Several experimental observations on this system have yet to find adequate theoretical explanation. Most striking is the observed relative intensity of the Ni-Ni peak to the Ni-Mn peak.

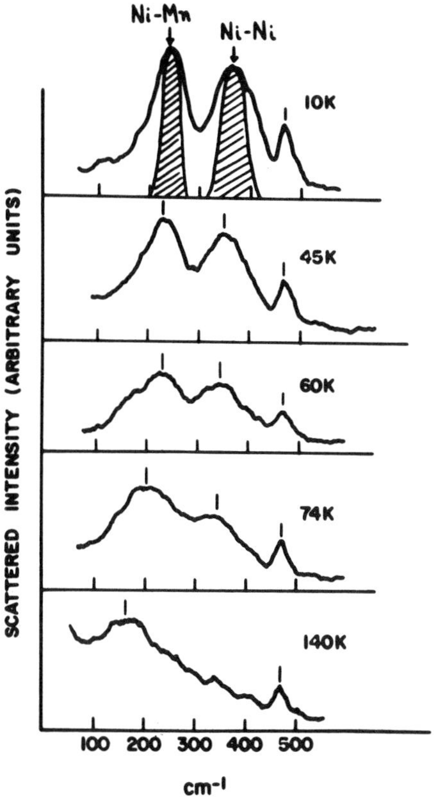

Fig. 10. Magnon pair spectra in the mixed two dimensional random antiferromagnet $Rb_2Mn_{.5}Ni_{.5}F_4$. The 475cm^{-1} feature is a phonon. The predictions of a simple Ising cluster model appear as shaded areas. $T_N \approx 66K$. After Ref. 27.

This ratio is too large by nearly a factor of three if the usual excited state exchange mechanism applies for magnon pair light scattering. Second, the scattering cross section of the more energetic Ni-Ni pair falls off much more rapidly with increasing temperature than the Ni-Mn peak. Third, even though the system is compositionally disordered the temperature renormalization of the pair mode frequencies is quantitatively the same as in the pure system, K_2NiF_4. Finally the low temperature linewidth appears to have two simply additive contributors, one due to disorder the other to magnon interactions. The latter is a temperature independent additive.

IV. EXPERIMENTAL RESULTS ON FOUR POINT DENSITY CORRELATION FUNCTIONS

A. Simple Fluids

The possibility of depolarized inelastic light scattering in simple monatomic fluids was recognized only relatively recently and arose from two rather different starting points. Both can be crudely viewed as extensions of second-order Raman scattering, so familiar in crystalline solids, to gases and liquids from roughly the single particle and collective viewpoints respectively. The latter view due to Halley[28] and Stephen[29] bears close analogies with the magnon-pair spectra discussed above, has met with success in treating roton-pair scattering in liquid helium and will be discussed below.

The former view, which we now discuss, due to Levine and Birnbaum[30] considers the dynamic polarizability of a pair of a pair of interacting (colliding) atoms so that

$$P^{\alpha\beta}(\vec{r},t) = \int d^3r'\rho_1(\vec{r}',t)\beta(\vec{r}',r)\rho_2(\vec{r}-\vec{r}',t).$$

Where $\rho_i(rt) = c\delta(\vec{r}-\vec{r}_i)\delta(t-t_i)$. During the binary collision process the initially isolated particles of polarizability α_o approach each other and distort their total polarizability $P^{\alpha\beta}$ from the unperturbed value, $2\alpha_o$. This polarizability pulse endures for a time of order $\tau_D \approx r_o/v_o$ where r_o is the atomic radius ($r_o \sim \sigma$ parameter for a Lennard-Jones potential) and v_o the relative atomic velocity. The spectrum produced by this collision process covers an extent in frequency of order $\sim \tau_D^{-1}$.

Several authors have considered the intermolecular scattering from uncorrelated binary collisions for a monatomic gas in thermal equilibrium.[31] If one treats the particle trajectories in the "turning point" approximation then it is easy to show that $P^{\alpha\beta}(t)$ is Lorentzian in t, so that an exponential frequency spectrum of the form $I_2(\nu) = I^o e^{-\nu/\Delta}$ results. Details of the interatomic potential and manner of thermally averaging do not alter the main qualitative conclusions of the binary collision model:

1) The spectrum is approximately exponential in shape with a characteristic frequency $\Delta = V_{TH}/r_o$.

2) The characteristic frequency Δ is independent of density (since it depends upon the duration τ_D of a collision rather than the time τ_c between collisions).

3) Δ varies as $\sqrt{T}$ through $v_{TH} = \sqrt{kT/m}$.

4) The integrated spectral intensity (I^o) varies quadratically with density, since $P^{\alpha\beta}$ increases with the probability of having two particular simultaneously present within an interaction volume.

The original observations of intermolecular light scattering by McTague and Birnbaum[32] were carried out in gaseous Kr and Xe at densities between 10 and 100 amagats (i.e. relatively dilute). Their experiments confirmed the major features of the above model in the case where $\tau_c >> \Delta^{-1} = \tau_D$.

The shortcomings of this model were soon brought to light, however, with the initial observations[33] of intermolecular light scattering in a simple liquid. In particular the spectrum of liquid argon at $T \approx 100K$ in equilibrium with its vapor revealed striking departures from the above predictions. Fig. 11 shows that for essentially the same temperature $\Delta_{gas} \simeq 6.5cm^{-1}$, while $\Delta_{liq} \simeq 20.0cm^{-1}$. Further the ρ^2 predicted density increase in scattered intensity falls short by more than an order of magnitude in relating the total intensities of the liquid and gas spectra ($\rho_{liq} = 800$ amagats; $\rho_{gas} = 13$ amagats.) While the breakdown of the isolated binary collision model at liquid densities (where $\tau_c \approx \tau_D$) is hardly surprising, the persistence of the exponential spectral lineshape presented some puzzles.

As is well known the branch of collective excitations representing density fluctuations in liquid He is well defined out to very large wave vector, including $K_r \simeq \pi/\sigma$, the famous "roton minimum". Light scattering from pairs of these excitations is dominated by contributions from the high density of states portions, exactly as in the case of magnon pairs discussed earlier. Theoretical calculations by Halley[28] and by Stephen[29] along these lines were carried out for liquid helium at very low temperatures. The

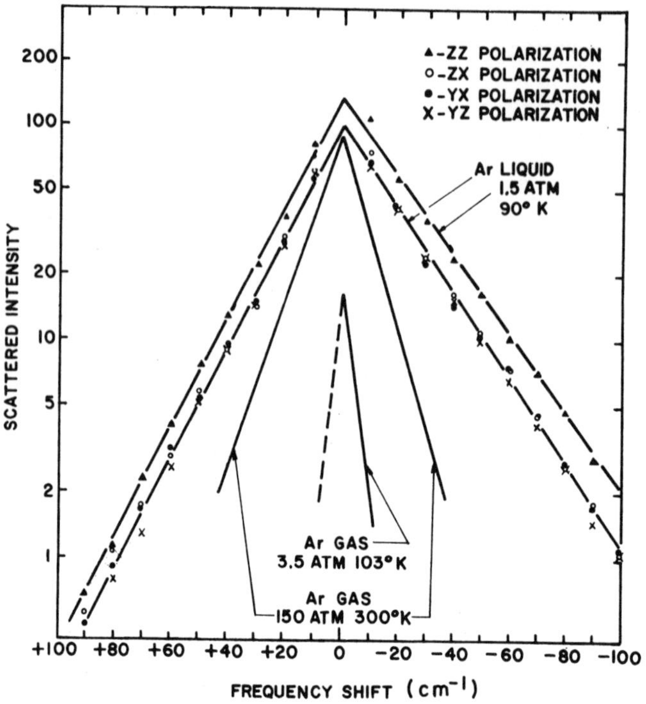

Fig. 11 Lineshapes of intermolecular light scattering spectra in liquid and gaseous argon. The gas spectra at 103K are not drawn to the same intensity scale as the others. After Ref. 33.

observed spectra(34) showed that (i) the roton pair energy is shifted down slightly from twice the energy of a single roton and (ii) the sharp spectral structure which distinguish the helium spectrum from those the classical rare gas liquids is rapidly smeared out as the liquid temperature is increased. Point (i) implies appreciable roton-roton interaction, as in the case of magnon pairs. This interaction has been the subject of considerable theoretical and experimental effort over the past few years, and recently agreement has become good enough to give confidence that roton pair spectral lineshapes can be used to measure the roton-roton interaction. Detailed comparisons of the observed spectral lineshapes with various theoretical calculations are made elsewhere in this volume.

The intermolecular spectra observed in the classical simple liquids can evidentally not be described in terms of either binary collisions between atoms or as due to pairs of weakly interacting but otherwise well defined short wavelength collective modes. In particular realistic calculation of $P^{\alpha\beta}$ has remained an elusive goal. McTague et al. have argued(33) that the formal approach which Stephen applied to liquid He can be generalized to fluids wherein the short wavelength density fluctuations are not well-defined collective modes. The sum over all pairs of such over-damped excitations will produce a spectrum centered at zero frequency of approximately the correct width. The assumed factorization of the four point correlation function is a worse approximation for liquid Ar than for superfluid He, but even so, the liquid Ar spectrum calculated this way and using the neutron measured $S(\vec{q},\omega)$ differed by only about 20% from the observed spectrum.(33) Gelbart(31) has pointed out some additional terms of importance left out of Stephen's original expression relating $n^{(4)}$ (1234) to lower order density correlation functions. These are discussed in his contribution to this volume.

Two routes of experimental attack on the problem of relating liquid state dynamics to intermolecular spectra in classical liquids have been employed. The first starts from the dilute gas point of view and follows the evolution of short time dynamics toward and beyond liquid state densities.(35) The second begins from the low temperature solid and examines the temperature evolution of the elementary excitations (phonons) up to and beyond the melting transition.(36) As we shall see, these experiments have suggested that the high frequency dynamical view of the liquid state as a highly anharmonic or "sloppy" solid is more convenient than that of an ultra dense gas. A number of experimental results have emerged which form benchmarks for the as yet unwritten theory of the dynamics of the dense fluid.

To bridge the gap between the liquid and dilute gas spectra it was desirable to gather spectra over a wide range of density and temperature, while varying them independently. Fleury, Daniels and Worlock(35) studied Ar, Kr, and Ne for temperatures between 30 and 300K and densities between 100 and 1800 amagats. The densities ranged from the dilute gas limit to as much as 50% in excess of the triple point liquid density. Figure 12 shows the spectra at 300K in Ar for several densities. ($\rho_{TP} \cong 800$ amagat). At constant temperature the characteristic frequency $\Delta(\rho,T)$ clearly increases dramatically with increasing ρ. In addition for higher densities, departure of the spectral lineshape from a single exponential becomes significant. Figure 13 summarizes the observed ρ and T dependence of Δ for fluid Ar.

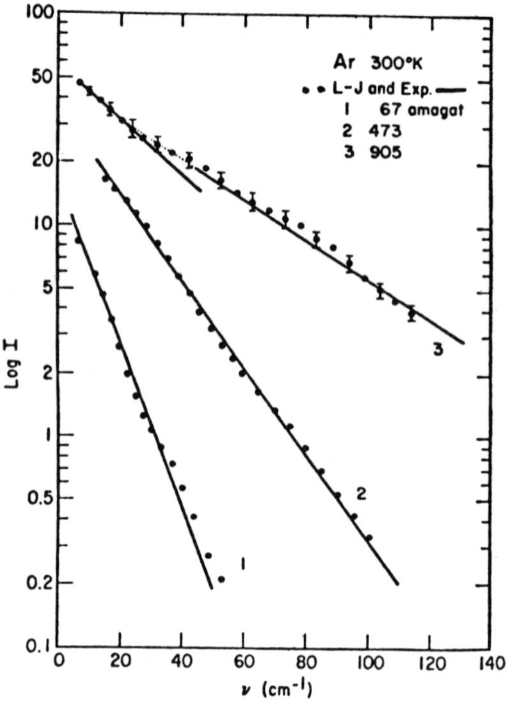

Fig. 12 Comparison of observed[35] intermolecular scattering lineshapes (solid lines) for argon at 300K and various densities (1) 67 amagat, (2) 473 amagat and (3) 905 amagat with results of molecular dynamics calculations. Overall intensity scales adjusted at each density. After Ref. 37.

The origin of the two-time behavior at high densities is not clearly understood and its nature appears to depend upon temperature, as evident from 300K spectra on Ne (which corresponds to 1033K spectra in Ar) where the sense of departure from a pure exponential lineshape is reversed.[35] These experiments demonstrate that the spectral lineshapes evolve smoothly with ρ and T and, somewhat surprisingly, that over a very wide range can be quantitatively characterized by the following simple empirical expression:[36]

$$\Delta(\rho,T) = \frac{3}{2\pi c}\left(\frac{\varepsilon}{m\sigma^2}\right)^{1/2}\left(\frac{kT}{\varepsilon}\right)^{1/2}\left[1+\left(\frac{2\sigma^3}{m}\right)^2\rho^2\right] \qquad (9)$$

where c is the speed of light; m,σ,ε are the atomic mass, Lennard-Jones distance and energy parameters, respectively. This equation has not been derived theoretically, but adequately describes experimental observation on Ar, Kr, Xe and Ne covering $0.7 < kT/\varepsilon < 4$ and $0.01 < \rho\sigma^3/m < 1$. It also represents perhaps the most general statement to date of the extension of the principle of corresponding states to high frequency (10^{-13} sec) dynamics.

Although there have been no formal theoretical calculations to explain the above experiments, a series of molecular dynamics calculations by Alder, Strauss and Weiss[37] have produced remarkably detailed agreement with the observed lineshapes (although not the intensities). In fact, the points in Fig. 16 are the results of their calculations for the experimental conditions indicated. Further their calculations for Ar at ∿1000K agree very well with the scaled neon 300K data.[35] Alder et al.[37] used the dipole-induced-dipole tensor:

$$T_{ij}^{xy} = \frac{1}{r_{ij}^3}\left[\delta_{xy} - (3x_{ij}y_{ij}/r_{ij}^2)\right] \tag{10}$$

to calculate the autocorrelation function

$$S^{xy}(\tau) = \left\langle \left[\sum_{i\neq j} T_{ij}^{xy}(0)\right]\left[\sum_{i\neq j} T_{ij}^{xy}(\tau)\right]\right\rangle. \tag{11}$$

$\vec{r}_{ij}$ is the distance between two particles. At high densities the time dependence of spatial arrangements for triplets and quartets, as well as pairs of particles must be averaged to calculate $S^{xy}(\tau)$. The MD calculations assumed the particles to interact via the standard Lennard-Jones potential, and were carried out for systems of up to 864 particles with different temperatures and densities. Some exploration was also done of the dependence of $S^{xy}(\tau)$ on the intermolecular potential. The spectral lineshapes (Fourier transforms of $S(\tau)$) particularly at low densities or high temperatures are practically indistinguishable for L-J, hard sphere and square well potentials, and agree within quoted errors with the experimental results. Some differences in absolute intensity are predicted among the three potentials, but these differences are small compared to the difference between the calculated and observed intensities. Although reasonably accurate values for the absolute scattered intensities have since been reported[36] for liquid Ar, Kr, and Xe, the calculations have yet to produce sufficiently

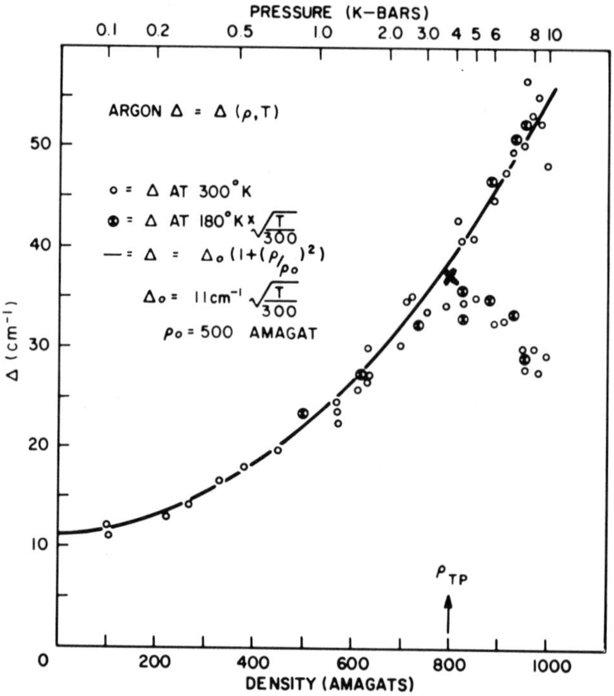

Fig. 13 Density and temperature dependences of characteristic frequencies (Δ) in argon. Δ_1 and Δ_2 are slopes in the low (0-50cm^{-1}) and high (>50cm^{-1}) regions of the spectra. Open circles and data at 300K. Closed circles are data at 180K, scales by $T^{1/2}$. X is the triple point liquid scaled by $T^{1/2}$.

accurate absolute intensities to permit distinguishing among the different intermolecular potentials on that basis (e.g. for liquid argon the L-J potential is calculated(37) to give an intensity ∿2.6 times greater than the hard sphere potential).

IV. B. Simple Solids

Experiments undertaken(36) to assess the relationships between the intermolecular light scattering in the simple liquids and the second order Raman (two phonon) spectra in the crystalline solids are summarized in Figs. 14 and 15. In Fig. 14 is shown the

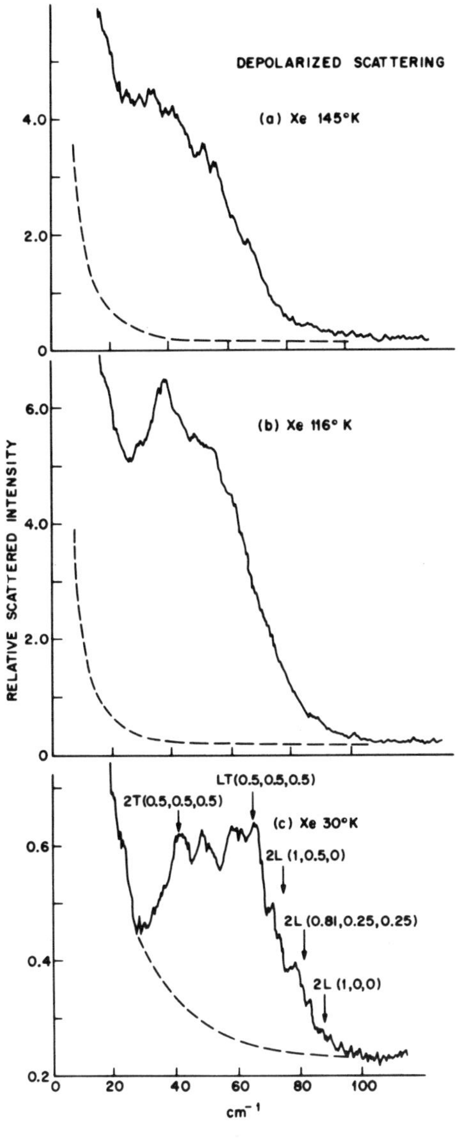

Fig. 14 Second order-Raman spectra in solid Xe. Dashed curve represents instrumental background. Arrows in (c) indicate calculated frequencies of two phonon features at various critical points in the fcc Brillouin zone. After Ref. 36.

second-order Raman spectrum from solid Xe. The fcc rare gas solids have only one atom/unit cell and therefore no first order Raman spectrum. At low temperatures, the second-order spectrum is weak but clearly exhibits structure associated with phonon pairs from various points in the Brillouin zone. Spectra in solid Ar and solid Kr are similar in appearance provided their frequencies scaled according to the material's Debye frequency. With increasing temperature the second-order spectrum gradually loses its sharp structure, with the 2TA peaks persisting to higher temperatures than the 2LA or LA-TA peaks. In addition, the relatively sharp high frequency cutoff (at $88cm^{-1}$ for Xe) becomes smeared out and eventually assumes a nearly exponential shape.

In Fig. 15 the solid and liquid spectra(36) are compared at the triple point. Spectra are shown with no change in intensity or frequency scales between the liquid and the solid. Note that as far as the high frequency dynamics ($\omega \gtrsim 2\omega_D$) are concerned the liquid and solid are indistinguishable. At low frequencies, however, the liquid spectra continue to increase exponentially while the solid spectra level off, so that the integrated intensities in the liquids are all much greater than in the corresponding solids.

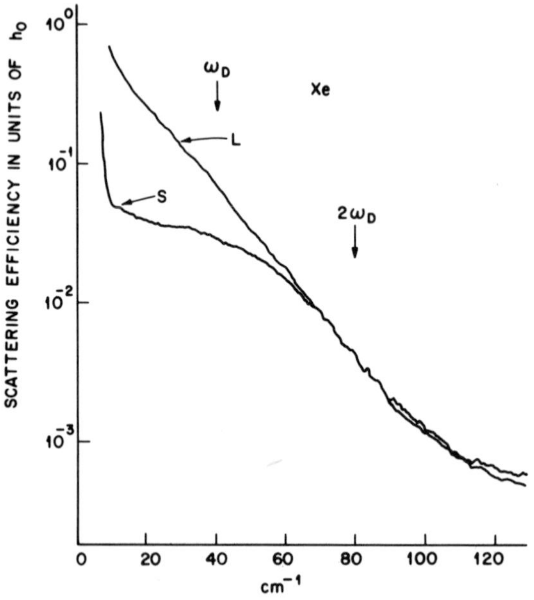

Fig. 15 Comparison of second-order Raman spectra in Xe just below (S) and just above (L) the melting temperature. Note logarithmic intensity scale. After Ref. 36.

In Table II various aspects of the second-order spectra for the condensed phases of the classical rare gases Ar, Kr and Xe are listed and experiment is compared with theory. The "calculated" characteristic frequency, Δ, is obtained from the corresponding states formula Eq. (9). The calculated total scattered intensity h_T for the solid is obtained from a finite temperature generalization[36] of the theory of Werthamer, Gray and Koehler.[38]

$$h_T = \frac{N}{V}\left(\frac{\alpha_o}{a^3}\right)^4 \left(\frac{\omega_i}{c}\right)^4 a^6 Z^2 \left(\frac{\hbar}{m\omega_D a^2}\right)^2 \left(\frac{n_i^2+2}{3}\right)^2 f\left(\frac{\hbar\omega_D}{kT}\right) \qquad (12)$$

where α_o is the atomic polarizability, a the lattice parameter, Z the number of nearest neighbors, ω_i the laser frequency, n_i the optical refractive index and $f(x) \doteq (1-e^{-x})^{-2}$. In view of the crude arguments used in this finite temperature extrapolation we would not expect Eq. (12) to be quantitatively valid to better than about a factor of 2. The considerably better agreement with experiment indicated in Table II is probably fortuitous.

Nevertheless, the agreement between experiment and theory regarding both intensity and frequency range leaves little doubt that the observed spectra in the classical rare gas solids are essentially completely described by two phonon processes. The smooth evolution of the high frequency spectra upon melting also supports the view that in simple classical fluids the spectra can be attributed to scattering from pairs of highly damped density fluctuations. The large increase in spectral intensity at low frequencies for the liquid relative to the solid accounts for all of the rather large difference in integrated intensities between the two phases. However this striking difference has not yet

Table II

	ω_D (cm^{-1})	$\Delta^{calc}_{liq.}$ (cm^{-1})	$\Delta^{obs}_{liq.}$ (cm^{-1})	h_T^{obs}(liq.) ($cm^{-1}Sr^{-1}$)	h_T^{obs}(sol.) ($cm^{-1}Sr^{-1}$)	h_T^{calc}(sol.) ($cm^{-1}Sr^{-1}$)
Ar	62	22.1	22	18×10^{-10}	4.6×10^{-10}	4.6×10^{-10}
Kr	46	16.7	17	83×10^{-10}	9.4×10^{-10}	6.9×10^{-10}
Xe	40	14.4	14	276×10^{-10}	33.6×10^{-10}	21.4×10^{-10}

received a satisfactory theoretical explanation. Perhaps some insight into this problem can be gained by nothing that h_T is related to the nonlinear optical susceptibility[39] which is known to receive contributions from "electronic" and "atomic configurational" mechanisms. The similarities between liquid and solid spectra at high frequencies may suggest that the "electronic" mechanism dominates there; while the "atomic configuration" contribution is more important at low frequencies and would naturally differ between the liquid and solid states.

Experimental studies of higher order spectra have been conducted in the condensed phases of ^3He, ^4He and their mixtures by Surko and Slusher.[40] While several aspects of the spectra are similar to those just discussed for the classical materials, there are some important differences. We have already discussed the sharp structure associated with roton pairs in superfluid ^4He, and now concentrate on the more general features seen over wide ranges of temperature, density and isotropic composition. Typical spectra are shown in Fig. 16 for solid and liquid ^4He. Except for a diminished spectral intensity at low frequencies these spectra appear at first glance very similar to those for Xe in Fig. 14. That is, the high frequency lineshapes are essentially identical in the liquid and solid states and there is an appreciable increase in low frequency intensity for the liquid, relative to the solid. As seen by the solid curve in Fig. 16, the observed peak in the spectrum corresponds rather well to that in the two phonon density of states (shown as the solid curve). Its persistence into the liquid state is a strong argument for the existence of relatively well-defined short wavelength phonon modes in the liquid.[40]

A striking difference between the helium spectra and those of the classical materials is the appreciable <u>additional intensity</u> observed at frequencies well above the predicted two phonon cutoff. Within a phonon type of description such intensity would require four-, or even six-phonon processes. Surko and Slusher have argued[40] that this additional high frequency scattering can successfully be viewed instead as due to creation of pairs of nearly free particle excitations. In He single free particle excitations at large energies (>20cm^{-1}) and momenta (>2.5Å^{-1}) have been observed by neutron scattering, and obey the dispersion relation $\varepsilon_k = \hbar^2k^2/2m^*$, with m^* nearly equal to the mass of a bare helium atom. By taking the final single particle states at wave vector k to be distributed about $\hbar^2k^2/2m^*$ with a Gaussian of width $\hbar k\Gamma$, they calculated the spectral response associated with excitation of pairs (at $\vec{k}$ and $-\vec{k}+\vec{q}$) summed over $\vec{k}$. The value of Γ was adjusted to give the best fit (Γ = 8.7 K/Å^{-1}) and the results compared with experiment in Fig. 17. The overall intensity scales are adjusted.[40] The success of this approach indicates that

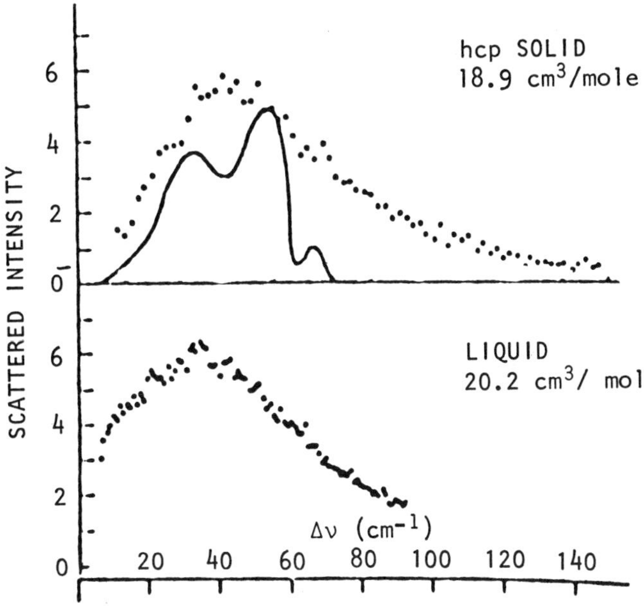

Fig. 16 Light scattering spectra of solid and liquid helium (points) at 3-4K. Solid curve is calculated[38] two phonon Raman spectrum. After Ref. 40.

pairs of both single particle and collective (phonon) excitations contribute significantly to the light scattering spectra in the condensed phases of helium. For the heavier rare gas solids and liquids the single particle pairs are much less important, presumably because for these materials the ratio of binding energy, E_b, to Debye frequency, $\hbar\omega_D$ is much greater than in He. That is, the single particle pairs should contribute for frequencies $\hbar\omega > 2E_b$ a spectral component of the form $I_{SPP}(\omega) = Ce^{-\omega/\omega_D}$. Surko and Slusher estimate[40] that for argon, C is $\simeq 15$ times the maximum of the two phonon intensity but since $E_b = 715°K$ and $\omega_D = 93°K$, the ratio $I_{SPP}/I_{2Ph} \sim 15\, e^{-1430/93} \approx 2\times 10^{-6}$ is negligible indeed.

To complete the discussion of higher order light scattering in simple systems, we consider the case of liquid and solid hydrogen. Although a diatomic molecule, H_2 at low temperatures

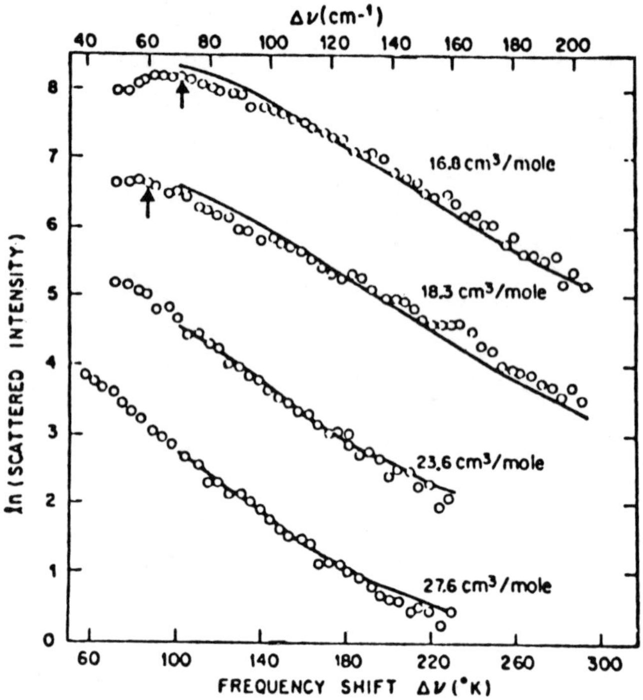

Fig. 17 Semilog plot of light scattering spectra in ^4He at large frequency shifts for different molar volumes. Upper two curves are for h.c.p. solid, lower two are for liquid ^4He. Data at each molar volume displaced vertically for display. Solid lines are calculated for scattering from pairs of free particles. After Ref. 40.

can be prepared in either the para (J=0) or ortho (J=1) configurations (or as arbitrary mixtures of the two). Thus it can be treated as a fluid of spherically symmetric particles in the pure para state. The intermolecular spectra(41) from liquid parahydrogen, fall not surprisingly in general appearance between the purely exponential spectra of the classical liquids and the sharply peaked roton pair spectrum of superfluid helium. In fact as seen in Fig. 18 the spectrum looks quite similar to those in normal liquid helium, thus supporting the existence of relatively well-defined collective phonon modes of short wavelength in para hydrogen.(42) The addition of ortho-hydrogen (J=1) even up to 75% concentrations, does not alter this behavior. The main effect of such addition is to introduce an additional depolarized quasi-elastic component due to $J = 1 \rightarrow 1$ transitions. These transitions correspond to orientation fluctuations which are absent in the spherical para-H_2 liquid. The unique ability afforded by liquid hydrogen to mix chemically identical spherical and anisotropic molecules at liquid densities can be exploited to examine the microscopic dynamics of the intermolecular interactions. This has been done through a careful study of rotational Raman line-shapes as functions of ortho/para ratio and temperature. See Ref. 43 for details.

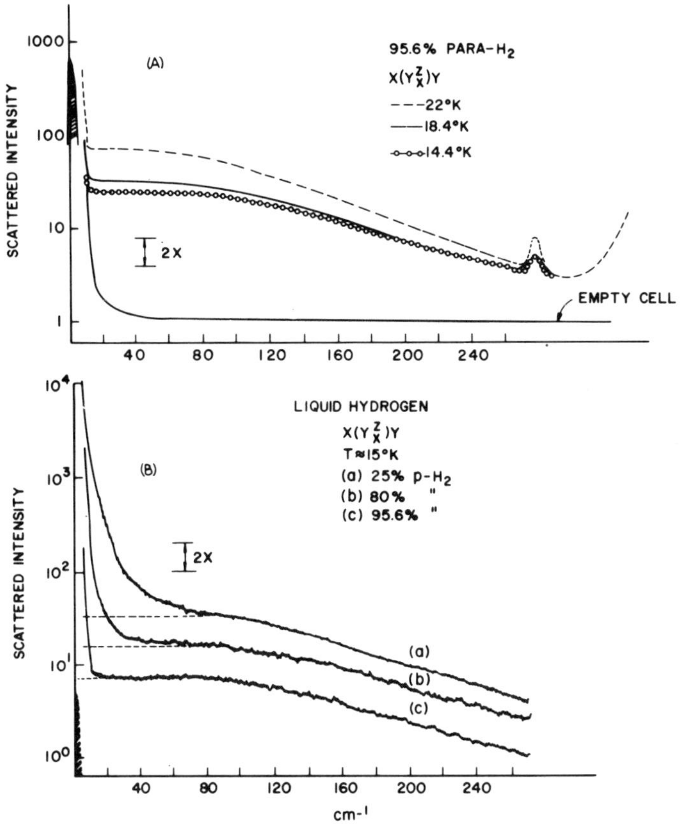

Fig. 18 Depolarized intermolecular scattering in liquid hydrogen. (A) Temperature variation in 95.6% para hydrogen. (B) variation with ortho-para composition at constant temperature (15K). Spectra are displaced vertically for display. After Ref. 41.

The case of solid (hcp) para hydrogen is similar to that of hcp ^{4}He, with the exception that the single particle pairs play less of a role. Fig. 19 compares the observed spectrum in solid H_2 to the calculated two phonon prediction (dashed curve). The correspondence is quite good, even including the dip in the vicinity of 120cm^{-1}. At low temperatures in ortho-hydrogen the broad J = 1 → 1 continuum seen in the liquid develops structure corresponding to quantized states called librons. Examination of these features has proved a powerful probe of the order-disorder transition in ortho hydrogen and para deuterium, and is discussed by Prof. Silvera in this volume.(44)

It is clear, in conclusion, that higher order light scattering in the condensed phases of simple molecular systems has revealed several new phenomena and has provided extensive quantitative data on the short time, small scale dynamics of simple solids and liquids. The spectra in solids can be semiquantitatively understood as due to pairs of collective (phonon) and

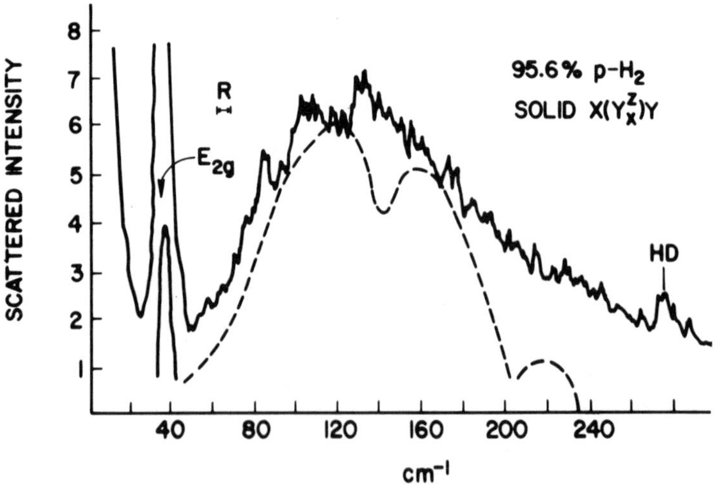

Fig. 19 Raman spectrum of solid hcp para hydrogen at 11K. One phonon scattering (Eg) is shown reduced by 10X. Two phonon spectrum calculated lineshape is given by dashed curve. After Ref. 41.

single particle excitations, with the latter being of practical importance only when the ratio of the binding energy to the Debye frequency is not too large. The theoretical situation in the dense fluids and liquids is much less satisfactory. With the exception of the roton pair region in superfluid ^{4}He, there is not even a semiquantitatively satisfactory theory for higher order spectra. While complete solution of this problem will require a full theory of the dynamics of the liquid state, and may therefore not be immediately forthcoming, theorists at least have a number of quantitative experimental benchmarks regarding the dynamics of simple fluids which did not exist even a few years ago.

V. HIGHER ORDER CORRELATION FUNCTIONS AND PHASE TRANSITIONS

It is well known that insofar as phase transitions and critical phenomena are concerned, the fundamental quantities of interest are the space and time dependent correlation functions involving fluctuations in the order parameter: $\psi(\vec{r},t) \equiv \psi_o + \delta\psi(\vec{r},t)$. Singular behaviors of the macroscopic order parameter $\psi_o = \langle\psi(\vec{r},t)\rangle$, the dynamic susceptibility $\chi_\psi(\vec{r},t) = \langle\psi(\vec{r},t)\psi(0,0)\rangle$, and other simply related functions are now relatively well understood for a wide variety of transitions thanks to a fruitful interplay between experiment and theory over recent years. In many cases scattering experiments can directly measure the Fourier transform of the dynamic susceptibility and thus provide relatively complete information on the dynamics of the critical fluctuations near T_c. For example, critical slowing down is manifested in the divergence of $\chi_\psi(q_c,\omega)$ as $\omega \to 0$, so that the signature in the scattering spectrum is the collapse of spectral weight to zero frequency, provided of course that the scattering probe couples linearly to $\psi(\vec{r},t)$. Light scattering from density fluctuations at the liquid gas critical point, and neutron scattering from simple antiferromagnets provide familiar examples.

However, there are many important cases, particularly with light scattering, where linear coupling to the order parameter is symmetry forbidden. This is in fact the case for all symmetry changing transitions, i.e. those in which the critical mode is not Raman active on both sides of T_c. In such cases (including the familiar $SrTiO_3$ cubic-tetragonal transition, the ferroelectric transition in perovskite crystals, etc.) $P^{\alpha\beta}(\vec{r},t) = a_2\psi^2(\vec{r},t)$, so that the important correlation function determining the scattered spectrum is fourth order: $\langle\psi(\vec{r},t)\psi(\vec{r},t)\psi(0,0)\psi(0,0)\rangle$.

The processes discussed earlier in this paper correspond to simultaneous creation or destruction of pairs of elementary excitations (that is, to sum processes). More important for critical phenomena are the corresponding <u>difference</u> processes, in which an excitation of wave vector $\vec{k}$ is created, while a second excitation on the same branch at wave vector $\vec{q}-\vec{k}$ is simultaneously destroyed. When the relevant excitations are phonons, this pair mode is often referred to as a "phonon density fluctuation".[46] The spectral lineshape from this process is determined by weighted k space averages of the excitations' group velocities, k-dependent lifetimes, and interaction effects. For an undamped dispersionless excitation, its density fluctuation spectrum would be a delta function at zero frequency, provided interaction effects are ignored. Thus the second-order difference Raman scattering can in principle provide very low frequency spectral response, which may contribute significantly to, or even solely determine the spectrum of fluctuations near T_c.

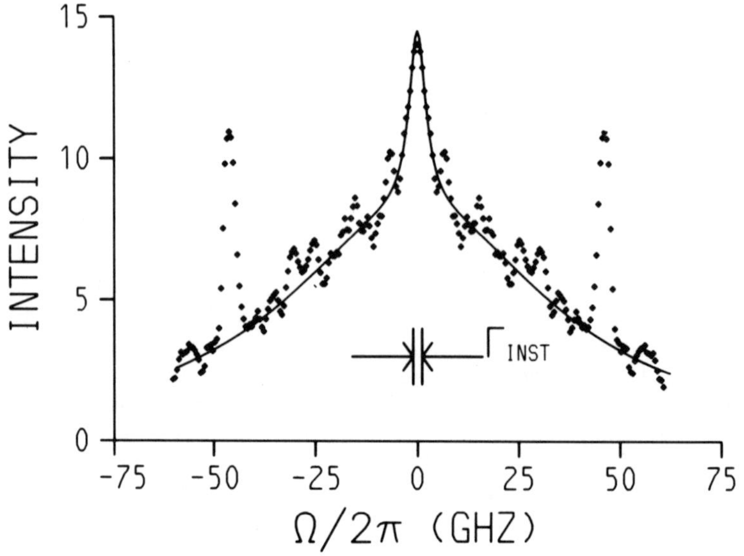

Fig. 20 Polarized quasielastic and Brillouin spectrum of $KTaO_3$ at 182K. Solid line is sum of two Lorentzians of full width at half maximum 5.0 and 80 GHz. The LA Brillouin components appear at ±47.5 GHz. After Ref. 46.

It has recently been possible to observe(46) light scattering directly from phonon density fluctuations in crystals. Fig. 20 shows the low frequency spectrum of $KTaO_3$ (O_h^1 symmetry, no allowed first order Raman spectrum) which exhibits two central peaks in addition to the familiar Brillouin scattering from acoustic phonons. The narrower peak is due to entropy fluctuations, made visible by an enhanced Landau-Placzek ratio.(45) The broader one is due to phonon density fluctuations. Fig. 21 shows the temperature dependence of its intensity and linewidth. As expected no singularities appear, inasmuch as $KTaO_3$ exhibits no phase transition in the temperature range covered. However these results clearly demonstrate that dynamic "central peaks" in crystalline solids can arise from higher order correlation functions.

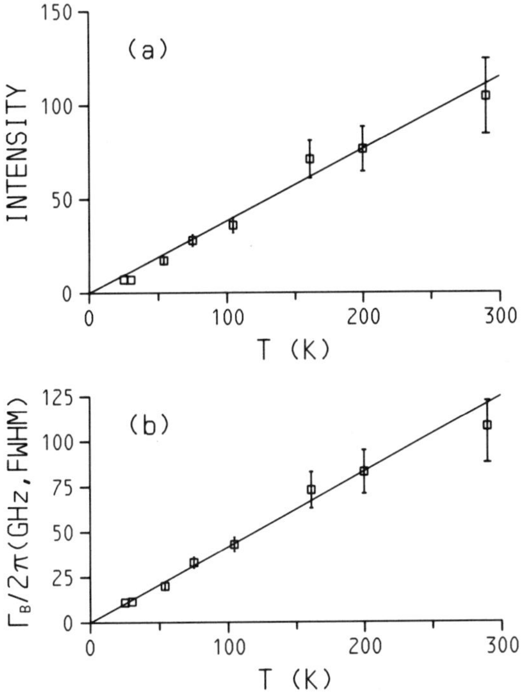

Fig. 21 Temperature dependence of (a) integreted intensity and (b) linewidth of the broad central component in $KTaO_3$, shown in Fig. 20. This feature has been attributed to phonon density fluctuations. After Ref. 46.

In the case of $SrTiO_3$, which exhibits a continuous phase transition from O_h^1 to D_{4h}^{14} symmetry at 107K, a singular dynamic central peak has been detected in the depolarized light scattering spectrum.(47) Figure 22 illustrates its development with $|T-T_c|$ as well as its interaction with the transverse acoustic phonons. Courtens(48) has explored theoretically the phonon density fluctuations arising from the Brillouin zone boundary soft mode above T_c, and has emphasized their relation to the fourth order correlation function. The spectral lineshape for this process has not yet been calculated in any such system, but it is clearly an important unsolved theoretical problem in critical dynamics. Similar observations have been reported at the uniaxial ferroelectric transition in lead germanate.(49) It was emphasized

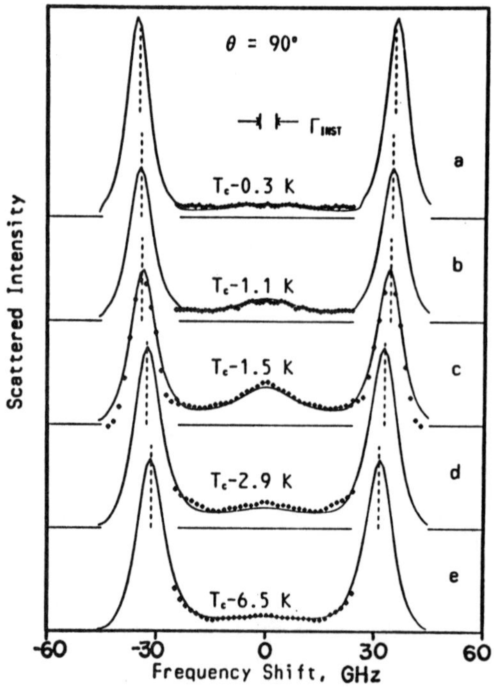

Fig. 22 Brillouin spectrum of $SrTiO_3$ showing emergence of a singular dynamic central peak associated with the cubic-tetragonal phase transition. T_c = 107K. After Ref. 47.

there that at present there is no general theory for the critical exponent describing the total scattered intensity - although when $a_2^{\alpha\beta}\psi^2$ corresponds to the energy density, the divergence should be described by the specific heat exponent, α.

It now appears likely that the origin of many of the "central peak" phenomena reported in neutron and light scattering studies of structural phase transitions(50) receives contributions from such phonon density fluctuations, although the precise influence of defects on critical behavior has not been fully explored.(51)

Although theoretical evaluation of fourth order dynamic correlation functions in phonon systems is extremely difficult, it is hoped that the recent availability of detailed experimental line-shape and intensity information will stimulate greater theoretical efforts in these directions.

REFERENCES

1. See for example, R. Loudon, Adv. in Phys. 17, 243, 1968.

2. P. A. Fleury, Proc. Int'l. Conf. on Magnetism, Vol. 1, Nauka, Moscow, 1974, p. 80.

3. T. Moriya, J. Appl. Phys. 39, 1042 (1968).

4. P. A. Fluery and R. Loudon, Phys. Rev. 166, 514 (1968).

5. P. A. Fleury, S. P. S. Porto, L. E. Cheesman and H. J. Guggenheim, Phys. Rev. Letters 17, 84, 1966.

6. S. R. Chinn, R. W. Davies, H. J. Zeiger, A.I.P. Conf. Proceedings #5 Magnetism and Magnetic Materials, ed. by C. D. Graham and J. J. Rhyne 1972, p. 317.

7. Y. A. Popkov, V. I. Fomin and B. V. Beznosikov, JETP Lett. 11, 264, 1970.

8. P. A. Fleury, Int'l. J. Magnetism, 1, 75, 1970.

9. P. Moch et al., "Light Scattering in Solids", edited by M. Balkanski, Flammarion, Paris, 1971, p. 138.

10. R. M. McFarlane, Phys. Rev. Lett. 25, 1454, 1970.

11. R. V. Pisarev, P. Moch and C. Dugautier, Phys. Rev. B7, 4185, 1973.

12. P. A. Fleury, Phys. Rev. 180, 591, 1969.

13. R. J. Elliott and M. F. Thorpe, J. Phys. C2, 1630, 1969, and this volume.

14. M. F. Thorpe, Phys. Rev. B2, 2690, 1970.

15. J. B. Parkinson, J. Phys. C2, 2012, 1969.

16. P. A. Fleury and H. J. Guggenheim, Phys. Rev. Lett. 24, 1346, 1970.

17. P. A. Fleury in Ref. 9, p.151; R. Loudon, J. Phys. C3, 872, 1970; and M. G. Cottam, J. Phys. C8, 1933, 1975.

18. C. R. Natoli and J. Ranninger, J. Phys. C6, 345, 1973.

19. U. Balucani and V. Tognetti, P.R. B8, 4247, 1973, and Rivista del Nuovo Cimento 6, 39, 1976.

20. Bohnen, C. R. Natoli and J. Ranninger, J. Phys. C7, 947, 1974.

21. A. van der Pol et al., Solid State Comm. 19, 177, 1976.

22. U. Balucani and V. Tognetti, Phys. Rev. B_, 1977 in press.

23. S. Lovesey and R. Meserve, J. Phys. C6, 79, 1973.

24. M. Buchanan, et al., J. Phys. C5, 2011, 1972.

25. P. A. Fleury, W. Hayes and H. J. Guggenheim, J. Phys. C8, 2183, 1975.

26. J. Als Nielsen, R. J. Birgeneau, H. J. Guggenheim, G. Shirane, Phys. Rev. B12, 4963, 1975.

27. P. A. Fleury and H. J. Guggenheim, Phys. Rev. B12, 985, 1975.

28. J. W. Halley, Light Scattering Spectra of Solids, edited by G. B. Wright, Springer-Verlag, New York, 1969, p. 177

29. M. J. Stephen, Phys. Rev. 187, 279, 1969.

30. H. B. Levine and G. Birnbaum, Phys. Rev. Lett. 20, 439, 1968.

31. For a review see W. M. Gelbart, Adv. Chem. Phys. 26, 1, 1974.

32. J. P. McTague, and G. Birnbaum, Phys. Rev. Lett. 21, 661, 1968 and Phys. Rev. A3, 1376, 1971.

33. P. A. Fleury and J. P. McTague, Optics Comm. 1, 164, 1969 and J. P. McTague, P. A. Fleury and D. B. DuPre, Phys. Rev. 188, 303, 1969.

34. T. J. Greytak and J. Yan, Phys. Rev. Lett. 22, 987, 1969.

35. P. A. Fleury, W. B. Daniels and J. M. Worlock, Phys. Rev. Lett. 27, 1493, 1971.

36. P. A. Fleury, J. M. W. Worlock and H. L. Carter, Phys. Rev. Lett. 30, 591, 1973.

37. B. J. Alder, H. L. Strauss and J. J. Weiss, J. Chem. Phys. 59, 1002, 1973.

38. N. R. Werthamer, R. L. Gray and T. R. Koehler, Phys. Rev. B2, 4199, 1970.

39. R. W. Hellwarth, A. Owyoung and N. George, P.R. A4, 2342, 1971.

40. C. M. Surko and R. E. Slusher, Phys. Rev. B13, 1095, 1976.

41. P. A. Fleury and J. P. McTague, Phys. Rev. Lett. 31, 914, 1973.

42. K. Carneiro, M. Nielsen, and J. P. McTague, Phys. Rev. Lett. 30, 481, 1973.

43. P. A. Fleury and J. P. McTague, Phys. Rev. A12, 317, 1975.

44. I. Silvera, this proceedings.

45. See for example, R. K. Wehner and R. Klein, Physica 62, 161, 1972; and G. J. Coombs and R. A. Cowley, J. Phys. C6, 121, 1973.

46. K. B. Lyons and P. A. Fleury, Phys. Rev. Lett. 37, 161, 1976.

47. K. B. Lyons and P. A. Fleury, Solid State Comm. 23, 477, 1977.

48. E. Courtens, Phys. Rev. Lett. 37, 1584, 1976.

49. P. A. Fleury and K. B. Lyons, Phys. Rev. Lett. 37, 1088, 1976.

50. See for example, "Anharmonic Lattices, Structural Transitions and Melting", edited by T. Riste, Noordhoff, Leiden, 1974.

51. B. I. Halperin and C. M. Varma, P.R. B14, 4030, 1976.

Part III
Liquids

SECOND-ORDER LIGHT SCATTERING BY CLASSICAL FLUIDS I: COLLISION INDUCED SCATTERING

William M. Gelbart

Department of Chemistry, University of California

Los Angeles, California 90024

INTRODUCTION

In this and the following lecture I shall deal with two important examples of higher-than-first-order light scattering from classical gases and liquids. The term "first-order" means here that (1) light has been scattered only once by the sample (i.e., doubly- and multiply-scattered waves make negligible contributions to the detached signal), and (2) the polarizabilities of the molecules are not distorted through interaction with the electronic charge clouds of neighboring molecules. Virtually all discussions in the journal and monograph literature[1] have dealt only with the first-order scattering of light, e.g. critical point fluctuations and sound propagation via Rayleigh-Brillouin spectra, molecular reorientation in liquids via Raman line shapes, and macromolecular weight distributions, conformational changes and diffusion.

In all these first-order examples the amplitude of the scattered field is linear in the molecular polarizabilities ($\underset{\approx}{\alpha}$) and the density fluctuations ($\Delta\rho$). Correspondingly, the observed cross sections are quadratic in the $\underset{\approx}{\alpha}$'s and the $\Delta\rho$'s. For example, the Rayleigh-Brillouin spectrum[1] of an atomic fluid is proportional (as we shall see below) to $\alpha^2 \langle \Delta\rho_{\underset{\sim}{K}} \Delta\rho_{-\underset{\sim}{K}} \rangle$ where: $\Delta\rho_{\underset{\sim}{K}} = \int d\underset{\sim}{r}\, \exp(i\underset{\sim}{K}\cdot\underset{\sim}{r})\Delta\rho(\underset{\sim}{r}) = \sum_i \exp(i\underset{\sim}{K}\cdot\underset{\sim}{R}_i)$; $\underset{\sim}{K}$ is the difference between the propagation wavevectors[1] for the incident and scattered light; $\underset{\sim}{R}_i$ is the position of the ith atom; and the brackets denote an equilibrium (canonical ensemble) average over all $\underset{\sim}{R}_i, \cdots \underset{\sim}{R}_N$.

In my lectures I shall treat second-order light scattering processes for which the observed cross sections involve α^4 (rather than α^2) and a four- (rather than two-) point correlation of the density fluctuations. In the first lecture I discuss the collision-induced scattering of light by classical gases and liquids. Here the incident light is scattered only once before being detected, but the polarizabilities of the molecules are distorted through interaction with the electronic charge clouds of their neighbors. Thus each scattered field amplitude is pairwise additive in the particle positions, giving rise to an observed intensity which is proportional to a four-point density correlation function. In the case of double light scattering at gas-liquid critical points (second lecture) only undistorted electronic charge clouds are involved, but each incident wave is scattered twice. Again, then, the electric field amplitude at the detector is pairwise additive (quadratic) in the density fluctuations.

After a brief outline of light scattering theory, I present in Section III a "quick and dirty" discussion of first-order spectra; a more rigorous discussion is included in the second lecture. The generalized hydrodynamics for an atomic fluid are treated in Section IV, so that multiple excitation contributions to first-order spectra can be described in terms of the system's approximate collective modes. Second-order spectra are discussed in Section V where a detailed theory of collision-induced optical effects is presented. In deriving the expressions relevant to the second order scattering of light by a classical fluid I shall not attempt to be rigorous. Instead my discussion will be heuristic, with references given where appropriate to the more formal theory available in the literature.

II. GENERAL LIGHT SCATTERING THEORY

In order to lay an adequate foundation for the later discussion of the new effects mentioned above, it is necessary first to treat in detail the more familiar first-order scattering of light.

Consider a classical gas or liquid made up of molecules whose dimensions are small compared with the wavelength of optical radiation. A linearly polarized light wave with electric field

$$\underset{\sim}{E}_0(\underset{\sim}{r},t) = \underset{\sim}{E}_0 \exp(i\underset{\sim}{k}_0\cdot\underset{\sim}{r} - i\omega_0 t) \tag{1}$$

is incident on this sample. Then from Maxwell's equations it follows[2-4] that

$$\underset{\sim}{E}_S(\underset{\sim}{r},t) = \underset{\sim}{E}_0(\underset{\sim}{r},t) + \int d\underset{\sim}{r}'\int dt'\ \underset{\approx}{T}(\underset{\sim}{r}-\underset{\sim}{r}',t-t')\cdot\underset{\sim}{P}(\underset{\sim}{r}',t') \tag{2}$$

Here $\underset{\sim}{E}_S(\underset{\sim}{r},t)$ is the local electric field at the point $\underset{\sim}{r}$ in the fluid at time t, and $\underset{\sim}{P}(\underset{\sim}{r}',t')$ is the polarization (induced dipole moment) at $\underset{\sim}{r}'$ at t'. The tensor $\underset{\approx}{T}$ is defined by[2,3]

$$\underset{\approx}{T}(\underset{\sim}{R},t) = \frac{c}{2\pi}\int_{-\infty}^{+\infty} dk\ \exp(-ikct)\underset{\approx}{T}_k(\underset{\sim}{R}) \tag{3}$$

where

$$\underset{\approx}{T}_k(\underset{\sim}{R}) = \begin{cases} (\underset{\sim}{\nabla}\underset{\sim}{\nabla} + k^2\underset{\approx}{I}) \left[\dfrac{\exp(ikR)}{R}\right], & R > \sigma = 0^+ \\ 0, & \text{otherwise} \end{cases} \quad (4)$$

$\underset{\approx}{T}(\underset{\sim}{r}-\underset{\sim}{r}',t-t')$ acts on the dipole moment $\underset{\sim}{P}(\underset{\sim}{r}',t')$ at $\underset{\sim}{r}',t'$ to give the electric field generated by it at the "neighboring point" $\underset{\sim}{r},t$. [The restriction $R > \sigma$ in Eq. (4) arises since a dipole does not polarize itself (we neglect the radiation reaction term[5]); $\sigma = 0^+$ corresponds to our implicitly making the "point-atom" assumption.] Thus Eq. (2) states that the local electric field at an arbitrary point in the fluid is equal to the incident field there plus the fields from the dipole moments induced throughout the rest of the sample.

Now we make the phenomenological assumption that $\underset{\sim}{P}(\underset{\sim}{r},t)$ can be written as

$$\underset{\sim}{P}(\underset{\sim}{r},t) = \underset{\approx}{\alpha}(\underset{\sim}{r},t)\rho(\underset{\sim}{r},t)\cdot\underset{\sim}{E}_s(\underset{\sim}{r},t) \quad (5)$$

where $\underset{\approx}{\alpha}$ is the molecular polarizability tensor, and

$$\rho(\underset{\sim}{r},t) = \sum_i \delta[\underset{\sim}{r} - \underset{\sim}{R}_i(t)] \quad (6)$$

is the instantaneous number density at $\underset{\sim}{r},t$. [$\underset{\sim}{R}_i(t)$ is the center-of-mass position of the ith molecule at time t.] Substituting Eq. (5) into Eq. (2) we obtain the following, <u>closed</u> equation for $\underset{\sim}{E}_s$:

$$\underset{\sim}{E}_s(1) = \underset{\sim}{E}_0(1) + \int d(2)\, \underset{\approx}{T}(1-2)\cdot\underset{\approx}{\alpha}(2)\rho(2)\underset{\sim}{E}_s(2) \quad (7)$$

Here "1" $\equiv \underset{\sim}{r},t$ and "2" $\equiv \underset{\sim}{r}',t'$.

If we attempt to solve Eq. (7) directly by iteration, we obtain $\underset{\sim}{E}_s$ as a power series in $\alpha\rho$, the polarization per unit volume. This can be shown,[6] however, to lead to divergences in the corresponding expansion for the intensity. These difficulties are avoided by recognizing[4-10] that light propagation takes place inside the fluid with a wavevector different from $\underset{\sim}{k}_0$. The uniform dielectric "background" can be accounted for by "renormalizing" the dipole tensors so that they include the effects of propagation through the medium <u>to all orders in $\alpha\rho$</u>. This has been carried out in full detail by Bedeaux and Mazur[5] and by Felderhof,[4] and describes correctly the scattering of light from the density <u>fluctuations</u> $\Delta\rho$ (rather than the total densities ρ).

Let us write then the total density at $\underset{\sim}{r},t$ as the sum of a uniform part ρ_0 and a fluctuating part $\Delta\rho$:

$$\rho(\underset{\sim}{r},t) = \rho_0 + \Delta\rho(\underset{\sim}{r},t) \qquad [\rho_0 \equiv \langle\rho(\underset{\sim}{r},t)\rangle = \frac{N}{V}] \tag{8}$$

Similarly, following Stephen,[9] we divide the field $\underset{\sim}{E}_s$ into a part $\underset{\sim}{E}_s^{(0)}$ and $\Delta\underset{\sim}{E}_s$. $\underset{\sim}{E}_s^{(0)}$ is the electric field which would propagate (without scattering) in the sample if its density were truly uniform: it satisfies

$$\underset{\sim}{E}_s^{(0)}(1) = \underset{\sim}{E}_0(1) + \alpha\rho_0\int d(2)\underset{\approx}{T}(1-2)\underset{\sim}{E}_s^{(0)}(2) \tag{9}$$

Eq. (9) is simply Eq. (7) with $\rho \to \langle\rho\rangle$ and $\underset{\approx}{\alpha} \to \alpha$, and its solution is well known. $\underset{\sim}{E}_s^{(0)}$ propagates in the sample with a velocity c/n where the refractive index n is related to the isotropic part (α) of $\underset{\approx}{\alpha}$ via the Lorentz-Lorenz equation[2-3]

$$\frac{n^2 - 1}{n^2 + 2} = \frac{4\pi}{3}\alpha\rho_0 \tag{10}$$

In order to simplify the ensuing discussion, we assume that n is close enough to unity so that the small differences between $\underset{\sim}{E}_0$ and $\underset{\sim}{E}_s^{(0)}$ can be neglected. (As a result, all our cross sections will be in error by multiplicative factors of $(n^2+2)/3$, etc.) Then it follows from Eqs. (7) and (8) that the scattered field $\Delta\underset{\sim}{E}_s$ satisfies the closed equation

$$\Delta\underset{\sim}{E}_s(1) = \int d(2)\underset{\approx}{T}(1-2)\cdot\underset{\approx}{\alpha}(2)\Delta\rho(2)\underset{\sim}{E}_0(2) + \int d(2)\underset{\approx}{T}(1-2)\cdot\underset{\approx}{\alpha}(2)\Delta\rho(2)\Delta\underset{\sim}{E}_s(2) \tag{11}$$

Solving Eq. (11) by iteration we obtain $\Delta\underset{\sim}{E}_s$ as a power series in $\alpha\Delta\rho$:

$$\Delta\underset{\sim}{E}_s = \Delta\underset{\sim}{E}_s^{(1)} + \Delta\underset{\sim}{E}_s^{(2)} + \cdots \tag{12}$$

where

$$\Delta\underset{\sim}{E}_s^{(1)}(1) = \int d(2)\underset{\approx}{T}(1-2)\cdot\underset{\approx}{\alpha}(2)\Delta\rho(2)\underset{\sim}{E}_0(2) \tag{13}$$

and

$$\Delta\underset{\sim}{E}_s^{(2)}(1) = \int d(2)\int d(3)\underset{\approx}{T}(1-2)\cdot\underset{\approx}{\alpha}(2)\Delta\rho(2)\underset{\approx}{T}(2-3)\cdot\underset{\approx}{\alpha}(3)\Delta\rho(3)\underset{\sim}{E}_0(3) \tag{14}$$

For reasons which shall become apparent soon we refrain from referring to $\Delta\underset{\sim}{E}_s^{(1)}$ and $\Delta\underset{\sim}{E}_s^{(2)}$ as the "singly" and "doubly" scattered fields - they are more properly associated with "first" and "second"-order optical response. Indeed $\Delta\underset{\sim}{E}_s^{(2)}$ contains <u>both</u> the collision-induced <u>and</u> double light scattering effects.

III. FIRST-ORDER SPECTRA

Let's look first at $\Delta\underset{\sim}{E}_s^{(1)}$ and recover the familiar Rayleigh-Brillouin and first-order Raman spectra. The power spectrum associated with the scattered field is given by[1,11]

$$I(\omega) = \int_{-\infty}^{+\infty} dt\ e^{i\omega t}\ H(t) \tag{15}$$

where

$$H(t) = \langle \Delta \underset{\sim}{E}_s(\underset{\sim}{R},t)\Delta \underset{\sim}{E}_s^*(\underset{\sim}{R},0)\rangle \tag{16}$$

Now, using the definitions of $\underset{\approx}{T}(1\text{-}2)$ and $\underset{\sim}{E}_0(2)$ given by Eqs. (3) and (1), respectively, Eq. (13) for $\Delta\underset{\sim}{E}_s^{(1)}$ can be written out as [for "1" = $\underset{\sim}{R}$,t and "2" = $\underset{\sim}{r}'$,t']

$$\Delta\underset{\sim}{E}_s^{(1)}(\underset{\sim}{R},t) = \tag{17}$$

$$\int d\underset{\sim}{r}' \int_{-\infty}^{+\infty} dt' \frac{c}{2\pi}\int_{-\infty}^{+\infty} dk e^{-ikc(t-t')} \underset{\approx}{T}_k(\underset{\sim}{R}-\underset{\sim}{r}')\cdot\underset{\approx}{\alpha}(\underset{\sim}{r}',t')\Delta\rho(\underset{\sim}{r}',t')\underset{\sim}{E}_0 e^{i\underset{\sim}{k}_0\cdot\underset{\sim}{r}'} e^{-i\omega_0 t'}$$

If we take the detector to lie on the y-axis (the sample is at the origin) so that $\underset{\sim}{R} = R\hat{y}$, and if we recognize that $|\underset{\sim}{R}| \gg |\underset{\sim}{r}'|$ for all $\underset{\sim}{r}'$ in the illuminated volume, then $|\underset{\sim}{R}-\underset{\sim}{r}'| \approx R-\hat{R}\cdot\underset{\sim}{r}'$ and

$$\underset{\approx}{T}_k(\underset{\sim}{R}-\underset{\sim}{r}') \approx \frac{k^2\exp[ik(R-\underset{\sim}{r}'\cdot\hat{R})]}{R}\begin{bmatrix}1 & 0 & 0\\ 0 & 0 & 0\\ 0 & 0 & 1\end{bmatrix} \tag{18}$$

The dyadic $(\underset{\sim}{I}-\hat{R}\hat{R})$ -- recall $\hat{R} = \hat{y}$ -- on the hand-hand side of Eq. (18) acts on $\underset{\sim}{E}_0 = E_0\hat{z}$ and preserves its z-component, $E_0\hat{z}$. Thus, if the incident field is z-polarized, the scattered field to <u>first order</u> will have a z-component

$$E_{sz}^{(1)}(\underset{\sim}{R},t) =$$

$$\frac{E_0}{2\pi c^2 R}\int d\underset{\sim}{r}' \int_{-\infty}^{+\infty} dt' \alpha_{zz}(\underset{\sim}{r}',t')\Delta\rho(\underset{\sim}{r}',t') e^{i\underset{\sim}{k}_0\cdot\underset{\sim}{r}'} e^{-i\omega_0 t'} \tag{19}$$

$$\int_{-\infty}^{+\infty} d\bar{\omega}\bar{\omega}^2 \exp[i\bar{\omega}(t' - t + \frac{R}{c} - \frac{\underset{\sim}{r}'\cdot\hat{R}}{c})]$$

Here we have simply changed the dummy integration variable in Eq. (17) from k to $\bar{\omega}/c$. The integral over $\bar{\omega}$ can be rewritten as

$$-2\pi \frac{d^2}{dt'^2}\delta[t' - \tau], \quad \tau = t - \frac{R}{c} + \frac{\underset{\sim}{r}'\cdot\hat{R}}{c}$$

so that the integral over t' becomes

$$\int_{-\infty}^{+\infty} dt' \exp(-i\omega_0 t')\alpha_{zz}(\underset{\sim}{r}',t')\Delta\rho(\underset{\sim}{r}',t') \frac{d^2}{dt'^2} \delta[t' - \tau]$$

$$= \frac{d^2}{dt'^2} [\exp(-i\omega_0 t')\alpha_{zz}(\underset{\sim}{r}',t')\Delta\rho(\underset{\sim}{r}',t')]\Big|_{t'=\tau} \tag{20}$$

$$= \alpha_{zz}(\underset{\sim}{r}',t')\Delta\rho(\underset{\sim}{r}',t') \frac{d^2}{dt'^2} \exp(-i\omega_0 t')\Big|_{t'=\tau}$$

The last line follows from the fact that the molecular polarizability and local density fluctuation are essentially constant over the period ($\approx 10^{-15}$ sec) of $\exp(-i\omega_0 t)$ (see, for example, the discussion on page 35 of reference 1). Thus the scattered field at the detector can be written through first order as

$$\Delta E_{sz}^{(1)}(\underset{\sim}{R},t) \approx \frac{E_0}{R}\frac{\omega_0{}^2}{c} e^{-i\omega_0 t}\exp(i\frac{\omega_0}{c}R)\int d\underset{\sim}{r}' e^{i\underset{\sim}{k}_0\cdot\underset{\sim}{r}'} e^{-i\underset{\sim}{k}\cdot\underset{\sim}{r}'}$$
$$\Delta\rho(\underset{\sim}{r}',t - \frac{|\underset{\sim}{R}-\underset{\sim}{r}'|}{c})\alpha_{zz}(\underset{\sim}{r}',t - \frac{|\underset{\sim}{R}-\underset{\sim}{r}'|}{c}) \tag{21}$$

Here we have written $\underset{\sim}{k}$ for $\frac{\omega_0}{c}\hat{R}$, the propagation wavevector for the scattered light.

Substituting Eq. (21) into Eqs. (15)-(16) we have, for the first-order power spectrum

$$I_{zz}^{(1)} = \frac{|E_0|^2}{R^2}\alpha^2(\frac{\omega_0}{c})^4\int_{\infty}^{+\infty} dt \exp[i(\omega-\omega_0)t]\int d\underset{\sim}{r}'\int d\underset{\sim}{r}'' \exp[-i(\underset{\sim}{k}-\underset{\sim}{k}_0)(\underset{\sim}{r}'-\underset{\sim}{r}'')$$
$$\langle\Delta\rho(\underset{\sim}{r}',t - \frac{|\underset{\sim}{R}-\underset{\sim}{r}'|}{c})\Delta\rho(\underset{\sim}{r}'', - \frac{|\underset{\sim}{R}-\underset{\sim}{r}''|}{c})\alpha_{zz}(\underset{\sim}{r}',t - \frac{|\underset{\sim}{R}-\underset{\sim}{r}'|}{c})\alpha_{zz}(\underset{\sim}{r}'', - \frac{|\underset{\sim}{R}-\underset{\sim}{r}''|}{c} \tag{22}$$

But the correlation function in Eq. (22) depends only on the <u>differenc</u> between the position and time arguments. Furthermore, if we neglect collisional distortion of polarizabilities and assume (reference 1, page 115) that the molecular orientations and center-of-mass positions are independent, the correlation function in Eq. (22) factors into a product of $\langle\Delta\rho(\underset{\sim}{r}'-\underset{\sim}{r}'',t)\Delta\rho(0,0)\rangle$ and $\langle\alpha_{zz}(\underset{\sim}{r}'-\underset{\sim}{r}'',t)\alpha_{zz}(0,0)\rangle$. Here we have also neglected $|\underset{\sim}{r}'-\underset{\sim}{r}''|$ compared with R. Finally, $\langle\alpha_{zz}(\underset{\sim}{r}'-\underset{\sim}{r}'',t)\alpha_{zz}(0,0)\rangle$ is identified in this approximation as $\langle\alpha_{zz}(\underset{\sim}{\Omega}(t),\{Q_i(t)\})\alpha_{zz}(\underset{\sim}{\Omega}(0),\{Q_i(0)\})\rangle$ where $\alpha_{zz}(\underset{\sim}{\Omega},\{Q_i\})$ denotes the zz-component of the polarizability tensor of a molecule whose orientation is $\underset{\sim}{\Omega}$ and whose internal vibrational displacements are $\{Q_i\}$. Then, writing $\Omega \equiv \omega-\omega_0$ and $\underset{\sim}{K} \equiv \underset{\sim}{k}-\underset{\sim}{k}_0$, the first-order power spectrum becomes

$$I_{zz}^{(1)} \simeq \frac{|E_0|^2}{R^2}\left(\frac{\omega_0}{c}\right)^4 V\int_{-\infty}^{+\infty} dt\; e^{i\Omega t} \langle\alpha_{zz}(t)\alpha_{zz}(0)\rangle S(\underset{\sim}{K},t) \tag{23}$$

where

$$S(\underset{\sim}{K},t) \equiv \int d\underset{\sim}{r}\; e^{-i\underset{\sim}{K}\cdot\underset{\sim}{r}} \langle\Delta\rho(\underset{\sim}{r},t)\Delta\rho(0,0)\rangle$$

is the usual, intermediate structure factor. Let $F(\Omega)$ and $S(\underset{\sim}{K},\Omega)$ denote the Fourier transforms of $\langle\alpha_{zz}(t)\alpha_{zz}(0)\rangle$ and $S(\underset{\sim}{K},t)$, respectively. Then from Eq, (23) it follows that $I_{zz}^{(1)}(\Omega)$ is a convolution of these two spectra:

$$I_{zz}^{(1)}(\Omega) = \frac{|E_0|^2}{R^2}\left(\frac{\omega_0}{c}\right)^4 V\int d\Omega' F(\Omega-\Omega') S(\underset{\sim}{K},\Omega') \tag{24}$$

Note that, in the case where the correlations between the molecular center-of-mass positions can be neglected, $S(\underset{\sim}{K},\Omega') \rightarrow \rho\delta(\Omega')$ and

$$I_{if}^{(1)}(\Omega) = \frac{|E_0|^2}{R^2}\left(\frac{\omega_0}{c}\right)^4 V\rho F(\Omega) \tag{25}$$

with

$$F(\Omega) \equiv \int_{-\infty}^{+\infty} dt e^{i\Omega t} \langle\alpha_{if}[\underset{\sim}{\Omega}(t),\{Q_i(t)\}]\alpha_{if}[\underset{\sim}{\Omega}(0),\{Q_i(0)\}] \tag{26}$$

(i and f denote the polarizations of the incident and scattered light). That is, the first-order scattering reduces to the (in this long wavelength ($\underset{\sim}{K}\rightarrow 0$) limit) familiar expression obtained[21] for the rotation-vibration Raman spectrum of a dilute molecular gas. In general, however, $I_{if}^{(1)}(\Omega)$ is a convolution -- see Eq. (24) -- of the $F(\Omega)$ and $S(\underset{\sim}{K},\Omega)$ spectra. In fact the situation is even more complicated -- this is because of the breakdown of the statistical independence assumed earlier for the molecular orientations and center-of-mass positions. A microscopic theory of these effects is presently being developed.[13] The coupling of orientations with the collective ($\underset{\sim}{K}$ dependent) modes of a liquid has, however, been treated quite extensively from a phenomenological (generalized hydrodynamic) point of view. This latter work is described in detail in Chapters 11 and 12 of the recent monograph by Berne and Pecora;[1] references to and discussion of the original papers are also provided there.

From this point on I shall restrict myself to fluids made up of atoms (or of molecules whose internal degrees of freedom can be neglected). This will enable us to focus on the differences between first- and second-order optical response, without the complications of rotational and vibrational motions. In the case of an atomic fluid, the fluctuating tensor $\underset{\approx}{\alpha}(r',t')$ in Eq. (17) becomes the constant scalar α. As a result $\Delta\underset{\sim}{E}_s^{(1)} || \underset{\sim}{E}_0$, i.e. the first-order scattering is completely polarized, and the power spectrum reduces to

$$I^{(1)}(\Omega) = \frac{|E_0|^2}{R^2}\left(\frac{\omega_0}{c}\right)^4 V\alpha^2 S(\underset{\sim}{K},\Omega) \tag{27}$$

where

$$VS(\underset{\sim}{K},\Omega) = \int_{-\infty}^{+\infty} dt\ e^{i\Omega t} \langle \rho_{\underset{\sim}{K}}(t)\ \rho_{-\underset{\sim}{K}}(0)\rangle \tag{28}$$

This representation of $S(\underset{\sim}{K},\Omega)$ follows immediately from Eqs. (15)-(16) and the fact that, for $\alpha_{zz}[\underset{\sim}{r}',t - (|\underset{\sim}{R}-\underset{\sim}{r}'|)/c] \to \alpha$ in Eq. (21), $\Delta E_{sz}^{(1)}(R,t)$ can be written as

$$\alpha\ \exp(-i\omega_0 t)\exp(i\frac{\omega_0}{c}R)\ \frac{E_0}{R}\left(\frac{\omega_0}{c}\right)^2 \rho_{\underset{\sim}{K}}\left(t - \frac{|\underset{\sim}{R}-\underset{\sim}{r}'|}{c}\right)$$

where

$$\rho_{\underset{\sim}{K}}(t) = \int dr\ \exp(-i\underset{\sim}{K}\cdot\underset{\sim}{r})\Delta\rho(\underset{\sim}{r},t) \tag{29}$$

Thus the first-order light scattering spectrum for an atomic fluid follows from the dynamics of the $\underset{\sim}{K}$th Fourier component of the density fluctuations.

IV. GENERALIZED HYDRODYNAMICS FOR AN ATOMIC FLUID

The simplest description of the $\rho_K(t)$ dynamics is gotten by solving the linearized hydrodynamic equations for an atomic fluid. Consider the usual dynamical variables defined by [here $\underset{\sim}{p}_i$ is the momentum of the ith atom and $V(R_{ij})$ is the potential energy of interaction between the ith and jth atoms]

$$\begin{aligned}
Q_1 &= \int d\underset{\sim}{r} e^{i\underset{\sim}{K}\cdot\underset{\sim}{r}}\rho(\underset{\sim}{r},t) = \int d\underset{\sim}{r} e^{i\underset{\sim}{K}\cdot\underset{\sim}{r}} \sum_i \delta[\underset{\sim}{r}-\underset{\sim}{R}_i(t)] = \sum_i e^{i\underset{\sim}{K}\cdot\underset{\sim}{R}_i(t)} \equiv \rho_{\underset{\sim}{K}}(t) \\
Q_2 &= \sum_i (\underset{\sim}{p}_i)_x \exp[i\underset{\sim}{K}\cdot\underset{\sim}{R}_i(t)] \equiv (\underset{\sim}{p}_{\underset{\sim}{K}})_x \\
Q_3 &= [\underset{\sim}{p}_{\underset{\sim}{K}}(t)]_y \\
Q_4 &= [\underset{\sim}{p}_{\underset{\sim}{K}}(t)]_z \\
Q_5 &= \sum_i\left[\frac{p_i^2}{2m_i} + \sum_{j\neq i} V(R_{ij})\right]\exp[i\underset{\sim}{K}\cdot\underset{\sim}{R}_i(t)] = e_{\underset{\sim}{K}}(t)
\end{aligned} \tag{30}$$

Then, as shown by Mori, and others[6,14] the $\underset{\sim}{Q}(t)$ obey the following equation:

$$\frac{d}{dt}\underset{\sim}{Q}(t) - i\underset{\approx}{\omega}\cdot\underset{\sim}{Q}(t) + \int_0^t ds\underset{\approx}{\phi}(t-s)\cdot\underset{\sim}{Q}(s) = \underset{\sim}{f}(t) \tag{31}$$

where

$$\underset{\approx}{\omega} = \langle\dot{\underset{\sim}{Q}}\ \underset{\sim}{Q}^*\rangle \cdot \langle\underset{\sim}{Q}\ \underset{\sim}{Q}^*\rangle^{-1} \tag{32}$$

is the matrix of "Euler coefficients";

$$\underset{\approx}{\phi}(t) = \langle f^* f(t)\rangle \cdot \langle\underset{\sim}{Q}\ \underset{\sim}{Q}^*\rangle^{-1} \tag{33}$$

is the matrix of "dissipative coefficients"; and

$$\underset{\sim}{f}(t) = \exp[(1-P)iLt](1-P)\dot{\underset{\sim}{Q}} \tag{34}$$

is the random force. Here P is the projection operator onto the space $\underset{\sim}{Q}$, i.e.,

$$PG \equiv \sum_i \frac{\langle Q_i\ G^*\rangle}{\langle Q_i\ Q_i^*\rangle} Q_i \tag{35}$$

for an arbitrary dynamical variable G. The brackets denote as usual a canonical ensemble average, and L is the Liouville operator defined by

$$iL \equiv \sum_i\left(\frac{\partial H}{\partial \underset{\sim}{p}_i} \cdot \frac{\partial}{\partial \underset{\sim}{R}_i} - \frac{\partial H}{\partial \underset{\sim}{R}_i} \cdot \frac{\partial}{\partial \underset{\sim}{p}_i}\right) \tag{36}$$

with H the hamiltonian for the system.

Eq. (31) is an exact identity which holds for any set $\underset{\sim}{Q}$ of dynamical variables. It simplifies considerably, however, if the set $\underset{\sim}{Q}$ contains all the "slow" variables. Then, $\underset{\sim}{f}$ -- which propagates in time wholly in the "unprojected" space $(1-P)\dot{\underset{\sim}{Q}}$ and hence has no component along $\underset{\sim}{Q}$ -- can be treated as a rapidly varying (compared to the $\underset{\sim}{Q}$ timescales) quantity. Correspondingly, its autocorrelation can be taken to be a Dirac delta function and

$$\int_0^t ds\underset{\approx}{\phi}(t-s)\cdot\underset{\sim}{Q}(s) \rightarrow \underset{\sim}{Q}(t)\cdot\underset{\approx}{\Gamma} \tag{37}$$

where

$$\underset{\approx}{\Gamma} = \lim_{s\to 0}\int_0^\infty dt\ e^{-st}\ \underset{\approx}{\phi}(t) \tag{38}$$

Thus Eq. (31) assumes the familiar Langevin form

$$\frac{d}{dt} Q_i(t) - i\sum_j \omega_{ij} Q_j(t) + \sum_j \Gamma_{ij} Q_j(t) = f_i(t) \tag{39}$$

with $f_i(t)$ playing the role of the random force associated with Q_i.

Note that Q_i's are coupled in Eq. (39) through the matrices $\underset{\sim}{\omega}$ and $\underset{\sim}{\Gamma}$. If we could construct linear combinations of them which were <u>un</u>coupled, i.e.

$$\frac{d}{dt} Q_i'(t) - [i\omega_i' - \Gamma_i']Q_i'(t) = 0 \tag{40}$$

then the solution of these equations would be trivial:

$$Q_i'(t) = Q_i'(0)\exp(i\omega_i' t)\exp(-\Gamma_i' t) \tag{41}$$

Furthermore, since -- by definition of the Liouville operator --

$$\frac{d}{dt} Q_i'(t) = iLQ_i'(t) \quad , \tag{42}$$

Eq (40) establishes that $Q_i'(t)$ is an eigenfunction of iL, with the real and imaginary parts of its eigenvalue corresponding to its decay rate and propagation frequency. (In writing Eq. (40) we have dropped the random force term appearing earlier.)

Using the notation of Kadanoff and Swift[15] let us write the eigenfunctions Q_i' and eigenvalues $i\omega_i' - \Gamma_i'$ of iL as $|\nu,\underset{\sim}{K}\rangle$ and S_ν:

$$iL|\nu,\underset{\sim}{K}\rangle = S_\nu|\nu,\underset{\sim}{K}\rangle \tag{43}$$

Here $\underset{\sim}{K}$ denotes the Fourier transform wavevector which characterizes each translationally invariant eigenstate of the system. Similarly, we write the original Q_i's as $|i,\underset{\sim}{K}\rangle$. By manipulating Eq. (43), e.g., by inserting $1 = 1-P + P = 1-P + \sum_i |i,\underset{\sim}{K}\rangle\langle i,\underset{\sim}{K}|$, we obtain the following coupled equations for the eigenfunction components $\langle j,\underset{\sim}{K}|\nu,\underset{\sim}{K}\rangle$:

$$\sum_j [S_\nu \delta_{ij} - L_{ij}(\underset{\sim}{K}) - U_{ij}(\underset{\sim}{K},S_\nu)] \langle j,\underset{\sim}{K}|\nu,\underset{\sim}{K}\rangle = 0 \tag{44}$$

Here

$$L_{ij}(\underset{\sim}{K}) \equiv \langle i,\underset{\sim}{K}|iL|j,\underset{\sim}{K}\rangle \tag{45}$$

and

$$U_{ij}(\underset{\sim}{K},S_\nu) = \langle i,\underset{\sim}{K}|iL(1-P) \frac{1}{S-(1-P)iL(1-P)} (1-P)iL|j,\underset{\sim}{K}\rangle \tag{46}$$

Note that Eqs. (44)-(46) are equivalent to Eqs. (31)-(34); one set is simply the Laplace transform of the other.

The eigenvalues associated with iL can now be obtained by setting equal to zero the determinant of $S\delta_{ij}-L_{ij}-U_{ij}$ and solving for S. The S-dependence of U_{ij} can be neglected in the same approximation that we dropped earlier the memory effects in the dissipative term of Eq. (31). Note that Eq. (44) differs from the usual $\sum_j[S\delta_{ij}-L_{ij}]\langle j|\nu\rangle = 0$ which arises in more familiar eigenvalue problems; this is because the $\{\langle j,\underset{\sim}{K}|\}$ do not form a complete set -- instead, (1-P) is "left over".

Now we are in a position to solve for the collective modes of a simple (atomic) fluid. They correspond to the special linear combinations $(\{\langle j|\nu\rangle\}_{j=1,2,\ldots,5})$ of the conserved quantities defined by Eqs. (30) which satisfy Eq. (44). Suppose we take the $\underset{\sim}{K}$ appearing in the Fourier-transformed densities of Eq. (30) to lie along the z-axis. Then, as shown by Kadanoff and Swift,[15] two eigenfunctions are given approximately by $(\underset{\sim}{p}_{\underset{\sim}{K}})_x$ and $(\underset{\sim}{p}_{\underset{\sim}{K}})_y$, with eigenvalues which reduce in the hydrodynamic limit $(\underset{\sim}{K},\underset{\sim}{S}\to 0)$ to $S = \eta(0,0)K^2/m$. Here $\eta(0,0)$ is the macroscopic shear viscosity and m is the uniform mass density of the fluid. For arbitrary ($\neq 0$) $\underset{\sim}{K}$ and S, the damping rate of these transverse momentum modes is given by the solution to

$$S = \eta(\underset{\sim}{K},S)K^2/m \tag{47}$$

where

$$\eta(\underset{\sim}{K},S)K^2 = \langle(\underset{\sim}{p}_{\underset{\sim}{K}})_y iL(1-P)\frac{1}{S-(1-P)iL(1-P)}(1-P)iL(\underset{\sim}{p}_{\underset{\sim}{K}})_y\rangle/k_BT = U_{33}/k_BT \tag{48}$$

Another approximate eigenfunction of the system is given by a linear combination having the form $A\rho_{\underset{\sim}{K}} + Be_{\underset{\sim}{K}}$. It corresponds to a nonpropagating (heat flow) entropy mode whose damping rate S satisfies

$$S = \lambda(\underset{\sim}{K},S)K^2/mC_p(\underset{\sim}{K}) \tag{49}$$

Here $\lambda(\underset{\sim}{K},S)$ is defined by an expression identical to Eq. (48) but with $(\underset{\sim}{p}_{\underset{\sim}{K}})_y$ replaced by the $A\rho_{\underset{\sim}{K}}+Be_{\underset{\sim}{K}}$; $\lambda(\underset{\sim}{K},S)$ and $C_p(\underset{\sim}{K})$ reduce in the $\underset{\sim}{K},S\to 0$ limit to the macroscopic thermal conductivity and heat capacity at constant pressure. The final two approximate eigenfunctions (first-order in the conserved densities) are propagating modes formed from the longitudinal momentum $[(\underset{\sim}{p}_{\underset{\sim}{K}})_z]$ and pressure $[C\rho_{\underset{\sim}{K}}+De_{\underset{\sim}{K}}]$ fluctuations. Their eigenvalues are given by

$$S_\pm = \pm iKc(\underset{\sim}{K}) + \frac{1}{2}K^2D(\underset{\sim}{K},S_\pm) \tag{50}$$

where

$$D(\underset{\sim}{K},S) = [\frac{4}{3}\eta(\underset{\sim}{K},S) + \xi(\underset{\sim}{K},S) + \lambda(\underset{\sim}{K},S)(\frac{1}{C_v(\underset{\sim}{K})} - \frac{1}{C_p(\underset{\sim}{K})})]/m \tag{51}$$

$c_s(\underset{\sim}{K}\to 0)$, the $\underset{\sim}{K}\to 0$ limit of the propagation velocity, is the macroscopic speed of sound, and $4/3\eta(\underset{\sim}{K},S)+\xi(\underset{\sim}{K},S)$ is defined by an equation identical to Eq. (48) but with $(p_{\underset{\sim}{K}})_{x \text{ or } y} \to (p_{\underset{\sim}{K}})_z$. $D(0,0)$ is the low frequency sound attenuation coefficient.

Note that all the generalized transport coefficients appearing above have the same form. Defining

$$X \equiv (1-P)\frac{1}{S-(1-P)iL(1-P)}(1-P)$$

we can write the shear viscosity, for example, as

$$\eta(\underset{\sim}{K},S)K^2 = \frac{1}{k_BT} \langle (p_{-\underset{\sim}{K}})_y iL\ X\ iL(p_{\underset{\sim}{K}})_y \rangle$$

Recall that P corresponds to a projection operator which acts on an arbitrary dynamical variable to give its components along the hydrodynamic densities $\{|j,\underset{\sim}{K}\rangle\}_{j=1,2,\ldots,5}$. (1-P) brings in all that is "left over" after these particular linear modes have been projected out. What <u>is</u> "left over"? First, there are short-wavelength rapidly-varying fluctuations. Second, there are the slowly varying dynamical variables which are not included in the $\{|j,\underset{\sim}{K}\rangle\}_{j=1,2,\ldots,5}$. The most important such variables are the <u>bilinear</u> products of the $|j,\underset{\sim}{K}\rangle$'s. A mode such as $Q_1'(\underset{\sim}{K}-\underset{\sim}{K}')Q_2'(\underset{\sim}{K}')$ has eigenvalues $S_{\nu_1}(\underset{\sim}{K}-\underset{\sim}{K}')+S_{\nu_2}(\underset{\sim}{K}')$, i.e.

$$iL|\nu_1,\underset{\sim}{K}-\underset{\sim}{K}'\rangle|\nu_2,\underset{\sim}{K}'\rangle = \{S_{\nu_1}(\underset{\sim}{K}-\underset{\sim}{K}') + S_{\nu_2}(\underset{\sim}{K}')\}|\nu_1,(\underset{\sim}{K}-\underset{\sim}{K}')\rangle|\nu_2,\underset{\sim}{K}'\rangle$$

and will be slowly varying if both $\underset{\sim}{K}$ and $\underset{\sim}{K}'$ are small, Thus

$$X = X_{\text{background}} + \sum_{(\nu,\nu')} \frac{d\underset{\sim}{K}'}{(2\pi)^3} \frac{|\nu,\underset{\sim}{K}-\underset{\sim}{K}'\rangle|\nu',\underset{\sim}{K}'\rangle\langle\nu',-\underset{\sim}{K}'|\langle\nu,\underset{\sim}{K}'-\underset{\sim}{K}|}{S-[S_\nu(\underset{\sim}{K}-\underset{\sim}{K}') + S_{\nu'}(\underset{\sim}{K}')]}$$

$X_{\text{background}}$ refers here to the contributions of short-wavelength, rapidly varying fluctuations to the 1-P projections in X. These will not be sensitive, for example, to small temperature changes near T_{critical}; they give rise to the "background" value of the transport coefficients. The bilinear terms in X, however, give rise to the "critical" part. In the case of the shear viscosity, say, the dominant term in X_{critical} arises from the bilinear variable involving the entropy mode -- this term gives the contribution to the decay rate of the K-transverse momentum mode $[(p_{\underset{\sim}{K}})_y]$ due to its breakup into two entropy (heat flow) modes of wavevectors $\underset{\sim}{K}-\underset{\sim}{K}'$ and $\underset{\sim}{K}'$. The contribution of trilinear variables has also been explicitly evaluated,[16] and similar calculations of the critical behavior of the remaining transport coefficients, (λ and ξ, etc.) have also been carried out.[17] This approach constitutes the mode-mode coupling theory of collective motion of dynamics and is discussed in the lectures of Mazenko and Oppenheim, and in the seminar by Kapral.

If we now think of $\rho_{\underset{\sim}{K}}$ as a linear combination of the approximate L-eigenfunctions -- only those with eigenvalues (49) and (50) contribute -- we see that this density fluctuation varies with time according to

$$\rho_{\underset{\sim}{K}}(t) \approx I\exp[-(\lambda K^2/mC_p)t]+J\exp[-(DK^2/2)t][\exp(ic_sKt)+\exp(-ic_sKt)] \tag{52}$$

Substituting this expression into Eq. (28) for $S(\underset{\sim}{K},\Omega)$ we find that $I^{(1)}(\Omega)$ -- see Eq. (27) -- consists of three peaks: one centered at $\Omega=0$ with width $\lambda K^2/mC_p$ (this arises from the nonpropagating heat-flow mode); and two centered at $\Omega=\pm c_sK$ with width $DK^2/2$ (corresponding to the damped but propagating sound waves). This constitutes the familiar Rayleigh-Brillouin triplet structure which appears as the first-order spectrum of atomic fluids.

Even within the first-order light scattering approximation, however, the above result is not necessarily adequate. Consider again the basic (Langevin) Eq. (39). By multiplying both sides by $Q_i^*(0)$ and performing an equilibrium (canonical ensemble) average, we obtain the following equation for the correlation functions:

$$\frac{d}{dt}\langle Q_i(t)Q_i^*(0)\rangle - i\sum_j\omega_{ij}\langle Q_j(t)Q_i^*(0)\rangle + \sum_j\Gamma_{ij}\langle Q_j(t)Q_i^*(0)\rangle = 0 \tag{53}$$

Recall that this equation is valid only if the set $\{Q_i\}$ includes all the slowly varying variables -- otherwise memory effects are important. Thus only if the five hydrodynamic densities [see Eq. (30)] exhaust this slowly varying set will $\langle\rho_{\underset{\sim}{K}}(t)\rho_{-\underset{\sim}{K}}(0)\rangle$ be given by the sum of three exponentials as in Eq. (52). Otherwise, if for example, bilinear products of these variables have also to be included, the coupling equations (53) include many new terms and the solutions for the $\langle Q_i(t)Q_i^*(0)\rangle$ take on new form. Keyes and Oppenheim[18] have shown in particular how the transverse momentum autocorrelation function $\langle(\underset{\sim}{p}_{\underset{\sim}{K}})_y(t)(\underset{\sim}{p}_{-\underset{\sim}{K}})_y(0)\rangle$ owes the existence of its "long-time tail" to the contribution of bilinear variables. An explicit treatment of the $\langle\rho_{\underset{\sim}{K}}(t)\rho_{-\underset{\sim}{K}}(0)\rangle$ has not been carried out as thoroughly, although discussions of the $\underset{\sim}{K}$,S-dependence of the Rayleigh linewidth are scattered throughout the literature.[19] In any case, the contributions of product modes (or, equivalently, "multiple excitations" -- see below) to the first-order spectrum $S(\underset{\sim}{K},\Omega)$ need to be studied further.

V. SECOND-ORDER SPECTRA

Let us now return to the second-order term $\Delta\underset{\sim}{E}_s^{(2)}$ in the scattered field: it is given by Eq. (14). Substituting this result into Eqs. (15)-(16) and carrying out manipulations identical to those involved in Eqs. (18) through (21), we find that the f-component (f=x, y, or z) of $\Delta\underset{\sim}{E}_s^{(2)}$ at the detector can be written as

$$(\Delta \underset{\sim}{E}_s^{(2)})_f(\underset{\sim}{R},t) = \frac{\alpha^2}{R}\left(\frac{\omega_0}{c}\right)^2 \int d\underset{\sim}{r}' \Delta\rho(\underset{\sim}{r}',t - |\underset{\sim}{R}-\underset{\sim}{r}'|/c) \int d\underset{\sim}{r}''$$

$$[\underset{\approx}{T}_{k_0}(\underset{\sim}{r}'-\underset{\sim}{r}'')]_{fz} \Delta\rho(\underset{\sim}{r}'',t - |\underset{\sim}{R}-\underset{\sim}{r}'|/c - |\underset{\sim}{r}'-\underset{\sim}{r}''|/c) \tag{54}$$

$$E_0 \exp(ik_0 \cdot r'')\exp[(-i\omega_0)\{t - \frac{|\hat{R}-\underset{\sim}{r}'|}{c} - \frac{|\underset{\sim}{r}'-\underset{\sim}{r}''|}{c}\}]$$

[Recall that we have taken $\underset{\sim}{\alpha}(\underset{\sim}{r},t)\to\alpha$ (since we are treating monatomic fluids) and $\underset{\sim}{E}_0(\underset{\sim}{r},t) = E_0 z \exp(i\underset{\sim}{k}_0\cdot\underset{\sim}{r})\exp(-i\omega_0 t)$.] Note that $\alpha\rho(\underset{\sim}{r}',t')[\underset{\approx}{T}_{k_0}(\underset{\sim}{r}'-\underset{\sim}{r}'')]\alpha\rho(\underset{\sim}{r}'',t'')\cdot\underset{\sim}{E}_0\cdot\exp(i\underset{\sim}{k}_0\cdot\underset{\sim}{r}'')\exp(-i\omega_0 t'')$ is the dipole moment induced at $\underset{\sim}{r}'$ at time t' due to the electric field arising from the dipole moment induced at $\underset{\sim}{r}''$ at time t'' by the incident $\underset{\sim}{E}_0$. The times t' and t'' differ of course by the time $|\underset{\sim}{r}'-\underset{\sim}{r}''|/c$ that it takes for light to travel from one point to the other. The propagator tensor $\underset{\approx}{T}_{k_0}(\underset{\sim}{R}-\underset{\sim}{r}')$ which acts on this dipole moment to give $\Delta\underset{\sim}{E}_s^{(2)}$ at $\underset{\sim}{R}$ at time $t' + |\underset{\sim}{R}-\underset{\sim}{r}'|/c$ has "disappeared" through its asymptotic representation, Eq. (18).

Because two density fluctuations have been excited by the incident light wave, $\Delta\underset{\sim}{E}_s^{(2)}$ corresponds to a second order Raman process.[21] The prediction of such an effect in a monatomic fluid was first made by Halley,[22] who suggested it should be observable in liquid helium. After its observation by Greytak and Yan,[23] the theory was clarified further by Stephen.[9] Note that each contribution to the $\Delta\underset{\sim}{E}_s^{(2)}$ in Eq. (54) is proportional to a product of two density fluctuations. Thus the second-order light scattering spectrum, given by

$$\int_{-\infty}^{+\infty} dt \, \exp(i\omega t)\langle\Delta\underset{\sim}{E}_s^{(2)}(\underset{\sim}{R},t)\Delta\underset{\sim}{E}_s^{(2)*}(\underset{\sim}{R},0)\rangle$$

will depend directly on the four-point correlation function for the density fluctuations. Very little is known about these functions. Suppose we assume, however, that it can be factored into a symmetrized product of the familiar two-point correlation functions $\langle\Delta\rho(\underset{\sim}{r}_1 t_1)\cdot\Delta\rho(\underset{\sim}{r}_2 t_2)\rangle$ -- this approximation is discussed further in the second lecture. Then $I_{fz}^{(2)}(\omega)$ reduces to

$$I_{fz}^{(2)}(\omega) = V\frac{\alpha^4}{R^2}\left(\frac{\omega_0}{c}\right)^4 \frac{6}{5}E_0^2[1+\frac{1}{3}(\hat{f}\cdot\hat{z})^2]\left(\frac{1}{2\pi}\right)^4 \int d\underset{\sim}{k}' d\omega' g^2(\underset{\sim}{k}')S(-\underset{\sim}{k},\Omega-\omega')S(\underset{\sim}{k}',\omega') \tag{55}$$

Here $g(\underset{\sim}{k})$ is defined by the Fourier transform of $\underset{\approx}{T}_k$ according to

$$\underset{\approx}{g}(\underset{\sim}{k}) \equiv \int d\underset{\sim}{r} \, \exp(-i\underset{\sim}{k}\cdot\underset{\sim}{r})\underset{\approx}{T}_k(\underset{\sim}{r}) = g(\underset{\sim}{k})[3\hat{k}\hat{k} - \underset{\approx}{I}] \tag{56}$$

and $\hat{f}$ is as before the unit vector along the final polarization direction.

In obtaining the structure factor for liquid helium, Stephen[9] has included only the contributions from single excitations. That is, in the language of our earlier discussion, the appropriate hydrodynamic equations have been linearlized in the density and momentum fluctuations so that the resulting eigenmodes propagate independently --without coupling to higher order product modes. [Actually, an empirical multiplicative factor has been included[9] to correct for these multiple excitation contributions to $S(\underset{\sim}{k},\omega)$.[20]] Large wave-vector (roton) modes are able to appear in the light scattering spectrum as long as $\underset{\sim}{k}-\underset{\sim}{k}' \approx k_{optical} \approx 0$. Thus the momenta of the two excitations created by the incident field will have nearly equal mangitudes and opposite directions. Most of the scattering involves $\approx 10°K$ rotons near the $\approx 2Å^{-1}$ minimum in the energy vs. momentum dispersion curve: there is a sharp peak in the observed[23] $I^{(2)}(\omega)$ spectrum at $\omega_0-\omega = 2E_{\substack{roton\\minimum}} \approx 20°k$. For more recent developments, see the seminar by Murray.

In calssical liquids, for which the dispersion curves exhibit no sharp structure, the second-order spectrum associated with Eq. (55) is expected to be essentially structureless. This is indeed the case for the line-shapes observed for liquid argon, krypton and xenon.[24] Because the first-order scattered field $\Delta\underset{\sim}{E}_s^{(1)}$ is completely polarized ($||\underset{\sim}{E}_0$) in the case $\underset{\sim}{\alpha}\to\alpha$, the observation of a <u>de</u>polarized ($\hat{f}\neq\hat{z}$) spectrum establishes ambiguously the presence of second-order effects. Nevertheless, because of the structureless dispersion curves for the classical liquids, the interpretation of $I^{(2)}(\omega)$ in terms of a solid-like second-order <u>Raman</u> effect is less compelling than in the case of low-temperature helium. This point is discussed further in the original paper of McTague et al.[24] and in the lectures of Fleury.

A. Integrated (over frequency) Second-Order Intensity

Consider now the integrated (over frequency) second-order intensity:

$$I_{fz}^{(2)} = \int_{-\infty}^{+\infty} d\omega\, I_{fz}^{(2)}(\omega) = 2\pi\langle |\{\Delta\underset{\sim}{E}_s^{(2)}(\underset{\sim}{R},0)\}_f|^2\rangle \tag{57}$$

Using Eq. (54) for $\{\Delta\underset{\sim}{E}^{(2)}(\underset{\sim}{R},0)\}_f$, and neglecting $|\underset{\sim}{r}'-\underset{\sim}{r}''|$ compared with $|\underset{\sim}{R}-\underset{\sim}{r}'|$, and $|\underset{\sim}{R}-\underset{\sim}{r}''|$ compared with $\underset{\sim}{R}$, in the time arguments of the density fluctuations, we have immediately that

$$I_{fz}^{(2)} = 2\pi\frac{\alpha^4}{R^2}E_0^2\left(\frac{\omega_0}{c}\right)^4 \int d\underset{\sim}{r}_1\int d\underset{\sim}{r}_2\int d\underset{\sim}{r}_3\int d\underset{\sim}{r}_4 [\underset{\approx}{T}_{k_0}(\underset{\sim}{r}_1-\underset{\sim}{r}_2)]_{fz}{}^*[\underset{\approx}{T}_{k_0}(\underset{\sim}{r}_3-\underset{\sim}{r}_4)]_{fz} \tag{58}$$

$$\times\langle\Delta\rho(\underset{\sim}{r}_1)\Delta\rho(\underset{\sim}{r}_2)\Delta\rho(\underset{\sim}{r}_3)\Delta\rho(\underset{\sim}{r}_4)\rangle \exp(i\underset{\sim}{k}_0\cdot\underset{\sim}{r}_2)\exp(-i\underset{\sim}{k}\cdot\underset{\sim}{r}_1)\exp(-i\underset{\sim}{k}_0\cdot\underset{\sim}{r}_4)\exp(i\underset{\sim}{k}\cdot\underset{\sim}{r}_3)$$

Expressing each density fluctuation as

$$\Delta\rho(\underset{\sim}{r}) = \sum_i \delta(\underset{\sim}{r}-\underset{\sim}{R}_i) - \rho$$

Eq. (58) can be rewritten in the f≠z case say (f=x) as

$$I^{(2)}_{depol} = 2\pi \frac{\alpha^4}{R^2} E_0^2 (\frac{\omega_0}{c})^4 \langle \sum_{i\neq j} T^d_{xz}(\underset{\sim}{R}_i-\underset{\sim}{R}_j) \sum_{k\neq l} T^d_{xz}(\underset{\sim}{R}_k-\underset{\sim}{R}_l)\rangle \qquad (59)$$

Here we have taken the long wavelength limit -- i.e., assumed that $2\pi/\underset{\sim}{k}_0 \simeq 5000$ Å greatly exceeds all positional correlation ranges -- so that the phase factors in Eq. (58) can be replaced by unity and

$$\underset{\approx}{T}_{k_0}(\underset{\sim}{R}) \xrightarrow{k_0R<<1} \frac{3\hat{R}\hat{R} - \underset{\approx}{I}}{R^3} \equiv \underset{\approx}{T}^d \qquad (60)$$

The ensemble average of the right hand side of Eq. (59), with T_{xz} defined by Eq. (60), has been evaluated by large-scale machine computation of the positional correlations. We shall discuss these results at the end of this lecture. First, though, we focus attention on the implicit assumption made earlier about the polarization of the atomic fluid by the incident electric field, i.e. the assumption that the atoms behave as polarizable "points". We shall show below how the second-order light scattering description is greatly improved by spreading the atomic polarizability density out through the atom in a physically suggestive way, instead of concentrating it all at the nucleus.

B. Collisional Polarizabilities

Consider again the classical electrostatic theory, according to which the local polarization (i.e., dipole moment density) $\underset{\sim}{\mu}(\underset{\sim}{r})$ in a system satisfies the equation [see Eqs. (5) and (7), with $\underset{\approx}{\alpha}(\underset{\sim}{r},t) \to \alpha(\underset{\sim}{r})$ and $\underset{\sim}{E}_0(\underset{\sim}{r},t) \to \underset{\sim}{E}_0$].

$$\underset{\sim}{\mu}(\underset{\sim}{r}) = \alpha(\underset{\sim}{r})\underset{\sim}{E}_0 + \alpha(\underset{\sim}{r})\int d\underset{\sim}{r}'\underset{\approx}{T}^d(\underset{\sim}{r}-\underset{\sim}{r}')\cdot\underset{\sim}{\mu}(\underset{\sim}{r}') \qquad (61)$$

Here $\alpha(\underset{\sim}{r})$ is the local polarizability density and $\underset{\sim}{E}_0$ is the constant applied electric field; $\underset{\approx}{T}^d(\underset{\sim}{r}) = (3\hat{r}\hat{r}-1)/r^3$ is the tensor which acts on a vector dipole moment (static in this case) at the origin to give the electric field arising from it at $\underset{\sim}{r}$. For our system we have in mind a pair of noble gas atoms a distance R apart. The induced dipole moment associated with the diatom is then

$$\underset{\sim}{\mu} = \int d\underset{\sim}{r}\underset{\sim}{\mu}(\underset{\sim}{r}) \equiv \alpha_{\parallel,\perp}(R)\underset{\sim}{E}_0 \qquad (62)$$

where $\parallel$ and $\perp$ refer to $\underset{\sim}{E}_0$ being alternately parallel and perpendicular to the internuclear axis. Approximating the polarizability density by the sum of isolated atom contributions (distortion effects are included later in the exchange and correlation terms) we write

$$\alpha(\underset{\sim}{r}) = \alpha_a(|\underset{\sim}{r} - \underset{\sim}{r}_a|) + \alpha_b(|\underset{\sim}{r} - \underset{\sim}{r}_b|) \qquad (63)$$

where $\underset{\sim}{r}_a$ and $\underset{\sim}{r}_b$ denote the positions of the nuclei.

Note that if we use the point dipole form

$$\alpha_a(\underset{\sim}{r}) = \alpha_0\delta(\underset{\sim}{r} - \underset{\sim}{r}_a) \qquad (64)$$

where α_0 is the atomic polarizability, then Eqs. (61)-(63) may be solved by iteration to give the familiar classical, dipole-induced-dipole (DID) results:[4]

$$\beta(R) \equiv \alpha_{\parallel}(R) - \alpha_{\perp}(R) = 6\alpha_0^2/R^3 + 6\alpha_0^3/R^6 + O(\alpha_0^4/R^9) \qquad (65)$$

and

$$\frac{1}{3}\mathrm{Tr}\underset{\approx}{\alpha}^{(2)}(R) \equiv \frac{1}{3}\mathrm{Tr}\underset{\approx}{\alpha}(R) - 2\alpha_0 \equiv \frac{\alpha_{\parallel}(R) + 2\alpha_{\perp}(R)}{3} - 2\alpha_0 = 4\alpha_0^3/R^6 + O(\alpha_0^4/R^9) \qquad (66)$$

Thus we can include overlap corrections by simply introducing a more reasonable distribution than Eq. (64) for $\alpha_a(\underset{\sim}{r})$, namely one in which the polarizability is spread out realistically through the atom.

Specifically, we use atomic, Hartree-Fock, finite electric field charge densities, divide the atom up into spherical shells, and take the dipole moment of each shell as a measure of the polarizability density at that distance from the nucleus. This procedure defines the approximation ($z = r\cos\theta$):

$$\alpha_a(\underset{\sim}{r}) = \frac{1}{E_0}\int_0^{2\pi} d\phi \int_0^{\pi} d\theta \sin\theta\, \rho(r\theta\phi/E_0) z \qquad (67)$$

Here $\rho(r\theta\phi/E_0)$ is the electron charge density of an atom which sits in a constant electric field, $E_0\hat{z}$. $\alpha_a(\underset{\sim}{r})$ is normalized so that

$$\int_0^{\infty} r^2 dr \alpha_a(\underset{\sim}{r}) = \alpha_0$$

If one uses Eq. (67) in Eqs. (61)-(63), then at large separations the $1/R^3$ DID results of Eqs. (65)-(66) are recovered while at shorter distances there are corrections due to overlap. More explicitly a single iteration of Eq. (61) leads to

$$\mathrm{Tr}\underset{\approx}{\alpha}^{(2)}(R) = 0 \qquad (68)$$

and

$$\beta(R) = 3\int d\underset{\sim}{r}_{1a}\int d\underset{\sim}{r}_{2b}\, \frac{3(\hat{r}_{12})_z(\hat{r}_{12})_z - 1}{r_{12}^3}\, \alpha_a(r_{1a})\alpha_b(r_{2b}) \tag{69}$$

where "a" and "b" refer to the two atomic centers and "1" and "2" to their electrons. Calculations of Eq. (69) for the light noble gases have been carried out and published.[25]

Suppose now, though, we proceed further by iterating a second time to obtain corrections of the form

$$\alpha_{\parallel}(R)_{corr} = \\ 2\sum_{i=x,y,z}\int d\underset{\sim}{r}''\int d\underset{\sim}{r}'\int d\underset{\sim}{r}\,\alpha_a(\underset{\sim}{r}'')[\underset{\approx}{T}(\underset{\sim}{r}''-\underset{\sim}{r}')]_{zi}\alpha_b(\underset{\sim}{r}')[\underset{\approx}{T}(\underset{\sim}{r}'-\underset{\sim}{r})]_{iz}\alpha_a(\underset{\sim}{r}) \tag{70}$$

with $\alpha_{\perp}(R)_{corr}$ the same except for x replacing z everywhere. The second order contributions to $\beta(R)$ and $Tr\underset{\approx}{\alpha}^{(2)}(R)$ are formed by taking the usual differences and sums of $\alpha_{\parallel,\perp}(R)_{corr}$. Eq. (70) reduces for large R to the $1/R^6$ DID terms of Eq. (65) and gives the proper overlap corrections at shorter distances. On top of these purely electrostatic $1/R^6$ terms, there is also a correlation contribution analogous to the London dispersion term in the intermolecular potential. This additional term, to a good approximation,[26] changes the asymptotic behavior of the trace from $4\alpha_0{}^3/R^6$ to

$$\frac{1}{3}\,Tr\underset{\approx}{\alpha}^{(2)}(R) \sim (4\alpha_0{}^3 + \frac{5}{9}\frac{\gamma C_6}{\alpha_0})/R^6 \tag{71}$$

where γ is the atomic hyperpolarizability and C_6 the $1/R^6$ van der Waal's coefficient. We make the assumption that the correlation contributions are reduced by overlap in the same way that the second-order electrostatic ones are.

Finally, exchange effects can be accounted for as follows. We take the total diatom wave function to be the antisymmetrized product of the two atomic, finite field, wavefunctions [A is the "antisymmetrizer"]:

$$\Psi = A\Psi_a\Psi_b \tag{72}$$

Then the incremental pair polarizability that results from exchange effects alone can be obtained from the dipole moment difference:

$$\mu - \mu 0 = \int d\underset{\sim}{r}[\rho(\underset{\sim}{r}) - \rho_a(\underset{\sim}{r}) - \rho_b(\underset{\sim}{r})]z \equiv \alpha_{\parallel,\perp}(R)E_0 \tag{73}$$

Here we have taken the electric field to lie along the z-axis with respect to which the interatomic axis may be $\parallel$ or $\perp$. $\rho(\underset{\sim}{r})$ is the total electron density -- no longer atom additive because of exchange - obtained from the normalized wavefunction in Eq. (72); ρ_a and ρ_b are the atomic electron densities. This "exchange" dipole moment can be expressed in terms of two-center orbital overlaps and dipole integrals, which we have evaluated[27] for He, Ne and Ar using the finite field HF atomic wavefunctions of Sitter.[28]

Typical results for the total $\alpha_{\parallel,\perp}(R)$ -- including overlap through second order, correlation, and exchange -- are shown (for helium) by the solid line in Fig. 1. We include for comparison the Hartree-Fock diatom results of O'Brien et al.[29a] (Δ's) and Fortune et al.[29b] (O's). The agreement is very good, and this is encouraging since the Hartree-Fock calculations[29] for the full diatom are no longer feasible for inert atoms heavier than helium.

Since no other accurate theoretical results are available for comparison in the case of the heavier noble gases, it is necessary to test our calculations directly against experiment. For example, if the second-order (depolarized) intensity given by Eq. (59) is expanded in powers of the density we find

$$I^{(2)}_{depol} = C\rho^2 \int_0^\infty dR\ R^2[\beta(R)]^2 \exp(-v(R)/k_BT) + O(\rho^3) \qquad (74)$$

Here: C is a multiplicative factor involving the incident intensity, distance to the detector, etc; v(R) is the interatomic pair potential and $\beta(R)$ is the "point" atom (DID) limit of $\alpha_\parallel(R) - \alpha_\perp(R)$. But because we have shown above that $\beta(R)$ should include overlap corrections which are nonnegligible for $R \lesssim R_e$ [where R_e is the minimum in v(R)], the ρ^2-term in $I^{(2)}_{depol}$ can be calculated by simply replacing $\beta_{overlap}(R)$ for $\beta_{DID}(R)$. The higher moments

$$\gamma_{2n} \equiv \int_{-\infty}^{+\infty} d\omega\ \omega^{2n}\ I^{(2)}_{depol}(\omega)$$

of the depolarized spectrum [Eq. (74) gives the 0th moment] can also be related[30] to $\beta(R)$ and v(R).

The first line of Table 1 shows that our room temperature calculated integrated intensity for argon -- normalized to the DID value -- falls nicely in the middle of the measured data. The second line predicts the measured value for $\Delta\nu_{1/2} = (\gamma_2/\gamma_0)^{1/2}/2\pi$, the half-width of the inelastic line shape. The third line shows $\gamma_4\gamma_0/(\gamma_2)^2$. For a purely gaussian spectrum, this quantity assumes the value 3; for an exponential it is 6 and for a Lorentzian it is infinite. [In all these calculations we have used for v(R) the very accurate MSV III potential of Parson et al.[31]]

There are at present no experimental values available for low density $\gamma_0(I^{(2)}_{depol})$ in the case of He or Ne. This is, of course, because the depolarized light scattering intensities scale roughly as α_0^4,

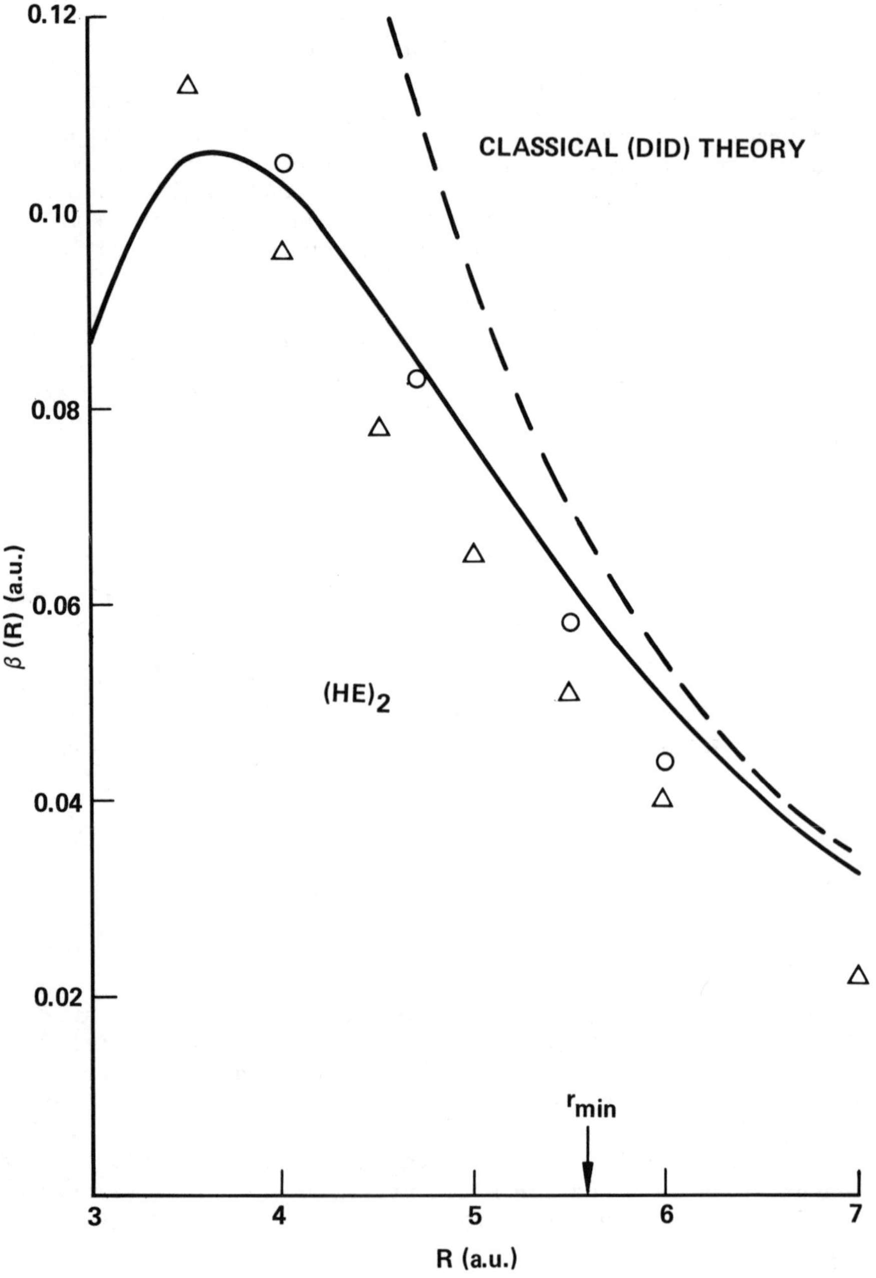
0.12
0.10
0.08
0.06
0.04
0.02
β (R) (a.u.)
CLASSICAL (DID) THEORY
(HE)2
r min
3
4
5
6
7
R (a.u.)

FIGURE 1

implying collisional anisotropy effects which are orders of magnitude smaller than those observed for Ar. Our calculations of $\beta(R)$ allow us to predict γ_0 values for He and Ne which are 0.72 and 0.73 smaller, respectively, than the corresponding DID values at 298°K.

Note that the ρ^2-$I^{(2)}_{depol}$ data only provide the area under $\beta^2(R)$ and hence do not allow for a direct look at the R-dependence of the collisional anisotropy.

TABLE I

Moments of (γ_n,n=0,2,4) of the low density depolarized spectrum of Ar, Calculated from $[\beta(R)]_{DID}$ and $[\beta(R)]_{Ref.\ 27}$.

	DID	Ref. 27	Experiment
I^{xz}_{TOT}	1	0.68	0.62 ±.06[a] 0.72 ±.11[b] (Kerr) 0.81 ± .12[c] 0.67 ± .03[d]
$\Delta\nu_{1/2} = \frac{1}{2\pi}\left(\frac{\gamma_2}{\gamma_0}\right)^{1/2}$	18.2 cm^{-1}	13.9 cm^{-1}	15.0 cm^{-1} [e]
$\frac{\gamma_0\gamma_4}{(\gamma_2)^2}$	11.5	7.2	7.3[f]

[a]J. P. McTague et al. J. de Phys. 33 C1-241 (1972).
[b]A. D. Buckingham and D. A. Dunmur, Trans. Faraday Soc. 64 1776 (1968).
[c]P. Lallemand, J. de Phys. 33 C1-257 (1972).
[d]M. Thibeau, G. C. Tabisz, B. Oksengorn and B. Vodar, J. Quant. Spectrosc. Radiat. Transfer 10 839 (1970).
[e]J. P. McTague and G. Birnbaum, Phys. Rev. Lett. 21 661 (1968).
[f]J. P. McTague, private communication.

Ideally, we would want to measure directly v(R) in a strong (≈10^7v/cm) electric field by the same extremely accurate differential scattering methods[31] used to determine the ordinary pair potentials -- but this experiment is still many years around the corner. From the excellent agreement between our calculations and both the Hartree-Fock diatom results and the measured ρ^2-$I^{(2)}_{depol}$ data, it appears that the collisional polarizabilities discussed above are quite accurate.

A much harder way to test the theoretical $\beta(R)$'s is to calculate $I^{(2)}_{depol}$ at higher densities and to compare the results with the corresponding laboratory data. To study the room temperature noble gases at high density requires of course that the sample be squeezed to very high pressures. The resulting strain-induced birefringence in the cell windows significantly distorts the measured depolarization. On the theoretical side the difficulty is that no approximations (e.g., density expansions) can be employed to simplify the evaluation of the ensemble average appearing in Eq. (59). Nevertheless several machine computations have been carried out.[32-34] Except for the study by Pinski and Campbell[34] -- which tests integral equation approximations for the pair-pair correlations contributing to the liquid depolarization -- these calculations use the molecular dynamics method (discussed in detail in the lectures of Rahman) and fall into two classes: those[32a,c] studying the density dependence of $I^{(2)}_{depol}$ and those[32b,33] trying to account for the observed $I^{(2)}_{depol}(\omega)$ line shapes.

The molecular dynamics studies of $I^{(2)}_{depol}(\omega)$ line shapes have shed some light on the nature of correlated collisions in the liquid. It is found,[33] for example, that the angular -- rather than radial -- correlations involving the interatomic separations are most important in accounting for the observed line shape. No isolated binary collision model, regardless of the form of $\beta(R)$, gives a realistic spectrum. With three- and four-body correlations included, however, the results do not depend strongly on $\beta(R)$.

The integrated intensities studies, on the other hand, all confirm that use of the DID $\beta(R)$ over estimates the depolarization, the discrepancy increasing with density. This is consistent with the ρ^2-results of Table I and the fact that the neglected overlap corrections decrease $\beta(R)$ significantly for $R \lesssim R_e$ (cf. Fig. 1). Using the full collisional anisotropy calculated by Oxtoby and Gelbart, or the empirically fit $\beta(R) = (6\alpha^2/R^3) - (A/R^n)$ of Levine and Birnbaum,[35] brings down the theoretical estimates of $I^{(2)}_{depol}$. Because of the subtle cancellation effects involving the two- ($ij=kl$), three ($j=l, i\neq k$) and four- ($j\neq l, i\neq k$) body contributions to the average in Eq. (59) -- each one is many orders of magnitude larger than their sum -- the machine computations are quite trickly to interpret. Also, as mentioned earlie the relevant experimental data are scarce and of inadequate accuracy. Nevertheless, we stress that the $\beta(R)|_{calculated}$'s discussed earlier descrive completely the collision-induced depolarized intesities observed for liquids and high pressure gases. The four-point density fluctuation correlation function "simply" needs to be evaluated carefully enough, as considered in the lectures of Rahman and the seminar by Pinski. As stressed by Fleury, though, it would be very useful to develop crude analytical models of the many-body structure/dynamics which could provide a conceptual underpinning for the measured changes in intensity and line shape.

REFERENCES

1. For a most recent and comprehensive account of light scattering theory and its applications, see Dynamic Light Scattering, by B. J. Berne and R. Pecora (John Wiley, 1976).

2. J. D. Jackson, Classical Electrodynamics (John Wiley, 1962).

3. M. Born and E. Wolf, Principles of Optics (Pergamon, 1964).

4. B. U. Felderhof, Physica 76 486 (1974).

5. D. Bedeaux and P. Mazur, Physica 67 23 (1973).

6. D. W. Oxtoby, unpublished Ph.D. dissertation (entitled "I. Optical Properties of Simple Fluids. II. Mode Coupling Theory of Critical Transport"), University of California, Berkeley, 1975.

7. P. Mazur, Advan. Chem. Phys. 1 309 (1958).

8. R. K. Bullough, Phil. Trans. Roy. Soc. (London) A258 387 (1965).

9. M. J. Stephen, Phys. Rev. 187 279 (1969).

10. D. W. Oxtoby and W. M. Gelbart, J. Chem. Phys. 60 3359 (1974).

11. G. B. Benedek, in Statistical Physics, Phase Transitions and Superfluidity, Vol. 2, M. Chrétien et al., eds. (Gordon and Breach, 1968).

12. A. Ben-Reuven and N. D. Gershon, J. Chem. Phys. 51 893 (1969).

13. see, for example, T. Keyes and B. Ladanyi, Mol. Phys. 6 1685 (1976).

14. H. Mori, Progr. Theoret. Phys. 33 423 (1965).

15. L. P. Kadanoff and J. Swift, Phys. Rev. 166 89 (1968).

16. F. Garisto and R. Kapral, J. Chem. Phys. 64 3826 (1976).

17. K. Kawasaki, Ann. Phys. (NY) 61 1 (1970).

18. T. Keyes and I. Oppenheim, Phys. Rev. A7 1384 (1973).

19. See, for example, the discussion and references given in H. L. Swinney and D. L. Henry, Phys. Rev. A8 2586 (1973).

20. See A. D. B. Woods in Quantum Fluids, ed by D. F. Brewer (North-Holland, 1966) for neutron scattering evidence of multiple excitation contributions to S(K,Ω) for liquid He.

21. M. Born and K. Huang, Dynamical Theory of Crystal Lattices (Oxford, 1968).

22. J. W. Halley, Phys. Rev. 181 338 (1969).

23. T. J. Greytak and J. Yan, Phys. Rev. Lett. 22 987 (1969).

24. J. P. McTague, P. A. Fleury and D. B. DuPré, Phys. Rev. 188 303 (1969).

25. D. W. Oxtoby and W. M. Gelbart, Mol. Phys. 29 1569 (1975).

26. A. D. Buckingham, Trans. Faraday Soc. 52 1035 (1956).

27. D. W. Oxtoby and W. M. Gelbart, Mol. Phys. 30 535 (1975).

28. R. E. Sitter, Jr., and R. P. Hurst, Phys. Rev. A5 5 (1972).

29. (a) E. F. O'Brien, V. P. Gutschick, V. McKoy and J. P. Mctague, Phys. Rev. A8 690 (1973); (b) P. J. Fortune, P. R. Certain and L. W. Bruch, Chem. Phys. Lett. 27 233 (1974).

30. P. Lallemand, J. Phys. Paris, 33 C1-257 (1972).

31. J. M. Parson, P. E. Siska and Y. T. Lee, J. Chem. Phys. 56 1511 (1972).

32. (a) B. J. Alder, J. J. Weis and H. L. Strauss, Phys. Rev. A7 281 (1973); (b) B. J. Alder, H. L. Strauss and J. J. Weis, J. Chem. Phys. 59 1002 (1973); and (c) J. Beers, B. J. Alder and H. L. Strauss, to be published.

33. B. J. Berne, M. Bishop and A. Rahman, J. Chem. Phys. 58 2696 (1973).

34. F. J. Pinski and C. E. Campbell, to be published, and seminar by Pinski in this volume.

35. H. Levine and G. Birnbaum, J. Chem. Phys. 55 2914 (1971).

SECOND-ORDER LIGHT SCATTERING BY CLASSICAL FLUIDS II:

DOUBLE LIGHT SCATTERING BY CRITICAL FLUIDS

William M. Gelbart

Department of Chemistry, University of California

Los Angeles, California 90024

I. INTRODUCTION

In this lecture I shall treat the double (and higher-order multiple) scattering of light by critical fluids. We shall discuss the basic physics behind this "new" effect and how much (how little?) it tells us about the correlations of density fluctuations in these many-body systems. The first predictions of the dependence of double light scattering intensities on the temperature, incident wavelength and sample size were given by Oxtoby and Gelbart.[1,2] These predictions were confirmed by the experiments of Reith and Swinney.[3] A more general and rigorous theory was then developed by Boots, Bedeaux and Mazur,[4-6] and several other experiments were performed.[7-9]

First, in Section II, I shall present the original discussion of Oxtoby and Gelbart. Consistent with the heuristic approach followed in my first lecture, this presentation makes few claims to rigor. Instead the idea is to lay a simple groundwork for later discussion of the Reith and Swinney experiments (Section III) and of the formal theory of Boots, et al. (Section IV).

II. SIMPLE THEORY OF DOUBLE SCATTERING

Recalling Eq. (I.54) for the second-order scattered field $\Delta \underset{\sim}{E}_s^{(2)}$, substituting its x-component into Eq. (I.16) and noting that $[\Delta \underset{\sim}{E}_s^{(1)}]_x = 0$ for $\underset{\sim}{E}_0 || \hat{z}$, we can write the zx-depolarized intensity (integrated over final frequency ω) as in Eq. (I.58):

$$I^{zx} = \frac{\alpha^4}{R^2} k_0^4 |E_0|^2 \int_{V_S} d\underset{\sim}{r}_1 \int_{V_I} d\underset{\sim}{r}_2 \int_{V_S} d\underset{\sim}{r}_3 \int_{V_I} d\underset{\sim}{r}_4 \exp[i\underset{\sim}{k}_0 \cdot (\underset{\sim}{r}_2 - \underset{\sim}{r}_4)] \quad (1)$$

$$\exp[-i\underset{\sim}{k} \cdot (\underset{\sim}{r}_1 - \underset{\sim}{r}_3)][\underset{\approx}{T}_{k_0}(\underset{\sim}{r}_1 - \underset{\sim}{r}_2)]_{xz}[\underset{\approx}{T}_{k_0}(\underset{\sim}{r}_3 - \underset{\sim}{r}_4)]^*_{xz} \langle \Delta\rho(\underset{\sim}{r}_1)\Delta\rho(\underset{\sim}{r}_2)\Delta\rho(\underset{\sim}{r}_3)\Delta\rho(\underset{\sim}{r}_4)\rangle$$

Here α is the isolated atom polarizability, R the distance to the detector, $\underset{\sim}{k}_0$ and $\underset{\sim}{k}$ the propagation wavevectors for the incident and scattered light ($\underset{\sim}{k}_0 || \hat{x}$ and $\underset{\sim}{k} || \hat{y}$), $|E_0|^2$ the initial intensity, and $\Delta\rho(\underset{\sim}{r})$ the instantaneous density fluctuation at $\underset{\sim}{r}$:

$$\sum_{i=1}^{N} \delta(\underset{\sim}{r} - \underset{\sim}{R}_i) - \frac{N}{V} \equiv \Delta\rho(\underset{\sim}{r}) \quad .$$

$$\underset{\approx}{T}_{k_0}(\underset{\sim}{r}) \equiv \exp(ik_0 r)\left[\left(\frac{3}{r^3} - \frac{3ik_0}{r^2} - \frac{k_0^2}{r}\right)\hat{r}\hat{r} + \left(\frac{k_0^2}{r} + \frac{ik_0}{r^2} - \frac{1}{r^3}\right)\underset{\approx}{I} \right] \quad (2)$$

acts on an oscillating ($\omega_0 = ck_0$) dipole moment to give the resulting electric field at a vector distance $\underset{\sim}{r}$ away. V_S denotes the volume of the sample "seen" by the detector, while V_I is the illuminated volume. These volumes arise from the fact that -- see Eq. (I.54) -- each second-order scattered field corresponds to: (1) the incident field $\underset{\sim}{E}_0$ polarizing a density fluctuation at $\underset{\sim}{r}_2$ [$\underset{\sim}{E}_0(\underset{\sim}{r}_2)] \neq 0$ iff $\underset{\sim}{r}_2 \epsilon V_I$] followed by (2) this induced moment giving rise [through $\underset{\approx}{T}_{k_0}(\underset{\sim}{r}_1 - \underset{\sim}{r}_2)$] to a field at $\underset{\sim}{r}_1$ which (3) radiates light seen by the detector [iff $\underset{\sim}{r}_1 \epsilon V_S$]. Again, we have neglected $|\underset{\sim}{r}_2 - \underset{\sim}{r}_1|$ compared with $|\underset{\sim}{R} - \underset{\sim}{r}_1|$, and set $|\underset{\sim}{R} - \underset{\sim}{r}_1| \simeq \underset{\sim}{R}$.

Note that in the $k_0 \to 0$ limit, the exponential phase factors in Eq. (1) can be replaced by unity and the dipole propagator reduces [see Eq. (2)] to $(3\hat{r}\hat{r} - \underset{\approx}{I})/r^3$. The resulting $I^{xz}(k_0 \to 0)$ is associated with the collision-induced depolarized light-scattering intensity discussed at the end of the previous lecture. Near the critical point of an atomic fluid, however, $k_0 r$ can no longer be neglected compared with 1, since the correlation distances are no longer small compared with the wavelength ($\approx 1/k_0$) of light. In this case all terms in Eqs. (1) and (2) must be retained.

It is the presence of $\langle \Delta\rho(1)\Delta\rho(2)\Delta\rho(3)\Delta\rho(4)\rangle$ in I^{zx} which originally prompted the suggestion[10,11] that second-order light scattering might provide new information about three- and four-particle correlations in critical fluids. As we shall see, however, the double scattering intensity yields instead "only" a new probe of the two-particle correlations.

It is natural at this point to decompose the 4-point correlation function into its disconnected and connected parts:[12]

$$<\Delta\rho(1)\Delta\rho(2)\Delta\rho(3)\Delta\rho(4)> = \tag{3}$$
$$<\Delta\rho(1)\Delta\rho(2)><\Delta\rho(3)\Delta\rho(4)>+<13><24>+<14><23>+<\Delta\rho(1)\Delta\rho(2)\Delta\rho(3)\Delta\rho(4)>_c$$

The connected ("c") correlation function is what remains after subtracting off all possible products of lower-order connected correlation functions involving the same density fluctuations. [Since $<\Delta\rho> = 0$, the 2- and 3-point correlation functions, $<\Delta\rho(1)\Delta\rho(2)>$ and $<\Delta\rho(1)\Delta\rho(2)\Delta\rho(3)>$, naturally contain no disconnected contributions.] The importance of this decomposition is that the connected correlation functions are short-ranged, equalling zero when any pair of arguments is separated by a macroscopic distance. Thus $<1234>_c$ will contribute to the depolarized light scattering intensity only when all four atoms are within a correlation length ξ of one another; ξ (to be discussed below) is small compared to the dimensions of V_S and V_I. The disconnected terms in Eq. (3), on the other hand, will contribute to I^{xz} for configurations in which the pairs of atoms are separated by macroscopic distances. Thus they are expected to dominate the depolarized intensity, by a factor of $\approx V/\xi^3$ where V is the volume of fluid;[12b] the $<1234>_c$ contribution can be neglected.

In order to evaluate Eq. (1) we substitute Eq. (3) for $<\Delta\rho(1)\Delta\rho(2)\Delta\rho(3)\Delta\rho(4)>$, with the connected term neglected. The $<13><24>$ term gives, for example, the following contribution to I^{zx}:

$$\frac{\alpha^4}{R^2} k_0{}^4 |E_0|^2 \int_{V_S} d\underset{\sim}{r}_1 \int_{V_I} d\underset{\sim}{r}_2 \int_{V_S} d\underset{\sim}{r}_3 \int_{V_I} d\underset{\sim}{r}_4 \exp[i\underset{\sim}{k}_0\cdot(\underset{\sim}{r}_2-\underset{\sim}{r}_4)]\exp[-i\underset{\sim}{k}\cdot(\underset{\sim}{r}_1-\underset{\sim}{r}_3)]$$
$$[\underset{\approx}{T}_{k_0}(\underset{\sim}{r}_1-\underset{\sim}{r}_2)]_{xz}[\underset{\approx}{T}_{k_0}(\underset{\sim}{r}_3-\underset{\sim}{r}_4)]^*_{xz}<\Delta\rho(\underset{\sim}{r}_1)\Delta\rho(\underset{\sim}{r}_2)><\Delta\rho(\underset{\sim}{r}_3)\Delta\rho(\underset{\sim}{r}_4)> \tag{4}$$

Now change variables to $\underset{\sim}{r}_{12}$, $\underset{\sim}{r}_{34}$, $\underset{\sim}{r}_2$ and $\underset{\sim}{r}_4$. In order to "decouple" these integrations it is convenient to express $u(r) \equiv <\Delta\rho(\underset{\sim}{r})\Delta\rho(\underset{\sim}{0})>$ as the Fourier transform of the static structure factor:

$$u(r) = \frac{1}{(2\pi)^3\rho}\int d\underset{\sim}{k}\,\exp(-i\underset{\sim}{k}\cdot\underset{\sim}{r})[S(k)]$$
$$= \frac{1}{(2\pi)^3}\int d\underset{\sim}{k}\,\exp(-i\underset{\sim}{k}\cdot\underset{\sim}{r})\left[\frac{4\pi\xi_0}{k^2+1/\xi^2} + \frac{S_{sr}(k)}{\rho}\right] \tag{5}$$

Here the first term in square brackets is the contribution to S(k) from the long-range (Ornstein-Zernicke[13]) part of u(r):

$$u_{\ell r}(r) = u_{oz}(r) = \xi_0\exp(-r/\xi)/r \tag{6}$$

ξ is the familiar "long-range correlation length"; ξ_0 is ≈ 1 Å and is almost independent of temperature for $T \approx T_c$. $S_{sr}(k)$ is the contribution to S(k) from the short-range part of u(r); the structure of $u_{sr}(r)$ is dominated by the nearest-neighbor peak and is expected to be sensibly constant throughout the critical ($T \approx T_c$) region. Substituting Eq. (5) into Eq. (4), and integrating over $\underset{\sim}{r}_2$ and $\underset{\sim}{r}_4$, we obtain

$$\frac{\alpha^4}{R^2} k_0{}^4 |E_0|^2 \rho^4 \frac{V_I}{(2\pi)^3} \int d\underset{\sim}{k}_1 |g(\underset{\sim}{k}_1)|^2 \left[\frac{4\pi\xi_0}{|\underset{\sim}{k}_1+\underset{\sim}{k}|^2+1/\xi^2} + \frac{1}{\rho} S_{sr}(|\underset{\sim}{k}_1+\underset{\sim}{k}|)\right]$$
$$\left[\frac{4\pi\xi_0}{|\underset{\sim}{k}_1+\underset{\sim}{k}_0|^2+1/\xi^2} + \frac{1}{\rho} S_{sr}(|k_1+k_0|)\right] \tag{7}$$

where

$$g(\underset{\sim}{k}_1) = \int_{V_S} d\underset{\sim}{r}\; \exp(i\underset{\sim}{k}_1\cdot\underset{\sim}{r})[\underset{\approx}{T}_{k_0}(\underset{\sim}{r})]_{xz} \tag{8}$$

In evaluating $|g(\underset{\sim}{k}_1)|^2$, Oxtoby and Gelbart chose a sample in which a small illuminated volume V_I is located at the center of a large, approximately spherical, observed volume V_S. [More realistic geometries have been treated subsequently[3,5-6,8b,14] but such improvements are not important for our present discussion.] They found that $|g(\underset{\sim}{k}_1)|^2$ behaves under the $\underset{\sim}{k}_1$ integration as

$$|g(\underset{\sim}{k}_1)|^2 = (\hat{k}_1\cdot\hat{x})^2(\hat{k}_1\cdot\hat{z})^2[h(k_1) + 8\pi^3 k_0{}^2 R_s \delta(k_1-k_0)] \tag{9}$$

Here R_s is the "radius" of V_S, i.e., the dimension of the sample volume seen by the detector. $h(k_1)$ is a slowly varying function of k_1, independent of R_s, and is given in Ref. 1.

Eight terms result when Eq. (7), with Eq. (9) substituted for $|g(\underset{\sim}{k}_1)|^2$, is multiplied out. The term involving the product $h(k_1)S_{sr}(\underset{\sim}{k}_1+\underset{\sim}{k})S_{sr}(\underset{\sim}{k}_1+\underset{\sim}{k}_0)$ corresponds to the short-range correlation contribution associated with the $1/r^3$ terms in $\underset{\approx}{T}_{k_0}$ [cf. Eq. (2)] -- we identify this as the depolarized light scattering intensity arising from collision-induced effects. The remaining seven terms have been calculated either analytically or numerically,[1] and one of them is found to be dominant over the others by many orders of magnitude. The one important term arises from the product of the R_s-dependent term in Eq. (9) and the two long-range structure factors. Since the $8\pi^3 k_0{}^2\cdot R_s\delta(k_1-k_0)$ term in Eq. (9) comes from the $1/r$ term in $\underset{\approx}{T}_{k_0}$, the contribution involving the product $8\pi^3 k_0{}^2 R_s\delta(k_1-k_0)S_{\ell r}(\underset{\sim}{k}_1+\underset{\sim}{k}_0)$ $S_{\ell r}(\underset{\sim}{k}_1+\underset{\sim}{k})$ is identified as the I^{zx} arising from true double scattering effects. For recall that $T_{k_0}(\underset{\sim}{r})\cdot\underset{\sim}{p}$ can be rewritten as

$\underset{\sim}{E}$ (at $\underset{\sim}{r}$ due to dipole oscillating at ck_o at origin)

$$= \exp(ik_0 r)(\frac{k_0{}^2}{r})[\hat{r}x\underset{\sim}{p})x\hat{r}] + \exp(ik_0 r)(\frac{1}{r^3} - \frac{ik_0}{r^2})[3\hat{r}(\hat{r}\cdot\underset{\sim}{p}) - \underset{\sim}{p}] \quad (10)$$

thereby making explicit the fact that the electric field is transverse only in the "far ($k_0 r >> 1$) zone".[15] That is, the true radiation term arising from $1/r$ survives only with the long-range correlations. Thus we have

$$I^{xz} \approx I^{zx}_{CI} + I^{zx}_{DS} \quad (11)$$

where I^{zx}_{CI} is the collision-induced depolarized light scattering intensity discussed at the end of the previous lecture, and

$$I^{zx}_{DS} \approx \frac{\alpha^4}{R^2} k_0{}^4 |E_0|^2 \rho^4 \frac{V_I}{2\pi} \xi_0 \; a^2 R_s \int_0^{\pi} d\theta_1 \int_0^{2\pi} d\phi_1 \frac{\sin^3\theta_1 \cos^2\theta_1 \sin^2\phi_1}{[1 + a\sin\theta_1\cos\phi_1][1 - a\sin\theta_1\sin\phi_1]} \quad (12)$$

with

$$a \equiv 2k_0{}^2/[2k_0{}^2 + (1/\xi)^2] \quad (13)$$

Eq. (12) follows from Eq. (7) upon integrating the $8\pi^3 k_0{}^2 R_s \delta(k_1 - k_0)$ $S_{\ell r}(\underset{\sim}{k}_1 + \underset{\sim}{k}) S_{\ell r}(\underset{\sim}{k}_1 + \underset{\sim}{k}_0)$ term over k_1.

The fact that real double scatterings take place, i.e., that some of the light waves scattered by atoms in the incident beam do not get out of the cell without being scattered again, means of course that the detector measures contributions to I^{zx} from atoms polarized outside the path of the incident beam. Thus the observed intensities no longer depend only on the illuminated volume V_I, but also on the size of the part of the sample which would otherwise be "dark". In particular we have shown [cf. Eq. (1])] how for the case of a spherical V_S with radius R_s, I^{zx}_{DS} is proportional to R_s. We shall also see that I^{zx}_{DS} varies with temperature, in the critical region, as $(k_0\xi)^4$ where ξ is the long-range correlation length. The background, collision-induced effects (I^{zx}_{CI}), on the other hand, arise from short-range correlations only and thus depend solely on local structure in the illuminated volume.

Recall that Eq. (4), with which we started, includes only the $<13><24>$ term in Eq. (3). The contributions from the remaining disconnected terms, however, can be shown to involve only the "$h(k_1)$"-term in $|g(\underset{\sim}{k}_1)|^2$ -- no "$R_s\delta(k_1 - k_0)$" terms survive. This is because factors such as $<12><34>$ do not allow $|\underset{\sim}{r}_1 - \underset{\sim}{r}_2|$ and $|\underset{\sim}{r}_3 - \underset{\sim}{r}_4|$ to become macroscopic -- hence they kill off the 1/r-terms in $\underset{\approx}{T}_k(\underset{\sim}{r}_1-\underset{\sim}{r}_2)$ and $\underset{\approx}{T}_k(\underset{\sim}{r}_3-\underset{\sim}{r}_4)$. Similarly, the connected term $<1234>_c$ vanishes unless $\underset{\sim}{r}_1$, $\underset{\sim}{r}_2$, $\underset{\sim}{r}_3$ and $\underset{\sim}{r}_4$ are all within a distance ξ ($<<V_S^{1/3}$) of each other, thereby squelching the double scattering contributions.

For $k_0\xi$ not too large (say, $\lesssim 0.3$) Eq. (12) for I_{DS}^{zx} can be shown[1] to increase monotonically with $k_0\xi$ according to

$$I_{DS}^{zx} \approx \frac{\alpha^4}{R^2} k_0{}^4|E_0|^2 \frac{64\pi^3}{15} \rho^4 V_I\xi_0{}^2R_s(k_0\xi)^4 \tag{14}$$

This follows directly from Eq. (12) upon expanding its integrand in powers of a, retaining only the leading term, and analytically evaluating the angular integrations.

Since we shall discuss experimental data for xenon in the next section, it is convenient to present first an approximate value of I_{CI}^{zx} for xenon near its critical point. We can get a rough estimate of the non-critical depolarization by extrapolating the low density results of Eq. (I.74) to ρ_c. We find [putting $\beta(r) = \beta_{DID}(r) = 6\alpha^2/r^3$]

$$I_{CI}^{zx} \approx \frac{\alpha^4k_0{}^4|E_0|^2V_{IS}}{R^2} \rho_c{}^2 \cdot \frac{24\pi}{5} \int_0^\infty dr\ \exp[-v(r)/k_BT_c]r^{-4} \tag{15}$$

Here V_{IS} is the portion of the illuminated volume seen by the detector. The integral in Eq. (15) has been evaluated using an accurate pair potential v(r) for xenon,[16] with the result

$$I_{CI}^{zx} \approx \frac{\alpha^4k_0{}^4|E_0|^2V_{IS}}{R^2} \rho_c{}^2(1.5\times10^{23}\text{cm}^{-3}) \tag{16}$$

This estimate is almost certainly an upper bound on the true non-critical depolarization; there are two effects which tend to lower it. First of all, in writing down Eq. (15) we have used the point dipole approximation. This is valid at large distances, but, as we have seen in the previous lecture, overlap at shorter distances will reduce the polarizability anisotropy and thus the low density light scattering by a factor of between 1/2 and 1. Secondly, we have extrapolated the low density slope to the critical region. At higher densities the local fluid environment becomes more uniform, on the average, so the depolarized intensity will be lower than the extrapolated low density slope. For <u>argon</u> at room temperature it is found[17] that I^{zx} at ρ_c is smaller by a factor of about 1/2 than the value obtained by extrapolating the low density expression [Eq. (15)]. Combining these numbers, we find for the non-critical depolarization,

$$I_{CI}^{zx} \approx \frac{\alpha^4k_0{}^4|E_0|^2V_{IS}}{R^2} \rho_c{}^2(6\times10^{22}\ \text{cm}^{-3}) \tag{17}$$

A more accurate estimate can be obtained (see below) by measurements carried out at ρ_c and at temperatures far from T_c, but this is not necessary for our present purposes.

Recalling Eq. (11), we can now add Eqs. (14) and (17) to obtain

$$I^{zx}_{total} \approx \frac{\alpha^4}{R^2} k_0{}^4 |E_0|^2 \rho_c{}^2 V_I \{(6\times10^{22}\ cm^{-3}) + \frac{64\pi^3}{15} \rho_c{}^2 \xi_0{}^2 R_s (k_0\xi)^4\} \tag{18}$$

Since second-order scattering is still a small effect compared with fist-order scattering, I^{zz}_{total} can be well-approximated by Eq. (I.27) from the previous lecture:

$$I^{zz}_{total} \approx I^{zz}_{first\text{-}order} \approx \frac{\alpha^2}{R^2} k_0{}^4 |E_0|^2 VS(K)$$

$$\xrightarrow[\text{critical point}]{K\xi \lesssim 0.3} \approx \frac{\alpha^2 k_0{}^4}{R^2} |E_0|^2 \rho_c{}^2 V_I 4\pi\xi_0\xi^2 \tag{19}$$

Thus we have, for $k_0\xi \lesssim 0.3$ (implying $K\xi \lesssim 0.3$ since $K \equiv |\underset{\sim}{k}-k_0| \lesssim k_0$)

$$\Delta = \frac{I^{zx}}{I^{zz}} \approx \frac{\alpha^2}{4\pi\xi_0} \{\frac{6\times10^{22}\ cm^{-3}}{\xi^2} + [\frac{64\pi^3}{15} \rho_c{}^2 \xi_0{}^2 R_s k_0{}^4] \xi^2\} \tag{20}$$

Note that ξ is the only quantity in Eq. (20) which is sensitive to the temperature in the critical region. Consider an experiment[3] in which xenon is prepared at its critical density[18] ($\rho_c = 5.1\times10^{21}$ molecules/cm^3) and T approaches T_c from above. Then, as the critical point is approached and ξ becomes larger, Δ will initially <u>decrease</u>. This corresponds to I^{zx} being dominated by I^{zx}_{CI} -- which stays constant [cf. Eq. (17)] -- while I^{zz} is increasing with ξ^2 [Eq. (19)]. For large enough ξ^2, however, the double scattering term in I^{zx} takes over, increasing as ξ^4 [Eq. (14)]; thus Δ should increase with ξ^2 in this region. The exact position and shape of the minimum will depend of course on the actual values of ρ_c, ξ_0, R_s and k_0. Using the ρ_c value cited above, $\xi_0 = 0.65\times10^{-8}$ cm (see discussion in Ref. 1), $R_s = 0.233$ cm and $k_0 = 2\pi/(4.579\times10^{-5}$cm), we find the Δ vs. ΔT curve shown in Fig. 1. The Δ vs. ξ expression given by Eq. (20) has been displayed in Fig. 1 as a Δ vs. ΔT plot by employing the experimentally determined relation

$$\xi = 1.8\ \text{Å}(\Delta T/T_c)^{-0.57} \tag{21}$$

for xenon.

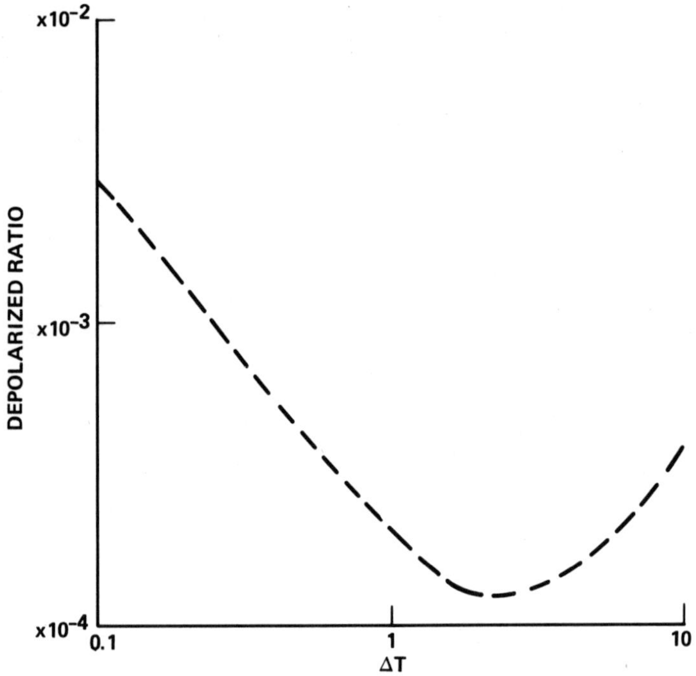

FIGURE 1

III. EXPERIMENT, AND A STILL SIMPLER THEORY

Reith and Swinney[3] have performed experiments to test specifically the predictions described in the preceding section. Depolarization ratio measurements were made for a sample of xenon at the critical density, over the temperature range $0.05 \leq T-T_c \leq 10°C$. Most of the measurements were performed with an argon-ion laser with $\lambda_0 = 4579$ Å, but at $T-T_c = 0.496°C$ data were also obtained at 5145 Å (argon laser) and at 6328 Å (He-Ne laser). The incident light from the laser was linearly polarized perpendicular to the scattering plane and the light scattered at 90° passed through a lens and rotatable linear polarizer. The lens formed an image of the sample on a pair of crossed adjustable slits. The horizontal slit was adjusted so that the observed sample heights ranged from 0.261 to 1.490 mm; the

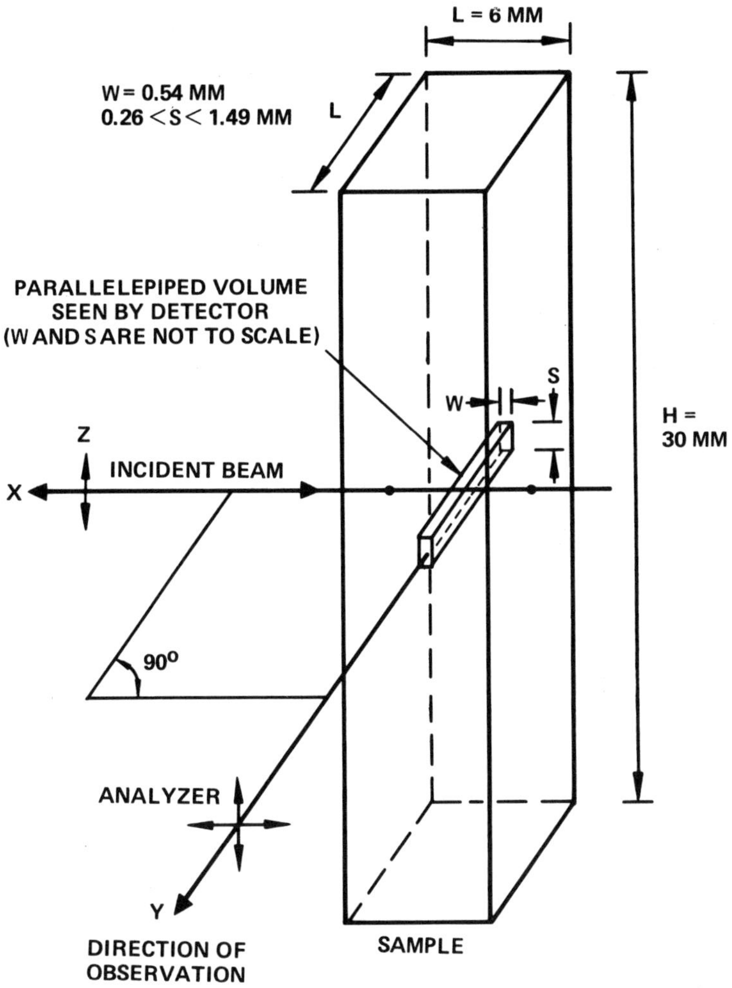

FIGURE 2

vertical slit which determined the width W of sample seen by the detector, was held fixed at W = 0.54 mm. Light passing through the crossed slit (S×W rectangle) was imaged by a second lens onto the cathode of a photomultiplier tube which measured its intensity. The experimental geometry is shown schematically in Fig. 2. The collimated, focused beam passing through the sample had a diameter of ≈0.21 mm.

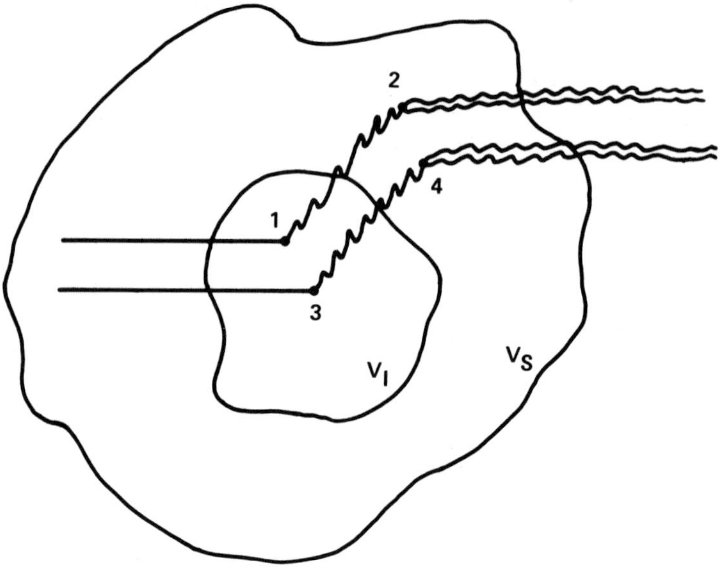

FIGURE 3A

Although the experimental geometry of Reith and Swinney is rather different from that assumed in the earlier theoretical discussion, the theory can be tested qualitatively by associating $2R_S$ with S. In particular Reith and Swinney found a Δ vs. ΔT curve which has the predicted shape (see Fig. 1), with the minimum lying between one and two degrees from T_c. Furthermore Δ was found to increase linearly with S and $k_0{}^4$, in agreement with our Eq. (20). Before discussing the experimental results in more detail, however, it is convenient to present first the simpler theory of double scattering suggested by Reith and Swinney.

Recall from the preceding section that the dominant contribution to I_{DS}^{zx} comes from the disconnected term $\langle\Delta\rho(\underset{\sim}{r}_1)\Delta\rho(\underset{\sim}{r}_3)\rangle\langle\Delta\rho(\underset{\sim}{r}_2)\Delta\rho(\underset{\sim}{r}_4)\rangle$ in the 4-point correlation function. This contribution involves the interference between two doubly scattered waves, the first involving the successive polarization of particles "1" and "2" [cf. the $\underset{\approx}{T}_{k_0}(\underset{\sim}{r}_1-\underset{\sim}{r}$ in Eq. (4)] and the second involving "3" and "4" [cf. $\underset{\approx}{T}_{k_0}(\underset{\sim}{r}_3-\underset{\sim}{r}_4)$]. This event is depicted schematically in Fig. 3(a), where the ——, ~~~~, ≈≈≈≈ lines denote the incident, singly- and doubly- scattered fields, respectively. <13><24> forces the initial polarizations at $\underset{\sim}{r}_1$ and $\underset{\sim}{r}_3$ to take place within a correlation length of one another,

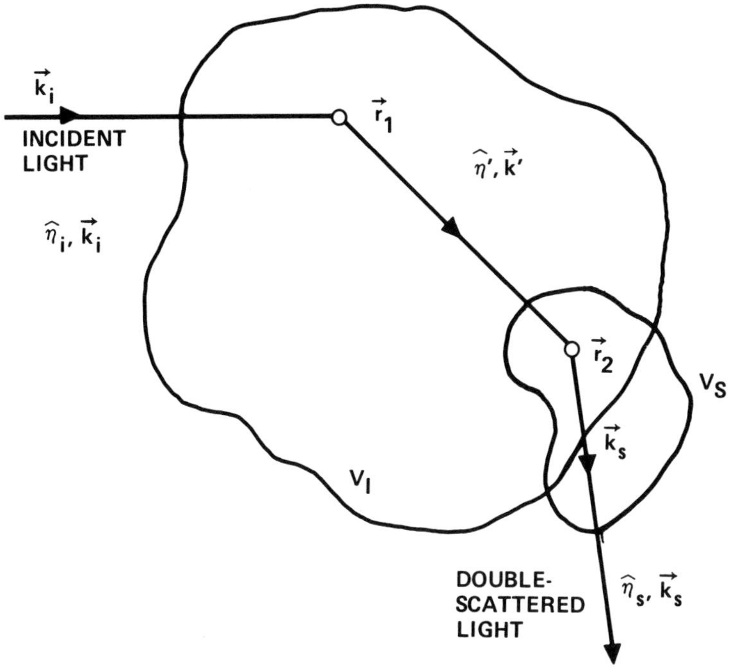

FIGURE 3B

and similarly for the final (second) polarizations at $\underset{\sim}{r}_2$ and $\underset{\sim}{r}_4$. The $\underset{\sim}{r}_1$-$\underset{\sim}{r}_2$ and $\underset{\sim}{r}_3$-$\underset{\sim}{r}_4$ separation can in principle be small (i.e., $\lesssim\xi$) but these configurations are much less likely (by at least the factor of $\xi/V_S{}^{1/3}$ mentioned earlier) than larger separation ones. Thus the two pairs 1-2 and 3-4 are essentially uncorrelated, and the double scattering event can be regarded as successive -- <u>and independent</u> -- single scatterings.

Reith and Swinney have exploited this conclusion by calculating the depolarized intensity in terms of the familiar single scattering cross sections. [Recall from our discussion (in the first lecture) of Stephen's second-order Raman theory[20] that the collision-induced depolarized intensity has also been approximated via a factorization of the four-point correlation function -- for the evaluation of collision induced (as opposed to double) light scattering cross sections, however, this simplification is far less satisfactory.] Consider the geometry for successive and independent single scatterings shown schematically in Fig. 3(b). The incident light is polarized along $\hat{\eta}_i$ and propagates with wavevector $\underset{\sim}{k}_i$, illuminating the volume V_I. The light scattered at the point $\underset{\sim}{r}_1$ in V_I, with polarization $\hat{\eta}'$ and wave-vector $\underset{\sim}{k}'$, is scattered again at $\underset{\sim}{r}_2$ with $\hat{\eta}_s$ and $\underset{\sim}{k}_s$. The direction of the polarization unit vectors $\hat{\eta}'$ and $\hat{\eta}_s$ are determined by the classical electrodynamics describing the "far" field produced by a dipole:[15]

$$\hat{\eta}' = [\hat{\eta}_i - \hat{k}'(\hat{\eta}_i\cdot\hat{k}')]/[1 - (\hat{\eta}_i\cdot\hat{k}')^2]^{1/2} \tag{22}$$

$$\hat{\eta}_s = \frac{\hat{\eta}_i - \hat{k}'(\hat{\eta}_i\cdot\hat{k}') + \hat{k}_s[(\hat{\eta}_i\cdot\hat{k}')(\hat{k}_s\cdot\hat{k}') - (\hat{k}_s\cdot\hat{\eta}_i)]}{\{1 - (\hat{\eta}_i\cdot\hat{k}')^2 - [(\hat{\eta}_i\cdot\hat{k}_s) - (\hat{k}_s\cdot\hat{k}')(\hat{k}'\cdot\hat{\eta}_i)]^2\}^{1/2}} \tag{23}$$

In general, the intensity of light singly scattered with $\hat{\eta}'$ and $\underset{\sim}{k}'$ from a volume V is given by Eq. (19) [we treated there the $\hat{\eta}_i\cdot\hat{k}' = 1$ case]

$$I(\hat{\eta}_i,\underset{\sim}{k}_i,\underset{\sim}{k}') = \alpha^2 k_0^4 \frac{I_i V}{R^2}\, |\hat{\eta}_i x \underset{\sim}{k}'|^2 S(|\underset{\sim}{k}_i - \underset{\sim}{k}'|) \tag{24}$$

$$\xrightarrow[\text{critical point}]{|\underset{\sim}{k}_i-\underset{\sim}{k}|\xi < 0.3} \frac{\alpha^2 k_0^4 I_i V}{R^2}\, \rho^2\, 4\pi\xi_0\xi^2 |\hat{\eta}_i x \underset{\sim}{k}'|^2 \equiv \frac{I_i V}{R^2}\,\sigma \tag{25}$$

where $I_i = |E_0|^2$ from before and $\hat{\eta}'$ is given by Eq. (22)

$$\sigma(\hat{\eta}_i,\underset{\sim}{k}') = 4\pi|\hat{\eta}_i x \underset{\sim}{k}'|^2\, \alpha^2 k_0^4\, \rho^2\xi_0\xi^2 \equiv |\hat{\eta}_i x \underset{\sim}{k}'|^2 \sigma_0 \tag{26}$$

is the familiar "Rayleigh ratio", the fraction of I_i scattered per unit volume into unit solid angle at $\hat{k}'$. Thus, referring to Fig. 3(b) we can write

$$dI(\underset{\sim}{r}_2) = \frac{I_i(\underset{\sim}{r}_1)d\underset{\sim}{r}_1}{|\underset{\sim}{r}_1-\underset{\sim}{r}_2|^2}\,\sigma(\hat{\eta}_i,\underset{\sim}{k}') = \begin{array}{l}\text{intensity incident on } d\underset{\sim}{r}_2\\ \text{at } \underset{\sim}{r}_2\text{, due to scattering}\\ \text{by volume } d\underset{\sim}{r}_1 \text{ at } \underset{\sim}{r}_1\end{array} \tag{27}$$

and

$$d^2I(\underset{\sim}{R}) = \frac{dI(\underset{\sim}{r}_2)d\underset{\sim}{r}_2}{R^2}\,\sigma(\hat{\eta}',\underset{\sim}{k}_s) = \begin{array}{l}\text{intensity at detector } (\underset{\sim}{R})\\ \text{due to scattering of } dI(\underset{\sim}{r}_2)\\ \text{by volume } d\underset{\sim}{r}_2 \text{ at } \underset{\sim}{r}_2\end{array} \tag{28}$$

Then

$$I_{DS}^{zx} = \int_{V_I} d\underset{\sim}{r}_1 \int_{V_S} d\underset{\sim}{r}_2\, d^2I(\underset{\sim}{R}) = \frac{1}{R^2}\int_{V_I} d\underset{\sim}{r}_1 \int_{V_S} d\underset{\sim}{r}_2\, \frac{I_i(\underset{\sim}{r}_1)}{|\underset{\sim}{r}_1-\underset{\sim}{r}_2|^2}\,\sigma(\hat{\eta}',\underset{\sim}{k}_s)\sigma(\hat{\eta}_i,\underset{\sim}{k}') \tag{29}$$

Substituting from Eqs. (22), (23) and (26), and specializing to the case of 90°-scattering, we obtain [with σ_0 defined by Eq. (26)]

$$I_{DS}^{zx}(90°) = \frac{\sigma_0{}^2}{R^2} \int_{V_I} d\underset{\sim}{r}_1 \int_{V_S} d\underset{\sim}{r}_2 \; I_i(\underset{\sim}{r}_1) \frac{x_{21}^2 z_{21}^2}{r_{21}^6} \tag{30}$$

Dividing by the single scattering (SS) polarized intensity

$$I_{SS}^{zz}(90°) = \frac{\sigma_0}{R^2} \int_{V_{IS}} d\underset{\sim}{r} I_i(\underset{\sim}{r}) \tag{31}$$

(where V_{IS} is the illuminated volume seen by the detector), the depolarization ratio becomes

$$\Delta = \sigma_0 \frac{\int_{V_I} d\underset{\sim}{r}_1 \int_{V_S} d\underset{\sim}{r}_2 I_i(\underset{\sim}{r}_1) \frac{x_{21}^2 z_{21}^2}{r_{21}^6}}{\int_{V_{IS}} d\underset{\sim}{r} \; I_i(\underset{\sim}{r}_1)} \tag{32}$$

Recalling Fig. 2, we see that V_I is a small cylindrical volume that pierces the center of the thin rectangular volume (V_S) seen by the detector. The dimensions of V_S are L×W×S, while the total sample-cell dimensions are L×L×H. The convergence of the focused beam is sufficiently small so that the beam diameter d_I can be taken as a constant throughout its length L in the sample. Hence the laser beam, which has a Gaussian profile, has an intensity in the sample given by

$$I_i(\underset{\sim}{r}) = I_0 \exp[-2(y^2 + z^2)/d_I^2] \tag{33}$$

Then, using $d_I \ll (V_S)^{1/3}$ (which is in fact the case here since $d_I \approx 0.21$mm and $L \approx 6.0$ mm), substitution of Eq. (33) into Eq. (32) yields

$$\Delta = \sigma_0 g S \tag{34}$$

where the dimensionless factor g, given by

$$g \equiv \frac{1}{SW} \int_{-L/2}^{+L/2} dx_1 \int_{-W/2}^{+W/2} dx_2 \int_{-L/2}^{+L/2} dy_2 \int_{-S/2}^{+S/2} dz_2 \; \frac{x_{21}^2 \; z_{21}^2}{r_{21}^6} \tag{35}$$

depends only on the V_S-geometry and is independent of the sample.

The integral in Eq. (35) has been evaluated numerically by Reith and Swinney with the result that g varies only slowly over the range of sample heights used [$0.261 \text{ mm} \leq S \leq 1.490$ mm]: $g = 0.77 \pm 0.01$. In the limit $L \to \infty$ and $W \to 0$, the integral can be evaluated analytically, giving $g = \pi/4 = 0.78$. Bray and Chang[14] have shown further that $g = \pi/4 - 1/6(1+\pi/2)\varepsilon^2 + \sigma(\varepsilon^4)$ where $\varepsilon \equiv S/L<<1$. Thus for S/L as big as one-tenth, the S-dependence of g is at most one percent, thereby establishing the linear dependence of Δ on S [cf. Eq. (34)].

Recalling now the form of our earlier estimate [cf. Eq. (17)] of the collision-induced contribution to I^{zx}, i.e.,

$$I^{zx}_{CI} = \frac{k_0{}^4 I_0 V_{IS}}{R^2}\, \alpha^4 \rho_c{}^2 C \tag{36}$$

and writing the polarized intensity as [cf. Eqs. (24)-(25)]

$$I^{zz}_{SS} = \frac{k_0{}^4 I_0 V_{IS}}{R^2}\, \alpha^2 \rho_c{}^2 4\pi\xi_0 \xi^2 \tag{37}$$

we have for the total Δ:

$$\Delta \equiv \frac{I^{zx}_{CI}}{I^{zz}_{SS}} + \sigma_0 g S = \frac{C'}{\xi^2} + \sigma_0 g S \tag{38}$$

where

$$C' \equiv \alpha^2 C/4\pi\xi_0 \tag{39}$$

is essentially independent of temperature and k_0. [C was estimated earlier to be 6×10^{22} cm^{-3}, but we shall treat it henceforth as an experimentally determined quantity.] Note that the choice $g=(16/15)(\pi/4)$ makes Eq. (38) reduce exactly to the Eq. (20) derived by Oxtoby and Gelbart for the special case of spherical $V_S>>V_I$.

Depolarization ratio data obtained for xenon at the critical density near the critical point are shown in Fig. 4 for four different values of S, the height of the sample seen by the detector. In agreement with the prediction of Oxtoby and Gelbart, as the critical point is approached, there is a transition in the depolarization ratio behavior from a regime where collision-induced depolarization is dominant to a regime where depolarization due to double scattering is dominant. The qualitative agreement between these data and the predictions of Oxtoby and Gelbart is particularly striking if Fig. 4 is compared with Fig. 1. It is clear from Fig. 4 that at $T-T_c = 10°C$, where the depolarization ratio is essentially independent of sample height, that double scattering is not important compared to the scattering from anisotropic collision clusters. As the critical point is neared, the depolarization ratio decreases because of the rapid increase in the polarized scattering intensity (critical opalescence). At $T-T_c$ = 2°C the depolarized intensity is larger for the larger sample heights

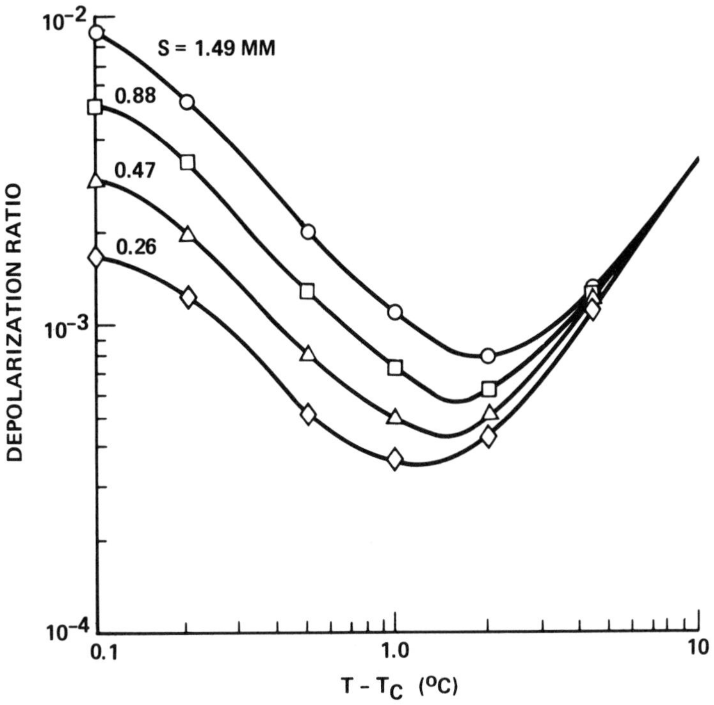

FIGURE 4

indicating that multiple scattering is already contributing significantly to the depolarization. Closer to the critical point the depolarization ratio begins to increase; here the behavior of Δ is expected to be given by the second term in Eqs. (20) and (38). We now make a quantitative comparison between these experimental results and the theoretical predictions.

The depolarization ratio measured for three different wavelengths at $T-T_c = 0.496°C$ is shown in Fig. 5. The linear dependence of the ratio on the height S, predicted by Eq. (38), is obeyed by these data within the experimental uncertainty. Moreover, within the experimental uncertainty the slopes of the curves are proportional to $(k_0^4)\lambda_0^{-4}$, as predicted. The S=0 intercepts of the three curves in Fig 5 are approximately equal, as predicted by Eq. (38), but the difference between the intercepts is slightly greater than the experimental uncertainty (which includes an uncertainty of 0.03 mm in S). The difference in intercepts is probably due to some strain-induced birefringence in the glass sample cell wall.

Depolarization data obatined for several temperatures at fixed wavelength are shown in Fig. 6. Again the measured ratios vary

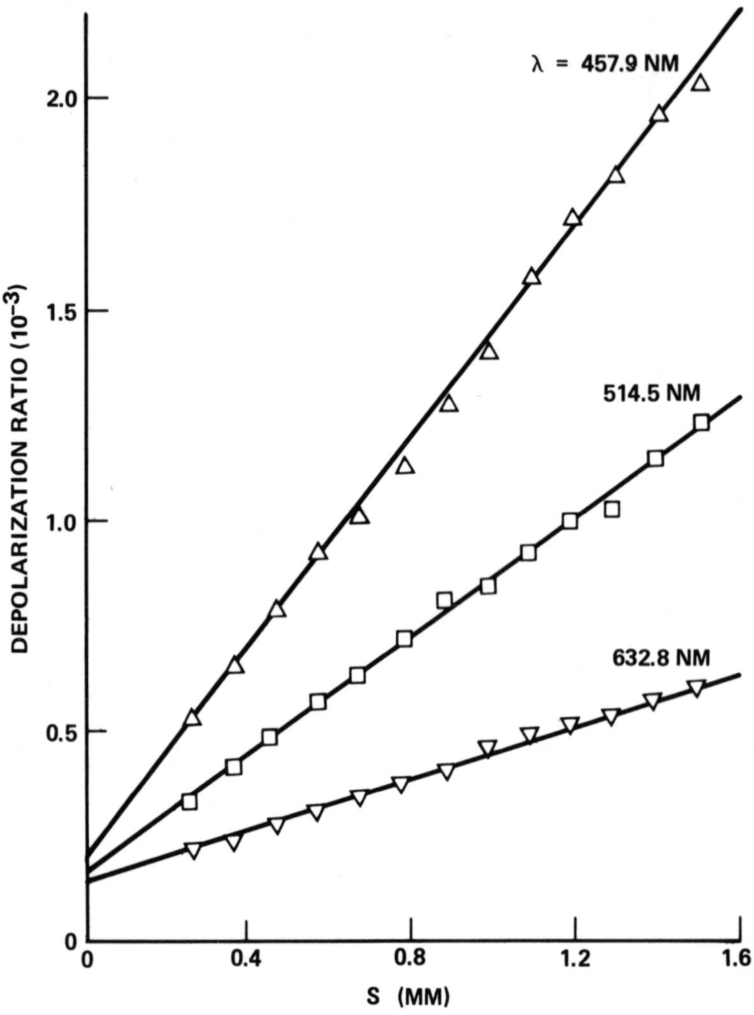
λ = 457.9 NM
514.5 NM
632.8 NM
DEPOLARIZATION RATIO (10^{-3})
2.0
1.5
1.0
0.5
0
0
0.4
0.8
1.2
1.6
S (MM)

FIGURE 5

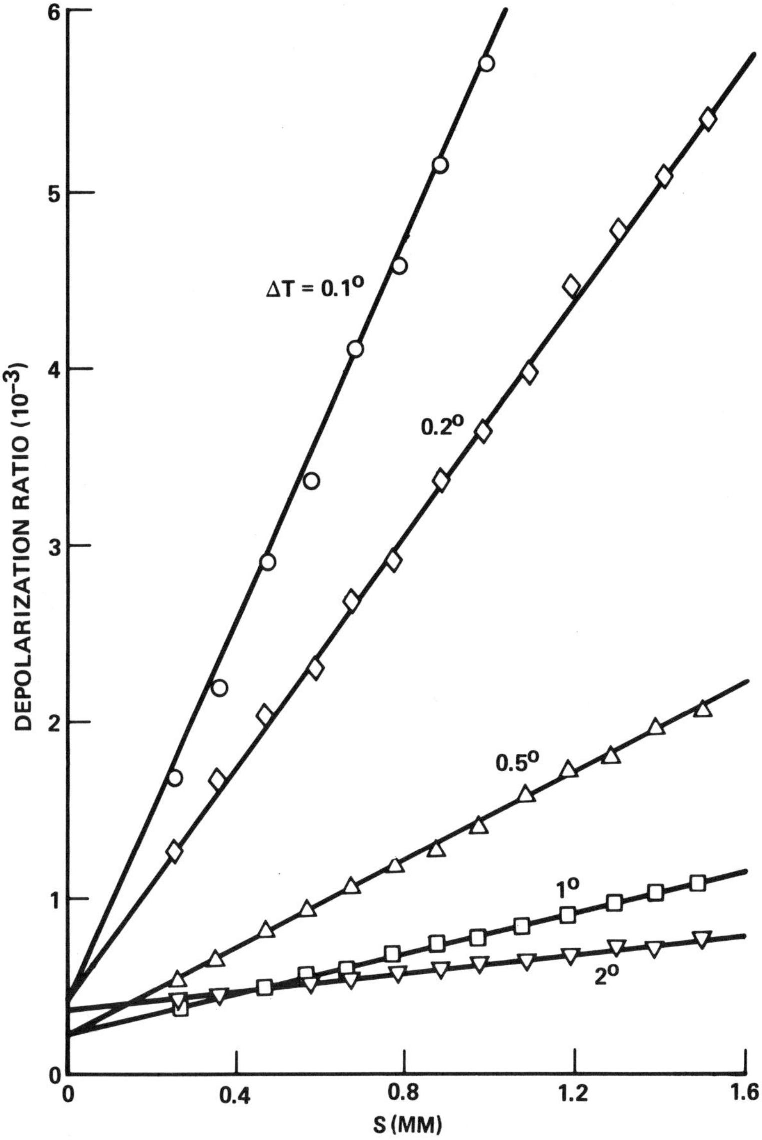

6
5
4
3
2
1
0
DEPOLARIZATION RATIO (10^{-3})
ΔT = 0.1°
0.2°
0.5°
1°
2°
0
0.4
0.8
1.2
1.6
S (MM)

FIGURE 6

linearly with S within the experimental uncertainty. The rapid increase in the slopes of the curves as the critical point is approached is clearly illustrated by the figure. Since all quantities in the theoretical expression [cf. Eq. (38)] for the slope are known from independent measurements and calculations, not only the temperature dependence -- but also the actual magnitude -- of the measured slopes can be compared with theory. The theoretical slope is given by[3]

$$g\sigma_0 = (0.77)\sigma_0 = (8.7\times10^{-6})(1+\frac{\Delta T}{T_c})(\frac{\Delta T}{T_c})^{-1.21}\ \text{cm}^{-1} \qquad (40)$$

The observed temperature dependence of the slopes are found to agree well with this expression, but the magnitudes are discrepant by about 30% -- almost certainly due to the uncertainties in the data which enter in Eq. (40) for σ_0.

From the above discussion it should be clear that the phenomenon of double scattering is qualitatively understood in the case of light depolarization by critical fluids. In particular it has been established that the measurement of $\Delta \equiv I^{zx}/I^{zz}$ near the critical point does not provide any information about connected (irreducible) 4-point correlations. Instead the double scattering has been seen to involve successive and independent single scattering events. Thus the depolarization ratio is linear in σ_0 [cf. Eq. (38)], the differential cross section ("Rayleigh ratio") for single scattering. σ_0, recall, is proportional -- through a factor of $\alpha^2 k_0^4 I_0 V_{IS}/R^2$ [cf. Eq. (24)] -- to the all important static structure factor S(k) which, in turn, is proportional to the isothermal compressibility, etc. Since

$$I_{SS}^{\text{polarized}} = \frac{I_i V_{IS}}{R^2}\,\sigma_0$$

σ_0 has commonly been obtained from absolute intensity determinations of $I_{SS}^{\text{polarized}}$. But, as is well appreciated by experimentalists, the measurement of absolute scattering intensities is extremely difficult.[21] Thus Reith and Swinney[3,22] have suggested that σ_0 be determined instead from the ratio of double- to single-scattering intensities measured in critical fluids or in solutions of macromolecules whose dimensions are not small compared with the wavelength of light. Since ratios of scattering intensities can almost always be measured more accurately and easily than absolute cross sections, this suggestion should provide a valuable new method for learning about σ_0.

Throughout the discussion in this lecture we have neglected the double scattering contributions to the polarized intensity. This is valid as long as the double scattering contributions are indeed negligible compared to those from single scattering. [In the case of the depolarized intensity, of course, the single scattering contributions vanish identically and the double scattering appears as the leading term.] Over the temperature range considered earlier,

$\Delta \leqslant 0.01$ and this means that I^{zz}_{DS} ($\approx I^{zx}_{DS}$)$<<I^{zz}_{SS}$, so that approximating I^{zz} by I^{zz}_{SS} introduces less than one percent error in Δ. Nevertheless, even though $I^{zz}_{DS}(90°)$ amounts to such a small fraction of $I^{zz}_{total}(90°)$, the angular distribution of I^{zz}_{total} can be significantly distorted by double scattering contributions. This possibility was first suggested by Oxtoby and Gelbart,[2] who showed that small double scattering contributions account for easily observable downward curvature in the usual Ornstein-Zernicke-Debye plots of $1/I^{zz}_{total}$ vs. $\sin^2(\theta_s/2)$. (Here θ_s is the scattering angle.) Neglect of this effect leads to spurious inferences about the pair (2-point) correlation function for critical fluids. We shall return to this point in the next section, but only after first discussing the rigorous theory of multiple light scattering due to Boots, Bedeaux and Mazur.[4-6]

IV. RIGOROUS MULTIPLE SCATTERING THEORY

Recall that throughout the above discussion -- and in the first lecture as well -- we have been careless about factors of n (the refractive index) and $(\varepsilon+2)/3$, etc. Thus we have neglected, for example, differences between $\underset{\sim}{T}_{k_0}$ -- defined by Eq. (2) -- and the properly renormalized propagator mentioned earlier [see references 4-6, 8-10 from the first lecture]. Being "quick" and "dirty" in this way has allowed us to expose most simply the mechanisms of second-order light scattering processes. All calculated cross sections, however, will be off by "local field" correction factors, e.g., powers of n and $(\varepsilon+2)/3$, etc. Also, attenuation of the incoming and outgoing light has been neglected. Similarly, we did not attempt any systematic treatment of the triple- and higher-order multiple light scattering intensities. These important improvements are contained in the work of Boots, Bedeaux and Mazur.[4-6]

Boots, et al. combine the Maxwell equation in a fluctuating medium,

$$[\underset{\sim}{k}\underset{\sim}{k} - k^2 + \omega^2] \cdot \underset{\sim}{e}(\underset{\sim}{k},\omega) = -\omega^2 \underset{\sim}{p}(\underset{\sim}{k},\omega) \tag{41}$$

with the constructive relation

$$\underset{\sim}{p}(\underset{\sim}{r},t) = (\underset{\approx}{\boldsymbol{\varepsilon}}_b - 1) \cdot \underset{\sim}{e} \tag{42}$$

to obtain the wave equation for the $\underset{\sim}{k},\omega$ Fourier component of the fluctuating electric field $\underset{\sim}{e}$:

$$[\underset{\sim}{k}\underset{\sim}{k} - k^2 + \omega^2 \underset{\approx}{\boldsymbol{\varepsilon}}] \cdot \underset{\sim}{e} = -\omega^2[(\underset{\approx}{\boldsymbol{\varepsilon}}_b - \underset{\approx}{\boldsymbol{\varepsilon}}) \cdot \underset{\sim}{e}] \equiv -\omega^2 \Delta \underset{\sim}{p} \tag{43}$$

Here $\underset{\approx}{\boldsymbol{\varepsilon}}$ is the macroscopic dielectric tensor defined by

$$\langle \underset{\sim}{p} \rangle \equiv (\underset{\approx}{\boldsymbol{\varepsilon}} - 1) \cdot \langle \underset{\sim}{e} \rangle \tag{44}$$

where $\underset{\sim}{p}$ is the fluctuating polarization density and the brackets denote as usual a canonical ensemble average. Note that $\underset{\approx}{\varepsilon}$ is not equal to $<\underset{\approx}{\varepsilon}_b>$ where $\underset{\approx}{\varepsilon}_b$ is the fluctuating dielectric tensor defined by Eq. (42). Solving for the scattered field [$\underset{\sim}{e}_s \equiv \underset{\sim}{e} - <\underset{\sim}{e}>$] at the detector, the intensity of the scattered light with polarization $\hat{u}$ and frequency ω is found to be

$$I_{\hat{u}}(\underset{\sim}{R},\omega)2\pi\delta(\omega-\omega') \equiv \hat{u} \cdot <\underset{\sim}{e}_s(\underset{\sim}{R},\omega)\underset{\sim}{e}_s^*(\underset{\sim}{R},\omega')> \cdot \hat{u} \tag{45}$$

$$= \frac{\omega^4}{(4\pi R)^2}\int d\underset{\sim}{r}'\int d\underset{\sim}{r}''\ \hat{u} \cdot \underset{\approx}{M}(\underset{\sim}{r}',\omega/\underset{\sim}{r}'',\omega') \cdot \hat{u}\ \exp[-\alpha(\omega)d_s]\ \exp[i\underset{\sim}{k}_s \cdot (\underset{\sim}{r}''-\underset{\sim}{r}')]$$

where $\underset{\approx}{M}$ is the autocorrelation function involving the polarization fluctuations defined by Eq. (43):

$$\underset{\approx}{M}(\underset{\sim}{r},\omega/\underset{\sim}{r}',\omega') \equiv <\Delta\underset{\sim}{p}(\underset{\sim}{r},\omega)\Delta\underset{\sim}{p}^*(\underset{\sim}{r}',\omega)> \tag{46}$$

d_s is the average of the distances from $\underset{\sim}{r}'$ and $\underset{\sim}{r}''$ to the surface of the sample and $\alpha(\omega)$ is the extinction coefficient, i.e., the factor $\exp[-\alpha(\omega)\ d_s]$ describes the attenuation of the light in the sample.

$$\alpha(\omega) \equiv 2\omega\ \mathrm{Im}\ n(\omega) \tag{47}$$

where $n(\omega)$ is the complex refractive index defined by

$$n^2(\omega) \equiv \varepsilon_T(\frac{\omega n}{c},\omega) \tag{48}$$

with the transverse dielectric constant defined in turn by (since the system is isotropic)

$$\underset{\approx}{\varepsilon}(\underset{\sim}{k},\omega) = \varepsilon_T(\underset{\sim}{k},\omega)[\underset{\approx}{I} - \hat{k}\hat{k}] + \underset{\approx}{\varepsilon}_L(\underset{\sim}{k},\omega)\hat{k}\hat{k} \tag{49}$$

Finally, the propagation wavevector for the scattered light is given by $\underset{\sim}{k}_s \equiv \omega \mathrm{Re}\, n(\omega)\hat{R}/c$.

Now the $\Delta\underset{\sim}{p}$-correlation function appearing in Eq. (45) must be expressed in terms of correlation functions involving $\Delta\rho$, where

$$\Delta\rho \equiv \sum_i \delta[\underset{\sim}{r}-\underset{\sim}{R}_i(t)] - \frac{N}{V} \tag{50}$$

is the density fluctuation discussed earlier. This expansion of $\underset{\approx}{M}$ in powers of $\Delta\rho$ is carried out by Boots et al.[4] who show that

$$\underset{\approx}{M} = \underset{\approx}{M}_{1,1} + \underset{\approx}{M}_{1,2} + \underset{\approx}{M}_{1,3} + \underset{\approx}{M}_{2,2} + \mathcal{O}(\Delta\rho^5) \tag{51}$$

Here the sum of the subscripts on each $\underset{\approx}{M}_{ij}$ corresponds to the order of $\Delta\rho$. More explicitly they find

$$\underset{\approx}{M}_{1,1} = \frac{\alpha_0^2}{3^4} \langle [(\underset{\approx}{\boldsymbol{\varepsilon}}+2)\cdot\Delta\rho(\underset{\approx}{\boldsymbol{\varepsilon}}+2)\cdot\langle \underset{\sim}{e}\rangle][(\underset{\approx}{\boldsymbol{\varepsilon}}+2)\cdot\Delta\rho(\underset{\approx}{\boldsymbol{\varepsilon}}+2)\cdot\langle \underset{\sim}{e}\rangle]^*\rangle \tag{52}$$

and

$$\underset{\approx}{M}_{1,2} = -\frac{\alpha_0^3}{3^4}(\boldsymbol{\varepsilon}_0+2)^4\langle[\Delta\rho\langle\underset{\sim}{e}\rangle][\Delta\rho\underset{\approx}{K}\cdot\Delta\rho\langle\underset{\sim}{e}\rangle]^*\rangle + \begin{matrix}\text{hermitian}\\ \text{conjugate}\end{matrix} \tag{53}$$

Here

$$\boldsymbol{\varepsilon}_0 = 1 + \frac{\alpha_0\rho}{1 - \frac{\alpha_0\rho}{3}} \tag{54}$$

and

$$\underset{\approx}{K} = \frac{1}{9}(\underset{\approx}{\boldsymbol{\varepsilon}}+2)\cdot\{\underset{\approx}{F}_{\boldsymbol{\varepsilon}} - \frac{1}{\underset{\approx}{\boldsymbol{\varepsilon}}+2}\}\cdot(\underset{\approx}{\boldsymbol{\varepsilon}}+2) \tag{55}$$

is the renormalized dipole tensor, e.g., in the low density limit ($\underset{\approx}{\boldsymbol{\varepsilon}} \to 1$)

$$\underset{\approx}{K} \to \underset{\approx}{F}_1 - \frac{1}{3} \equiv \underset{\approx}{T} \tag{56}$$

where $\underset{\approx}{T}$ is the familiar propagator arising in low density scattering theory:

$$\underset{\approx}{F}_1 \equiv \lim_{\underset{\approx}{\boldsymbol{\varepsilon}}\to 1}\{\underset{\approx}{F}_{\boldsymbol{\varepsilon}}\} = \lim_{\underset{\approx}{\boldsymbol{\varepsilon}}\to 1}\left\{\frac{\omega^2}{[\underset{\sim}{k}\underset{\sim}{k} - k^2 + (\omega^+ io^+)^2\underset{\approx}{\boldsymbol{\varepsilon}}(\underset{\sim}{k},\omega)]}\right\} \tag{57}$$

($\underset{\approx}{F}_{\boldsymbol{\varepsilon}}$ is the retarded wave propagator in the medium.)

$$\underset{\approx}{M}_{1,3} =$$

$$\frac{\alpha_0^4(\boldsymbol{\varepsilon}_0+2)^4}{3^4}\langle[\Delta\rho\langle\underset{\sim}{e}\rangle][\{\Delta\rho\underset{\approx}{K}\cdot\Delta\rho\underset{\approx}{K}\cdot\Delta\rho - \Delta\rho\underset{\approx}{K}\cdot\langle\Delta\rho\underset{\approx}{K}\Delta\rho\rangle - \langle\Delta\rho\underset{\approx}{K}\cdot\Delta\rho\rangle\underset{\approx}{K}\cdot\Delta\rho\}\langle\underset{\sim}{e}\rangle]^*\rangle$$

$$+ \text{ hermitian conjugate} \tag{58}$$

and

$$\underset{\approx}{M}_{2,2} =$$

$$\frac{\alpha_0^4(\boldsymbol{\varepsilon}_0+2)^4}{3^4}\langle[\{\Delta\rho\underset{\approx}{K}\cdot\Delta\rho - \langle\Delta\rho\underset{\approx}{K}\cdot\Delta\rho\rangle\}\langle\underset{\sim}{e}\rangle][\{\Delta\rho\underset{\approx}{K}\cdot\Delta\rho - \langle\Delta\rho\underset{\approx}{K}\cdot\Delta\rho\rangle\}\langle\underset{\sim}{e}\rangle]^*\rangle \tag{59}$$

When $\underset{\approx}{M}_{1,1}$ is substituted into Eq. (45) a single scattering intensity is obtained which is identical to Einstein's[23] except for a minor correction describing attenuation in the sample. This difference arises from the fact that, before and after being scattered, the light passes through a fluctuating medium described by $\underset{\approx}{\boldsymbol{\varepsilon}}$ rather than through an average density ($\boldsymbol{\varepsilon}_0$) sample, i.e., $\underset{\approx}{\boldsymbol{\varepsilon}} \to \boldsymbol{\varepsilon}_0$ rids of this difference.

To evaluate the higher order $\underset{\approx}{M}_{i,j}$'s the "Gaussian approximation"[4] is introduced:

$$\langle \Delta\rho(\underset{\sim}{k}_1,\omega_1)\Delta\rho(\underset{\sim}{k}_2,\omega_2)\Delta\rho(\underset{\sim}{k}_3,\omega_3)\rangle = 0 \tag{60}$$

$$\begin{aligned} &\langle \Delta\rho(1)\Delta\rho(2)\Delta\rho(3)\Delta\rho(4)\rangle = \\ &\langle \Delta\rho(1)\Delta\rho(2)\rangle\langle \Delta\rho(3)\Delta\rho(4)\rangle + \langle 13\rangle\langle 24\rangle + \langle 14\rangle\langle 23\rangle \end{aligned} \tag{61}$$

This is equivalent to neglecting the connected contributions to the 3- and 4-point correlation functions as we did earlier, following Eq. (3). In this approximation

$$\underset{\approx}{M}_{1,2} = 0 \tag{62}$$

and

$$I^{1,3}_{\hat{u}}(\underset{\sim}{R},\omega) = \frac{\omega^4(\boldsymbol{\varepsilon}_0-1)^4}{(4\pi R)^2}\cos\theta I_0 V^{1,3}_S S(\underset{\sim}{k}_S-\underset{\sim}{k}_0,\omega-\omega_0) + \begin{matrix}\text{hermitian}\\ \text{conjugate}\end{matrix} \tag{63}$$

where $\cos\theta = \hat{u}\cdot E_0$, $\underset{\sim}{E}_0(\underset{\sim}{r},t) = \underset{\sim}{E}_0\exp(i\underset{\sim}{k}_0\cdot\underset{\sim}{r})\exp(-i\omega_0 t)$ is the incident field, and $V^{1,3}_S$ is a factor with the dimension of a volume.[6] $I^{1,3}$ is the intensity associated with interference between singly and triply scattered waves. Because of the $\cos\theta \equiv \hat{u}\cdot\hat{E}_0$ factor -- note in fact that $I^{1,3}$ is directly proportional to the single scattering intensity I^{pol}_{SS} -- it cannot contribute to depolarized scattering. Even though the single-triple (1,3) terms are the same order in $\alpha_0\Delta\rho$ as the double scattering contributions (2,2), it can be shown that they are negligible in comparison. This is because in double scattering the first scattering takes place anywhere in the illuminated volume V_I and the second anywhere in the observed volume V_S. In the single-triple case, on the other hand, all the scatterings must take place in V_{IS} (the illuminated volume seen by the detector: $V_{IS} = V_I \times V_S$), just as in the single scattering case.

In the Gaussian approximation, the double scattering intensity reduces to [note relation to Eq. (I.55) from first lecture]

$$R^2 I^{2,2}(\hat{R},\hat{u},\omega) = \frac{I_0}{2\pi}\left[\frac{\omega_0{}^2\alpha_0\rho_0|n^2+2|^2}{36\pi}\right]^4 \int_{-\infty}^{+\infty} d\omega' \int_{V_I} d\underset{\sim}{r}_1 \int_{V_S} d\underset{\sim}{r}_2$$

$$\{\exp[-\alpha(d_0 + d_{21} + d_s)]/d_{21}^2\} S(\underset{\sim}{k}-k_0\ \hat{\Omega}_{21},\omega-\omega') \tag{64}$$

$$S(k_0\hat{\Omega}_{21} - \underset{\sim}{k}_0,\omega'-\omega_0)[\hat{u}\cdot(1 - \hat{\Omega}_{21}\hat{\Omega}_{21})\cdot\hat{u}_0]^2$$

where

$$\hat{\Omega}_{21} \equiv (\underset{\sim}{r}_2-\underset{\sim}{r}_1)/|\underset{\sim}{r}_2-\underset{\sim}{r}_1| \equiv \frac{(\underset{\sim}{r}_2-\underset{\sim}{r}_1)}{d_{21}} \tag{65}$$

and $\underset{\sim}{k}$ and k_0 are as usual given by $(\omega_0 \mathrm{Re}n(\omega)/c)\hat{R}$ and $(\omega_0 \mathrm{Re}n(\omega)/c)\hat{k}_0$. d_0 ($\tilde{d}_s$) is the distance between $\underset{\sim}{r}_1$ ($\underset{\sim}{r}_2$) and the point where the incident (scattered) light enters (leaves) the sample. Integrating Eq. (64) over ω, and dividing the result by

$$R^2 I^{pol} \simeq I_0\left[\frac{\omega_0{}^2\alpha_0\rho_0|n^2+2|^2}{36\pi}\right]^2 S(\underset{\sim}{k}-\underset{\sim}{k}_0)(\hat{u}\cdot\hat{u}_0)^2 \int_{V_S} d\underset{\sim}{r}\ \exp[-\alpha(d_0+d_s)] \tag{66}$$

gives the depolarization ratio Δ. Using the Clausius-Mossotti equation

$$\alpha_0\rho_0 = \frac{3(n^2-1)}{n^2+1} \tag{67}$$

and assuming that the vertical diameter of V_I is sufficiently small so that both $I^{1,1}$ and $I^{1,2}$ are linear in this dimension, Δ assumes the form[6]

$$\Delta = \left[\frac{\omega_0{}^2(\varepsilon-1)(\varepsilon+2)}{12\pi}\right]^2 \left\{\frac{S(\sqrt{2}k_0)4\sinh(\alpha x_S)\sinh(\alpha y_I)}{\alpha^2}\right\}^{-1}$$

$$\int_{-d/2}^{+d/2} dz \int_{-X_S}^{X_S} dx_2 \int_{-Y_I}^{Y_I} dy_1 \int_{-X_I}^{X_I} dx_1 \int_{-Y_S}^{Y_S} dy_2\ \frac{z^2}{d_{21}^6}(x_1-x_2)^2 S(k_0\{2[1-\frac{y_1-y_2}{d_{21}}]\}^{1/2}) \tag{68}$$

$$S(k_0\{2[1+\frac{x_1-x_2}{d_{21}}]\}^{1/2})\exp[-\alpha(x_1+y_2+d_{21})]$$

Here $d_{21} = [(x_1-x_2)^2 + (y_1-y_2) + z^2]^{1/2}$ and d is the vertical diameter of the observed volume. $\pm X_S$, $\pm Y_S$ and $\pm X_I$, $\pm Y_I$ are the positions of the boundaries of the observed and illuminated volumes, respectively. Here the coordinate system has been centered at the common origin of V_S and V_I, with the x-axis in the direction of the beam and the z-axis perpendicular to the scattering plane.

If the attenuation in the sample is sufficiently small (i.e., $1/\alpha$ is large compared to ℓ, a typical value of X_S and Y_S) then the depolarization ratio given by Eq. (68) satisfies[6]

$$\Delta = \Delta_0 d[1 + \mathcal{O}(d^2/\ell^2)] \quad , \quad \alpha\ell << 1 \tag{69}$$

Here [writing κ_T for the isothermal compressibility: $\kappa_T = 4\xi_0\xi^2/k_BT$][13]

$$\Delta_0 = \frac{1}{4}\,\pi\left[\frac{\omega_0{}^2(\varepsilon-1)(\varepsilon+2)}{12\pi}\right]^2 k_BT\kappa_T \tag{70}$$

$$= \frac{1}{4}\,\pi\,\sigma_0$$

where σ_0 is the "Rayleigh ratio" defined earlier [cf. Eq. (26)]. Thus the double scattering terms in Eqs. (20) and (38) are reproduced [$d \leftrightarrow S, 2R_S$]. Also, as before, the general result for Δ [cf. Eq. (68)] must be evaluated numerically in the case of particular experiments.

Boots et al. were concerned with explaining the depolarization data of Trappeniers, Michels and Huijser.[7] In this experiment, Δ was measured for CO_2 at the critical density over a temperature range between 0.02 and 0.7% above T_c. The incident beam propagates along the x-axis, with a rectangular cross section 4.5 mm along the horizontal ($\hat{y}$) direction and 0.35 mm along the vertical ($\hat{z}$); its polarization lies along the vertical direction. The light scattered along the y-axis is detected, the observable volume having the shape of a parallelepiped with variable vertical dimension (d). Using the appropriate experimental values for the $(X,Y)_{S,I}$ appearing in Eq. (68) and the relevant critical parameters for CO_2, Δ has been evaluated numerically by Boots et al.[6] Agreement with experiment over the range $0.02°C \lesssim \Delta T \lesssim 0.7°C$ is excellent, suggesting negligible contributions from: anisotropic collision clusters; triple- and higher- multiple scattering; and beam deflection by gravity-induced density gradients.

Finally we return to the point, mentioned earlier (end of Section III) concerning multiple light scattering contributions to I^{pol}. Assuming that $I^{pol} \approx I_{\text{single scattering}}$ we have [cf. Eq. (66), neglecting attenuation]

$$I^{pol} \approx I\left(\frac{n^2+2}{3}\right)^4 \frac{\omega_0{}^4\alpha_0{}^2\rho_0{}^2}{(4\pi R)^2}\, S(|\underset{\sim}{k}-\underset{\sim}{k}_0| \approx 2k_0\sin\frac{\theta_S}{2})$$

$$= \frac{I_0\left(\frac{n^2+2}{3}\right)^4 \frac{\omega_0{}^4\alpha_0{}^2\rho_0{}^2}{4\pi R^2}\,\xi_0\xi^2}{1 + [2k_0\sin\frac{\theta_S}{2}]^2\xi^2} \tag{72}$$

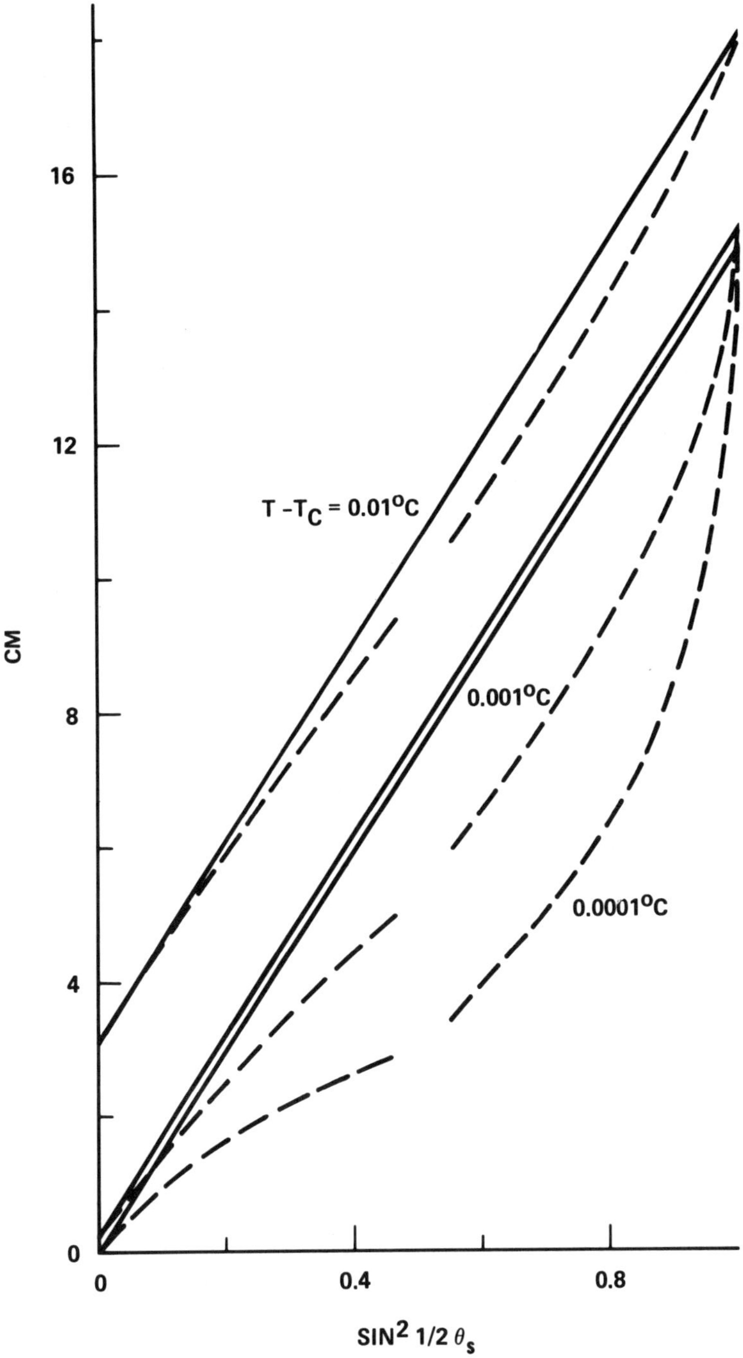

FIGURE 7

Here θ_s is the scattering angle and we have used the fact that $k \approx k_0$ to write $|\underset{\sim}{k}-\underset{\sim}{k}_0| \approx 2k_0\sin(\theta_s/2)$. Thus a plot of $1/I^{pol}$ vs. $\sin^2(\theta_s/2)$ should give a straight line whose slope gives the temperature dependence of ξ^2; this is the famous Ornstein-Zernicke-Debye (OZD) plot.[13] Eq. (72) for I^{pol} assumes, however, that: (1) only single scattering need be considered and (2) the 2-point density correlation function, whose Fourier transform gives S(k), behaves "classically",[13] i.e. it falls off at large separations as

$$\xi_0\, e^{-r/\xi}/r^{1+\eta}$$

where $\eta=0$. For η different from zero (as suggested, for example, by recent Ising model calculations[13]) the dependence of $1/I_{SS}$ on $\sin^2(\theta_s/2)$ is no longer linear and curvature at small θ_s is expected for the OZD plots of sufficiently critical systems. Correspondingly, measurements of this curvature can in principle provide a means for determining η. However, gravity-induced density gradients can also[24] give rise to deviations from Eq. (72). Furthermore, nonlinearity in the OZD plots is not observed until ΔT is so small that double scattering contributions to I^{pol} need to be considered. This was first done by Oxtoby and Gelbart, using the simple theory described in Section II. More recently Boots et al.[6] have applied their more rigorous formulation of multiple scattering to this particular problem.

Figure 7 shows theoretical plots of $1/(I_{SS}^{pol})$ (solid lines) and $1/(I_{SS}^{pol} + I_{DS}^{pol})$ (dashed curves) vs. $\sin^2(\theta_s/2)$ for the particular experimental geometry appropriate to a CO_2 study by White and Maccabee.[25] For $\Delta T = .001°C$, I_{DS} contributes as much as 30% of I^{pol} at certain angles; thus for $\Delta T < 0.001°C$, higher-order multiple scatterings need to be considered as well. Even at 0.01°C, though, the double scattering contributions have a nonnegligible effect on the OZD plots. The details of the deviations from linearity are found[6] to be very sensitive to the attentuation and geometry effects not included in the original calculations.[2]

Finally, I should mention that triple and higher-order multiple scatterings can be treated analogously to double scatterings, once it is recognized that successive single scatterings are independent. Using the Gaussian approximation mentioned earlier for the factorization of the many-body correlations, and a diagrammatic book-keeping device, Boots et al.[6] have succeeded in deriving a general expression for the nth-order multiple scattering intensity. Explicit evaluation of even the triple scattering cross section, however, is not yet feasible.

We have concentrated throughout this lecture on the case of double light scattering from simple fluids near their critical points. This is because the relevant theory and experiment have attracted so much recent interest. It is important to note, however, that the phenomenon of multiple light scattering is quite universal and is by no means restricted to critical fluids. As discussed in Metiu's seminar, for example, double (and higher-order) scattering effects are important in the study of spinodally decomposing systems, where dielectric fluctuations grow spontaneously. Also, multiple scattering is commonly observed for samples consisting of particles (e.g. macro-molecules, water droplets) whose dimensions are comparable with the wavelength of light.

REFERENCES

1. D. W. Oxtoby and W. M. Gelbart, J. Chem. Phys. 60 3359 (1974).

2. D. W. Oxtoby and W. M. Gelbart, Phys. Rev. A10 738 (1974).

3. L. A. Reith and H. L. Swinney, Phys. Rev. A12 1094 (1975).

4. H. M. J. Boots, D. Bedeaux and P. Mazur, Physica 79A 397 (1975).

5. H. M. J. Boots, D. Bedeaux and P. Mazur, Chem. Phys. Lett. 34 197 (1975).

6. H. M. J. Boots, D. Bedeaux and P. Mazur, Physica 84A 217 (1976).

7. N. J. Trappeniers, A.-C. Michels and R. H. Huijser, Chem. Phys. Lett. 34 192 (1975).

8. (a) D. Beysens, A. Bourgou and H. Charlin, Phys. Lett. 53A 236 (1975); (b) D. Beysens and G. Zalczer, Phys. Rev. A15 765 (1977).

9. Y. Garrabos, K. Tufeu and B. LeNeindre, C. R. Acad. Sci. (Paris) 282 313 (1976).

10. H. L. Frisch and J. McKenna, Phys. Rev. 139 A68 (1965).

11. R. D. Mountain, J. Phys. (Paris) 33 C4-265 (1972).

12. See, for example, (a) V. Korenman, Phys. Rev. A2 449 (1970); (b) J. Swift, Annals of Phys. 75 1 (1973).

13. H. E. Stanley, Introduction to Phase Transitions and Critical Phenomena (Oxford University Press, New York, 1971).

14. A. J. Bray and R. F. Chang, Phys. Rev. A12 2594 (1975).

15. J. D. Jackson, Classical Electrodynamics (Wiley, New York, 1966).

16. J. A. Barker, R. O. Watts, J. K. Lee, T. O. Schafer and Y. T. Lee, J. Chem. Phys. 61 3081 (1974).

17. M. Thibeau, B. Oksengorn and B. Vodar, J. Phys. (Paris) 29 287 (1968); and C. R. Acad. Sci. (Paris) 265 722 (1967).

18. A. B. Cornfield and H. Y. Carr, Phys. Rev. Lett. 29 28 (1972).

19. M. Giglio and G. B. Benedek, Phys. Rev. Lett. 20 1145 (1969).

20. M. J. Stephen, Phys. Rev. 187 279 (1969).

21. See, for example, the discussion in E. R. Pike, W. R. M. Pomeroy and J. M. Vaughan, J. Chem. Phys. 62 3188 (1975).

22. L. Reith and H. L. Swinney, Opt. Comm. 17 111 (1976).

23. A. Einstein, Ann. der Physik 33 1275 (1910).

24. O. Splittorff and B. N. Miller, Phys. Rev. A9 550 (1974).

25. J. A. White and B. S. Maccabee, Phys. Rev. A11 1706 (1975).

NUMERICAL CALCULATIONS IN CLASSICAL LIQUIDS

A. Rahman

Argonne National Laboratory

Argonne, Illinois 60439

Preliminary Considerations

We are concerned with a system of N particles with masses m_i, i=1, ..., N which interact with a potential V which is a continuous function of the coordinates of the particles namely $\underline{r}_i$, i=1, ..., N. The extension to include a system of rigid bodies as well will be mentioned later. The classical equations of motion of the N particles are

$$m_i \ddot{\underline{r}}_i = -\underline{\nabla}_i V \quad , \quad i=1, \ldots, N,$$

so that the knowledge of all the $\underline{r}_i$ and $\dot{\underline{r}}_i$ at any time t_o is in principle enough to obtain the positions and velocities at any other time t.

It is not the purpose here to discuss the analytical problems related with the solution of such systems of second-order, non-linear, coupled, ordinary differential equations. It is also not the purpose here to discuss the practical problems related with the numerical solution of such equations from the point of view say of orbit theory where one is required to calculate the motion of a few particles with prescribed accuracy.

We are interested in that class of problems of classical statistical mechanics which can be studied by solving for the motion of N interacting particles where N is of the order of 10^2 to 10^4. However, it should be borne in mind that if the potential V is such that by elementary algebraic manipulation the N-body problem can be transformed to N 1-body (i.e., 1-coordinate) problems, then again our interest in that problem will be much reduced; two obvious examples are the ideal gas (V=0) and the harmonic vibrations about a stable configuration (V = a quadratic form in displacements away from the stable configuration). There is, however, an intermediate situation where V contains a "perturbation" which couples the N 1-body problems; for example in a harmonic system the N normal modes may be coupled by a small anharmonic part in V. We shall not be concerned with the special questions that will naturally arise in such a context. Thus, to summarize, we are restricted here to the statistical mechanics of N strongly interacting particles.

It is obvious that modern computers are essential for this kind of work. Hence, in the following presentation, we shall often use terminology which is relevant only for the writing of computer programs; on some occasions it will appear as if the logical outline of a computer program is being discussed. (A warning is most important at this point; with the advent of parallel processors, vector machines and so on, new ways of organizing the computation will become necessary.)

For starting a computational project of this kind the very first question is to decide on the value of N, the size of the system. Eventually the cost of computation has to be considered; in any case, it is good practice to start with a 'small' system, say of 100 particles, to build up the operating machinery.

At this point, we shall assume that the potential function V is not a purely repulsive potential between the particles; this means that it is possible for the system to be stable when it is an isolated system of N particles. For purely repulsive potentials, it is necessary to introduce a confining mechanism of one sort or another. This will be discussed below when we consider periodic boundary conditions. At present we are considering an isolated system of N particles interacting with the potential V.

The initial positions and velocities can be chosen in a variety of ways some of which are the following:

i) positions on a regular array of points of suitable geometry and a distribution of velocities suitably randomized.

ii) positions as above with suitably randomized small displacements and all zero velocities or chosen as in i) above.

iii) randomly chosen positions avoiding situations of extremely large potential energy and velocities zero or as in i) above. Close overlap between particles is bound to produce a large potential energy; this can be avoided either by a trial and error procedure or, after all the positions have been chosen, by expanding the system so that the worst overlap ceases to be an overlap.

iv) positions and velocities from a previous calculation is an obvious procedure which hardly needs to be mentioned.

We shall see below how, starting in one fashion or another, one can manipulate the system to bring it to a desired thermodynamic state. In the case of the system under consideration (N particles and no boundaries), the only quantity which can be prescribed is the total energy E (it is understood that the total linear and angular momenta have been made to vanish when the initial positions and velocities were being chosen). Since $E=V+\Sigma m_i \dot{\underline{r}}_i^2/2$ the reduction of all $\dot{\underline{r}}_i$ by a factor $\alpha<1$ will reduce the energy E and vice versa for $\alpha>1$. The choice of α and the frequency with which it is applied will vary according to circumstance.

At this point, we introduce the temperature scale for measuring the kinetic energy. Writing $\Sigma m_i \dot{\underline{r}}_i^2/2 = 3Nk_BT(t)/2$ we shall say that the system has a certain temperature T at time t to imply that is has total kinetic energy at time t equal to $3Nk_BT(t)/2$. The average of T(t) over a sufficiently long dynamical run will be referred to as the temperature of the system for that run.

Using this terminology, the reduction (or increase) in E, the total energy, is achieved by 'cooling' (or 'heating') the system. From our understanding of statistical mechanics we expect that the equation of state of such a system, T(E), is the dependence on E of the average temperature T of the system over a long enough time t; it is more usual to think of this relation as E(T). It is also obvious that if T(t) at any moment t is high enough the system will dissipate by a process of evaporation or explosion depending on the circumstances. As against heating, the cooling of the system has limits on two accounts. Firstly, the factor α put equal to zero is the ultimate in reducing the temperature at time t to zero; however immediately afterwards the temperature will start rising by the conversion into kinetic energy of some of the potential energy of the system at t; thus drastic temperature reduction cannot be achieved in one step; secondly, as the system is cooled it gets more and more sluggish and takes longer and longer to sample the configuration space available to it.

Choice of the Algorithm

The problem is to convert the differential equations into a set of difference equations which enables us to go from time t to $t+\Delta t$ with a suitably chosen Δt. We shall consider later the value to be chosen for Δt.

In the literature that has grown over the last fifteen years or so, the simplest algorithm is that used by Verlet[1] and his collaborators at Orsay, France. Using q(t) to represent any one of the 3N coordinates which describe the system completely, we have

$$q(t+\Delta t) = q(t) + \Delta t\dot{q}(t) + 1/2(\Delta t)^2\ddot{q}(t) + \ldots$$

$$q(t-\Delta t) = q(t) - \Delta t\dot{q}(t) + 1/2(\Delta t)^2\ddot{q}(t) - \ldots$$

therefore

$$q(t+\Delta t) = 2q(t) - q(t-\Delta t) + (\Delta t)^2\ddot{q}(t) + O(\Delta t^4)$$

But, if m is the mass the acceleration of the coordinate q(t) is given by

$$\ddot{q}(t) = -m^{-1}\partial V/\partial q = a_q(t) \qquad \text{, say.}$$

Obviously $a_q(t)$ can be calculated from the knowledge of all the coordinates at time t i.e., of all the q(t). Thus the only information that needs to be carried in memory is $q(t-\Delta t)$ and q(t). Using all the q(t) we first calculate all the $a_q(t)$ and having done that all the $q(t+\Delta t)$ from above. Thus for N particles one needs 3N locations for each of $q(t-\Delta t)$, q(t) and $a_q(t)$. (Notice that the $q(t+\Delta t)$ do not need 3N extra locations even though the way of writing the equation for $q(t+\Delta t)$ above gives that impression).

It will be noticed that the new coordiantes $q(t+\Delta t)$ have been obtained without reference to the velocities at time t. To get the velocities at time t, so as to calculate the kinetic energy at that time, the simplest procedure is to calculate

$$(q(t+\Delta t) - q(t-\Delta t))/2\Delta t$$

as a measure of the velocity at time t.

Whatever the algorithm used, the process of calculating $\partial V/\partial q$ can obviously be combined with that of calculating V itself so that at the end of the process of going from t to $t+\Delta t$ we have the kinetic and the potential energy at time t.

A predictor-corrector type algorithm which has been used by various authors can be found in a report by Gear.[2] This algorithm makes use of the derivatives of q up to a predetermined order, say 5, for example, and all these derivatives need to be carried in the memory. Let us denote q by q_o, $\dot{q}\Delta t$ by q_1, $\ddot{q}(\Delta t)^2/2!$ by q_2 and so on. The first step is to predict the values of all the derivatives at $t+\Delta t$ using a Taylor expansion. At $t+\Delta t$ let p_i be the predicted values of q_i. We have

$$
\begin{aligned}
p_o &= q_o + q_1 + q_2 + q_3 + q_4 + q_5 \\
p_1 &= q_1 + 2q_2 + 3q_3 + 4q_4 + 5q_5 \\
p_2 &= q_2 + 3q_3 + 6q_4 + 10q_5 \\
p_3 &= q_3 + 4q_4 + 10q_5 \\
p_4 &= q_4 + 5q_5 \\
p_5 &= q_5
\end{aligned}
$$

Hence from p_2 we can get the predicted value of the acceleration at $t+\Delta t$.

Using p_o the predicted positions at $t+\Delta t$ we can calculate the accelerations in predicted positions; let $a(p_o)$ denote this value and let $\tilde{p}_2$ denote $a(p_o)(\Delta t)^2/2!$. It is the difference $\Delta = \tilde{p}_2 - p_2$ which allows us to get corrected values c_i for all the predicted values p_i as follows:

$$c_i = p_i + f_{i2}^{(5)}\Delta \qquad i = 0,1,\ldots,5$$

where the magic numbers $f_{i2}^{(5)}$ are 3/16, 251/360, 1, 11/18, 1/6, 1/60, respectively for $i = 0,1,\ldots,5$. In $f_{ij}^{(k)}$, j denotes the order of the differential equations being solved, (k) the order of the highest derivative being used in the algorithm and i goes from 0 to k. The reason for choosing three indices for these magic numbers is that the values of $f_{ij}^{(k)}$ depend on all three indices.

It is important to note that for clarity of presentation we have used the symbols p and c. In fact, the only memory locations needed are for the q's and for the acceleration $a(p_o)$ mentioned above.

The predictor-corrector method has the advantage that one can use the differences $c_o - p_o$ and $c_1 - p_1$ to put bounds on the acceptable error. The predictor-corrector loop can be repeated if necessary by using the corrected values as the predicted values and going back to the calculation of $a(c_o)$ and then Δ would be $\tilde{c}_2 - c_2$, $\tilde{c}_2$ being $a(c_o)\times(\Delta t)^2/2!$.

This algorithm has other possibilities too; it allows us to solve a coupled set of 1st and 2nd order differential equations (e.g. the Newton-Euler equations of motion for rigid bodies) in the same manner as explained above.[3] As for the problem of initiating the calculation, a practical scheme would be to put all the derivatives from 2 upwards equal to zero; in a few steps of Δt the algorithm itself generates appropriate values. More sophisticated methods of starting a calculation can also be used but in the context of statistical mechanics the 'error' committed in the first few Δt has no significance at all for the final results stretching over thousands of Δt.

Calculation of the Forces

This being the most time consuming part of the whole calculation, particular attention has to be paid so as to make it as efficient as possible. Let us assume that the potential V is given to be a sum of 2-body potentials depending on the pair distance alone.

$$V = \sum_i \sum_{j>i} \phi_{ij}(r_{ij})$$

The force on i due to j is

$$-\nabla_i \phi_{ij}(r_{ij}) = -(r_{ij}^{-1} \frac{d\phi_{ij}}{dr_{ij}})\mathbf{r}_{ij}$$

where $\mathbf{r}_{ij} = \mathbf{r}_i - \mathbf{r}_j$. Thus, we need, in analytic or tabulated form, the two functions ϕ_{ij} and $r_{ij}^{-1}\phi'_{ij}$ for all pairs ij. Note that we could drop the indication ij on ϕ_{ij} if all particles were identical. Also, if ϕ_{ij} depends on even powers of r only (the famous Lennard-Jones 6-12 potential for example) then $-r_{ij}^{-1}\phi'_{ij}$ also has only even powers which makes for a great saving since no square root calculation needs to be performed. However for any function ϕ_{ij} one can use this advantage (of not having to calculate square roots) by tabulating the function not as a function of r but of r^2. Eventually each advantage has to be balanced between memory requirements for finely meshed tables (to avoid involved interpolation arithmetic) against the cost of interpolating in a not so finely meshed table.

The calculation of the forces thus proceeds by programming the double sum as indicated above with j>i (rather than j≠i and a 1/2 to go with it), calculating $\mathbf{r}_{ij}$, its square r_{ij}^2, the value of ϕ'_{ij} and $r_{ij}^{-1}\phi'_{ij}$ and finally the three numbers $(r_{ij}^{-1}\phi'_{ij})\mathbf{r}_{ij}$ which gives the components of the force on i due to j. The same values with changed signs give the force on j due to i. The contribution to the virial from this interaction is obviously $(r_{ij}^{-1}\,\phi'_{ij})\times r_{ij}^2$. Thus we get

$$\mathbf{f}_i = -\sum_{j\neq i} \nabla_i \phi_{ij}(r_{ij}) \qquad i = 1,\ldots,N$$

$$\Xi = -\sum_i \sum_{j>i} \nabla_i \phi_{ij}(\mathbf{r}_{ij})\cdot\mathbf{r}_{ij} \equiv -\sum_i \mathbf{r}_i \cdot \sum_{j\neq i} \nabla_i \phi_{ij}$$

While r_{ij}^2 is available to make these calculations, it is obviously useful to evaluate other properties of the system which depend on the distance. The most important is the pair correlation function g(r). We shall come back to this function after considering the removal of the surface through periodic boundary conditions. Note that if the system of N particles is composed of ν different kinds of particles there will be $\nu(\nu-1)/2$ different pair correlations.

Truncation of the Potential and Lists of Interacting Neighbours

In many cases it may be worthwhile working with a truncated potential. Symbolically we write

$$\phi_{ij}(r) = \phi_{ij}(r_c) \quad \text{for} \quad r \geq r_c$$

The most famous example of this is the truncated Lennard-Jones potential made famous by the perturbation theory of liquids. In that case one takes r_c to be the value ('bottom of the well') at which $\phi'(r)=0$; this makes the truncation smooth in the derivative of ϕ as well. However, when investigating the effect of the attractive part of the potential we may still decide to truncate the potential at a suitable large value.

In either case, the following procedure invented by Verlet is useful when dealing with moderately large systems. For very small systems (N=100 to 200) such procedures are not of any use, and for very large systems (N=5000 to 10,000) further elaboration of the procedure becomes necessary. This will be dealt with later.

Unless one is dealing with very dilute systems (for which molecular dynamics becomes a doubtful method of investigation) one can state that for several Δt of molecular dynamics the neighbours up to distance r_c will be, in large majority, unchanged. A few move out of range (i.e., r_c) and a few will move within range. Thus, if at any moment we construct a list of neighbours not up to r_c but up to r_c+S, where S denotes a 'skin' thickness, then for several Δt after that moment, we need to consult only this list (and not the whole system) to identify neighbours up to r_c and to throw away those beyond r_c.

A little consideration will show that there is an optimum balance between the value of S and the number of Δt for which the list, once made, may be used. We shall not dwell on this. However, we shall mention the fact that the list array should be a one dimensional array (and a very long one too!) with an auxiliary array of dimension N which is a marker array providing indicators for the part of the list relevant for each of the N particles. The following diagram will make this obvious.

# of element of list	1	2	3	.	.	.	.	.	.	.	L
content of register	a	b	c	.	.	.	.	.	.	.	
# of element of markers	1	2	3	.	.	.	N				
content of registers	λ_1	λ_2	λ_3	.	.	.	λ_N				

L is the number of locations reserved for the list ('dimension' of list), N is the number of particles. a,b,c,...are the identities of particles, i.e., a b c, are any of the tags 1 to N and are modified every time the list is remade. λ_1=0 (always), λ_2 is that element of the total list where the sub-list of neighbours interacting with 1 ends and at λ_2+1 the sub-list of neighbours interacting with 2 starts.

Depending on various cost factors in a particular computing organization, it may be worthwhile breaking up the list into blocks and storing them on peripheral devices the blocks being read into fast memory as the need arises. We shall not consider this detail here.

Periodic Boundary Conditions

We shall consider the usual periodic boundary conditions (pbc) which enable us to extend a parallelopiped box to infinity by an integral number of translations in the three directions which determine the shape and size of the box. It is customary, while working on liquids, to use a cubic box; however, when dealing with the dynamics of crystals of non-cubic symmetry it is more convenient to work in a non-orthogonal system and to calculate the squares of distances, for example, with appropriate cross terms. However, we shall not consider this further assuming that the N particles are confined in a cubic box of length L repeated indefinitely by translations in the three orthogonal directions.

Pbc have the merit of removing the surface effects in a mathematically well defined manner. Intuitively one can conclude that for potential functions which are short ranged compared to half the box size the effect of pbc will be rather small. For long range potentials like the Coulomb interaction this may not be so and it appears that systematic analysis of such effects has not yet been made. The work done so far on the one component plasma and on molten salts using 200 to 1000 particles has given no indication of systematic effects arising out of pbc. Recently Mandell[4] was able to conclude from his molecular dynamics work on small systems using the (6-12) potential that pbc induce certain 3-body correlations and systematic effects can also be seen in the value of the virial.

In visualizing a system with pbc it may be convenient to think in terms of a 'box' with walls and particles 'entering' from one face when 'leaving' from the opposite face. However, since the 'box' can be drawn anywhere in space, as long as it has the correct size and is never tilted, one can always think of any particle as being at the center of the 'box' rather than at one face or another. Both ways of visualizing the pbc are of course equally valid. When dealing with a short range potential and a range of interaction r_c ($<L/2$) the two pictures will appear in a hand-drawn sketch as:

The shaded region is a sphere of radius r_c. In a computer program with coordinates given with respect to a fixed origin (usually taken at the center of the 'box') it is the picture on the left which will be in operation. While checking to see if two particles i, j are within range each component x_{ij}, y_{ij}, z_{ij} of $\underset{\sim}{r}_{ij} = \underset{\sim}{r}_i - \underset{\sim}{r}_j$ has to be tested separately; if $|x_{ij}| < r_c$ one proceeds to y_{ij} but otherwise one takes $\tilde{x}_{ij} = x_{ij} - L[x_{ij}/|x_{ij}|]$ as the x coordinate difference and the test is made again. This is equivalent to slicing a plate (of thickness 2 x r_c), perpendicular to the x axis before looking at y and z. After completing this operation for y the sum of squares is tested against r_c^2 and the z coordinate is tested only if this sum is $\leq r_c^2$. Finally the pair is accepted for interaction if the distance squared is $\leq r_c^2$. It should be noted that the sign $[x_{ij}/|x_{ij}|]$ can be determined without arithmetic by simply remembering whether i is in the left or the right half of the box. Also, for L much larger than $2r_c$ (say $3r_c$ or more) the slicing can be made more efficient by including, after the test $|x_{ij}| < r_c$, the additional test $|x_{ij}| > L-2r_c$ before applying the translation and testing $\tilde{x}_{ij}$ against r_c.

When making the list of interacting neighbours one needs to use r_c+S, S being the 'skin'. It is possible to incorporate the signals for the operation (0 or $\pm$ 1) $\times$ L which gives $\tilde{x}_{ij}$ out of x_{ij} in the computer word of the list array itself; this is because, say with 16 bits in a word, there is enough room to accommodate this signal for all three directions as well as the identity of the particle. However, the overhead in decoding the composite contents of the computer word has to be carefully considered.

When a particle coordinate goes beyond $\pm$ L/2 it has to be reset so as to bring it back into the box from the opposite face. This operation simply recognizes that the image is already in the box and transfers the particle tag to its image. However, when a listing process is being used in the manner just described in the last paragraph the resetting process has to be carried through only at the moment of preparing the list and not after every time step.

Recent Developments in Molecular Dynamics Techniques

Very recently a new method has been developed for expediting molecular dynamics calculations with continuous pari interactions. This is called 'Multiple Time Step Methods in Molecular Dynamics; by the authors Streett, Tilderley and Saville.[5] This depends on the observation that the force on particle i can be broken up into a 'primary' force from its close heighbours (the first shell) and a 'secondary' force from the rest. The latter varies more slowly and hence can be developed as a suitable Taylor series by keeping its several higher time derivatives in memory. Thus the search for a large number of interacting neighbours is dropped in favour of the higher time derivatives of the secondary force. Thus the primary force is calculated at every step and the secondary force and its derivatives only after several Δt.

Some significant developments for the study of assemblies of rigid molecules have recently taken place. Evans[6] has shown that Cayley-Klein parameters can be used with advantage in place of Euler angles. From a completely different view point, Berendsen, Ciccotti and Ryckaert[7] have developed practical methods of introducing rigidity as constraints; these authors' main concern is the development of methods for dealing with long chain molecular dynamics whereas the method of Evans is applicable strictly to completely rigid molecules.

Multiple Particle Correlation Functions

The direct determination of the triplet correlation function has been found to be feasible using configurations generated by the molecular dynamics or the Monte Carlo method. Since most of the calculations up to now have used a pair-wise additive interaction, the determination of the pair correlation has become a routine matter which is incorporated, without extra cost, into the main body of the calculation. The determination of higher order correlation functions then becomes a large additional task.

To determine the pair correlation function, g(r), we note that, by definition, for a system of N particles occupying volume V,

$$\langle n_{r,r+\Delta r}\rangle = (4\pi r^2 \Delta r\ N/V)\ g(r) \quad ,$$

where the left side indicates the average (over configurations and all N particles of the system) of the number of neighbours found at distance r, $r+\Delta r$ from a particle. Some advantage can be gained by eliminating the distance in favour of its square $s = r^2$, since

$$\langle n_{s,s+\Delta s}\rangle = (2\pi s^{1/2} \Delta s\ N/V)\ g(s) \quad .$$

Thus the count on the left hand side is made in terms of squares of distances and hence no square root need be taken in the main body of the counting process. In either form, as mentioned above, this counting process adds only a trivial amount of extra arithmetic and memory locations; for multiple component systems the above statements can be easily generalized.

In determining 3-particle correlation functions the use of the distance square leads to a simple formula for the six-dimensional volume element.

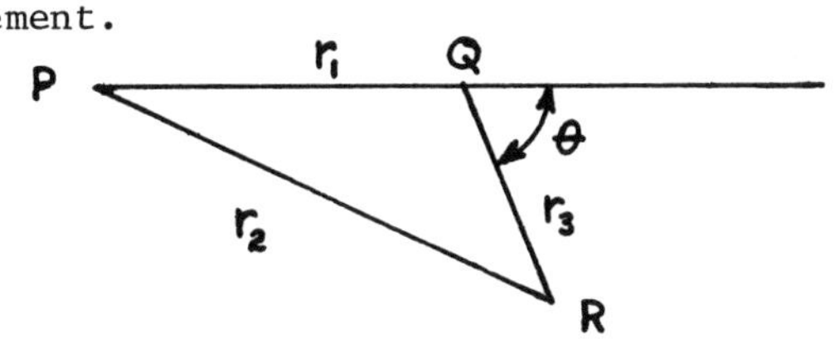

The volume element is $(4\pi r_1^2 dr_1) \times (2\pi \sin\theta\, d\theta\, r_3^2 dr_3)$. But $\cos\theta$ is given by $(r_2^2 - r_1^2 - r_3^2)/2r_1r_3$ and hence the variation of the angle θ gives $d(\cos\theta) = r_2dr_2/r_1r_3$ and the volume element becomes $\pi^2 d(r_1^2)d(r_2^2)d(r_3^2)$. Thus in this case even the normalization constant can be expressed entirely in terms of the square of distances.

The Lennard-Jones system is the one most thoroughly studied for the structure of the three particle correlation function. At high densities large deviations from the superposition approximation of Kirkwood have been found. For a detailed account of the problem of triplet-correlations and their role in determining the thermodynamic functions see Raveche and Mountain[8]. A more complex situation exists for example in connection with the thermodynamic properties of mixtures of molten salts which we shall describe below.

The conformal ionic solution theory[9] considers the mixture of two liquids, called A-X and A-Y, containing N particles of A, N_1 of X and N_2 of Y, $N_1 + N_2 = N$. Let $X_i = N_i/N$. For convenience the presentation is limited to a binary mixture. The particles denoted by A will sometimes be referred to as anions and X and Y as cations.

Let the partition function of the mixture be denoted by Z_m. Then

$$Z_m \times (N!N_1!N_2!) = I_m \equiv \int e^{-\beta\{\sum_{c=1}^{N_1} \phi_c(g_1) + \sum_{c=1}^{N_2} \phi_c(g_2) + \text{ the 'rest'}\}} d\tau$$

where each ϕ_c represents $\sum_{a=1}^{N} \phi_{rep}(g, r_{ca})$ i.e., only the short range repulsive energy of cation C with all the anions. The conformal ionic solution theory is based on the assumption that the particles X and Y have a short range repulsive potential with the particles A, namely ϕ_{rep}, which apart from its dependence on distance, contains a size parameter g; X and Y only differ in their parameters g. In other words the 'rest' of the potential function does not change in going from one system to the other. For $N_i = N$ the mixture reduces to the pure liquid A-X or A-Y which have the partition function Z_i

$$Z_i \times (N!N!) = I_i \equiv \int e^{-\beta\{\sum_{c=1}^{N} \phi_c(g_i) + \text{the 'rest'}\}} d\tau$$

We would like to evaluate

$$\Delta A_m^e = -kT\,\{\log I_m - X_1 \log I_1 - X_2 \log I_2\}$$

where I's indicates the configurational integrals already defined.

When $g_1 = g_2 = 1$ both liquids and hence the mixture reduce to a reference liquid, making $I_m = I_1 = I_2 = I_0$ (say) and ΔA_m^e vanishes.

Hence we shall expand ΔA_m^e in powers of $g_i - 1 = \gamma_i$, i = 1, 2. We therefore need derivatives of $\log I_m$ etc. with respect to the g_i's; for this the derivatives of ϕ_c with respect to g will be needed; we shall write $\psi_c \equiv d\phi_c/dg$. Also U_m will denote the potential functions for the mixture and $(U_m)_{g_1=g_2=1}$ will be denoted by U_o, the potential for the reference liquid. The 1st order terms in ΔA_m^e are

$$-kT\{\sum_{i=1}^{2} \gamma_i (\frac{\partial}{\partial g_i} \log I_m)_{g_1=g_2=1} - \sum_{i=1}^{2} X_i \gamma_i (\frac{\partial}{\partial g_i} \log I_i)_{g_i=1}\}$$

Now

$$(\frac{\partial}{\partial g_i} \log I_m)_{g_1=g_2=1} = -\beta(I_m^{-1} \int \sum_{c=1}^{N_i} \psi_c(g_i) e^{-\beta U_m} d\tau)_{g_1=g_2=1}$$

$$= -\beta N_i \langle c \rangle$$

where

$$\langle c \rangle = \int (\psi_c)_{g=1}\, e^{-\beta U_o} d\tau / \int e^{-\beta U_o} d\tau$$

Similarly $(\frac{\partial}{\partial g_i} \log I_i)_{g_i=1} = -\beta N \langle c \rangle$ and hence as can be seen by simple algebra the 1st order terms in ΔA_m^e vanish.

The second order results are obtained in a similar fashion. In the formal Taylor expansion these are:

$$-kT\left\{\frac{1}{2}\sum_i\sum_j\left(\frac{\partial^2\log I_m}{\partial g_i\partial g_j}\right)_1\gamma_i\gamma_j - \frac{1}{2}\sum_i X_i\left(\frac{\partial^2\log I_i}{\partial g_i^2}\right)_1\gamma_i^2\right\},$$

and reduce to a particularly simple form (after considerable algebraic manipulation). The result is

$$(2kT)^{-1}\;(<cd> - <c>^2)\;N^2X_1X_2\;(g_1-g_2)^2$$

where $c \neq d$. (For a multicomponent mixture the term $X_1X_2(g_1-g_2)^2$ becomes $\sum_i\sum_{j>i} X_iX_j(g_i-g_j)^2$). In the above expression $<cd>$ denotes the average value (for $c \neq d$) of $(\psi_c\psi_d)_{g=1}$ in the reference salt, i.e., with the potential U_o just as $<c>$ denotes the average of $(\psi_c)_{g=1}$ as already indicated.

The above expression can be written in an alternative form. Let Ψ denote $(\sum_{c=1}^{N}\psi_c)_{g=1}$. Then obviously $<\Psi>=N<c>$ and $(N-1)<c>^2 = <\Psi>^2/N-<c>^2$. We also have $(N-1)<cd> = <c[\Psi-(\psi_c)_{g=1}]>$ $= <\Psi^2>/N-<c^2>$. This gives $(N-1)[<cd>-<c>^2] = [<\Psi^2>-<\Psi>^2]/N -$ $[<c^2>-<c>^2]$. This shows how the sign of the second order term depends on the fluctuation of the repulsive interaction of each cation with all the anions (i.e., of $c \equiv \psi_c$) and on the fluctuation of the sum of all such repulsive interactions. Going back to the simple form $<cd>-<c>^2$ given above, since the repulsive interaction and its derivative is short ranged, we can expect $<cd>\neq<c><d>=<c>^2$ only for closely occurring cations (those sharing the same anion?) and the difference $<cd>-<c>^2$ should be appreciable only for such pairs of cations. This argument suggests that the results will not depend on N the size of the system. This needs to be investigated.

Let us consider the structure of the two averages $<cd>$ and $<c>^2$.

$$<c> = \sum_{a=1}^{N}\int\left(\frac{d\phi_{rep}(g,r_{ca})}{dg}\right)_{g=1} e^{-\beta U_o}\,d\tau\Big/\int e^{-\beta U_o}d\tau$$

$$= N \times \text{average of } \left(\frac{d\phi_{rep}(g,r_{ca})}{dg}\right)_{g=1}$$

in the reference system. This is a pair property and depends on the pair correlation between a and c type particles in the reference system.

However (with the normalization constant suitably put in),

$$<cd> = \int \sum_{a=1}^{N} \left(\frac{d\phi_{rep}(g,r_{ca})}{dg}\right)_{g=1} \sum_{b=1}^{N} \left(\frac{d\phi_{rep}(g,r_{db})}{dg}\right)_{g=1} e^{-\beta U_o} d\tau$$

depends on the triplets of the type (cd,a) for c ≠ d and the pair-pair correlation of the type (cd,ab) c ≠ d, a ≠ b; here c,d are of one kind and a,b of the other.

At present the only method available to use this theory to test against experimental values is computational; the averages calculated for a reference liquid can be used in the above theory to predict the mixing properties. Any disagreement can arise out of two different causes: the inadequacy of the potential functions to describe the departures from the reference salt potential function or the inadequacy of the above perturbation theory to converge. The latter point however can be tested by doing computer calculations on a suitable model system. This has not yet been done.

As far as the adequacy of the potential is concerned, calculations have shown that the quantity $<cd>-<c>^2$ is indeed sensitive to the repulsive potential between the cations and anions.[9] If integral equation techniques can be developed (see Pinski's contribution to this volume[10]) to predict the higher order particle correlations for such systems they will indeed be very useful.

From the computational point of view the time correlations which have received most attention are those related with self-diffusion and inelastic neutron scattering. Both these are contained in the wave-vector dependent correlation function

$$N^{-1} \sum_i \sum_j \exp[i\underset{\sim}{k} \cdot (\underset{\sim}{r}_i(o) - \underset{\sim}{r}_j(t))] \quad .$$

The i = j terms correspond to self-diffusion and the function in its totality (i.e., including i = j as well as the i ≠ j terms) corresponds to density fluctuations.

For many of the discussions relevant to this meeting we are concerned with correlation functions of the type <Q(0)Q(t)> where the Q depends on the properties of pairs of particles. For example the tensor $\sum_{i \neq j} \underset{\approx}{\alpha}(\underset{\sim}{r}_{ij})$, $\underset{\approx}{\alpha}(\underset{\sim}{r})$ being the polarizability tensor of a pair at separation $\underset{\sim}{r}$ is of interest in light scattering experiments. In all such cases where Q depends on the position of pairs we need to know the behaviour of 3 types of correlations namely (ij, ij), (ij, ik), (ij, kℓ); in NMR experiments for example only the 1st of the three is involved. Thus, instead of a particle-particle correlation, one is concerned with a pair-pair correlation function.

The scheme used by Pinski and Campbell[11] for the structural and dynamical pair-pair correlations will be described below. In a molecular dynamics calculation using a pair-wise additive potential function for the system the collection of distance statistics (for getting g(r)) is a simple matter. Suppose a pair i,j with vector separation $\underset{\sim}{r}_{ij}$ has been located (j being the minimum image of i, i.e. it is inside the molecular dynamics 'box' if i is thought of as being at the center of the 'box'). The contribution of this vector $\underset{\sim}{r}_{ij}$ to the six quantities of the type $x_\alpha x_\beta/r^2$ can be added on to six arrays reserved for the purpose. The arrays obviously consist of 'bins' corresponding to the value of $r^2(=s)$ and a suitable mesh size Δs. If working with r rather than s is decided upon then of course the bin size will be some suitable Δr.

At the end of the dynamical step, in addition to positions, velocities and other suitable information for time correlations analysis, these six arrays can also be recorded on the magnetic tape or other suitable (large data storage) device. Obviously the arrays will have to be cleared to make them ready for the information at the next Δt. It is obvious that the three arrays containing x^2/r^2, y^2/r^2, z^2/r^2 when added will give the statistics necessary for the pair correlation function.

Consider for example the time correlations of the quantity

$$Q(t) = \sum_i \sum_{j\neq i} \left[\frac{1}{r_{ij}^3} + f(r_{ij})\right] (x_{ij}\, y_{ij}/r_{ij}^2)$$

Then, apart from the problem of resolution which can always be handled by a judicious choice of a sufficiently small bin size,

$$Q(t) = \sum_{\text{bins}} \left[\frac{1}{r^3}\right]_{\text{bin}} \left(\frac{xy}{r^2}\right)_{\text{bin}} + \sum_{\text{bins}} [f(r)]_{\text{bin}} \left(\frac{xy}{r^2}\right)_{\text{bin}}$$

Thus, for the same dynamics run, one can evaluate rather easily, the effect of various functions f(r) on the value of $\langle Q(o)Q(t)\rangle$ and in particular on that of $\langle Q^2\rangle$.

The results of a molecular dynamics calculation on a Lennard-Jones system of 500 particles at $\rho^* = 0.397$ and $T^* = 2.50$ are given below. The quantity T_{xy} to be calculated for each configuration is [following the work of Oxtoby and Gelbart[12]]

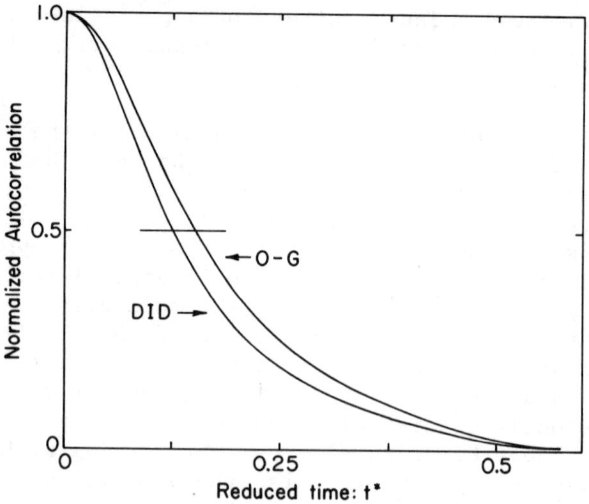

Fig. 1. Effect of the Oxtoby-Gelbart overlap term on the polarizability auto-correlation of a Lennard-Jones system (at $T^* = 2.5$, $\rho^* = 0.397$). At t=0 the DID value 0.0229 of $\langle T_{xy}^2 \rangle$ reduces to 0.0145 with the inclusion of the O-G term. In argon $t^* = 0.5$ is about 10^{-12} sec. The slower decay in time will lead to a 20% sharpening of the frequency spectrum i.e., by about 4 cm-1.

$$T_{xy} = \sum_i \sum_{j \neq i} \frac{a_o^3 \beta(r_{ij})}{\alpha_o^2} x_{ij} y_{ij} / r_{ij}^2$$

$$= \beta_o' \sum_i \sum_{j \neq i} \left[\left(\frac{\sigma}{r_{ij}} \right)^3 - \alpha \exp(-\gamma r_{ij}/\sigma) \right] \frac{x_{ij} y_{ij}}{r_{ij}^2}$$

where $\beta_o = 6a_o^3/\sigma^3 = .02262$, $\alpha = 605.62$, $\gamma = 7.741$ (from Oxtoby and Gelbart). The manner of writing T_{xy} obviously makes it dimensionless since a_o is the Bohr radius β the pair polarizability anisotropy and α_o the atomic polarizability. The value of $\langle T_{xy}^2 \rangle$ is found to be 2.29×10^{-2} without the term in α and 1.45×10^{-2} with it. Obviously the term $\langle T_{yz}^2 \rangle$ and $\langle T_{zx}^2 \rangle$ are also calculated and the average of the three is quoted above. The calculation of the time correlation $\langle T_{xy}(\tau) T_{xy}(t+\tau) \rangle_\tau$ is made by reading through the stored values at various times and using as many values of τ as possible. It is seen from figure 1 that the inclusion of α makes the decay less rapid (due to the decrease in the importance of the nearby neighbours whose effect varies more rapidly). The half-width in time increases by about 20% which on transformation gives a spectrum which is only about 4 cm^{-1} sharper than the pure DID width of about 20 cm^{-1}.

A similar calculation can be made for liquid He at T = 0 using a Jastrow variational wave function. Pinski and Campbell[11] have used the function $\Psi = \prod_{i>j} \exp[u(r_{ij})]$, $u = (b\sigma/r)^5/2$, b = 1.17 and σ = 2.556 A for He at a number density $\rho = .364\sigma^{-3}$. Using a 500 particle system the value of $\langle T_{xy}{}^2 \rangle$ is found to be .070 without the Oxtoby-Gelbart term and .048 when it is included.

References

1. L. Verlet, Phys. Rev. 159, 98 (1967)
2. a) C. W. Gear, ANL Report #7126, Argonne National Laboratory (1966)
   b) C.W. Gear, 'Numerical Initial Value Problems in Ordinary Differential Equations'. (Pretice Hall, Englewood Cliffs, N.J. 1971)
3. A. Rahman and F.H. Stillinger, J. Chem. Phys. 55, 3336 (1971)
4. M. Mandell, J. Stat. Physics, 15, 299 (1976)
5. Streett, Tilderley and Saville (preprint)
6. Evans (preprint to be published in J. Chem. Phys.)
7. J.P. Ryckaert, G. Ciccotti, H.J.C. Berendsen (to be published) (see W.F. Van Gunsteren and H.J.C. Berendsen in CECAM workshop report on 'Models for Protein Dynamics', CECAM, 91405, Orsay, France, 1976)
8. H. Raveche, R. Mountain in 'Progress in Liquid State Physics', Edited by C. Croxton, John Wiley & Son, London
9. a) M.L. Saboungi and M. Blander, J. Chem. Phys. 63, 212 (1975)
   b) M.L. Saboungi and A. Rahman, J.Chem. Phys. 66, 2773 (1977)
10. F. J. Pinski: this volume
11. F.J. Pinski and C.E. Campbell (preprint)
12. D.W. Oxtoby and W.M. Gelbart., Mol. Phys. 29, 1569 (1975)

OPTICAL POLARIZATION IN MOLECULAR DIELECTRIC FLUIDS

Abraham Ben-Reuven

Chemistry Department, Tel-Aviv University

Tel-Aviv, Israel

The characteristics of optical polarization and light scattering in molecular fluids differ from those of simple atomic fluids in several ways:

(1) The shape of the molecules (if they are not nearly spherical) is extremely important in forming short-range correlations between orientations and positions and among orientations, or even long-range orientational order (as in liquid crystals).

(2) The complexity of Raman line spectra (owing to electronic and vibrational energy-level structure, and also to rotational structure in gases and in "quantum" liquids) introduces the problem of dealing with the variation of the scattering amplitudes and the correlation functions with the internal quantum numbers. An illustration to the complexity of these effects is presented here by D. Frenkel [1].

(3) The richness of intermolecular-force effects in collision-induced light scattering, compared to the simple dipole-dipole interaction of the Stephen approach to atomic fluids [2]. All kinds of higher-multipole and short-range forces contribute to collision-induced scattering, causing a variety of phenomena such as inducing symmetry-forbidden Raman transitions.

In dealing with light scattering between well-defined internal states, we generally distinguish between nonresonant scattering, in which only the initial and final states are in resonance through interaction with the radiation field (ordinary Rayleigh and Raman spectra) and resonance scattering, in which there is also an intermediate state in resonance with the two others (resonance Raman or resonance fluorescence). In the former (nonresonant) case we, furthermore, distinguish between single-molecule scattering and two-molecule scattering; the latter includes both collision-induced scattering at short intermolecular distances and double scattering at long distances, as in the Stephen formula. Single and double scattering in condensed matter involve, respectively, two-point and four-point correlations in space and time. Resonance scattering, on the other hand, is treated as a succession of two dipolar transitions occurring at different times, and therefore involves four-point correlations even in single scattering.

SINGLE-MOLECULE NONRESONANT SCATTERING

Single scattering of light is described by a single-molecule scattering amplitude consisting of a matrix element of the polarizability-density operator. Although this is a second-order optical process, the dynamics of the intermediate states in the Kramers-Heisenberg polarizability away from resonance is irrelevant, and the scattering process may be described as a single "event", or point in space and time (just as in the atomic case), with a corresponding two-point correlation function. Orientational correlations enter the expressions through the fact that the polarizability $\overleftrightarrow{\alpha}$ is generally a second-rank tensor, and not a scalar, in configuration space. In "classical" fluids (with unquantized rotation), Rayleigh scattering is related to a time correlation function of a spatial Fourier component of the polarizability density,

$$A_k(t) = \sum_{B=1}^{N} (\vec{n}_i \cdot \overleftrightarrow{\alpha}_B(t) \cdot \vec{n}_s) \exp[i\vec{k} \cdot \vec{r}_B(t)] \tag{1}$$

where $\vec{n}_i$, $\vec{k}_i$, $\vec{n}_s$, and $\vec{k}_s$ are, respectively, the polarizations and wavevectors of the incident and scattered fields, and $\vec{k} = \vec{k}_i - \vec{k}_s$. Here both orientations $\overleftrightarrow{\alpha}_A(t)$ and positions $\vec{r}_A(t)$ are time dependent. The removal of geometry-dependent factors and the classification of the various orientation-position correlations in isotropic fluids can be done by employing symmetry arguments [5]. The dynamics of orientational correlations and their coupling to hydrodynamic modes have been subject to extensive experimental and theoretical investigations, and to numerous reviews of which we refer to the latest [4].

In Raman scattering (including rotational transitions in "quantum" fluids) one should reckon, in addition, with the internal-state dependence of the scattering amplitudes $\langle a|\alpha|b\rangle$, and its effect on the correlation functions. As the amplitudes form a doubly-labeled basis in the sets of internal quantum numbers, the proper self energy is a tetradicly-labeled supermatrix in this basis (in the same fashion as dipole-absorption line spectra [5]).

TWO-MOLECULE SCATTERING

In order to introduce two-molecule (collision-induced and double) scattering, one can employ a many-body two-point polarizability density according to the scheme of Samson et al. [6]. The basic idea of this scheme is to replace the one-molecule dipole-dipole polarizability density (ω_i implying the limit ω_i+i0),

$$\overleftrightarrow{A}_o(\vec{r}) = -\sum_B [\vec{\mu}_B, (\omega_i - L_B)^{-1}\vec{\mu}]\delta(\vec{r}-\vec{r}_B) \equiv \sum_B \overleftrightarrow{\alpha}_B \delta(\vec{r}-\vec{r}_B) \qquad (2)$$

by the many-body expression

$$\overleftrightarrow{A}(\vec{r}_1,\vec{r}_2) = -[\vec{M}(\vec{r}_1), (\omega_i - L)^{-1}\vec{M}(\vec{r}_2)] \qquad (3)$$

where

$$\vec{M}(\vec{r}) = \sum_B \vec{\mu}_B \delta(\vec{r}-\vec{r}_B) \qquad (4)$$

is the electric-dipole density. Here L_B is the quantum Liouville operator of the single molecule, $L_B X = (1/\hbar)\ [H_B, X]$, whereas L is the full quantum-electrodynamic Liouville operator, including all intermolecular Coulomb and radiative couplings.

Retaining only dipole-dipole couplings, regarding only nonpolar fluids, and considering only one- and two-molecule contributions,

$$A^{\alpha\beta}(\vec{r}_1,\vec{r}_2) = A_o^{\alpha\beta}(\vec{r}_1)\delta(\vec{r}_1-\vec{r}_2) - A_o^{\alpha\gamma}(\vec{r}_1)F^{\gamma\delta}(\vec{r}_1-\vec{r}_2)A_o^{\delta\beta}(\vec{r}_2) \qquad (5)$$

where $\overleftrightarrow{F}(\vec{r})$ is the fully-retarded dipole propagator [2,6], and the summation convention over Cartesian components (Greek characters) is implied. Eq. (5), which is equivalent to Stephen's formula [2], is derived at the cost of assuming that (a) one considers only matrix elements $\langle a|\alpha|b\rangle$ such that $\omega_{ab}/\omega_i \ll 1$, and that (b) certain corrections of the order of v/c (where v is the molecular velocity) can be discarded. This equation comprises both collision-induced effects at short distances ($\omega r/c \ll 1$) and double scattering at larger distances ($\omega r/c > 1$) [2].

Interactions in molecular fluids, are rarely ever adequately accounted for by dipolar forces, and other contributions to the hamiltonian in Eq. (3) should be reckoned. Samson et al. [6] consider higher-order multipolar contributions to collision-induced scattering. Their approach is valid for small, rather rigid, molecules at distances where overlap is negligible. (At shorter distances other types of interactions, as well as the finite size of the molecular charge distributions should be reckoned). The contributions to Eq. (3) owing to quadrupolar interactions include couplings between A_o and the dipole-quadrupole polarizability densities

$$C_o^{\alpha,\beta\gamma}(\vec{r}) = -\hbar^{-1} \sum_B [\mu_B^\alpha, (\omega_i - L_B)^{-1} q_B^{\beta\gamma}] \delta(\vec{r}-\vec{r}_B)$$
$$C'^{\alpha\beta,\gamma}_o(\vec{r}) = -\hbar^{-1} \sum_B [q_B^{\alpha\beta}, (\omega_i - L_B)^{-1} \mu_B^\gamma] \delta(\vec{r}-\vec{r}_B) \qquad (6)$$

with $U^{\alpha\beta\gamma} = \nabla^\alpha F^{\beta\gamma}$ as the propagator. If, in addition, one considers molecules with permanent multipole moments, one should add terms coupling the molecular dipole density, Eq. (4), and the quadrupole density,

$$Q^{\alpha\beta}(\vec{r}) = \sum_B q_B^{\alpha\beta} \delta(\vec{r}-\vec{r}_B) \qquad (7)$$

to a third-order hyperpolarizability density

$$B_o^{\alpha\beta\gamma} = \hbar^{-2} \sum_B [\mu_B^\alpha, (\omega_i - L_B)^{-1} [\mu_B^\beta, (\omega_i - L_B)^{-1} \mu_B^\gamma]] \delta(\vec{r}-\vec{r}_B) \qquad (8)$$

which plays a major role in second-harmonics generation.

A careful study of the relevant contributions to Eq. (3) leads to the proper analysis of a variety of collision-induced phenomena. For example, the dipole-quadrupole polarizabilities, Eq. (6), have the right symmetry necessary to induce the forbidden A_{1g} to F_{1u} and F_{2u} vibrational Raman transitions in SF_6 [7]. Also, second-harmonics generation in atomic fluids can provide direct information on three-particle (six-point) correlations [8].

RESONANCE SCATTERING

The practice of treating single-molecule nonresonant scattering as a single point should be modified in the presence of an intermediate resonance state, owing to the long time delay forced by the resonance. A more suitable practice, in this case, is to treat the scattering as a two-point process. A procedure for dealing with intermediate resonances (in a form applicable to dilute systems such as gases or impurities) was developed recently for single-molecule time-resolved and frequency-resolved resonance fluorescence [9], based on a solution of the density-matrix equations of motion (using a tetradic scattering-matrix formalism). The upshot of this procedure is the realization that the scattering process is to be treated as a sequence of two electric-dipole transitions.

In dense media, this implies that two-point amplitudes (or four-point correlations) should be reckoned in single scattering with intermediate resonance, each carrying a dipole amplitude. In this respect, resonance scattering is similar to consecutive coherent absorption of two photons. A fuller investigation of this problem will open the way to the analysis of higher-order optical effects with intermediate resonance levels.

REFERENCES

[1] D. Frenkel, this volume.

[2] W. M. Gelbart, this volume; Adv. Chem. Phys. 26, 1 (1974); M. J. Stephen, Phys. Rev. 187, 279 (1969).

[3] R. Pecora and W. A. Steele, J. Chem. Phys. 42, 1872 (1965); A. Ben-Reuven and N. D. Gershon, J. Chem. Phys. 51, 893 (1969).

[4] D. R. Bauer, J. I. Brauman, and R. Pecora, Ann. Rev. Phys. Chem. 27, 443 (1976); M. W. Evans, Adv. Mol. Relax. Inter. Proc. 10, 203 (1977).

[5] U. Fano, Phys. Rev. 131, 259 (1963).

[6] R. Samson, R. A. Pasmanter, and A. Ben-Reuven, Phys. Rev. A 14, 1224 and 1238 (1976).

[7] W. Holzer and R. Ouillon, Chem. Phys. Lett. 24, 589 (1974); R. Samson and A. Ben-Reuven, J. Chem. Phys. 65, 3586 (1976).

[8] R. Samson and R. A. Pasmanter, Chem. Phys. Lett. 25, 405 (1974).

[9] S. Mukamel, A. Ben-Reuven, and J. Jortner, Phys. Rev. A 12, 947 (1975); S. Mukamel and A. Nitzan, J. Chem. Phys. 66, 2462 (1977).

ROTATIONAL RELAXATION OF SOLUTE MOLECULES IN DENSE NOBLE GASES

Daan Frenkel

Department of Chemistry, University of California

Los Angeles, CA 90024

Broadening of the pure rotational transitions of a dissolved molecule cannot occur if this molecule is embedded in a medium that is always perfectly isotropic. The aim of this seminar is to indicate that there is a relation between rotational linebroadening and the anisotropy of local density fluctuations around the rotating molecule. It is in fact possible to derive an expression which quantitatively relates rotational linewidths to certain 4-point density correlation functions. Rotational linewidth measurements may therefore be used to obtain information about higher order density correlation functions. Most of the topics mentioned in this seminar have been discussed in a number of publications (1, 2,3), to which the reader is referred for more details.

The interaction between a linear molecule, dissolved in a noble gas fluid, and the atoms of that fluid may be written as,

$$V(\vec{1}_\mu;\vec{R}_1,\ldots,\vec{R}_N) = \sum_{i=1}^{N}\sum_{\ell=0}^{\infty} v_\ell(\vec{R}_i)(4\pi/2\ell+1)\sum_{m=-\ell}^{\ell} Y_{\ell m}(\vec{1}_\mu)Y^*_{\ell m}(\vec{1}_{R_i}) \quad , \quad 1.$$

where $\vec{1}_\mu$ is the unit vector along the axis of the linear molecule, R_i is the distance of the i-th atom to the molecule, and $v_\ell(r)$ is the r-dependent part of the ℓ-th term in the Legendre polynomial expansion of the atom-molecule intermolecular potential. In eqn.1 pairwise additivity of intermolecular interactions has been assumed. Defining $\rho(R)$ as the number density at distance R from the molecule, eqn.1 may be rewritten in the following way:

$$V(\vec{1}_\mu) = \sum_{\ell,m} (4\pi/2\ell+1)Y_{\ell m}(\vec{1}_\mu)\int d^3\vec{R}\; v_\ell(R)\rho(\vec{R})Y^*_{\ell m}(\vec{1}_R) \quad . \quad 2.$$

Equation 2 clearly shows that the perturbation acting on the molecule depends on the size, range and spherical symmetry of the local density fluctuations around the dissolved probe molecule. Throughout the following, it will be assumed that we are dealing with a very dilute solution (one linear rotor, dissolved in N noble gas atoms, with N>>1). The problem we are interested in, is the relation between the molecular motions in the host fluid, and the rotational linewidths of the dissolved (quantized) rotor. As a first approximation, one may neglect the effect of the anisotropy of the atom-molecule interaction on the time evolution of the translational coordinates in the fluid; i.e. one assumes that, as far as translational motions are concerned, the molecule behaves as a spherical particle which interacts with its neighbours through a potential $v_o(r)$. If the translational degrees of freedom may be treated classically, the above assumption allows one to use, for instance, standard Molecular Dynamics techniques (4) to compute the trajectories of the particles in the system (5). Supposing, for the moment, that the time-dependence of the translational coordinates (and hence the time-dependence of the fluctuating local density $\rho(\vec{r})$) is known, we may compute the, now explicitly time-dependent, anisotropic interaction $V(\vec{1}_\mu)$, using eqn.2. This time-dependent interaction, added as a perturbation to the free rotor hamiltonian of the molecule, determines the time evolution of the rotational degrees of freedom. Given $V(\vec{1}_\mu;t)$ one may, in principle, solve the Schrödinger equation for the rotational motion of the molecule and thereupon compute, for instance, the dipole moment correlation function of the rotor. The shape of the rotational far infrared spectrum of the molecule is related, through Fourier transform, to the molecular dipole moment correlation function. Hence, knowledge of the time dependent local density fluctuations in a fluid allows us, in principle, to determine the shape of the rotational far infrared spectrum of a weakly anisotropic probe molecule. In practice, however, explicit solution of the time dependent Schrödinger equation is not feasible. One should therefore look for a simple, approximate way to compute the dipole moment correlation function , given $V(\vec{1}_\mu;t)$.

Space does not permit us to discuss how such an approximate expression for the dipole moment correlation function may be obtained (see however refs.2,3). We just state the result that, by making a few assumptions that seem plausible for weakly anisotropic quantized rotors, one may obtain an explicit relation between rotational linewidths, and time correlation functions of the following form:

$$g_\ell(t) \equiv \iint d\vec{r}\, d\vec{r}\,' \langle\rho(\vec{r};0)\rho(\vec{r}\,';t)\rangle v_\ell(r)v_\ell(r')P\,(\vec{1}_r\cdot\vec{1}_{r'}) \quad . \qquad 3.$$

Where $\langle\rho(\vec{r};0)\rho(\vec{r}';t)\rangle$ is the correlation function of the density at distance $\vec{r}$ from the probe at time 0 and the density at distance $\vec{r}'$ at time t. The r-dependent part of the ℓ-th term in the Legendre polynomial expansion of the intermolecular potential, $v_\ell(r)$,determines the range of density fluctuations that may influence the rotational motion (this range is typically of the order of one molecular diameter). The Legendre polynomial $P_\ell(\vec{1}_r.\vec{1}_{r'})$ in eqn.3 guarantees that only those density fluctuations that have the same symmetry as the leading terms of the intermolecular potential, have a non-negligible effect on the rotational motion.

The Fourier-Laplace transform of $g_\ell(t)$ is denoted by $G_\ell(\omega)$:

$$G_\ell(\omega) \equiv \int_0^\infty \exp(-i\omega t)\; g_\ell(t)\; dt \quad . \tag{4.}$$

For reference sake, we give the explicit relation between the linewidth $\Gamma_{j\to j'}$ of a rotational transition between two levels j and j', and the function $G_\ell(\omega)$ defined above

$$\Gamma_{j\to j'} = \hbar^{-2} \sum_{\ell=1}^{\infty} [K(j,j',\ell)G_\ell(0) + \sum_{j''\neq j} M(j,j'',\ell)G_\ell(\omega_{jj''}) + \sum_{j''\neq j'} M(j',j'',\ell)G_\ell(\omega_{j''j'})\;]/(2\ell+1) \tag{5.}$$

The coefficients $K(j,j',\ell)$ and $M(j,j'',\ell)$ may be expressed in terms of 3-J and 6-J symbols (2,3). It is essential to note that the linewidth Γ, is determined by the values of the $G_\ell(\omega)$ at frequencies that correspond to level spacings in the unperturbed rotor. This fact is important, because it limits the number of probe molecules that may be used in an actual experiment to investigate local density fluctuations in fluids. Clearly, as $G_\ell(\omega)$ is determined by the density-density correlation function $\langle\rho(\vec{r};0)\rho(\vec{r}';t)\rangle$, it will be non-zero only in the frequency domain characteristic of translational motions in fluids. If one wishes to study $G_\ell(\omega)$ experimentally, one should therefore select a probe molecule that has thermally populated rotational transitions precisely in this frequency range. In the far infrared linewidth measurements described in ref.6, $HC\ell$ was used to probe local density fluctuations in argon, at different fluid densities. $HC\ell$ is a suitable probe molecule in this case, because its rotational transitions cover the entire relevant frequency range. For an analysis of these experimental data in terms of local density fluctuations, and a comparison with the results of Molecular

Dynamics calculations on the same system (5), the reader is referred to refs.2 and 3.

The correlation function $\langle\rho(\vec{r};0)\rho(\vec{r}';t)\rangle$ appearing in eqn.3, is actually related to a 4-point density correlation function:

$$\langle\rho(\vec{r};0)\rho(\vec{r}';t)\rangle = \iint d\vec{R}\, d\vec{R}'\, \langle n_P(\vec{R};0)n_H(\vec{R}+\vec{r};0)n_P(\vec{R}';t)n_H(\vec{R}'+\vec{r}';t)\rangle \quad 6.$$

where n_P and n_H are the probe- and host number densities respectively. Similar 4-point density correlation functions play a role in collision induced dipolar absorption in noble gas mixtures, in which case the quantity corresponding to $g\ell(t)$ (eqn.3), is the correlation function of the collision induced dipole:

$$\langle\vec{\mu}_{ind}(0)\cdot\vec{\mu}_{ind}(t)\rangle = \iiiint \mu(r)\mu(r')\, P_1(\vec{1}_r\cdot\vec{1}_{r'})\,\cdot \quad 7.$$

$$\cdot\ \langle n_A(\vec{R};0)n_B(\vec{R}+\vec{r};0)n_A(\vec{R}';t)n_B(\vec{R}'+\vec{r}';t)\rangle\, d\vec{R}\, d\vec{R}'\, d\vec{r}\, d\vec{r}'$$

n_A and n_B denote the number densities of atoms A and B respectively. The correlation function of the collision induced polarizability distortion, that plays a role in collision induced depolarized light scattering in monatomic fluids, may similarly be expressed in terms of 4-point density correlation functions. We have therefore at our disposition a number of experimental techniques that probe different (more or less complementary) parts of 4-point density correlation functions in fluids. Obviously, none of the experiments mentioned above will provide us with sufficient information to allow unambiguous determination of a 4-point density correlation function. However, as these experiments probe different angular and radial parts of 4-point density correlation functions, it is to be expected that, combining these different experimental techniques, it will be easier to discriminate between the various theoretical models for 4-point density correlation functions (7,8,3) that have been proposed.

This work is part of the research of the Foundation for Fundamental Research of Matter (FOM) and was made possible by financial support from the Netherlands Organization for Pure Research (Z.W.O.).

REFERENCES

1. J.van der Elsken,D.Frenkel,Faraday discussions of the Chemical Society,No.11,125 (1977)
2. D.Frenkel, thesis, Amsterdam 1977.
3. D.Frenkel,J.van der Elsken, J.Chem.Phys.,to be published.
4. see Dr.Rahmans' lecture notes. (this volume).
5. D.Frenkel,J.van der Elsken,Chem.Phys.Lett. 40,14 (1976).
6. D.Frenkel,D.J.Gravesteijn,J.van der Elsken,Chem.Phys.Lett. 40, 9 (1976).
7. M.J.Stephen,Phys.Rev. 187,279 (1969).
8. see Frank Pinski's contribution. (this volume).

SPINODAL DECOMPOSITION: AN OUTLINE

Horia Metiu

Department of Chemistry, University of California

Santa Barbara, California, U.S.A.

If we attempt to dissolve phenol in water, for example, at room temperature, we shall not succeed. Small droplets of a phenol-water solution of concentration c_1 will float in a solution of phenol-water of concentration c_2. If we heat up the liquid the droplets disappear at a certain temperature T_0. We shall refer to this phenomenon as mixing. Once the temperature is above T_0, we can attempt to cool the system and repeat the experiment in reverse. If, for simplicity, we assume that nucleation does not exist, we see that the mixture separates into droplets whenever the system is cooled below T_s. This phenomenon, of unmixing of two liquids upon cooling, is called spinodal decomposition; T_s is the spinodal temperature. If we repeat these experiments for different proportions, c, of phenol and water and plot $T_0(c)$ and $T_s(c)$, we obtain the phase diagram shown in Fig.1. In the spinodal region, located below $T_s(c)$, the system separates

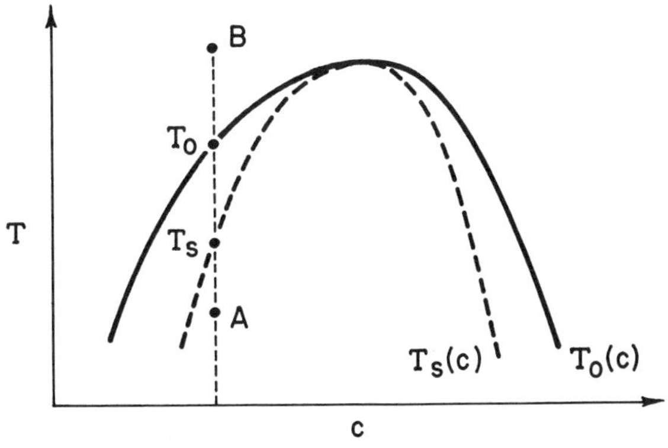

Fig.1. Phase diagram for mixing of a binary mixture.

spontaneously into droplets; above $T_o(c)$ the system is homogeneous. For the mixing experiment we start with a system consisting of droplets in the state A (Fig.1) and instantaneously increase the temperature to that corresponding to B. In this state the droplets are no longer stable and the material moves out of them to create a homogeneous state. The kinetics of this process is described by the diffusion equation. In the case of the spinodal decomposition the system is initially homogeneous and will spontaneously start to form droplets. This is possible only if some material is spontaneously transported from a region of lower to one of higher concentration. Such "counter-diffusion" can be described by the usual diffusion equation if the diffusion coefficient is negative. Therefore, the two related phenomena, the mixing and the spinodal decomposition, can be described by a unique, generalized diffusion equation, if we allow the diffusion coefficient to become negative inside the spinodal region. The counter-diffusion and hence the negativity of the diffusion coefficient are related to the thermodynamic instability of the homogeneous system at temperatures located below the spinodal; an initial inhomogeneity, created, for example, by a fluctuation, will evolve away from the unstable, homogeneous state and counter diffuse to make the system even more inhomogeneous.

The generalized diffusion equation, describing the mixing and the spinodal decomposition, is

$$\partial c(\underset{\sim}{r},t)/\partial t = M \nabla^2 \{\delta F/\delta c(\underset{\sim}{r},t)\} + R(\underset{\sim}{r},t) \quad . \tag{1}$$

Here M is a positive constant and $R(\underset{\sim}{r},t)$ is a Gaussian random function whose statistical properties are characterized by

$$\langle R(\underset{\sim}{r},t)\rangle = 0 \tag{2}$$

$$\langle R(\underset{\sim}{r},t)\, R(\underset{\sim}{r}',t')\rangle = -2\, M\, \nabla^2_{\underset{\sim}{r}}\, \delta(\underset{\sim}{r}-\underset{\sim}{r}')\, \delta(t-t') \quad . \tag{3}$$

F is the free energy of the inhomogeneous system[1] given by

$$F(c(\underset{\sim}{r})) = \int d\underset{\sim}{r}\, \{f(c(\underset{\sim}{r})) + K(\nabla c(\underset{\sim}{r}))^2\} \quad ; \tag{4}$$

K is a positive constant and $f(c(\underset{\sim}{r}))$ is the homogeneous contribution to the free energy; f is not the free energy density of a real, homogeneous system.[2]

The first to use this generalized diffusion equation (with R = 0) for an analysis of the spinodal decomposition were Hillert[3a] and Cahn.[3b] The random function was added by Cook,[4] to simulate the role of the thermal fluctuations. A more general formulation was proposed by Langer[5] who has used a master equation for the probability $P(c(\underset{\sim}{r}),t)$ that at a certain time t the system has the concentration profile $c(\underset{\sim}{r})$. The probability that the system changes its concentration from $c(\underset{\sim}{r})$ to $c'(\underset{\sim}{r})$, in a given time interval τ, is taken[5] to be:

$$W(c \to c') \propto \exp\{-\iint d\underset{\sim}{r}\, d\underset{\sim}{r}'\; c(\underset{\sim}{r})c(\underset{\sim}{r}')\; \Delta(\underset{\sim}{r},\underset{\sim}{r}')\}\; \exp\{-\tfrac{\beta}{2}[F(c') - F(c)]\}. \tag{5}$$

This formula favors spontaneous concentration changes that will diminish the free energy. The Gaussian term, with a large, positive definite kernel Δ, cuts off excessive changes in concentration (at the given time scale τ) even if they are thermodynamically favored. A microscopic study of a simple Ising model[6] suggests that such a formula for the transition rate W can be obtained if a microscopic master equation is coarse-grained in time. However, in general, one should consider eq.(5) to be a phenomenological postulate. Using a path integral method one can show[7] that Cahn's equation (eq.(1) with $R \equiv 0$) represents the most probable evolution of the system. If fluctuations around the most probable evolution are allowed, one obtains eqs.(2)-(3). These results are valid within a first order perturbation scheme[7] in which $\Delta(\underset{\sim}{r},\underset{\sim}{r}')^{-1}$ is a small parameter (i.e., small time scale). Calculations to second order yield more complex equations.[8]

To use the equations (1)-(4) we perform the functional derivative in (1) and expand $\partial f(c)/\partial c$ in powers of $\delta c(\underset{\sim}{r},t) \equiv c(\underset{\sim}{r},t) - c_o$. We take c_o to be the concentration of the system at the moment of quench, $t = 0$, and assume that c_o is space independent and that the rate of quench is instantaneous. These are not realistic assumptions and they can be removed.[9] We shall use them here for simplicity. As a result, eq.(1) becomes:

$$\partial\delta c(\underset{\sim}{r},t)/\partial t = M\,\nabla^2\,\{f''(c_o) - K\,\nabla^2\}\;\delta c(\underset{\sim}{r},t)$$

$$+ M\,\nabla^2 \sum_{n=2}^{\infty} \frac{\overset{(n+1)}{f}(c_o)}{n!}\,[\delta c(\underset{\sim}{r},t)]^n + R(\underset{\sim}{r},t)\;. \tag{6}$$

For short times, $\delta c(\underset{\sim}{r},t = 0) = 0$ and we can neglect the sum in eq.(6), to obtain a diffusion equation. If we Fourier transform it with respect to $\underset{\sim}{r}$ the diffusion coefficient is given by $D(k) = M[f''(c) - k^2K]$; this has the qualitative features discussed in the introduction. If $f_o''(c) < 0$ the diffusion coefficient is negative for small k; hence, the long wavelength inhomogeneities, caused by the fluctuations (that is, by $R(\underset{\sim}{k},t)$), are amplified by this diffusion equation. If we define the spinodal region as the one in which the diffusion coefficient can be negative, then the linearized equation indicates that the spinodal region is defined by $f''(c_o) < 0$ and the spinodal line by $f''(c_o) = 0$. This corresponds to our intuitive feeling that negativity of the diffusion coefficient is related to some form of thermodynamic instability condition. Note however that taking into account the nonlinear terms changes the diffusion coefficient and the definition of the spinodal line.

The inhomogeneity created by the decomposition can be measured by light,[10a] X-ray,[10b] and neutron scattering.[10c] The intensity of the single scattering is given by $I(Q,t) = \int d\underset{\sim}{r}\, e^{i\underset{\sim}{Q}\underset{\sim}{r}} \langle \delta c(\underset{\sim}{r},t)\, \delta c(o,t)\rangle$, where Q is the scattering wave-vector and t is the time elapsed since the decomposition has started. At short times we can neglect the nonlinear terms in eq.(6) and compute easily I(Q,t). The formula obtained in this manner[11] predicts diffractive scattering which has been observed in all cases studied experimentally. More detailed predictions (such as the dependence on the scattering vector and time) are correct for some of the systems studied;[10,11] however, a number of strong deviations have been observed.[10,12] Possible reasons for these deviations are suggested below. They are all related to the neglect of the high order correlation functions which are the topic of this Institute.

a) Since below the spinodal the concentration fluctuations are enhanced arguments of the type used by Oxtoby and Gelbart[13,14] indicate that multiple scattering of light becomes important. One can show[15] that for systems with optically shperical molecules, depolarized scattering, as 90 degrees, measures the time evolution and the Q dependence of the four point concentration correlation function. Such measurements will allow a more detailed test of the theory. The intensity of the polarized measurements determine the sum of the single and double scattering contributions. Since the latter have a different time evolution and Q dependence some of the discrepancy between the measurements and the theory may be due to the fact that only single scattering contributions have been used so far for the interpretation of the data.

b) The nonlinear terms in eq.(6) are not negligible, for cases when the decomposition is fast. They can be included through perturbation theory in a manner very similar to that used by Zwanzig[16] for systems at equilibrium. The outcome of such calculation is that the "bare" diffusion coefficient appearing into the linearized equation is "dressed" by the nonlinear terms and becomes time dependent. The absolute value of the negative diffusion coefficient is diminished in time and the growth of the fluctuations slows down (as compared to the predictions of the linear equation). If we persist in defining the spinodal region as the part of the phase diagram for which the dressed diffusion coefficient is negative then we discover that the spinodal region changes in time.

c) Other contributions to the scattering intensity are due to the preparation of the sample,[8] rate of quench,[8] and contributions from the fact that the heat of mixing is not instantly removed from the system.[8]

The equation (6) is central to the statistical theory of the kinetics of phase transitions[12] (nucleation[17] and spinodal decomposition) and dynamic critical phenomena.[1] Spinodal decomposition happens to be the phenomenon in which the dynamical behavior predicted by this equation can be tested experimentally in more detail. We feel therefore that this process will play a central role in our understanding of the value and limitation of this basic equation.

References

[1] G. Mazenko, see the lecture in this volume.
[2] J. S. Langer, Physica 73, 61 (1974).
[3] M. Hillert, Acta Met. 9, 525 (1961).
J. W. Cahn, Acta Met. 9, 795 (1961); 10, 179 (1962).
[4] H. E. Cook, Acta Met. 18, 297 (1970).
[5] H. S. Langer, Ann. Phys. (N.Y.) 65, 53 (1971).
[6] H. Metiu, K. Kitahara and J. Ross, J. Chem. Phys. 63, 5116 (1975).
[7] H. Metiu, K. Kitahara and J. Ross, J. Chem. Phys. 64, 292 (1976); ibid., 65, 393 (1976).
[8] H. Metiu, unpublished.
[9] H. Metiu, Chem. Phys. Lett. (in print).
[10] We give only a few recent references:
a) A. J. Schwartz, J. S. Huang and W. I. Goldburg, J. Chem. Phys. 62, 1847 (1975); 63, 499 (1975); N. Wong and C. M. Knobler, J. Chem. Phys. 66, 4707 (1977).
b) R. Acuña and A. Bonfiglio, Acta Metal. 22, 399 (1974).
c) J. Vrijen and C. van Dijk in Fluctuations, Instabilities and Phase Transitions, ed. T. Riste (Plenum, New York, 1975).
[11] J. W. Cahn, Trans. Metal. Soc. AIME 242, 166 (1968).
[12] H. Metiu, K. Kitahara and J. Ross in Studies in Statistical Mechanics (in press).
[13] W. M. Gelbart, see lecture in this volume.
[14] D. W. Oxtoby and W. M. Gelbart, J. Chem. Phys. 60, 3359 (1974); Phys. Rev. A 10, 738 (1974); L. A. Reith and H. L. Swinney, Phys. Rev. A 12, 1094 (1975).
[15] H. Metiu and G. Korzeniewski, Chem. Phys. Lett. (submitted); H. Metiu, J. Chem. Phys. (submitted).
[16] R. Zwanzig in Statistical Mechanics, ed. S. A. Rice, K. F. Freed and J. C. Light (U. of Chicago, 1972).
[17] J. S. Langer and L. A. Turski, Phys. Rev. A 8, 3230 (1973).

MODE COUPLING CALCULATIONS OF CRITICAL PHENOMENA AND POLYMER DYNAMICS

Raymond Kapral

University of Toronto

Toronto, Ontario, Canada M5S 1A1

CRITICAL DYNAMICS

The origin of the anomalous behavior of transport properties near the critical point has been the subject of many investigations and the physical basis of such effects is well understood. Mode coupling theory[1] and more recently the dynamic renormalization group approach[2] have provided a mathematical framework in which such anomalies can be calculated. Actual calculations are frequently difficult to carry out since the mode coupling equations are complex and the level of description which is required is not always clear due to the lack of a well defined smallness parameter in many cases. In addition, experimental results are often obtained in a region where background effects are large. As a result of these problems, our knowledge of the anomalous behavior in some systems is not complete. For example, in the past there has been disagreement as to whether the shear viscosity of fluids and fluid mixtures was finite or divergent at the critical point. In order to resolve some of these questions, we have carried out a fairly extensive set of calculations for the Rayleigh linewidth and shear viscosity of Xe[3-5]. Our calculations included the lowest order vertex and frequency corrections, all background terms and non-Ornstein-Zernicke corrections.

In order to derive a set of self consistent equations for the Rayleigh linewidth and sheer viscosity which incorporate the lowest order vertex and frequency corrections, we consider the coupling among the linear, bilinear and trilinear variables of the set,

$$c_k, g_k, g_q c_{k-q}, c_q c_{k-q}, g_{q'} c_{q''} c_{k-q'-q''}, g_{q'} g_{q''} c_{k-q'-q''}, c_{q'} c_{q''} c_{k-q'-q''} \tag{1}$$

where g_k and c_k are the local momentum density and order parameter fluctuations, respectively. The equations of motion for these variables can be obtained by making use of Mori's identity. For example, the equation of motion for the viscous propagator (x is transverse to k)

$$G_g(k,t) = \langle g_k^x(t) g_{-k}^x \rangle \langle g_k^x g_{-k}^x \rangle^{-1}, \tag{2}$$

takes the form

$$\frac{\partial G_g(k,t)}{\partial t} = \frac{k^2 \eta^B}{\rho_o} G_g(k,t) + [2(2\pi)^3]^{-1} \int d\vec{q}\, i q^x (\chi_q^{-1} - \chi_{k-q}^{-1})$$

$$\langle c_q(t) c_{k-q}(t) g_{-k}^x \rangle. \tag{3}$$

Here η^B is the background viscosity and $\chi_k = \langle c_k c_{-k} \rangle$ is the static susceptibility. Similarly, the equation of motion for the nonlinear correlation function $\langle c_q(t) c_{k-q}(t) g_{-k}^x \rangle$ can be derived and is coupled to a higher order nonlinear correlation function, $\langle \vec{g}_{q'}(t) c_{q''}(t) c_{k-q'-q''}(t) g_{-k}^x \rangle$ etc. When the equation for the correlation function involving the trilinear variable is solved and used in order to find $\langle c_q(t) c_{k-q}(t) g_{-k}^x \rangle$ we obtain the following integral equation:

$$\langle c_q(t) c_{k-q}(t) g_{-k}^x \rangle$$

$$= \frac{k_B T}{\rho_o} i q^x (\chi_{k-q} - \chi_q) \int_o^t dt_1 G_c(q,t-t_1) G_c(k-q,t-t_1) G_g(k,t_1)$$

$$- \frac{k_B T}{\rho_o (2\pi)^3} \int_o^t dt_1 \int_o^{t_1} dt_2 \int d\vec{q}' (\vec{k}-\vec{q}) \cdot \vec{q} - \hat{q}' (\vec{q} \cdot \hat{q}') \quad G_g(q',t_1-t_2) \times$$

$$\times \{ \frac{\chi_q}{\chi_{q+q'}} G_c(q,t-t_2) G_c(k-q,t-t_1) G_c(k-q-q',t_1-t_2) +$$

$$+ \frac{\chi_{k-q}}{\chi_{k-q-q'}} G_c(q,t-t_1) G_c(k-q,t-t_2) G_c(q+q',t_1-t_2) \}$$

$$\langle c_{q'+q}(t_2) c_{k-q-q'}(t_2) g_{-k}^x \rangle . \tag{4}$$

where $G_c(k,t)$ is the propagator for the order parameter.

If eq. (4) is solved by iteration to first order and the result is used in eq. (3) we obtain an expression for the Laplace transform of the viscous propagator in the form

$$G_g(k,z) = [z + k^2\eta(k,z)/\rho_o]^{-1}, \tag{5}$$

with the generalized shear viscosity given by

$$\eta(k,z) = \eta^B + J_k^1(z) + J_k^2(z) . \tag{6}$$

Here $J_k^1(z)$ is the standard anomalous contribution,

$$J_k^1(z) = \frac{k_B T}{2(2\pi)^3 k^2} \int d\vec{q}\; (q^x)^2 \; \frac{(\chi_{k-q}-\chi_q)^2}{\chi_q \chi_{k-q}} \; [z + \Gamma_q(z) + \Gamma_{k-q}(z)]^{-1}, \tag{7}$$

and the lowest order vertex correction is given by,

$$J_k^2(z) = -\frac{\rho_o}{4k^2} \left(\frac{k_B T}{\rho_o (2\pi)^3}\right)^2 \int d\vec{q}d\vec{q}' \; (\chi_q^{-1}-\chi_{k-q}^{-1}) \; \frac{\chi_q}{\chi_{q+q}} + \frac{\chi_{k-q}}{\chi_{k-q-q'}}$$

$$\times \; (\chi_{k-q-q'} - \chi_{q+q'}) \; [z+q'^2\eta(q',z)/\rho_o]^{-1} \; (\vec{k}-\vec{q}) \cdot [\vec{q}-\vec{q}'(\vec{q}\cdot\hat{q}')] \tag{8}$$

$$[\vec{q}-(\vec{q}\cdot\hat{k})\hat{k}]\cdot(\vec{q}+\vec{q}')[z+\Gamma_q(z)+\Gamma_{k-q}(z)]^{-1}[z+\Gamma_{q+q'}(z)+\Gamma_{k-q'-q}(z)]^{-1} .$$

In these equations $\Gamma_k(z)$ is the Rayleigh linewidth; it can be found by an analogous procedure to that outlined above. The results has the form,

$$\Gamma_k(z) = \Gamma_k^B + I_k^1(z) + I_k^2(z) \tag{9}$$

Here again Γ_k^B is the background term, $I_k^1(z)$ is the anomalous contribution

$$I_k^1(z) = \frac{k^2 k_B T}{8\pi^2 \rho_o} \int d\vec{q} \left(1-(\hat{q}\cdot\hat{k})^2\right) \frac{\chi_{k-q}}{\chi_k} [z + q^2 \frac{\eta(q,z)}{\rho_o} + \Gamma_{k-q}(z)]^{-1}, \tag{10}$$

while $I_k^2(z)$ is a vertex correction whose explicit form is given by

$$I_k^2(z) = -k^2 \left(\frac{k_B T}{\rho_o (2\pi)^3} \right)^2 \int d\vec{q} d\vec{q}' (\chi_{q'}^{-1} - \chi_{q-q'}^{-1})(\chi_{q-q'} - \chi_{k-q})\chi_{q'}\chi_k^{-1}$$

$$\times [(\vec{k}-\vec{q})\cdot\hat{k} - \hat{k}\cdot(\vec{k}-\vec{q}')(\vec{k}-\vec{q})\cdot(\vec{k}-\vec{q}')|\vec{k}-\vec{q}'|^{-2}][z+q^2\eta(q,z)/\rho_o + \Gamma_{k-q}(z)]^{-1}.$$

$$\times \quad z+|\vec{k}-\vec{q}'|^2\eta(k-q',z)\rho_o + \Gamma_{q'}(z) \quad ^{-1} \quad z+\Gamma_{q'}(z)+\Gamma_{k-q}(z)+\Gamma_{q-q'}(z) \quad ^{-1} \tag{11}$$

The frequency corrections are not indicated above.

We have solved equations (6) and (9) self consistently including background terms in the temperature region $\varepsilon(T-T_c)/T_c=10^{-3}-10^{-5}$ where experimental results are available. We have also solved these equations at the critical point where background terms are zero by assuming $\Gamma_k = A_D k^{2+Z_D}$ and $\eta(k) = A_\eta k^{Z\eta}$ and solving the resulting integral equations for the exponents Z_D and Z_η.

The results of our calculations can be summarized as follows. Vertex and frequency corrections to linewidth are small (in the 1-2% range) in qualitative agreement with earlier studies.[6] It is important to retain the linewidth in the denominator of eq. (10); neglect of this term leads to errors in the range 2-10% for the linewidth and 10-20% for the viscosity. The shear viscosity is sensitive to a variety of contributions. In the temperature region of interest, the background linewidth provides a natural cut-off for the $\vec{q}$ interaction in eq. (3) and leads to a viscosity which diverges logarithmically in this region.[7] The magnitude of the shear viscosity also depends strongly on vertex and non-Ornstein-Zernicke corrections. Good agreement with experiment[8] is found in this region. As one approaches very close to the critical point, the background terms will go to zero and the non-locality of Γ_q will provide a cut-off.

This is born out by the calculation of Z_D and Z_η. Eq. (9) yields $Z_D = 1-Z_\eta$, and eq. (6) yields $Z_\eta = -0.07$. This result is in good agreement with renormalization group calculations[2] which yield $Z_\eta = -0.065$. Hence, as the critical point is approached, the viscosity begins to diverge logarithmically with a magnitude that depends sensitively on background terms. As one gets very close to the critical point this logarithmic divergence turns into a weak power law divergence, $\eta \sim |T-T_c|^{-0.044}$. This power law regime is probably not experimentally accessible for Xe.

POLYMER DYNAMICS

It is also possible to use eq. (9) to study the dynamics of a large polymer in dilute solution. The dynamic structure factor for a single chain in solution is

$$S(\vec{k},z) = \langle n_k(z)n_k\rangle = [z + \Gamma_k^p(z)]^{-1} \tag{12}$$

where now n_k is the Fourier transform of the local monomer density in the chain and $\Gamma_k^p(z)$ is the polymer linewidth function. The anomalous part of the linewidth of a fluid mixture becomes important near the critical point due to the long range of the concentration fluctuations. Similarly, the polymer static structure factor is long ranged for a large polymer due to the connectivity of the chain. Hence, one might expect that the contribution in eq. (9) will dominate the behavior for a large polymer. This is, indeed, the case. For example, the center of mass diffusion coefficient can be calculated from the small $\vec{k}$ limit of $\Gamma_k^p(z=0)$ in eq. (12) using eq. (9) with χ_k now the polymer static structure factor. The result is just the Kirkwood expression for D,

$$D = \frac{k_B T}{n^2} \sum_{\substack{i,j=1 \\ (i\neq j)}}^{n} \langle \hat{k}\cdot \underline{\underline{T}}(R_{ij})\cdot \hat{k}\rangle_p \quad , \tag{13}$$

where $\underline{\underline{T}}$ is the Oseen tensor and $\langle\cdots\rangle_p$ denotes an average over all configurations of the polymer chain. This procedure has been used to calculate the full k dependence of Γ_k^p for a variety of polymer models.[9] For large kS, where S is the radius of gyration of the polymer, we find $\Gamma_k^p \sim k^3$ in agreement with earlier work.[10]

We have also studied the dynamics of a polymer chain in a binary critical mixture where a hydrodynamic description of the solvent is not appropriate.[11] In this case, the Kirkwood form is modified to

$$D = \frac{k_B T}{n^2} \sum_{\substack{i,j=1 \\ (i \neq j)}}^{n} \int d\vec{r}\, f(\vec{r})\, \langle \hat{k} \cdot \underline{\underline{T}}(\vec{R}_{ij} - \vec{r}) \cdot \hat{k} \rangle_p, \qquad (14)$$

where $f(\vec{r})$ takes into account the non-locality of the solvent shear viscosity. These techniques have also been used to calculate the full friction tensor of a polymer chain in dilute solution.[12]

REFERENCES

1. K. Kawasaki in "Phase Transitions and Critical Phenomena" eds. C. Domb and M. S. Green (Academic Press, N. Y., 1976), p. 166.

2. P. C. Hohenberg and B. I. Halperin, Rev. Mod. Phys. 49, 435 (1977); G. F. Mazenko, this volume.

3. F. Garisto and R. Kapral, J. Chem. Phys. 63, 3560 (1975).

4. F. Garisto and R. Kapral, J. Chem. Phys. 64, 3826 (1976).

5. F. Garisto and R. Kapral, Phys. Rev. A14, 884 (1976).

6. S. M. Lo and K. Kawasaki, Phys. Rev. A5, 421 (1972); A8, 2176 (1973).

7. D. W. Oxtoby and W. M. Gelbart, J. Chem. Phys. 61, 2957 (1974).

8. H. J. Strumpf, A. F. Collings and C. J. Pings, J. Chem Phys. 60, 3109 (1974).

9. R. Kapral, D. Ng and S. G. Whittington, J. Chem. Phys. 64, 2957 (1974).

10. E. Dubois-Violette and P. G. de Gennes, Physics 3,181 (1967).

11. C. Duzy and R. Kapral, J. Chem. Phys. 66, 2887 (1977).

12. D. Ng, R. Kapral and S. G. Whittington, J. Phys. A, to appear.

TWO ROTON RAMAN SCATTERING IN SUPERFLUID He^4

Cherry Ann Murray

Department of Physics
Massachusetts Institute of Technology
Cambridge, Massachusetts 02139 U.S.A.

The possibility of using Raman scattering as a probe for studying the elementary excitations of superfluid helium was first suggested by Halley[1] in 1968. Raman scattering in helium is a second order process in which two excitations of nearly equal and opposite momenta are created in the liquid. The spectrum of scattered light at an energy loss E is a measure of the density of two excitation states with zero total wavevector and energy E.[1,2] The primary contribution comes from the roton region of the dispersion curve. At low temperatures this consists of a well defined line near $E = 2\Delta$, where Δ is the energy of the roton minimum. I will concentrate on the 2-roton light scattering. A discussion of the spectrum in other energy regions can be found in reference 3.

In the first observation of Raman scattering from He^4 by Greytak and Yan,[4] the scattering from the "maxons" (excitations near the maximum of the dispersion curve) was weaker than expected.[1,2] In order to explain this discrepancy Ruvalds and Zawadowsky[5] and Iwamoto[6] showed that even a weak interaction between excitations could greatly modify the 2-excitation density of states from the noninteracting case. By invoking an interaction between maxons, they could deplete the density of pair states in that region. Assuming the same interaction between pairs of rotons, they found an enhancement in the density of 2-roton states corresponding to a resonance or bound state with an energy for the pair falling below 2Δ.

In order to investigate the possibility of such a resonance, higher resolution measurements of the Raman spectrum were made by Greytak *et al.*[7] in the vicinity of the 2-roton line. Figure 1

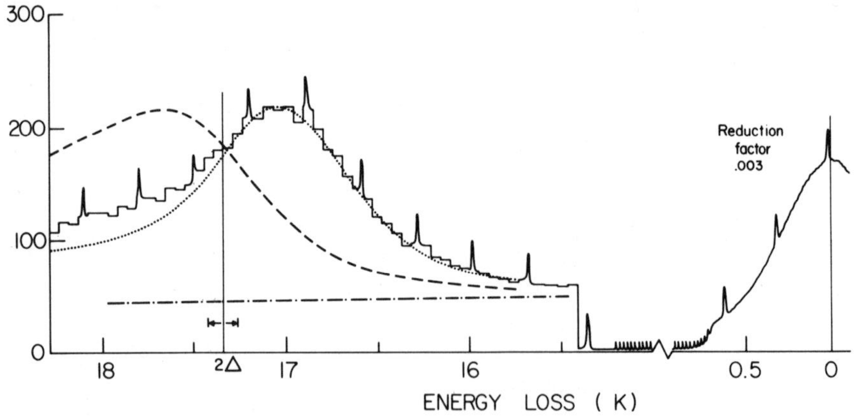

Fig. 1 Raman spectrum of He^4 at 1.2 K and SVP. The strong peak at zero energy shift is caused by Brillouin scattering and indicates the experimental profile (FWHM = .68 K). The dotted curve is a theoretical fit to the data based on a two roton bound state. The dashed curve corresponds to non-interacting rotons. The dot-dashed line is the background level. From ref. 7.

shows a typical experimental trace from those experiments taken at 1.2 K and saturated vapor pressure (SVP). The maximum in the scattering occurs at an energy shift less than 2Δ. This cannot be attributed to the finite line width of the rotons or the finite width of the spectrometer. Both of these effects would cause the maximum in the spectrum to occur at energy shifts greater than 2Δ. The theoretical prediction for the non-interacting case taking these effects into account is shown by the dashed line. In addition, a fit to the theory of Ruvalds and Zawadowsky with binding energy for the pair $E_B = .37 \pm .1$ K is shown by the dotted line. This value of E_B was obtained mainly from the difference of the position of the 2-roton peak and the value of 2Δ obtained by neutron scattering measurements of the dispersion curve by Cowley and Woods.[8] Details of the spectral line shape and width are obscured in this experiment by the instrumental width of the spectrometer (FWHM = .68 K), which was about twice the binding energy.

A separate measure of the binding energy can be obtained from the peak shape <u>alone</u>, independent of the neutron scattering measurements of Δ, if the instrumental width of the spectrometer is less than E_B. For this reason, and also to look for a finite zero temperature width of the 2 roton peak, a low temperature, extremely

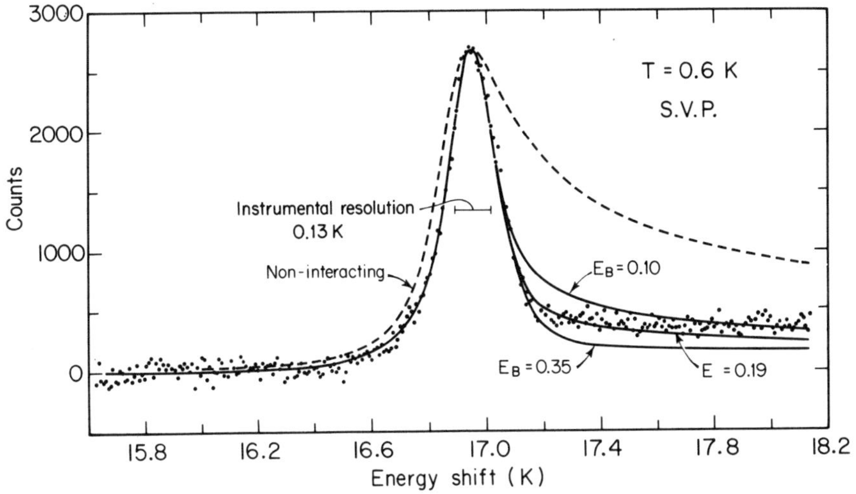

Fig. 2 Comparison of the two-roton Raman line shape at 0.6 K and SVP with various theoretical predictions. The dots are the data after background subtraction. All the theoretical curves have been calculated for an intrinsic full width 0.11 K for the pair. For the sake of line shape comparison each theoretical curve has been shifted and normalized so that its peak position corresponds to that of the data. From ref. 9.

high resolution experiment was carried out.[9] The data were obtained at a temperature of .6 K and SVP where the width of a single roton is negligible. The instrumental width of the flat Fabry-Perot spectrometer was .13 K (FWHM). Absolute frequency measurements were made with an accuracy of ±.03 K using an interferometric calibration scheme.[10] Thus a separate measurement of the frequency of the 2-roton peak was obtained along with the line shape data.

Figure 2 shows the comparison of the observed 2-roton line shape with various theoretical predictions. For the sake of line shape comparison, the theoretical spectra have been shifted so that all the peak maxima coincide. The measured peak is definitely more symmetric than the broken curve, which corresponds to the non-interacting model. Also the peak is 80% broader than the instrumental profile of the spectrometer. This corresponds to an intrinsic T = 0 width for the pair $\Gamma = .11 \pm .01$ K, which is due to the allowed decay of the roton pair into a pair of phonons of net zero momentum.[6] For the analysis of figure 2 we used a Lorentzian broadened interacting density of pair states. From the line shape of the peak alone, we obtain a best fit to the theory corresponding to $E_B = .22 \pm .07$ K.

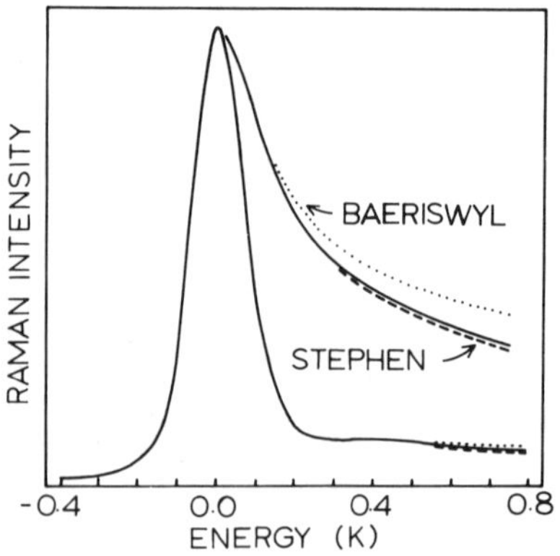

Fig. 3 Representative theoretical predictions for the two-roton Raman line shape folded with our highest resolution instrument. No intrinsic width for the pair has been included. The solid lines are calculated with a cutoff function constant in momentum space. In the case of Stephen,[2] the cutoff function is f(r) = θ(r-a) with a = 1.57 Å. For Baeriswyl,[14] f(r) = g(r), the pair correlation function for the liquid. The upper curves correspond to the non-interacting density of states, the lower to the case E_B = 0.35 K. All spectra have been normalized and shifted so that the peaks coincide.

The position of the 2-roton peak was found to be 16.97±.03 K, in excellent agreement with previous measurements.[7] We then used the measured peak position and our determination of E_B from line shape analysis of the peak to obtain a value for Δ. We suggested that the true value of Δ was ~.08 K smaller than 8.67±.04 K quoted by Cowley and Woods.[8] Recently a very high accuracy neutron scattering measurement of Δ was made by Woods *et al.*[11] They found Δ = 8.618±.009 K, in good agreement with our result.

Woerner and Stephen[12] have calculated the Raman spectrum including in addition to the roton-roton interactions an interaction between the continuum of 2-phonon states degenerate in energy with the 2-roton states. They find that the δ function bound state of Ruvalds and Zawadowsky is broadened into a resonance at T = 0 by the roton-phonon interaction. Both the energy shift and asymmetry of the 2-roton peak are affected by this interaction. As a result of their analysis, they estimate the 2-roton peak is located .26 K below 2Δ. Subtracting our measured peak position from twice the latest neutron value for Δ, we find .27±.05 K. This indicates an excellent consistency between the Raman and neutron scattering results.

The form of the coupling of light to the excitations in the theory of Stephen[2] has been very controversial, as small adjustments in the cutoff function used to avoid self-polarization of atoms can cause marked differences in the theoretical spectrum over large frequency ranges. However, in our most recent experiment the various forms for the cutoff which have been proposed[2,13,14] generate indistinguishable theoretical spectra in the narrow region around the roton peak. Figure 3 shows theoretical line shapes folded with our instrument for both the non-interacting and interacting densities of states, where various forms for the cutoff are used. Differences due to the variation of the weighting factor $Z^2(k)$ over this narrow frequency range are also negligible compared to the scatter in the data.

In conclusion, we have observed a slightly asymmetric peak in the Raman spectrum of He^4 which occurs at an energy shift 16.97± .03 K. This shift falls below twice the minimum energy for a single roton, 2Δ. The zero temperature line width of the peak was found to be .11±.01 K. Using our measurements of the line shape of the peak along with a model of roton-roton interactions we can make a determination of Δ. It is reassuring that our value for Δ obtained in this manner is in excellent agreement with recent neutron scattering results and the absolute position of the Raman peak which is model independent.

References

1. J.W. Halley, Bull. Am. Phys. Soc. 13, 398 (1968) and Phys. Rev. 181, 338 (1969)
2. M.J. Stephen, Phys. Rev. 187, 279 (1969)
3. T.J. Greytak, Physics of Quantum Fluids, 1970, Tokyo Summer Lectures in Theoretical and Experimental Physics. R. Kubo and F. Takano, editors (Tokyo, Syokabo)
4. T.J. Greytak and J. Yan, Phys. Rev. Lett. 22, 987 (1969)
5. J. Ruvalds and A. Zawadowsky, Phys. Rev. Lett. 25, 333 (1970)
6. F. Iwamoto, Prog. of Theoretical Phys. (Japan) 44, 1135 (1970)
7. T.J. Greytak, R. Woerner, J. Yan, and R. Benjamin, Phys. Rev. Lett. 25, 1547 (1970)
8. R.A. Cowley and A.D.B. Woods, Can. J. Phys. 49, 177 (1971)
9. C.A. Murray, R.L. Woerner, and T.J. Greytak, J. Phys. C: Solid State Phys. 8, L90 (1975)
10. R.L. Woerner and T.J. Greytak, Rev. Sci. Ins. 47, 383 (1976)
11. A.D.B. Woods, P.A. Hilton, R. Scherm, and W.G. Stirling, J. Phys. C: Solid State Phys. 10, L45 (1977)
12. R.L. Woerner and M.J. Stephen, J. Phys. C: Solid State Phys. 8, L464 (1975)
13. R.A. Cowley, J. Phys. C: Solid State Phys. 5, L287 (1972)
14. D. Baeriswyl, Phys. Lett. 41A, 297 (1972)

ROTON BOUND STATES IN LIQUID ^{4}He

Roger Hastings

North Dakota State University

Fargo, North Dakota 58102

The low lying excited states of liquid ^{4}He are long lived "quasiparticle" excitations which are described by the familiar phonon-roton dispersion curve.[1] Since the liquid is isotropic, the quasiparticle density of states is singular at the energy of the roton minimum Δ_R (and also at the energy of the maximum in the dispersion curve). Scattering processes in which quasiparticles decay into roton pairs are enhanced by the singular density of states, leading to a splitting of the dispersion curve into two branches.[2,3,4] The roton pair scattering amplitude is similarly enhanced, and an arbitrarily small attractive roton interaction is sufficient to form roton bound states.[3]

These dynamical features of ^{4}He, which are observed in neutron scattering[1] and light scattering[5,6,7,8] experiments, can be understood theoretically in terms of the following quasiparticle Hamiltonian:

$$H = E_0 + \sum_{\underset{\sim}{k}} \varepsilon(k)\alpha^\dagger_{\underset{\sim}{k}}\alpha_{\underset{\sim}{k}} + \frac{1}{2}\sum_{\underset{\sim}{k}_1,\underset{\sim}{k}_2,\underset{\sim}{k}_3} g_3(\underset{\sim}{k}_1,\underset{\sim}{k}_2,\underset{\sim}{k}_3)\{\alpha^\dagger_{\underset{\sim}{k}_1}\alpha^\dagger_{\underset{\sim}{k}_2}\alpha_{\underset{\sim}{k}_3} + \text{h.c.}\}$$

$$+ \sum_{\underset{\sim}{k}_1,\underset{\sim}{k}_2,\underset{\sim}{k}_3,\underset{\sim}{k}_4} g_4(\underset{\sim}{k}_1,\underset{\sim}{k}_2,\underset{\sim}{k}_3,\underset{\sim}{k}_4)\alpha^\dagger_{\underset{\sim}{k}_1}\alpha^\dagger_{\underset{\sim}{k}_2}\alpha_{\underset{\sim}{k}_3}\alpha_{\underset{\sim}{k}_4} \quad . \qquad (1)$$

E_o is the ground state energy, and $\varepsilon(k)$ is the non-interacting quasiparticle excitation energy. The three-point interaction term, g_3, describes the scattering of one quasiparticle into quasiparticle pair states, and leads to the splitting of the excitation spectrum into two branches. The four-point interaction, g_4, describes quasiparticle pair scattering and results in bound state formation.

The density correlation functions entering the neutron and light scattering intensities can be computed providing that a relationship between the density operator and the quasiparticle annihilation and creation operators, $\alpha_{\underset{\sim}{k}}$ and $\alpha^{\dagger}_{\underset{\sim}{k}}$, is specified. This relationship may be approximated by:[4]

$$\rho_{\underset{\sim}{k}} = \sqrt{S(k)}\;(\alpha_{\underset{\sim}{k}}+\alpha^{\dagger}_{-\underset{\sim}{k}}) \quad , \tag{2}$$

where $\rho_{\underset{\sim}{k}}$ is the Fourier transform of the density operator, and S(k) is the static structure factor. The Hamiltonian, Eq. (1), along with Eq. (2) can be derived through a transformation of the exact helium Hamiltonian to a representation in which the excited states are orthogonalized products of density operators acting on the ground state.[2,9] It can be shown that retention of the g_4 term in Eq. (1) is consistent with including terms which are quadratic in the α's in Eq. (2). However, in the following discussion we neglect quadratic terms in Eq. (2) as well as terms in H which are higher than quadratic order in the α's, and four-point terms in H which are not of the form included in Eq. (1).

Under the conditions discussed in Ref. 4, the exact ground state of Eq. (1) is given by $\alpha_{\underset{\sim}{k}}|g\rangle=0$. A glance at Eq. (1) shows that $|g\rangle$ is an eigenstate of $\tilde{H}$ with energy E_0. With this result and with the aid of Eq. (2), the density correlation functions entering the neutron and light scattering intensities can now be written down. The formal results are discussed in the references.[4,10] Bound state formation arises within the theory by approximating the four-density Raman scattering correlation function as a sum of ladder diagrams. This leads to an integral equation for the correlation function which is exact at zero temperature when the g_3 term in Eq. (1) is neglected.

Numerical calculations[4] of the Raman spectrum employ the following techniques: (i) the four-point interaction $g_4(\underset{\sim}{k}_1,\underset{\sim}{k}_2,\underset{\sim}{k}_3,\underset{\sim}{k}_4)$ is assumed to be an isotropic pair potential, i.e. $g_4(|\underset{\sim}{k}_1-\underset{\sim}{k}_3|)\times\delta(\underset{\sim}{k}_1+\underset{\sim}{k}_2+\underset{\sim}{k}_3+\underset{\sim}{k}_4)$; (ii) this potential is expanded in spherical harmonics, the $\ell=2$ component being relevant to the Raman scattering; (iii) the potential is further assumed to separate, i.e. $g_4^{\ell=2}(k,k')=g\eta(k)\eta(k')$, which allows the ladder sum integral equation to be solved analytically; (iv) $\eta(k)$ is taken to be a Lorentzian of width γ peaked around the roton wavevector; (v) one quasiparticle propagators appearing in the theory are related to the dynamic structure factor, and are taken from neutron scattering experiments; (vi) the matrix element coupling light to the helium atoms is taken as the dipole-induced dipole matrix element with a small r cut-off; (vii) the calculated Raman intensity is convoluted with an experimental resolution function. With γ and the constant g_4 taken as adjustable parameters, the calculated Raman spectrum gives a reasonable fit to the experimental spectrum for frequencies $10^{\circ}K<\hbar\omega<60^{\circ}K$. In particular the bound state peak in the experimental spectrum, split off below the two-roton continuum, is quantitatively accounted for by the theory.

Although the quasiparticle field theory described above gives a satisfactory description of the experimental Raman spectrum, there remain serious questions as to its applicability to liquid helium. The dynamic structure factor, as calculated within the theory,[4] fails to provide a quantitative description of neutron scattering measurements. A second and perhaps more serious criticism, is the failure of the theory to satisfy a hard sphere excluded volume condition.[11,12] The condition states that the four-density correlation function entering the Raman scattering intensity, when Fourier transformed to position space, must vanish for values of r smaller than the helium atom hard core diameter. The calculated non-interacting correlation function fails to satisfy this condition, and it can be shown that the interactions included in the ladder sum give no contribution to the Fourier transform. The static four-density correlation function can be computed exactly within the model of Eqs. (1) and (2), and it also fails to satisfy the condition.[11] This function has recently been computed numerically[13] using Monte Carlo techniques (the excluded volume condition is automatically satisfied in this calculation), and it is found that the field theory model calculation is missing a term which has structure for momenta near the roton momentum. Such a term must also be missing from the model calculation of the dynamic correlation function. Will the missing term have structure near or below the two-roton energy, thus modifying the roton bound state analysis? A possible improvement of the model would be to include quadratic terms in Eq. (2) and choose their coefficients in such a way that the excluded volume condition is satisfied.

REFERENCES

1. A. D. B. Woods and R. A. Cowley, Rept. Prog. Phys. 36, 1135 (1973).
2. H. W. Jackson, Phys. Rev. A 8, 1529 (1973).
3. A. Zawadowski, J. Ruvalds, and J. Solana, Phys. Rev. A 5, 399 (1972).
4. R. Hastings and J. W. Halley, Phys. Rev. A 10, 2488 (1974).
5. T. J. Greytak and J. Yan, Phys. Rev. Lett. 22, 987 (1969).
6. T. J. Greytak, R. Woerner, J. Yan, and R. Benjamin, Phys. Rev. Lett. 25, 1547 (1970).
7. C. A. Murray, R. L. Woerner and T. J. Greytak, J. Phys. C 8, L90 (1975).
8. C. A. Murray, proceedings of this conference.
9. C. Campbell, Progress in Liquid Physics, Croxton, ed. (Wiley, New York, 1977).
10. J. W. Halley lecture, proceedings of this conference.

11. P. Kleban, Phys. Lett. A 49, 19 (1974).
12. P. Kleban and R. Hastings, Phys. Rev. B 11, 1878 (1975).
13. F. Pinski and C. Campbell, to be published and F. Pinski, this volume.

INTEGRATED RAMAN INTENSITY IN ^{4}He

Frank J. Pinski†

School of Physics & Astronomy, Univ. of Minnesota

Minneapolis, Minnesota 55455

This seminar will explain a ground state calculation of the integrated Raman intensity in ^{4}He. This integrated intensity is related to a static correlation function which can be calculated from a new integral equation. The emphasis of this talk is placed on deriving this integral equation. The resulting correlation function is then used to calculate the integrated Raman intensity. In contrast to previous calculation[1,2], the excluded volume condition[3] is satisfied. The calculated intensity is 50% larger than the experimental result.[4]

Remember that Stephen[1] treated the four-ρ correlation function as a convolution of two-ρ pieces by eliminating U_4, the connected part. Specifically, the correlation function is decomposed by

$$\langle\rho_{\underset{\sim}{k}}(t)\rho_{-\underset{\sim}{k}}(t)\rho_{\underset{\sim}{k}'}(0)\rho_{-\underset{\sim}{k}'}(0)\rangle - \langle\rho_{\underset{\sim}{k}}\rho_{-\underset{\sim}{k}}\rangle\langle\rho_{\underset{\sim}{k}'}\rho_{-\underset{\sim}{k}'}\rangle$$
$$= [\delta_{\underset{\sim}{k},\underset{\sim}{k}'}+\delta_{\underset{\sim}{k},-\underset{\sim}{k}'}][\langle\rho_{\underset{\sim}{k}}(t)\rho_{-\underset{\sim}{k}}(0)\rangle]^2 + U_4(\underset{\sim}{k},\underset{\sim}{k}',t).$$

Note that although U_4 is only of order N and that the kept term is of order N^2, both these parts contribute to the intensity in the same order of N when the momentum sums are performed. An integral equation of the static correlation function $U_4(\underset{\sim}{k},\underset{\sim}{k}',t=0)$ can be obtained--but some preliminaries are necessary.

† Present address: Department of Physics, State University of New York at Stony Brook, Stony Brook, New York 11794.

For the ground state, a wave function can be calculated in some manner (for instance using variational methods). The normalization integral of this wave function can be used as a generating function. A set of parameters is introduced into the normalization integral

$$I\{C_{\underset{\sim}{k}}\} = \int |\psi|^2 \prod_{i \neq j} \exp\{\frac{1}{N} \sum_{\underset{\sim}{k}} C_{\underset{\sim}{k}} e^{i\underset{\sim}{k}\cdot\underset{\sim}{r}_{ij}}\} d^N\underset{\sim}{r}$$

with $C_{\underset{\sim}{k}}=C_{-\underset{\sim}{k}}$ to keep $I\{C_{\underset{\sim}{k}}\}$ real. The first derivative of $I\{C_{\underset{\sim}{k}}\}$ is related to $S(k)$, the liquid structure function, when all C_k's are zero. Likewise the four-ρ correlation function is related to the second derivative of $I\{C_{\underset{\sim}{k}}\}$,

$$\langle \rho_{\underset{\sim}{k}}\rho_{-\underset{\sim}{k}}\rho_{\underset{\sim}{\ell}}\rho_{-\underset{\sim}{\ell}}\rangle - \langle \rho_{\underset{\sim}{k}}\rho_{-\underset{\sim}{k}}\rangle\langle \rho_{\underset{\sim}{\ell}}\rho_{-\underset{\sim}{\ell}}\rangle = N^2 \frac{1}{I} \frac{\partial^2 I\{C_k\}}{\partial C_{\underset{\sim}{k}}\partial C_{\underset{\sim}{\ell}}}\bigg|_{\{C_k=0\}} .$$

The Jastrow approximation for the ground state can be expressed as a product of two-body functions, $\psi=\exp\{\Sigma_{i<j}u(r_{ij})\}$. The main consequence of introducing $u(r)$ is to keep the hard cores of the Helium atom from overlapping. The hypernetted chain approximation for the pair distribution function is a natural outgrowth of the Jastrow approximation:

$$\ln g(r,\{C_k\}) = u(r) + \frac{1}{N}\sum_{\underset{\sim}{k}} C_{\underset{\sim}{k}}[e^{i\underset{\sim}{k}\cdot\underset{\sim}{r}} + e^{-i\underset{\sim}{k}\cdot\underset{\sim}{r}}]$$

$$+ \frac{1}{N}\sum_{\underset{\sim}{k}} \frac{[S(k,\{C_k\})-1]^2}{S(k,\{C_k\})} e^{i\underset{\sim}{k}\cdot r}$$

where

$$g(r,\{C_k\})-1 = \frac{1}{N}\Sigma_{\underset{\sim}{k}}[S(k,\{C_k\})-1]e^{i\underset{\sim}{k}\cdot\underset{\sim}{r}} = \frac{1}{N}\sum_{\underset{\sim}{k}} \frac{\partial \ln I}{\partial C_{\underset{\sim}{k}}} e^{i\underset{\sim}{k}\cdot\underset{\sim}{r}} .$$

Taking derivatives of this equation (with respect to C_k), an integral equation is formed which can be simplified by making a Legendre expansion of U_4. Defining

$$U_4(\underset{\sim}{k},\underset{\sim}{q})/N = \Sigma_{\ell m}\hat{U}_4^{\ell}(k,q)Y_{\ell m}(\hat{k})Y^*_{\ell m}(\hat{q})$$

the integral equation separates into a set of decoupled equations, which for even ℓ can be written as

$$\frac{\hat{U}_4^{\ell}(k,q)}{S^2(k)} = 2\hat{S}^{\ell}(k,q) + \frac{1}{(2\pi)^3\rho}\int h^2 dh \hat{S}^{\ell}(h,k)[S^2(h)-1]\,\frac{\hat{U}_4^{\ell}(h,q)}{S^2(h)}$$

with

$$S(\underset{\sim}{k}+\underset{\sim}{q})-1 = \sum_{\ell m} \hat{S}^{\ell}(k,q)Y_{\ell m}(\hat{k})Y^*_{\ell m}(\hat{q}).$$

The projections of U_4 with odd ℓ are zero.

Using a choice[5] for S(k), the integral equation for $\ell=0$ and $\ell=2$ has been solved with the results displayed as a contour plot in figure 1. The symmetry that $\hat{U}_4^{\ell}(k,q)=\hat{U}_4^{\ell}(q,k)$ has been used, displaying only half of each contour. The only major feature of each contour is a negative peak along the diagonal at $2Å^{-1}$, the roton wavenumber.

These results could be used to calculate the integrated Raman intensity but problems arise in Fourier transforming $\beta(r)$, the anisotropic pair polarizability. The calculation can be done in r-space by noticing that g(r) is obtained from a functional derivative of the normalization integral (within the Jastrow approximation). The pair-pair correlation function P can be obtained by

$$(N\rho)^2 P(\underset{\sim}{r},\underset{\sim}{r}') = \sum_{i\neq j}\sum_{k\neq\ell} \langle\delta(\underset{\sim}{r}-\underset{\sim}{r}_{ij})\delta(\underset{\sim}{r}'-\underset{\sim}{r}_{k\ell})\rangle = \frac{4}{I}\,\frac{\partial^2 I}{\delta u(\underset{\sim}{r})\delta u(\underset{\sim}{r}')}\,.$$

A convenient function to define is

$$H(\underset{\sim}{r},\underset{\sim}{r}') = N[P(\underset{\sim}{r},\underset{\sim}{r}')-g(r)g(r')] - \frac{g(r)}{\rho}\,[\delta(\underset{\sim}{r}+\underset{\sim}{r}')+\delta(\underset{\sim}{r}-\underset{\sim}{r}')].$$

Again an integral equation can be obtained from the HNC approximation. By defining the Legendre projections of H, the decoupled equation for even ℓ projections can be written as

$$\hat{H}^{\ell}(x,y) = 2g(x)g(y)\hat{M}^{\ell}(x,y) + g(x)\rho\int h^2 dh\, \hat{M}^{\ell}(h,x)\hat{H}^{\ell}(h,y)$$

where

$$M(\underset{\sim}{x}+\underset{\sim}{y}) = \frac{1}{N}\sum_k\left[1-\frac{1}{S^2(k)}\right]e^{i\underset{\sim}{k}\cdot(\underset{\sim}{x}+\underset{\sim}{y})} = \sum_{\ell m}\hat{M}^{\ell}(x,y)Y_{\ell m}(\hat{x})Y^*_{\ell m}(\hat{y}).$$

Again the odd ℓ projections are zero.

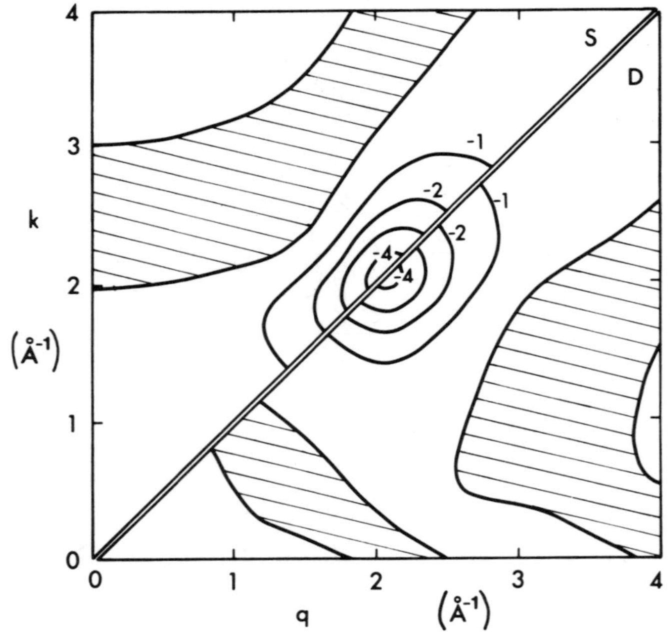

FIG. 1. Contour plots of $\ell = 0$ (S-wave) and $\ell = 2$ (D-wave) Legendre projections of $U_4(\vec{k},\vec{q})/N$ at 0.0218Å^{-3}. Shaded areas indicate $0 \le U_4^{\ell}/N < 0.5$.

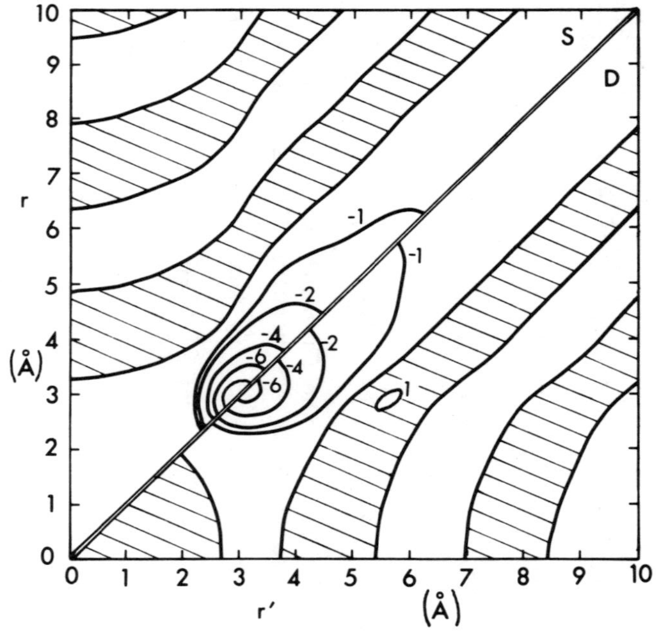

FIG. 2. Contour plots of $\ell = 0$ (S-wave) and $\ell = 2$ (D-wave) Legendre projections of $H(\vec{r},\vec{r}')$ at 0.0218Å^{-3}. Shaded areas indicate $0 \le H_\ell < 1$.

The solutions for the $\ell=0$ and $\ell=2$ projections are plotted in figure 2. A large negative peak exists along the diagonal at 3.0-3.4Å (the nearest neighbor peak in g(r) is at 3.3Å). In the remaining areas, H is small in comparison to this peak. Also if r or r' is smaller than 2Å, H is approximately zero. This is a consequence of the factor g(r)g(r') in the inhomogeneous term of the integral equation.

The $\ell=2$ projection of the pair-pair correlation function and the anisotropic pair polarizability are sufficient to calculate the total Raman intensity. Summing over polarizations the ratio of the 2nd order to the 1st order scattering intensity can be expressed as

$$\frac{I^{(2)}}{I^{(1)}} = \left(\frac{7}{3}\right) \frac{\pi}{15} \frac{N\rho^2}{S(k\to 0)} \int d\underset{\sim}{r}d\underset{\sim}{r}'\beta(r)\beta(r')Y_{22}(\hat{r})Y^*_{22}(\hat{r}')P(\underset{\sim}{r},\underset{\sim}{r}').$$

Using the results of the integral equation and the point dipole-induced-dipole (DID) form for β(r), the calculated ratio (.0039) is over twice as large as the experiment (.0016) of Greytak and Yan.[4] However, when corrections (calculated by Oxtoby and Gelbart[6] to take into account the finite size of the atoms) to β(r) are added, the calculated ratio is reduced to .0025. The pair-pair correlation function can be generated using Monte Carlo methods or molecular dynamics and similar results are obtained. Furthermore, for classical fluids, the line shape can be determined using molecular dynamics[7].

This work was done with C.E. Campbell at the University of Minnesota and supported in part by the National Science Foundation under grant DMR 76-14777. Both of us are indebted to J.W. Halley and P. Kleban for many helpful discussions.

REFERENCES

1. M. Stephen, Phys. Rev. 187, 279 (1969).
2. A.L. Fetter, J. Low Temp. Phys. 6, 487 (1972).
3. See the articles by R. Hastings and J.W. Halley in these proceedings.
4. T.J. Greytak and S. Yan, Phys. Rev. Lett. 22, 987 (1969).
5. C.C. Chang and C.E. Campbell, Phys. Rev. B15, 4238 (1977).
6. D.W. Oxtoby and W.M. Gelbart, Mol. Phys. 30, 535 (1975).
7. B.J. Alder, H.L. Strauss and J.J. Weis, J. Chem. Phys. 59, 1002 (1973); also see the article by A. Rahman in these proceedings.

RESPONSE FUNCTION OF THE SUPERCONDUCTING ORDER PARAMETER†

A. M. Goldman

School of Physics and Astronomy, Univ. of Minnesota

Minneapolis, Minn. 55455

As the order parameter of a superconductor is bilinear in electron field operators its response function is a four-point function of the field operators. Developments in recent years involving electron tunneling have led to the experimental determination of the imaginary part of the space and time Fourier transform of the response function of superconductors despite the order parameter being off-diagonal in the number representation and not having a classical laboratory conjugate field which couples to it.[1,2,3]

The superconductor under study is incorporated in an asymmetric tunneling junction in which the other electrode is a superconductor with a higher transition temperature. Then the imaginary part of the Fourier transform of the response function, $\chi''(\omega,q)$, is proportional to an excess current in the I-V characteristic of the junction with the frequency ω set by the bias voltage and the wave vector q determined by a magnetic field applied parallel to the plane of the junction. The geometry is shown in Fig. 1. Here δ is the oxide barrier thickness, λ_1 and λ_2 are the London penetration depths of the superconductors, labelled 1 and 2; T_{c1} and T_{c2} are the transition temperatures and d_1 and d_2 are the thicknesses. Usually $T_{c1} \approx T << T_{c2}$; $d_1, d_2 << \xi(T)$; $\lambda_2 < d_2$ and $\lambda_1 >> d_1$. The physical origin of the connection between the excess current and $\chi''(\omega,q)$ can be seen by modeling the pair tunneling between the two electrodes using the effective Hamiltonian

†Supported in part by USERDA under contract EY-76-S-02-1569.

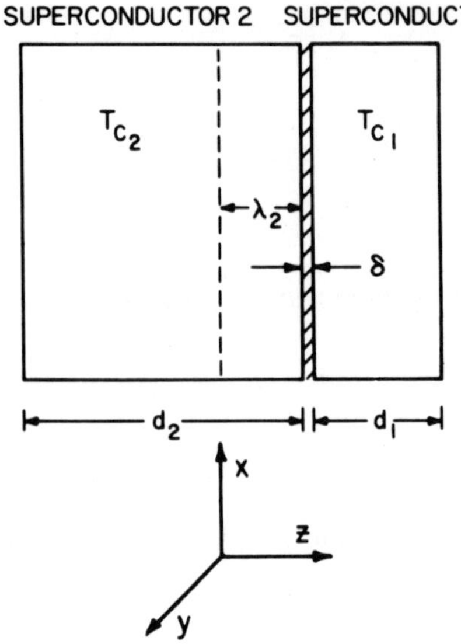

Figure 1: Schematic of an asymmetric junction. d_1 and d_2 are typically 10^{-5} cm and $\lambda_2 \sim 5 \times 10^{-6}$cm. Superconductor 2 is always Pb and superconductor 1 is Al or an alloy of Al and Er.

$$(1) \quad \mathcal{H}_I = \frac{\overline{C}}{d_1} e^{-i\omega t} \int_1 d^2 r e^{iq\chi} \hat{\Delta}(r) + h\cdot c$$

where $\omega = 2eV/h$ and $q = (2e/hc)\, H[\lambda_1 + d_2/2]$. The quantity $\hat{\Delta}$ is an operator for the order parameter of superconductor 1 and is given by $\hat{\Delta} = g\, \Psi_\uparrow \Psi_\downarrow$, where g is the coupling constant, and $\Psi_\uparrow$ and $\Psi_\downarrow$ are field operators for the destruction of spin-up and spin-down electrons. The constant $\overline{C}$ is given by

$$(2) \quad \overline{C} = \frac{\hbar}{e^2}(R_N A)^{-1} \ln \frac{4\, T_{c2}}{T_{c1}}$$

where R_N is the normal tunneling resistance and A is the area of the plane of the junction. From this effective Hamiltonian it can be shown[2] using linear response theory that there is an excess current due to pair tunneling of the form

$$(3) \quad \langle I_1 \rangle = \frac{4e\, |\overline{C}|^2 A}{\hbar d} \chi''(\omega, q) \,.$$

For $T > T_{c1}$ this is the only contribution from the tunneling of electron pairs. When $T < T_{c1}$, the quantity $\langle\hat{\Delta}\rangle_0 \neq 0$ and one finds in addition to the excess current, the usual Josephson current.

Experiments have been carried out on Pb-Al_2O_3-Al junctions and the order parameter susceptibility of Al above T_{c1} has been found to be in excellent agreement with that calculated from the time-dependent Ginzburg-Landau equation.[3] Extension of this procedure below T_{c1} is complicated by the ac and dc Josephson effects which are linear in the coupling constant of Eq. (1) and are thus much bigger than the current resulting from the induced gap which is proportional to $|C|^2$. The magnetic field dependence of the dc Josephson current can be used to get around this difficulty. One simply applies a field in the plane of a junction large enough to destroy the Josephson current, but below the critical fields of the electrodes. Alternatively, the field may be set at a value which biases the junction at one of the minima of the "diffraction pattern" plot of critical Josephson current vs. applied magnetic field.

An understanding of the data obtained below T_{c1} results if one employs a form of the fluctuation-dissipation theorem which relates $\chi''(\omega,q)$ to the Fourier transform of the order-parameter correlation function $S(\omega,q)$.[4] In the relevant limit, $\beta h\omega << 1$ and

$$(4) \quad S(\omega,q) = \frac{k_B T}{\pi} \frac{\chi''(\omega,q)}{\omega} .$$

Here the quantity $\beta = 1/k_BT$. Since χ'' is proportional to the excess current and ω is proportional to the voltage, $S(\omega,q)$ is proportional to the excess current divided by the dc voltage at which it is measured. In Fig. (2) $S(\omega,q)$ vs. ω is plotted at several temperatures for an aluminum film. A peak at the origin of $S(\omega,q)$ vs. ω results when the modes are diffusive. A peak at finite frequency is the signature of a propagating mode.

The propagating mode in this instance appears to be described by the theory of Schmid and Schön.[5] (See the dashed lines of Fig. (2).) In this theory fluctuations of both the real and imaginary parts of the order parameter are considered and the dispersion relation for the propagating mode is linear. The theory also considers the coupling of $\Delta(rt)$ to δf, the fluctuation in the quasiparticle distribution function. This coupling is the so-called "anomalous" term of the Gor'kov-Eliashberg equations.[6] In the linearized theory the real and imaginary parts of the fluctuations of $\Delta(rt)$ are decoupled and the "longitudinal" mode is always diffusive. The "transverse" mode involves a variation in the number of quasiparticles, can be excited by quasiparticle injection, and is a propagating mode at sufficiently high frequencies. The

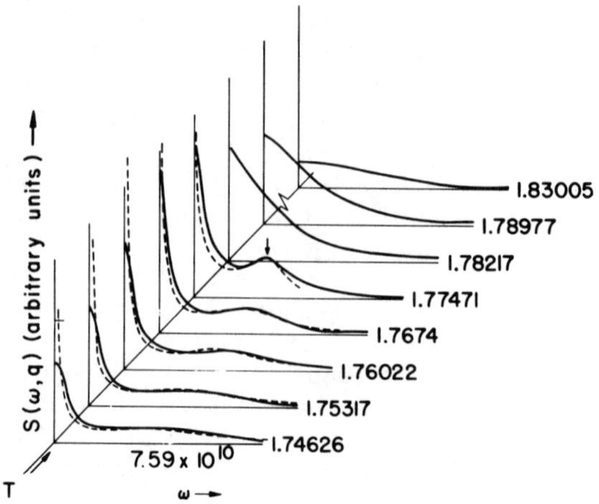

Figure 2: Structure factors at several temperatures in an Al film. This illustrates the appearance and disappearance of the propagating mode as a function of temperature.

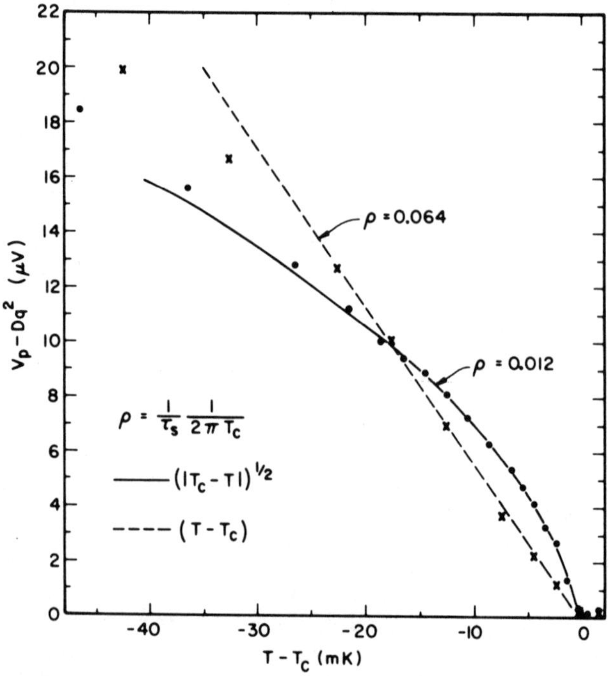

Figure 3: Longitudinal mode relaxation frequency for ρ = 0.012 and 0.064, corresponding to two different concentrations of the paramagnetic impurity Er.

charge density associated with the propagating transverse mode is small and as $\vec{k} \cdot \vec{j}_N = -\vec{k} \cdot \vec{j}_S$ the motion of supercurrent is counteracted by the normal current which is the source of the damping of the mode.[5] Propagation also occurs only when there is a gap in the single-particle excitation spectrum. Thus in the gapless regime of a superconductor doped with magnetic impurities the mode is not expected to propagate,[7] an effect which has been observed in our work on Al-Er films.[8]

A systematic investigation of the magnetic impurity dependence of $\chi''(\omega,q)$ for $T < T_{c1}$ is currently in process. In Fig. (3) we plot the relaxation frequency of the "longitudinal" mode corrected for the effect of finite q for two different depairing parameters; $\rho = 0.012$ for which the superconductor has a gap over the temperature range displayed and $\rho = 0.064$ for which a gapless regime extends 20 mK below T_c. In the gapless regime the "anomalous" term is switched off by the pair-breaking and the temperature dependence of the relaxation frequency changes from $(T_c-T)^{1/2}$ to (T_c-T) as is observed. A more detailed account of this work will be published elsewhere.[8]

In a current-carrying state the "longitudinal" and "transverse" modes are coupled and as the transport current approaches the critical current the modes are predicted to soften.[9] Such a result has been observed and will also be described in detail elsewhere.[8] The softened modes are significant in that they are believed to be the nucleation modes for phase-slip centers in weak links.

REFERENCES

1. R. A. Ferrell, J. Low Temp. Phys. 1, 433 (1969).
2. D. J. Scalapino, Phys. Rev. Lett. 24, 1052 (1970).
3. J. T. Anderson, R. V. Carlson, and A. M. Goldman, J. Low Temp. Phys. 8, 29 (1972).
4. R. V. Carlson and A. M. Goldman, J. Low Temp. Phys. 25, 67 (1976).
5. A. Schmid and G. Schön, Phys. Rev. Lett. 34, 941 (1975) and G. Schön, Thesis (1976) unpublished.
6. L. P. Gor'kov and G. M. Eliashberg, Zh. Eksp. Teor. Fiz. 54, 612 (1968) [Sov. Phys. JETP 27, 328 (1968)].
7. O. Entin-Wohlman and R. Orbach, to be published.
8. F. Aspen and A. M. Goldman, to be published.
9. V. Ambegaokar, Phys. Rev. Lett. 39, 235 (1977).

Part IV
Phonons

SECOND ORDER PHONON SPECTRA

Heinz Bilz

Max-Planck-Institut fuer Festkorperforschung

Stuttgart, Germany, Buesnauer Str. 171

1. Introduction: Phonons and their interactions with photons

In these lectures, the second order phonon spectra of solids are discussed as they are observed in infrared absorption and Raman scattering. The basic formalism for our discussion is given in the lectures by R. Stinchcombe on response theory and by W. Halley on perturbation theory.[1] Other useful information may be found in the books and reviews by Born and Huang,[2] Cochran and Cowley,[3] Maradudin et al.[4] Horton and Maradudin,[5] Birman,[6] and Bilz et al.[7] The lectures are essentially based on this last handbook article.

The dynamical properties of crystals determine the photon infrared absorption, inelastic neutron scattering and, to a large extent, inelastic photon scattering by phonons, i.e. Raman scattering. The interpretation of infrared and Raman spectra requires, therefore, an understanding of the basic features of lattice dynamics. Quantum theory of solids describes the crystal properties in terms of elementary excitations and their mutual interactions. Dynamical properties are represented by phonons (lattice vibrations) and their interactions mainly with other phonons (anharmonicity), electrons (electrophonon coupling), and photons (interactions with radiation). The emphasis of this article is on the interrelation between theory and experiment, i.e. on the microscopic or phenomenological, interpretation of experimental spectra.

At present, quantitative calculations of phonon spectra are usually restricted to diatomic crystals, especially those with structures like those of the alkali halides or germanium and its homologues, since our knowledge of the lattice vibrations is still rather poor for polyatomic and low-symmetry crystals.

The theory of lattice vibrations constitutes the background and the natural starting point for the discussion. There are recent reviews by Cochran and Cowley,[3] Ludwig,[8] Cochran,[9] Maradudin, et al.,[4] Sinha,[10] and several articles in Horton and Maradudin.[5]

An important part of the theory of lattice vibrations is the construction of models. A good example is the so-called shell model for phonons which describes the adiabatic harmonic electron-ion interaction in terms of localized charges and coupling constants. This provides a natural explanation of some long-range ion-ion forces in insulators in terms of induced dipole forces. Anharmonic extensions of this shell model and its modifications seem very desirable, and first attempts in this direction are discussed. Our approach is complementary to that by Birman.[11] He emphasizes the symmetry-related properties of crystals interacting with radiation fields (e.g. selection rules). We focus attention on the dynamical aspects: for example, the relative importance of cubic and quartic anharmonic coupling of phonons in certain absorption processes.

2. Model and microscopic theory of phonons

Lattice vibrations or their quantum-mechanical analogue, phonons, are excitations in a solid which are due to correlations between the displacements of the ions from their equilibrium positions. Microscopically, they originate from the Coulomb forces between electrons and nuclei but with a strong modification by the quantum-mechanical uncertainty law and the Pauli principle for the electrons. This leads to rather complicated effective forces between the ions, and many approximations (e.g. the adiabatic principle) and/or models (e.g. the rigid ion model) are usually introduced in order to simplify actual calculations of dispersion curves and properties. Usually, microscopic or macroscopic parameters have to be used which are fitted to experimental data, and very few a priori calculations exist where a satisfactory parameter-free description of lattice vibrations has been obtained. Thus, the determination of dispersion curves generally relies upon some experimental information. The most direct and, therefore, the most powerful method is inelastic neutron scattering which gives phonon frequencies of all wave lengths with a precision of a few percent.[12] The method sometimes fails for high-frequency modes or materials with strong incoherent scattering.

Infra-red and Raman spectra allow the determination of long-wave-length optic modes in crystals with an accuracy of more than 0.1%. The phonons with shorter wave-lengths appear in the second-order part of the spectra, which usually exhibit continuous bands and can be analyzed in a few simple cases in terms of combinations of single phonons. Therefore, in general, one needs additional information from other experimental sources, in particular from inelastic neutron scattering.

From the neutron or optical scattering data of a crystal, a set of force constants can be fitted more or less to the experimental data. With the force constants, the basis for calculations of one-, two-, or multi-phonon densities is established.

It is therefore of great importance for the analysis of phonon spectra to start with a well-grounded theoretical treatment of the lattice vibrations in the harmonic approximation. For a condensed survey of the present state of the theory of phonons in insulators, with emphasis on some important aspects of infra-red and Raman spectra, we refer the reader to articles by Cochran,[9] Sinha,[10] Sham,[13] and Bilz et al.[7,14]

3. Phonon-photon interaction

3.1 Charges and polarizabilities of ions and bonds

The treatment of phonons in terms of localized entities such as point charges, dipoles, polarizabilities, etc. and (except for the Coulomb forces) short-range nearest and weak second-nearest neighbor forces turns out to be quite successful because of the explicit consideration of electronic degrees of freedom. Formal long-range interionic forces are often due to short-range adiabatic electron-ion forces, and it is this aspect of the theory of lattice vibrations which allows for a description of phonon dispersion curves with very few parameters, for example, in a shell model or a bond charge model. The final justification of these different approaches is given by successful extension of these models to a treatment of phonon spectra observed in light absorption or scattering.

We are focussing attention on the simple case of diatomic solids. There exist obvious extensions to the case of more complex crystals. The basic distinction in the treatment of insulating solids seems to be that between ionic and covalent solids which is very common among chemists.[15,16] The main difference between these two proto-types is in the electron-ion interaction: In ionic solids, we may consider this interaction first in order to define (positively or negatively charged) ions; the crystal is then formed by the attractive forces between the ions and stabilized by the repulsive overlap forces between the valence electrons. In covalent solids, the forming of pair bonds is the decisive point which means that the originally independent atoms in a bond lose their individual character; here the attractive electron-ion interaction overcomes the repulsive ionic forces which leads to a hybridization of nearly degenerate atomic levels and consequently to an electronic charge accumulation between the atoms. In the discussion of phonons, this situation is reflected by the use of two different models: the shell model, on one side, where the different ionic and electronic multipoles are defined completely at the ionic lattice sites, and the bond charge model, on the other side, where the deformable part of the electronic charge density between the ions is treated as an independent entity often called a bond charge. In brief, we may say that in ionic solids (such as the alkali halides) the deformable part of the electronic charge density is localized at the anion lattice sites, while in covalent solids this charge density is localized between the ions. The fundamental distinction between 'ions' and 'bonds' is based on this difference.

So far, we have assumed that the third parameter in a solid, namely its 'metallicity' is relatively weak (as, e.g., in LiF or C). Metallicity describes delocalization of the bonding electrons which means that the promotion energy for an electronic charge transfer goes to zero. In such a case, for example, in a zero-gap crystal like grey tin or HgTe, both the simple ionic and covalent pictures are inadequate. Since here the screening of charges by the electrons dominates the situation, the most natural description is obtained by assuming a nearly free electron gas which keeps the positive ions together as in a neutral plasma.

The three characteristics of a solid, namely, covalency, ionicity and metallicity, are usually of different importance and behavior in different energy regimes. In the static energy limit where we do not consider any motion of the ions, we have a close relation between structure or stability and static ionic charges or bond charges. We may calculate, e.g., the cohesive energy in terms of those entities. If we consider the crystal in a low-frequency external field, as in infrared absorption, we have to introduce the corresponding dynamical entitites, i.e., transverse effective charges, etc. Only in the limiting case of weakly polarizable ionic crystals (LiF, etc.) static and dynamic charges are not very different while in strongly covalent systems, such as GaAs, the static and dynamic ionic charges may differ by an order of magnitude and show no interrelation whatsoever.[17] Finally, in purely covalent materials like diamond, infrared absorption is caused by so-called non-linear dipole moments which may be described in a model by using the electronic bond charges as effective charges. This part of the infra-red absorption is, therefore, more related to Raman scattering since here the fundamental interaction of the external light is with the electrons of the system, while the phonons come into play only by their weak modulation of the electronic polarizability.

Again, the description of the crystal polarizability is quite different in ionic as compared with covalent crystals. The linear polarizability showing up in optical phonons can be described in terms of individual ionic polarizabilities only in strongly ionic crystals where the effective field acting on the ions is close to the Lorentz-Lorenz field.

This ionic concept already becomes doubtful in crystals like MgO which still have the 'ionic' rocksalt structure but exhibit a strong delocalization of the outer $2p^6$ density of the oxygen ion. This becomes even more obvious if we analyse the non-linear Raman polarizabilities where the change of the crystal polarizability due to phonon modulation is measured. Contrary to what one might expect at first glance, this change is not a local property of well-defined and stable ions (Li^+, Na^+, Sr^{++}, etc., F^-, Cl^-, etc.) since here the outer filled electronic shells are rather rigid with respect to a change of their wave-functions. Therefore, the easiest way for those ionic solids to change their polarizability is by low-energy charge transfer from the anions to the cations which may be described in terms of interionic Raman polarizabilities, i.e., a real pair or overlap (but not 'bond') property. In covalent solids, on the contrary, the change of the polarizability of a bond is related to the dynamical change of the charge accumulation between the ions, i.e., a change of hybridization of the wave-functions. This cannot be achieved in terms of rigid bond charges but more conveniently by defining bond polarizabilities or (because of the high symmetry of the system) equivalent atomic deformabilities with intra-atomic non-linear polarizabilities. Somewhat surprisingly, this description can be carried through to the oxidic II-VI compounds including even those with sodium-chloride structure (e.g. MgO), where the 'bond' polarizability is now centered around the oxygen ion which dominates the Raman spectra with its intra-ionic non-linear deformability. This partially 'covalent' behavior of MgO and similar compounds is connected with the fact that O^{--} is not stable as a free ion. This means that the 2p charge density of O^{--} can relax in response to a displacement of neighboring ions. In crystals like $SrTiO_3$, where two of the oxygen neighbors (Ti) are very different from the other four (Sr), an important (linear and non-linear) anisotropy of the oxygen polarizability is observed which is intimately connected with the properties of the ferro-electric soft mode at q=0. We may therefore say that in $SrTiO_3$ and similar crystals, the O-Ti pairs are more covalent than the O-Sr pairs although the macroscopic dielectric properties are still isotropic due to the cubic symmetry of the crystals. In chain-like systems like Se and Te or other low-symmetry crystals, the anisotropy is showing up in the dielectric tensor and consequently in the optic phonons, infrared absorption, etc.

This introductory discussion may have shown how useful the chemical distinction between (predominantly) ionic and (strongly) covalent crystals is for a qualitative survey of the dynamical properties of solids even if the quantitative application of those concepts to a description of the measured entities is still restricted. There is, at present, a strong and increasingly successful effort going on to establish more quantitative links between the macroscopic, the microscopic and the model entities. The reader is referred, in the case of ionic solids, to the treatment of effective fields and effective charges as given by Fröhlich,[18] Szigeti,[19] Maradudin, et al.[14] and, for their interrelation to the model and microscopic theory, to the above-mentioned reviews by Cochran,[9] Sinha,[10] Sham,[13] and Bilz et al.[7,14] In the case of covalent crystals, we mention the treatments by Phillips,[16,20] by van Vechten,[21] and by Harrison.[22] The latter approach is based on papers by Hall,[23] Coulson et al.,[24] and an early attempt by Leman and Friedel.[25]

3.2 Theory of light absorption and scattering

The general theory of light absorption and scattering by solids is given in the lectures by Stinchcombe.[1] Here we focus attention on the dynamical aspects of second-order phonon spectra in specific (non-quantum) solids which are related in an obvious way to fourth-order displacement-displacement correlation functions.

The theory of infrared absorption and of Raman scattering describes processes where the absorption or scattering of light results in excitation of phonons in the crystals. There generally exists a marked distinction between the fundamental processes in light scattering and infrared absorption. Infrared absorption involves mainly the anharmonic interactions between phonons which are intimately connected with lattice stability, thermal expansion, lifetime of phonons, etc. Raman scattering, on the other hand, is due to non-linear electron-phonon coupling which produces changes of the crystal polarizability, non-linear susceptibilities, etc.

We note that infrared absorption in covalent crystals, such as diamond, concerns the electronic charges in the electronic dipole moments. These dipole transitions are directly seen in the electronic optical absorption but their details, which reflect the electronic band structure, are only important in the resonance Raman effect.

In the main part of our discussion, electronic excitations play only a "virtual" role as intermediate states. They dress the phonons by leading to polarizabilities, deformabilities, effective ionic charges and bond charges. The excitation energies for real electronic transitions in non-metals are high as compared to phonon energies. Here, in infrared absorption and non-resonant Raman scattering, the exciting photon energies are still low so that we are able to examine the above-mentioned concepts for adiabatic electrons when going to non-linear (phonon-phonon and electron-phonon) processes.

We begin with the non-relativistic Hamiltonian of the total system of radiation and matter[26]

$$H_c = H_r + H_m \tag{3.1}$$

where the Hamiltonian of the radiation H_r is

$$H_r = \frac{1}{8\pi} \int d^3\underline{r}\, (\underline{E}^2 + \underline{B}^2) \tag{3.2}$$

and the gauge invariant Hamiltonian of the crystal including its interacting with radiation reads

$$H_m = T_e(\underline{A}) + T_i(\underline{A}) + \tag{3.3}$$

$$+ U_e + U_i + U_{ei} \tag{3.4}$$

Here,

$$T_e(\underline{A}) = \frac{1}{2m} \sum_{\ell} (\underline{P}_\ell - \frac{e}{c} \underline{A})^2 \tag{3.5}$$

and

$$T_i(\underline{A}) = \sum_n \frac{1}{2M_n} (\underline{P}_n + \frac{eZ_n}{c} \underline{A})^2 \tag{3.6}$$

are the gauge invariant kinetic energies of the electrons and nuclei, respectively. The charges Z_n are the bare charges of nuclei in principle, but we shall later prefer to replace them by effective static ionic charges and to modify the interaction terms correspondingly. These terms denote

$$U_e = \frac{1}{2} \sum_{\ell,\ell'}{}' \frac{e^2}{|\underline{r}_\ell - \underline{r}_{\ell'}|} \tag{3.7}$$

the electronic interaction,

$$U_i = \frac{1}{2} \sum_{n,n'}{}' \frac{Z_n Z_{n'} e^2}{|\underline{r}_n - \underline{r}_{n'}|} \tag{3.8}$$

the ionic interaction, and

$$U_{ei} = -\sum_{\ell,n} \frac{Z_n e^2}{|\underline{r}_\ell - \underline{r}_n|} \tag{3.9}$$

that between electrons and ions.

The crystal excitations with gauge invariant momenta $\underline{p} - \frac{e}{c}\underline{A}$ and $\underline{p}_n + \frac{e}{c} Z_n \underline{A}$ are the so-called (excitonic and phonon-like) polaritions. These quasi-particles are very useful concepts in the harmonic approximation. For the general case, we prefer to separate the excitations in a crystal without the electromagnetic field from their interaction with radiation. We divide, therefore, H_m into a crystal part and an interaction term

$$H_m = H + H_{int} \tag{3.10}$$

with

$$H = T_e + T_i + U_e + U_i + U_{ei} \tag{3.11}$$

$$H_{int} = -\frac{e}{mc} \sum_{\ell} [\underline{A} \cdot \underline{P}_{\ell} + \frac{e}{2c} \underline{A}^2] + \tag{3.12}$$

$$+ \frac{e}{c} \sum_{n} \frac{Z_n}{M_n} [\underline{A} \cdot \underline{P}_n + \frac{Z_n}{2} A^2] \tag{3.13}$$

We note that we begin with the so-called radiation gauge ($\nabla \cdot \underline{A}=0, \phi=0$). Furthermore, we always work in the Schrödinger representation.

The single terms in the Hamiltonian H_{int} are not gauge invariant and therefore do not give matrix elements which are observables of the crystal. Since for most cases we can use a simple dipole approximation, we transform H_{int} with the help of a gauge transformation[26] into a more convenient form. With the introduction of the charge density

$$e\rho(r) = e[\sum_{\ell} \delta(\underline{r}-\underline{r}_{\ell}) - \sum_{n} Z_n \delta(\underline{r}-\underline{r}_n)] \tag{3.14}$$

and the charge current density

$$e\, j(\underline{r}) = \frac{e}{2m} \sum_{\ell} [\underline{P}_{\ell} \delta(\underline{r}-\underline{r}_{\ell}) + \delta(\underline{r}-\underline{r}_{\ell}) \underline{P}_{\ell}]$$

$$- \frac{e}{2} \sum_{n} \frac{Z_n}{M_n} [\underline{P}_n \delta(\underline{r}-r_n) + \delta(\underline{r}-r_n) \underline{P}_n] \tag{3.15}$$

we obtain for the interaction Hamiltonian

$$H'_{int} = -\int d^3\underline{r}[e\underline{r}\cdot\underline{E}(\underline{r},t)\rho(\underline{r})] \qquad (3.16)$$

$$-\int d^3\underline{r}\{\frac{e}{c}\underline{j}(\underline{r})\cdot[(\underline{r}\cdot \underline{grad})\underline{A}+\underline{r}x\underline{B}]\} \qquad (3.17)$$

$$+\int d^3\underline{r}\{\rho(\underline{r})[(\underline{r}\cdot \underline{grad})\underline{A}+\underline{r}x\underline{B}]^2\} \qquad (3.18)$$

H'_{int} contain, successively, the electric dipole interaction, (Eq. (3.16)), the electric quadrupole and magnetic dipole interactions, (Eq. (3.17)), and higher order corrections (Eq. (3.18)). The dipole approximation neglects all but the first term in H'_{int} which reads

$$H_{dip} = -\int d^3\underline{r}\ \underline{m}(\underline{r}) \cdot \underline{E}(\underline{r},t) \qquad (3.19)$$

with the dipole operator

$$\underline{m}(\underline{r}) = -e\underline{r}\rho(\underline{r}) \qquad (3.20)$$

and the linearly polarized field with frequency ω

$$\underline{E}(\underline{r},t) = \underline{E}(\underline{r})e^{-i\omega t} \qquad (3.21)$$

We remark that higher order (electric quadrupole, etc.) terms can be obtained in Eqs. (3.17) and (3.18) if necessary. We note that the full quantum-mechanical treatment of the coupled system of radiation and matter in the dipole approximation leads to the same result as the semi-classical treatment.[27]

3.2.1 Infrared Absorption - Formal Expressions

Infrared absorption may be described as the complex transverse response of the crystal to a (classical) external electric field. The situation is reminiscent of a system of one or more active oscillators with frequency-dependent dampings moving in a forced vibration with a single applied frequency. In the simpler crystals, only one or a very few active dispersion oscillators usually exist. This leads to a rather simple general structure of the spectra in terms of a few reflection or absorption bands. The interesting features appear in the secondary structure of these bands and give detailed information about the anharmonic coupling of phonons and related properties. There is, therefore, no advantage in transforming the total Hamiltonian at the very beginning into a system of many coupled phonon-photon modes (phonon-polaritons) which all give their individual contributions to the absorption spectra so that the useful resonance oscillator concept is lost.

To obtain the complex transverse response observed in infrared absorption, one can use a Kubo formula[1] for a system in thermal equilibrium. This microscopic approach to the linear response of a system to an external field has to be consistent with the macroscopic thermodynamical theory of susceptibilities, which defines these entities in terms of derivatives of thermodynamic potentials with respect to 'external' fields. The fact that both procedures usually lead to the same result is discussed in standard text books.[28] Here we summarize a few formulae for the dielectric susceptibility. It is useful to write the susceptibility explicitly in terms of matrix elements. We define eigenstates, n, of the isolated crystal with Hamiltonian H. According to the definitions of the foregoing section, and with $H_{nn}=\hbar\omega_n$, and $\rho_n = Z^{-1}\exp(-\beta\hbar\omega_n)$, χ can be written as the spectral representation of a retarded Green function

$$\chi_{\alpha\beta}(\underset{\sim}{q},\underset{\sim}{q}',\omega) = \frac{1}{\hbar V} G^r_\omega \left[m_\alpha(\underset{\sim}{q}) \,|\, m_\beta(\underset{\sim}{q}') \right]$$

$$= -\frac{1}{\hbar V} \sum_{nm} \frac{\rho_n - \rho_m}{\omega + \omega_n - \omega_m + i\varepsilon}\, m_\alpha(\underset{\sim}{q})_{nm}\, m_\beta(\underset{\sim}{q}')_{mn}$$

$$= \chi_{\alpha\beta}(-\underset{\sim}{q}',-\underset{\sim}{q}',-\omega) \tag{3.22}$$

Since the dipole moment density $\underset{\sim}{m}(\underset{\sim}{r})$ is a hermitian operator $\underset{\sim}{m}(\underset{\sim}{q})_{nm} = \underset{\sim}{m}(-\underset{\sim}{q})^*_{mn}$, for most purposes, it is sufficient to consider in χ only those elements with $\underset{\sim}{q}' = -\underset{\sim}{q}$, i.e.

$$\chi_{\alpha\beta}(\underset{\sim}{q},-\underset{\sim}{q},\omega) \equiv \chi_{\alpha\beta}(\underset{\sim}{q}\omega). \tag{3.23}$$

For example, in the case of a perfect lattice, it follows from lattice periodicity that the product of matrix elements $m_\alpha(\underset{\sim}{q})_{nm} m_\beta(\underset{\sim}{q}')_{mn}$ is only nonvanishing if $g + g'$ is equal either to zero or to a reciprocal lattice vector. In the infrared and even in the visible region, the wavelength of the radiation is very large compared with the lattice constant, so that for a perfect crystal attention can be restricted to $\underset{\sim}{q}' = -\underset{\sim}{q} \tilde{\ } 0$. If the product of matrix elements $m_\alpha(\underset{\sim}{q})_{nm} m_\beta(-\underset{\sim}{q})_{nm}$ is symmetric in the states n and m, one can write

$$\chi_{\alpha\beta}(\underset{\sim}{q}\omega) = \frac{1}{\hbar V}\sum_{\substack{nm \\ (\omega_n>\omega_m)}} (\rho_m-\rho_n) m_\alpha(\underset{\sim}{q})_{nm} m_\beta(-\underset{\sim}{q})_{mn} \frac{2\omega_{nm}}{\omega_{nm}^2-\omega^2-i\varepsilon\omega} = \chi_{\beta\alpha}(-\underset{\sim}{q}\omega) \quad (3.24)$$

For the diagonal elements ($\alpha=\beta$), which in the case of cubic crystals are the only important ones, one obtains for the imaginary part describing absorption

$$\mathrm{Im}\chi_{\alpha\alpha}(\underset{\sim}{q}\omega) = \frac{\pi}{\hbar V}(1-e^{-\beta\hbar\omega})\sum_{nm} \rho_n |m_\alpha(\underset{\sim}{q})_{nm}|^2 \delta(\omega+\omega_n-\omega_m) \quad (3.25)$$

The right-hand side of this equation can be expressed in terms of the Fourier transform of the dipole moment auto-correlation function

$$\int_{-\infty}^{\infty} dt e^{i\omega t} \langle m_\alpha(\underset{\sim}{q}t) m_\alpha(-\underset{\sim}{q}0)\rangle \equiv J_\omega\left(m_\alpha(\underset{\sim}{q})|m_\alpha(-\underset{\sim}{q})\right) \quad (3.26)$$

$$= 2\pi\sum_{nm} \rho_n |m_\alpha(\underset{\sim}{q})_{nm}|^2 \delta(\omega+\omega_n-\omega_m). \quad (3.27)$$

The function J represents the spectral density of the fluctuations of the dipole moment. For finite frequency the spectral density function follows from Eq. (3.25) as

$$J_\omega\left(m_\alpha(\underset{\sim}{q})|m_\alpha(-\underset{\sim}{q})\right) = 2\hbar V[1+n(\omega)]\mathrm{Im}\chi_{\alpha\alpha}(q\omega), \ (\omega\neq 0) \quad (3.28)$$

with

$$n(\omega) = (e^{\beta\hbar\omega}-1)^{-1} \quad (3.29)$$

Since the imaginary part of $\chi_{\alpha\alpha}(\underset{\sim}{q}\omega)$ is an odd function of frequency and, because of $1+n(-\omega) = -n(\omega)$, Eq. (3.28) may be written

$$J_\omega\left(m_\alpha(\underset{\sim}{q})|m_\alpha(-\underset{\sim}{q})\right) + J_{-\omega}\left(m_\alpha(\underset{\sim}{q})|m_\alpha(-\underset{\sim}{q})\right) = \quad (3.30)$$

$$= 2\hbar V \coth\left(\frac{\beta\hbar\omega}{2}\right)\mathrm{Im}\chi_{\alpha\alpha}(\underset{\sim}{q}\omega), (\omega\neq 0),$$

where

$$\coth\left(\frac{\beta h\omega}{2}\right) = 2n(\omega)+1 \quad (3.31)$$

The relation (3.30) is the fluctuation-dissipation theorem.[1]

These relations are widely used for the determination of optical constants from experimental data.[29]

Further spectral properties of the susceptibility can be derived from the Fourier transform of the commutator in the retarded Green function. In the diagonal case, $\alpha=\beta$, $\underset{\sim}{q}'=-\underset{\sim}{q}$, one obtains

$$\int_{-\infty}^{\infty} dt e^{i\omega t} \langle m_\alpha(\underset{\sim}{q}t) m_\alpha(-\underset{\sim}{q}0) - m_\alpha(-\underset{\sim}{q}0) m_\alpha(\underset{\sim}{q}t)\rangle = 2\hbar V \mathrm{Im}\chi_{\alpha\alpha}(\underset{\sim}{q}\omega) \tag{3.32}$$

This equation can be used as a starting point for discussing sum rules and moment expansions.[30]

A very useful sum rule is given by the following equation (in the specific case of a diatomic cubic crystal with one resonance frequency ω_{TO}):

$$\int_0^{'\infty'} \omega\varepsilon''(\omega)\,d\omega = \frac{2\pi^2}{\mu c}(e_T^*)^2 n \tag{3.33}$$

$$= \frac{\pi}{2}(\varepsilon_o - \varepsilon_\infty)\omega_{TO}^2 \tag{3.34}$$

This infrared dipole sum rule is completely analogous to the well-known electronic sum rule if one considers that $e_T^* = Z_T^*$ is the transverse effective ionic charge and n the number of positive charges per unit cell. The oscillator strength is essentially given by the difference between the total dielectric response ε_o in the static frequency limit minus the electronic response ε_∞ in this limit. Using the so-called Lyddane-Sachs-Teller relation we can write Eq. (3.34) in the following form

$$\int_0^{'\infty'} d\omega \frac{\varepsilon''(\omega)}{\varepsilon_\infty} = \frac{\pi}{2}(\omega_{LO}^2 - \omega_{TO}^2) = \frac{\pi}{2}\omega_{PL}^2 \tag{3.35}$$

$$\simeq \frac{\pi}{2}(\omega_{LO} + \omega_{TO})\Delta R \tag{3.36}$$

which shows that the frequency width of the reflection band, ΔR, approximately given by $\omega_{LO}-\omega_{TO}$, describes the integrated absorption reduced by ε_∞. The analogy with the electronic case is obvious if we substitute the electronic for the ionic plasma frequencies.

3.2.2. Raman scattering - Formal Expressions

The origin of phonon Raman scattering is a change of the crystal's polarizability due to a 'collision' of photons with electrons which on their part excite phonons via the non-linear electron-ion interaction. The frequency of the external light is usually high compared with the phonon frequencies. Therefore, the phonons cannot follow the frequency of the external light, i.e., the ions stay practically at rest while the electrons are polarized by the light. As an order of magnitude, we can state that the Raman cross section for ionic polarization is down by a factor of

$$\omega_{ion}^4/\omega_{el}^4 \sim (m_{el}/M_{ion})^2 \lesssim 10^{-6} \qquad (3.37)$$

as compared to the electronic Raman effect. This is equivalent to saying that the adiabatic approximation is usually very well fulfilled for Raman scattering. We note that the ionic Raman effect has so far not been seen experimentally.

For the remainder of our discussion we use the adiabatic approximation, i.e., we replace the crystal Hamiltonian H by its expectation value in the many-body electron ground state, the crystal potential. In the dipole approximation for the interaction Hamiltonian, H_{DIP}, the result of the inelastic light scattering can be described a transition polarizability amplitude $\underset{\sim}{P}(\omega,\omega')$. It consists of matrix elements of the dipole operator $\underset{\sim}{m}(\underset{\sim}{r})$ which depend on the position of $\underset{\sim}{r}_\ell$ of the ions. They can therefore be expanded into displacements of the ions from their equilibrium positions. This treatment leads to the polarizability theory of Raman scattering which is called the (generalized) Platzek approximation.[31,11]

Alternatively, one can use the polariton picture of Raman scattering which is an extension of the harmonic polariton treatment. This picture seems not to be very helpful for an interpretation of two or higher order quasi-particle excitations in a crystal, because these excitations involve preferentially phonons (or excitons) of short wavelengths for which the deviation of photon wave vector from zero is usually irrelevant.

A third approach emphasizes the analogy of Raman scattering with infrared absorption as a response function and uses the formalism of retarded Green functions for the calculations of the Raman cross section. This approach allows a systematic treatment of Raman scattering as a many-body-problem and may therefore be a useful further development of the theory.

We now summarize the results of the polarizability theory and of the Green function theory for Raman scattering. The complete system of radiation and matter is described by the Hamiltonians, Eqs. (3.1). The interaction H_{int} is used in the dipole gauge and the dipole approximation, Eq. (3.19). We assume that the scattering process can be described by transitions from well defined initial and final states of the system which are eigenstates of the 'free particle' part of the Hamiltonian H_o while the remainder V defines the different types of interaction in the system:

$$H = H_o + V \tag{3.38}$$

Here

$$H_o = H_r + H_e + H_{ph} \tag{3.39}$$

where

$$H_r|N\rangle = \hbar(N+1/2)\omega|N\rangle \tag{3.40}$$

defines the free photon states $|N\rangle$,

and $$H_e|m\rangle = [T_e + U_e + U_{ei}(R^o)]|m\rangle = E_m|m\rangle \tag{3.41}$$

corresponds to the electronic energy in the static approximation, while

$$H_{ph}|n_\lambda\rangle = (T_i + \underset{\sim}{\emptyset}_2)|n_\lambda\rangle = \hbar(n_\lambda + 1/2)\omega_\lambda|\tilde{n}_\lambda\rangle \tag{3.42}$$

describes the phonons (with quantum numbers $\lambda=(q,j)$ in the adiabatic-harmonic approximation

The interaction part is then

$$V \equiv H - H_o = H_{e,r} + H_{e,ph} + H_{ph,ph} \tag{3.43}$$

with

$$H_{e,r} \simeq H_{dip},$$

the electron-phonon coupling, $H_{e,ph}$, and the anharmonic potential,

$$H_{ph,ph} = \emptyset_{anh} \quad .$$

The cross section for Raman scattering may be described in terms of an effective interaction potential V^{eff}, which contains the photon-electron interaction as well as the electron-phonon interaction

$$\frac{d^2a}{d\Omega d\omega} = \frac{\omega_i \omega_s}{(2\pi)^2 c^4} \frac{q_s}{q_i} \int_{-\infty}^{\infty} dt e^{i\omega t} \langle i|V^{eff}(t)V^{eff}(0)|f\rangle_T \tag{3.44}$$

Here, $\langle .....\rangle_T$ denotes the thermal average at temperature T over all final states with $E=E_F$. $V^{eff}(t)$ is the time-dependent interaction operator in the Heisenberg representation. The final determination of the cross section needs the evaluation of V^{eff}. This operator determines the order of perturbation theory which is given by the number of intermediate states required for a transition from $|i\rangle$ to $|f\rangle$ with the interaction potential V in Eq. (3.43). The only part of V which leads to a creation or destruction of a photon is H_{dip}. Since it is linear in the electric field we need H_{dip} in second order for an inelastic scattering process where a photon with energy $\hbar\omega_i$ is destroyed while another photon with energy $\hbar\omega_s$ is created. The dipole operator $\underset{\sim}{m}(\underset{\sim}{r})$ is linear in the electronic and ionic coordinates $\underline{r}_\ell$ and $\underline{r}_n$ and can lead in a second order to elastic light scattering (with no net effect on the electron or phonon states). Since real transitions of electrons are forbidden, this is the only possibility of scattering in second order which determines the electronic polarizability ε_∞ of the crystal in the optic frequency regime.

In order to obtain inelastic scattering due to phonons we have to consider the electron-phonon or the phonon-phonon coupling. The latter one, in fact, can parallel the polarization process but cannot lead to a net result of phonon excitations or destructions in a low order process since there is no direct relation between the phonon-phonon coupling and the electron-phonon interaction. We therefore expect that the dominant term in phonon light scattering is due to $H_{e,ph}$. Raman scattering is, therefore, at least a third order process, if one-phonon processes are considered but of quartic or higher order for two or many-order phonon processes.

The general structure of third-order perturbation theory[32] gives

$$V_{if}^{eff} = \sum_{mm'} \frac{\langle i|V|m\rangle\langle m|V|m'\rangle\langle m'|V\ |f\rangle}{(E_m-E_i)(E_{m'}-E_i)} \tag{3.45}$$

A typical term of (3.45) is at T=0,

$$\left(V^{eff}_{if}\right)_{mm'} = \frac{\langle 0|H_{dip}|m\rangle\langle m|H_{e,ph}|m'\rangle \ \langle m'|H_{dip}|0\rangle}{(E_m-E_o)(E_{m'}-E_o-\hbar\Omega)} \tag{3.46}$$

$$= \frac{(2\pi\hbar\omega_i)^{1/2} \ d_{om} D_{mm'} \ d^*_{m'0} (2\pi\hbar\omega_s)^{1/2}}{(\omega_i+\omega_{om}-i\varepsilon)(\omega_s+\omega_{om'}-i\varepsilon)} \tag{3.47}$$

with the dipole matrix elements

$$ed_{om} = \langle 0|\underset{\sim}{m}(\underset{\sim}{r})\cdot\underset{\sim}{e}(\underset{\sim}{q})|m\rangle \tag{3.48}$$

calculated with equilibrium electronic wave functions and tensorial electron-phonon matrix elements

$$\underset{\sim}{D}_{mm'} = \langle m|H_{e,ph}|m'\rangle \tag{3.49}$$

the explicit form of which depends on the specific approximation used in the calculation. The cross section now becomes

$$\frac{d^2\sigma}{d\Omega_s \, d\omega_s} = \frac{\omega_i\omega_s^3}{c^4} \sum_{\alpha\beta,\alpha'\beta'} e^i_\alpha e^s_\beta \, i_{\alpha\beta,\alpha'\beta'} e^i_{\alpha'} e^s_{\beta'} \tag{3.50}$$

with the fourth rank tensor

$$i_{\alpha\beta,\alpha'\beta'}(\Omega) = \frac{(2\pi)^2\hbar}{\omega_i\omega_s} \left(V^{eff}\right)^i_{\alpha\beta} \left(V^{eff}\right)^{s\dagger}_{\alpha'\beta'} \delta(\Omega-\omega_i+\omega_s) \tag{3.51}$$

$$= \left[\sum_{mm'} \frac{d_{om,\alpha} \ D_{mm'} d_{m'o,\beta}}{(\omega_i+\omega_{om}-i\varepsilon)(\omega_s+\omega_{om'}-i\varepsilon)}\right] \left[\sum_{mm'} \frac{d_{om,\alpha'} \ D_{mm'} \ d_{m'0,\beta'}}{(\omega_i + \omega_{om}-i\varepsilon)(\omega_s+\omega_{om'}-i\varepsilon)}\right] \tag{3.52}$$

Here we have used the relation between the intensity of the electric field and the energy of the corresponding photon in a monochromatic light flux

$$\hbar\omega_{i,s} = \frac{1}{2\pi} \ | \ \underset{\sim}{E}^{i,s}(\omega)|^2 \ . \tag{3.53}$$

To make contact with the usual representation of the scattering tensor $\underline{i}$ in terms of transition polarizabilities P we go back from the representation of the interaction V in terms of matrix elements with equilibrium wave functions to those with adiabatic wave functions. Then, the dipole matrix elements are displacement-dependent and, instead of using the electron-phonon interaction $H_{e,ph}$ explicitly, we may use an expansion of the P's into ionic displacements. The transition rate now looks the same as in second order perturbation theory but becomes explicitly dependent on the ionic displacements

$$W_{if} = \frac{2\pi}{\hbar} \left| \underline{P}^{on} (\underset{\sim}{q}_i \underset{\sim}{q}_s) \right|^2 (2\pi\omega_i)(2\pi\omega_s)\delta(E_i - E_\ell) \tag{3.54}$$

where use has been made of Eq. (3.53), and the transition polaarizability is given in the dipole approximation. For the purpose of this section it is transformed with the help of a Fourier transform of the dipole operator into the form

$$P^{on}_{\alpha\beta}(\underset{\sim}{q}_i \underset{\sim}{q}_s) = \int d^3\underline{r} e^{i(\underset{\sim}{q}_i - \underset{\sim}{q}_s)\cdot \underline{r}} P_{\alpha\beta}(\underline{r}) \tag{3.55}$$

$$= \frac{1}{\hbar} \sum_m \left[\frac{\langle 0|m_\alpha(\underset{\sim}{q}_s)|m\rangle\langle m|m_\beta(\underset{\sim}{q}_i)|n\rangle}{-\omega_{mo} + \omega_i - i\varepsilon} - \frac{\langle 0|m_\alpha(\underset{\sim}{q}_i)|m\rangle\langle m|m(\underset{\sim}{q}_s)|n\rangle}{\omega_{mo} + \omega_s - i\varepsilon} \right]$$

This shows clearly the form of a second order perturbation theory with displacement-dependent dipole moments $\underline{m}$. Equivalently, we can replace the second order perturbation theory with the dipole interaction by a first order theory with an effective Hamiltonian:

$$V^{eff} = -\frac{1}{2} \int d^3\underline{r}\, \underline{E}(\underline{r})\, \underline{\chi}(\underline{r})\, \underline{E}(\underline{r}) \tag{3.56}$$

with the susceptibility

$$\chi(\underline{r}) = \frac{1}{V} \underline{P}(\underline{r}) \tag{3.57}$$

Considering the finite temperature of the crystal we use the density matrix to calculate thermal averages

$$\rho_n = Z^{-1} \exp(-\beta\hbar\omega_n) \tag{3.58}$$

with $\beta^{-1}=kT$, and ω_n denoting the different phonon energies. We obtain then the scattering tensor

$$i_{\alpha\beta,\alpha'\beta'}(\Omega, \underset{\sim}{Q}) = \sum_{nm} \rho_n P^{nm}_{\alpha\beta}{}^{*}(\underset{\sim}{Q}) P^{mn}_{\alpha'\beta'}(-\underset{\sim}{Q})\ \delta(\Omega-\omega_{nm}) \tag{3.59}$$

$$= \int e^{-irt} \langle P^*_{\alpha\beta}(\underset{\sim}{Q},t) | P_{\alpha'\beta'}(-\underset{\sim}{Q})\rangle \frac{dt}{2\pi} \tag{3.60}$$

The Raman scattering tensor is essentially the Raman dynamic form factor $S(\Omega,Q)$. The quantum-mechanical representation for $\underline{P}$ of Eq. (3.60) is related to the electronic susceptibility in a simple manner:

$$\sum_n \rho_n P_{\alpha\beta}(\omega,\underset{\sim}{q})_{nm} = \chi_{\alpha\beta}(\omega,\underset{\sim}{q}) \tag{3.61}$$

or

$$\langle \underline{P}(\omega,\underset{\sim}{q})\rangle = \underline{\chi}(\omega,\underset{\sim}{q}) \tag{3.62}$$

Since we are interested in the effect of photons, we used Platzek's approximation[2,31] and assumed that the electrons are in the ground state at the beginning and at the end of the interaction, so that the quantum numbers n and m denote merely the phonon states. The polarizability will therefore be expanded into phonon normal coordinates so that explicit formulae are obtained for one-phonon and multi-phonon scattering processes in crystals in various approximations.

Next we describe the Green function theory of Raman scattering, analogous to our earlier treatment of infrared absorption. We develop the theory here by replacing the dipole moments of the infrared case by polarizabilities or susceptibilities of the Raman case. The Raman cross section may be interpreted as a non-linear response function due to the fluctuations of the crystal's electronic polarizability. These fluctuations scatter the incident photon and may therefore be used to define a photon self-energy, the imaginary part of which is the finite photon lifetime due to Raman scattering. We begin with the retarded Green function of an incident photon with quantum numbers $\rho_i=(q_i,s_i)$ which may be defined by[7]

$$G_{t-t'}\left(A(\rho_i)\,|\,A(\overline{\rho}_i)\right) = i\Theta(t-t')\langle[A(\rho_i,t),\ A(\overline{\rho}_i,t')]_-\rangle \tag{3.63}$$

Here, normal coordinates $A(\rho)$ are introduced and the vector potential $\underline{A}(\underline{r})$ and the electric field $\underline{E}(\underline{r})$ are written as

$$\underline{A}(\underline{r}) = \frac{1}{V^{1/2}} \sum_{\rho} A(\rho)\underline{e}(\rho)e^{i\underset{\sim}{q}\cdot\underline{r}} + \text{c.c} \tag{3.64}$$

$$\underline{E}(\underline{r}) = -i\,\frac{1}{V^{1/2}} \sum_{\rho} A(\rho)\underline{e}(\rho)\,\frac{\omega_\rho}{c}\,e^{i\underset{\sim}{q}\cdot\underline{r}} + \text{c.c.} \tag{3.65}$$

$$A(\rho) = \left(\frac{2\pi\hbar}{\omega_\rho}\right)^{1/2} [a^\dagger(\rho) + a(\rho)]\ . \tag{3.66}$$

The phenomenological Hamiltonian of the radiation field in the nearly transparent frequency regime of the crystal has the form

$$H_r = \frac{1}{8\pi} \int d^3\underline{r} \sum_{\alpha\beta} \varepsilon_{\alpha\beta} E_\alpha(\underline{r}) E_\beta(\underline{r}) \tag{3.67}$$

$$= \hbar \sum_{\rho} \omega_\rho [a^\dagger(\rho)a(\rho)+1/2] \tag{3.68}$$

where $\varepsilon_{\alpha\beta}$ is the frequency-dependent dielectric tensor of the crystal in the equilibrium configuration, while the $a, a^\dagger$ are photon destruction and creation operators with the usual commutation rules. c is the light velocity in vacuum. The polarization vectors obey the condition $\underline{e}(\underset{\sim}{q}) = \underline{e}^*(\underset{\sim}{q})$, and the light frequency is determined by

$$\omega_\rho^2 = (cq)^2 [\sum_{\alpha\beta} e^*_\alpha(\rho)\varepsilon_{\alpha\beta}(\omega_\rho)e_\beta(\rho)]^{-1} \tag{3.69}$$

If the nuclei are allowed to vibrate their motion is represented by the anharmonic lattice Hamiltonian which includes Coulomb forces but no retardation. The effect of the phonons on the light propagation is here taken into account by introducing a space- and time-dependent electronic susceptibility or optical dielectric constant,

$$\varepsilon_{\alpha\beta}(\underline{r},t) = \varepsilon_{\alpha\beta} + 4\pi\delta\chi_{\alpha\beta}(\underline{r},t) \tag{3.70}$$

It is assumed that the fluctuations depend only on the displacement of the atoms from their equilibrium positions so that they can be expanded in terms of phonon normal coordinates. The fluctuation in Eq. (3.70) is related to a change in the energy of the phonon-photon system giving rise to the interaction Hamiltonian

$$H_{er} = -\frac{1}{2}\int d^3\underline{r} \sum_{\alpha\beta} \delta\chi_{\alpha\beta}(\underline{r})E_\alpha(\underline{r})E_\beta(\underline{r}) \tag{3.71}$$

$$= -\frac{\hbar}{2} \sum_{\rho_1\rho_2} \delta\chi_{\rho_1\rho_2} A(\rho_1)A(\rho_2)$$

Here the coefficient

$$\delta\chi_{\rho_1\rho_2} = 2\pi \sum_{\alpha\beta} \frac{(\omega_1\omega_2)^{3/2}}{c^2 q_1 q_2} e_\alpha(\rho_1)e_\beta(\rho_2)\delta\chi_{\alpha\beta}(-\underset{\sim}{q}_1,-\underset{\sim}{q}_2) \tag{3.72}$$

is expressed by the Fourier transform of the spatially varying fluctuation $\delta\underline{\chi}$. Then, the total Hamiltonian is

$$H = H_r + H_{er} + V \tag{3.73}$$

This Hamiltonian describes the phonon-photon system in an approximation suited to treating scattering of light at optical frequencies which are high compared to the phonon frequencies.

For the determination of the transition rate of light scattering processes the self-energy of the incident photon must be analysed. This is done by obtaining the equation of motion for the retarded Green function. We obtain for the Green function the following equation

$$\left\{ \frac{\partial^2}{\partial t^2} + \omega^2_{\rho_i} \right\} G_{t-t'}\left(A(\rho_i)|A(\tilde{\rho}_i)\right) =$$

$$2\omega_{\rho_i}\delta(t-t') + 2\omega_{\rho_i}\sum_{\rho_s} G_{t-t'}\left(\delta\chi_{\rho_i\rho_s} A(\rho_s)|\ A(\overline{\rho}_i)\right) \tag{3.74}$$

When the new Green functions on the r.h.s. are differentiated with respect to t' the result is

$$\left\{ \frac{\partial^2}{\partial t'^2} + \omega^2_{\rho_i} \right\} G_{t-t'}\left(\delta\chi_{\overline{\rho}_i\rho_s} A(\rho_s)|A(\overline{\rho}_i)\right) = \tag{3.75}$$

$$2\omega_{\rho_i} \sum_{\rho_{s'}} G_{t-t'}\left\{ \delta\chi_{\overline{\rho}_i\rho_s} A(\rho_s) \mid \delta\chi_{\overline{\rho}_i\rho_{s'}} A(\rho_{s'}) \right\} \tag{3.75}$$

because the $\delta\chi$ and $A(\rho_i)$ commute for equal time. Taking the Fourier transform of Eqs. (3.74) and (3.75) results in

$$G_\omega\left\{A(\rho_i)\,|\,A(\overline{\rho}_i)\right\} =$$

$$= G^{(0)}_{\rho_i}(\omega) + G^{(0)}_{\rho_i}(\omega) \sum_{\rho_s\rho_{s'}} G_\omega\left\{ \delta\chi_{\overline{\rho}_i\rho_s} A(\rho_s)\,|\,\delta\chi_{\rho_i\rho_{s'}} A(\rho_{s'})\right\} G^{(0)}_{\rho_i}(\omega) \tag{3.76}$$

where

$$G^{(0)}_{\rho_i}(\omega) = \left.\frac{2\omega_{\rho_i}}{\omega^2_{\rho_i} - (\omega - i\varepsilon)^2}\right|_{\varepsilon\to 0^+} \tag{3.77}$$

In the random phase approximation one arrives after some approximations at the cross section for Raman scattering.

$$\frac{d^2\sigma}{d\Omega d\omega_s} = \frac{\omega_i\omega_s^3}{2\pi c^3\varepsilon^{1/2}} \sum_{\substack{\alpha\beta\\ \alpha'\beta'}} e_\alpha(\rho_i)e_\beta(\rho_s)\; J_\Omega\left\{P_{\alpha\beta}(\underset{\sim}{Q})\,|\,P_{\alpha'\beta'}(\underset{\sim}{\tilde{Q}})\right\} e_{\alpha'}(\rho_i)e_{\beta'}(\rho_s) \tag{3.78}$$

with

$P_{\alpha\beta}(\underset{\sim}{Q}) = V\delta\chi_{\alpha\beta}(\underset{\sim}{Q})$. J_Ω is defined in equation (3.79) below.

This result expresses the scattering rate in terms of the susceptibility autocorrelation function as in (3.60). It must be emphasized, however, that we have calculated the radiation inside of the crystal whereas the intensity of the free, scattered radiation, as measured outside of the crystal, is more conveniently defined with respect to a different solid angle element, say $d\Omega_f$, which is related to $d\Omega_s$

inside of the crystal by the surface geometry and by the laws of refraction. Assuming, for simplicity, that the scattered ray passes the surface vertically, we have $d\Omega_s = \varepsilon^{1/2} d\Omega_f$.

To make the formal similarity of infra-red and Raman processes more apparent, we may use the relations between correlation and Green functions. One obtains for the spectral density function J_Ω:

$$J_\Omega(\delta\chi|\delta\chi) = 2[n(\Omega)+1] \; \mathrm{Im}\; G^r_\Omega \; (\delta\chi|\delta\chi) \tag{3.79}$$

$$J_{-\Omega} = \{n(\Omega)/[n(\Omega)+1]\} \; J_\Omega \tag{3.80}$$

where J_Ω are the Stokes and $J_{-\Omega}$ the anti-Stokes components of the Raman scattering and $G^r_\Omega \; (\delta\chi|\delta\chi)$ the retarded Green function of the polarizability $\underline{P}$ in the (ω,q) representation. The Raman tensor is, therefore, for the Stokes component

$$i_{\alpha\beta,\alpha'\beta'}(\underline{Q},\Omega) = [n(\Omega)+1] \; \mathrm{Im} G^r_\Omega(\delta\chi_{\alpha\beta}|\delta\chi_{\alpha'\beta'}) \tag{3.81}$$

A comparison of this equation with that for the susceptibility $\chi_{\alpha\beta}$ in Eq. (3.22) shows their formal similarity.

3.2.3. Infrared Absorption in Terms of Phonons

We assume that the changes of the crystal state can be described by a certain net number of phonons, with definite wave vector q and branch indices j, which have been excited in the crystal (absorption or Stokes scattering) or taken from its phonon resevoir (emission or anti-Stokes scattering). The electrons are always supposed to be again in their initial states after the interaction has taken place. It is well-known from quantum mechanics that such polarization processes may be satisfactorily approximated by a two-level system with an effective gap oscillator strength. This allows for a pseudo-classical description of the electronic excitations in terms of oscillators. This description may be extended to the non-linear couplings as they appear. e.g., in the nonlinear dipole moments. An important tool for the understanding of crystal excitations is the study of the change of external parameters such as temperature, pressure, and static fields. However, the discussion of effects induced by a variation of these parameters will not be covered in these lectures. The reader is referred to the existing literature[2].

The lattice normal vibrations are introduced in the adiabatic and harmonic approximations. Each phonon is characterized by a wave vector q and by a branch index j, and determined by its frequency $\omega_\lambda \equiv \omega(qj)$ and polarization vectors $\underline{e}(\kappa|\lambda)$:

$$\underline{U}(L) = (NM_\kappa)^{-1/2} \sum_\lambda \underline{e}(\kappa|\lambda)Q(\lambda)e^{i\underline{q}\cdot\underline{x}(L)}, \quad L\equiv(\ell,\kappa) \tag{3.82}$$

The normal coordinates $Q(\lambda)$ contain the frequency dependence $\exp(i\omega t)$.

Expression (3.82) describes plane lattice waves with complex amplitudes $Q(\lambda)$. Quantizing these harmonic vibrations, the lattice hamiltonian becomes

$$H = \hbar\sum_\lambda \omega_\lambda [a^\dagger(\lambda)a(\lambda) + 1/2] \tag{3.83}$$

The energy quanta $\hbar\omega_\lambda$ are the phonons; $a^\dagger(\lambda)$ and $a(\lambda)$ are the phonon creation and destruction operators respectively. Corresponding to the classical normal coordinates $Q(\lambda)$, phonon field operators

$$A(\lambda) = \left(\frac{2\omega_\lambda}{\hbar}\right)^{1/2} Q(\lambda) = a(\lambda) + a^\dagger(\lambda) \tag{3.84}$$

are used $[\lambda=(-q,j)]$. All properties of a crystal which are uniquely defined by the positions of the nuclei and which can be expanded in the nuclear displacements may be expressed by these phonon field operators. The following commutation relation will be used

$$[\dot{A}(\lambda),A(\lambda')] - -i2\omega(\lambda)\Delta(\underset{\sim}{q}+\underset{\sim}{q}')\delta_{jj'} \equiv -i2\omega_\lambda\delta_{\lambda\bar{\lambda}'} \tag{3.85}$$

The delta factor

$$\Delta(\underset{\sim}{q}) = N^{-1}\sum_\ell e^{i\underset{\sim}{q}\cdot\underline{x}(\ell)} \tag{3.86}$$

equals one if q is zero or a reciprocal lattice vector; in all other cases, $\Delta(\underset{\sim}{q})$ vanishes. The anharmonic part of the potential reads

$$H_A = \sum_{n\geqq 3} \frac{1}{n!} \sum_{L_1 \ldots L_n} \sum_{\alpha_1,\ldots,\alpha_N} \Phi_{\alpha_1,\ldots,\alpha_N}(L_1 \ldots L_N) U_{\alpha_1}(L_1) \ldots U_{\alpha_N}(L_N) \tag{3.87}$$

By this interaction Hamiltonian, processes are described in which three or more phonons interact simultaneously. The expansion (3.87) refers to the configuration with minimal potential energy. This configuration is defined by the conditions $\partial\Phi/\partial U_\alpha(L)=0$ for vanishing stresses.

With (3.84), the displacements of Eq. (3.82) can be written

$$U_\alpha(L) = \sum_\lambda \varepsilon_\alpha(\kappa|\lambda) A(\lambda) e^{i\underset{\sim}{q}\cdot\underline{x}(L)} \tag{3.88}$$

Here, for compactness, the vectors

$$\underline{\varepsilon}(\kappa|\lambda) \equiv \left(\frac{\hbar}{NM_\kappa 2\omega_\lambda}\right)^{1/2} \underline{e}(\kappa|\lambda) \tag{3.89}$$

are introduced. Substituting (3.82) and (3.84) in (3.87) we obtain

$$H_A = \sum_{n\geqq 3} \frac{\hbar}{n!} \sum_{\lambda_1\ldots\lambda_n} \phi_n(\lambda_1\ldots\lambda_n) A(\lambda_1)\ldots A(\lambda_n) \tag{3.90}$$

The Fourier transformed coefficients of the potential are given by

$$\phi_n(\lambda_1\ldots\lambda_n) = \frac{1}{\hbar} \Delta(\underset{\sim}{q}_1 + \ldots + \underset{\sim}{q}_n) \sum_{\substack{L_1 \ldots L_n \\ \alpha_1 \ldots \alpha_n}} \phi_{\alpha_1\ldots\alpha_n}(L_1 \ldots L_n) \times \tag{3.91}$$

$$\times\, \varepsilon_{\alpha_1}(\kappa_1|\lambda_1) \ldots \varepsilon_{\alpha_n}(\kappa_n|\lambda_n) \exp\{i[\underset{\sim}{q}_1\cdot x(L_1) + \ldots + \underset{\sim}{q}_n\cdot x(L_n)]\}$$

The delta factor expresses quasi-momentum conservation. The $\underset{\sim}{q}$-dependent coefficients are symmetric in the phonon indices λ.

Taking into account the phonon-phonon interaction H_A the lattice Hamiltonian is

$$H = \hbar \sum_\lambda \omega_\lambda [a^\dagger(\lambda)a(\lambda)+\tfrac{1}{2}] + \sum_{n\geqq 3} \frac{\hbar}{n!} \sum_{\lambda_1\ldots\lambda_n} \phi_n(\lambda_1\ldots\lambda_n) A(\lambda_1)\ldots A(\lambda_n) \tag{3.92}$$

Within the adiabatic approximation the crystal dipole moment can be given as a series in nuclear displacements:

$$\underline{M} = \underline{M}^{o} + \sum_{L\beta} \underline{M}_{1\beta}(L)U_{\beta}(L) + \frac{1}{2}\sum_{LL'}\sum_{\beta\gamma} M_{2\beta\gamma}(LL')U_{\beta}(L)U_{\gamma}(L') + \ldots \tag{3.93}$$

$\underline{M}^{o}$ denotes a constant dipole moment of the elementary cell. The linear dipole moment is the term of first order in the displacements; the coefficients $\underline{M}_{1\beta}(L)$ denote the formal charge tensor of the lattice particles. The higher-order dipole moments are physically explained by virtual electronic excitations during the displacements of an atom from its equilibrium position. These moments are connected with deformations of the electronic charge distribution. In the rigid ion model, only the linear dipole moment is present. In models like the simple shell model, non-linear dipole moments arise by means of anharmonic forces. For alkali halides these may be derived from a Born-Mayer type two-body potential, while for the diamond structure, because of the equality of the atoms, three-atom interactions must be considered.[33] The relation of the second-order dipole moment to the anharmonicity in infra-red absorption has been investigated, for example, by Szigeti,[33] Cowley,[34] Borik,[35] and Knohl.[36]

Substituting (3.88) into (3.93), the dipole moment for N elementary cells ($\lambda^{o} \equiv (\underline{0}j)$) is written:

$$\underline{M} = \underline{M}^{o}_{o} + \hbar\left[\sum_{j} \underline{M}_{1}(\lambda^{o})A(\lambda^{o}) + \frac{1}{2}\sum_{\lambda\lambda'} \underline{M}_{2}(\lambda\bar{\lambda}')A(\lambda)A(\bar{\lambda}')\ldots\right] \tag{3.94}$$

The representation of the lattice dipole moment in (3.94) allows the expansion of the susceptibility in terms of normal coordinates $A(\lambda)$.

The lattice part of the susceptibility (see equation (3.22)) is decomposed into a double series of retarded phonon Green functions by

$$G_t(M_\alpha|M_\beta) = \sum_{nm\geq 1} \frac{1}{n!m!} \sum_{\substack{\lambda_1\ldots\lambda_n \\ \lambda'_1\ldots\lambda'_m}} M_\alpha(\lambda_1\ldots\lambda_n)M_\beta(\lambda_i\ldots\lambda'_m)\, G_t(\lambda_1\ldots\lambda_n|\lambda'_1\ldots\lambda'_m) \tag{3.95}$$

where

$$G_t(\lambda_1\ldots\lambda_n|\lambda'_1\ldots\lambda'_m) = \tag{3.96}$$

$$i\Theta(t)\langle|[A(\lambda_1 t)\ldots A(\lambda_n t), A(\lambda_1' 0)\ldots A(\lambda_m' 0)]|\rangle$$

With the Fourier transforms of (3.95) and (3.96) the lattice susceptibility tensor

$$\chi^L(\omega) = \frac{1}{\hbar V}\sum_{nm\geq 1}\chi^{(nm)}(\omega) \tag{3.97}$$

is obtained where

$$\chi_{\alpha\beta}^{(nm)}(\omega) = \frac{1}{n!m!}\sum M_\alpha(\lambda_1\ldots\lambda_n)M_\beta(\lambda_1'\ldots\lambda_m')\; G_\omega(\lambda_1\ldots\lambda_n|\lambda_1'\ldots\lambda_m')$$

The phonon Green functions (3.96) are completely determined by the lattice Hamiltonian. They describe the phonons and the phonon-phonon interactions; the phonon-photon interaction is represented by the coefficients of the dipole moment, $M(\lambda_1\ldots\lambda_n)$. The electronic part of the susceptibility is given by,

$$(\lambda^o = (\underline{0}j))$$

$$P_{\alpha\beta}^{el} = P_{\alpha\beta}^{o} + \sum_j P_{\alpha\beta}(\lambda^o)\langle A(\lambda^o)\rangle + \frac{1}{2}\sum_{F,j_1,j_2} P_{\alpha\beta}(\lambda_1,\bar{\lambda}_2)\langle A(\lambda_1)A(\bar{\lambda}_2)\rangle + \ldots \tag{3.98}$$

where the temperature dependent part is represented by a series of thermal average values.

In the harmonic approximation, the phonon Green functions and the susceptibilities can be calculated exactly. We obtain for linear dipole moments (n=m=1 in Eq. (3.97)

$$G_\omega^{(0)}(\lambda|\lambda') = \lim_{\varepsilon\to 0^+}\frac{2\omega_\lambda}{\omega_\lambda^2 - (\omega+i\varepsilon)^2}\,\delta_{\lambda\lambda'}\,, \tag{3.99}$$

and the susceptibility of polar crystals with undamped dispersion oscillators is given by

$$\chi_{\alpha\beta}^{(1,1)} = \sum_{\lambda^o} M_\alpha(\lambda^o)M_\beta(\lambda^o) \frac{2\omega_{\lambda^o}}{\omega_{\lambda^o}^2 - (\omega+i\varepsilon)^2} . \qquad (3.100)$$

According to (3.100) the absorption spectrum consists of a series of sharp lines corresponding to the number of dispersion oscillators λ^o with frequencies ω_{λ^o}. In the simple case of an ionic cubic diatomic lattice (the transverse optical frequency ω_{TO} is the "Reststrahlen" - oscillator) the dipole moments are expressed by

$$M_\alpha(R)M_\beta(R) = \frac{(Ze)^2}{\mu\hbar 2\omega_{TO}} \delta_{\alpha\beta} \qquad (3.101)$$

and the susceptibility by

$$\chi_R^L(\omega) = \frac{(Ze)^2}{\mu V_o} \frac{1}{\omega_{TO}^2 - (\omega+i\varepsilon)^2} = \frac{1}{4\pi} \frac{\omega_{PL}^2}{\omega_{TO}^2 - (\omega+ i\varepsilon)^2} . \qquad (3.102)$$

Here, Z e is the ionic charge, V_o is the volume of the elementary cell, and μ is the reduced ionic mass.

For second-order dipole moments, theoretical investigations of the absorption were done first, by Born and Huang,[2] Lax and Burstein,[37] and Sizigeti.[38] Their results can be described by a retarded harmonic two-phonon Green function,

$$G_{2,\lambda_1\lambda_2}(\omega) = \frac{2(\omega_1+\omega_2)(1+n_1+n_2)}{(\omega_1+\omega_2)^2 - (\omega+i\varepsilon)^2} + \frac{2(\omega_1-\omega_2)(n_2-n_1)}{(\omega_1-\omega_2)^2 (\omega+i\varepsilon)^2} \qquad (3.103)$$

With this Green function, the susceptibility is obtained

$$\chi_{\alpha\beta}^{(2,2)}(\omega) = \frac{1}{2} \sum_{\lambda_1\lambda_2} M_\alpha(\lambda_1\lambda_2)M_\beta(\lambda_1\lambda_2) \sum_{+-} \frac{2(\omega_1\pm\omega_2)\left[(n_2+\frac{1}{2})\pm(n_1+\frac{1}{2})\right]}{(\omega_1\pm\omega_2)^2 - (\omega+i\varepsilon)^2}$$

(3.104)

The harmonic approximation for χ^L is of practical use only in nonionic crystals like germanium which do not contain infra-red active dispersion oscillators. Even in III-V compounds, or substances like SiC, the integrated oscillator strength of the absorption via the linear dipole moment is comparable with that of the non-linear dipole monent and dominates in the frequency region near the transverse optic frequency ω_{TO}. In that case, the frequency dependence of the absorption is strongly determined by the frequency-dependent damping function which might be understood as the inverse lifetime of the dispersion oscillator with $\omega=\omega_{TO}$. This damping is due to the anharmonic coupling of phonons and, therefore, is obtained by calculating all possible combinations of two and more phonons which contribute to a delay of the TO-mode (as far as they are consistent with conservation laws of energy, quasi-momentum, parity, etc.). This means that a representation of the retarded Green function $G_\omega(n|m)$ in an anharmonic approximation is thought adequate for the discussion of infra-red spectra. Such an approximation is derived from a solution of the equations of motion for G_ω. Since these functions are essentially products of phonon field operators $A(\lambda)$, the starting point is the equation of motion

$$\left(\frac{\partial^2}{\partial t^2} + \omega_\lambda^2\right) A(\lambda) = -2\omega(\lambda) \sum_{n \geqq 3} V^{(n)}(\lambda) \tag{3.105}$$

with

$$V^{(n)}(\lambda) = \sum_{\lambda_1 \ldots \lambda_{n-1}} \frac{1}{(n-1)!} \Phi_n(\lambda, \lambda_1 \ldots \lambda_{n-1}) A(\lambda_1) \ldots A(\lambda_{n-1})$$

Here, effects of static external fields, pressure, etc. are neglected.

From (3.105) the following equation of motion for the retarded one phonon Green function is obtained:

$$\left[\omega_{\lambda_1}^2 - (\omega+i\varepsilon)^2\right] G_\omega(\lambda_1|\lambda_2) = 2\omega_{\lambda_1} \delta_{\lambda_1 \bar{\lambda}_2} - \omega_{\lambda_1} \sum_{\lambda_3 \lambda_4} \Phi_3(\bar{\lambda}_1 \lambda_3 \lambda_4) G_\omega(\lambda_3 \lambda_4|\lambda_2) \tag{3.106}$$

in the case of the harmonic approximation, $V^{(n)} \equiv 0$.

The exact solution of (3.106) requires the determination of $G_\omega(\lambda\lambda'|\lambda'')$ which is coupled to higher order Green functions in an infinite set of equations of motion. In order to derive a practical solution, this system of equations has to be cut off at a certain order; in practice, this means that a certain Green function of definite order has to be approximated by lower order functions. This is equivalent to a certain diagrammatic expansion of thermal Green functions.[1]

The most important case for our discussion is that of a single dispersion oscillator. This case is realized in all diatomic crystals such as the alkali halides and the II - VI and III - V compounds and, furthermore, in triatomic crystals of CaF_2 structure, where the second dispersion oscillator is Raman active only.

In this case, Eq. (3.106) becomes

$$\left[\omega_R^2 - (\omega+i\varepsilon)^2\right] G_R(\omega) = 2\omega_R - \omega_R \sum_\lambda \Phi_3(R,\lambda\bar{\lambda}) G_\omega(\lambda\bar{\lambda}|R) \tag{3.107}$$

Here, the single "Reststrahlen" oscillator is denoted by R and we use the notation $G_\omega(R|R') = G_R(\omega)\delta_{RR'}$.

In order to get an approximately self-consistent solution, as discussed above, the self-energy of the dispersion oscillator is defined by introducing a Dyson equation:

$$G_R(\omega) = G_R^{(0)}(\omega)\left[1 - \Pi_R(\omega)\, G_R(\omega)\right], \tag{3.108}$$

Where $G_R^{(0)}(\omega)$ is given by

$$G_R^{(0)}(\omega) = \frac{2\omega_R}{\omega_R^2 - (\omega + i\varepsilon)^2} \tag{3.109}$$

From a systematic discussion of the relations of $G_R(\omega)$ to higher order Green functions, a representation of the self-energy $\Pi_R(\omega)$ may be developed in terms of anharmonic parts of the lattice potential up to infinite order. In practice, this discussion is mainly restricted to cubic and quartic anharmonicity. With (3.108) the Green function is

$$G_R(\omega) = \frac{2\omega_R}{\Omega_R^2(\omega) - (\omega + i\varepsilon)^2} \tag{3.110}$$

where

$$\Omega_R^2(\omega) = \omega_R^2 + 2\omega_R \Pi_R(\omega) \tag{3.111}$$

The complex self-energy Π_R is split into a real and an imaginary part,

$$\Pi_R(\omega) = \Delta_R(\omega) - i\Gamma_R(\omega) \tag{3.112}$$

As long as the temperature is not too high, $\Gamma_R(\omega)$ is small compared to ω_R. A quasi-particle can then be defined by the pole of $G_R(\omega)$ which has the excitation energy $\hbar\tilde{\omega}_{R'}$, where

$$\tilde{\omega}_R^2 \equiv \mathrm{Re}\,\Omega_R^2(\omega=\omega_R) \tag{3.113}$$

$$= \omega_R^2 + 2\omega_R \Delta_R(\tilde{\omega}_R) \tag{3.114}$$

$$= \tilde{\omega}^2(\omega)\,|\omega=\omega_R \tag{3.115}$$

A consistent solution of (3.114) needs a determination of $\Delta_R(\omega)$. Very often it is sufficient to assume $\Delta_R(\tilde{\omega}_R) \cong \Delta_R(\omega_R) \equiv \Delta_R^o$; the next approximation leads to

$$\Delta_R(\tilde{\omega}_R) \simeq \Delta_R^o \left[1 + \frac{\partial}{\partial\omega} \Delta_R(\omega)_{\omega=\omega_R}\right], \tag{3.116}$$

and so on. Eq. (3.113) defines a pseudo-harmonic frequency $\tilde{\omega}_R$ for the dispersion oscillator (R). It is a matter of convenience whether ω_R is supposed to be a "pure" harmonic frequency at T = 0 (which is advantageous if the interatomic potentials are known as e.g. for rare gas crystals) or rather is defined[39] as the quasi-harmonic $\omega_R(T)$ which includes the effect of thermal expansion of the crystal. The second approach is preferred when the real potentials are not known in crystals such as the alkali halides. However, information about cubic and quartic potential coefficients at a finite temperature can be obtained from certain experiments. Consequently, all parameters of the theory will be assumed to be quasi-harmonic entities.

The imaginary part of the self-energy $\Gamma_R(\omega)$ is a frequency-dependent damping function or inverse lifetime of the dispersion oscillator which determines the structure of the linewidth in the infra-red absorption for the dispersion oscillator with $\omega = \omega_{TO}$.

Replacing $G_R^{(o)}$ in the equation following (3.97) by $G_R(\omega)$, the complex lattice susceptibility of a dispersion oscillator is finally obtained for a diatomic cubic crystal:

$$\chi_R^L(\omega) = \frac{(Ze)^2}{\mu v_c} \frac{1}{\omega_R^2 + 2\omega_R \Delta_R(\omega) - \omega^2 - i2\omega_R \Gamma_R(\omega)} \tag{3.117}$$

The real and imaginary parts of the dielectric susceptibility including the electronic part $\chi^0 = \chi_\infty$, then are given by

$$\varepsilon'(\omega) + i\varepsilon''(\omega) = \varepsilon_\infty + \frac{(\varepsilon_0 - \varepsilon_\infty)\tilde{\omega}^2(0)}{\left[\tilde{\omega}^2(\omega) - \omega^2\right]^2 + 4\omega_R^2 \Gamma_R(\omega)} \left[\tilde{\omega}^2(\omega) - \omega^2 + i2\omega_R \Gamma_T(\omega)\right]. \tag{3.118}$$

This result is slightly different from a classical formula.[39]

$$\varepsilon^{cl}(\omega) = \varepsilon_\infty + \frac{(\varepsilon_0 - \varepsilon_\infty)\omega_{TO}^2}{(\omega_{TO}^2 - \omega^2)^2 + \omega_{TO}^2 \gamma_R^2} (\omega_{TO}^2 - \omega^2 + i\omega_{TO}\gamma_R), \quad \omega_R = \omega_{TO} \tag{3.119}$$

where ω_{TO} is the experimental frequency and γ_R is a damping constant which is usually fitted to experimental data near ω_{TO}.

The damping function or inverse lifetime of a dispersion oscillator (R) is determined by the different decay modes for the corresponding phonon with $\omega = \omega_{TO}$. Only cubic and quartic anharmonicity up to second order in the interaction potentials are considered. This seems to be appropriate for the bulk of experimental data, but cases exist where higher order terms are necessary.

The results for a single dispersion oscillator may be summarized as follows:

$$\Pi_R(\omega) = \left[\frac{1}{2}\sum_{\lambda_1}\Phi_4(R,R_1,\lambda_1,\bar{\lambda}_1)-\sum_{R'}\Phi_3(R,R,R')\quad V_3(R',\lambda_1,\bar{\lambda}_1)\right](2n_\lambda+1)$$

$$-\frac{1}{2}\sum_{\lambda_1\lambda_2}|\Phi_3(R,\lambda_1,\lambda_2)|^2 G_{2,\lambda_1,\lambda_2}(\omega)$$

$$-\frac{1}{6}\sum_{\lambda_1,\lambda_2,\lambda_3}|\Phi_4(R,\lambda_1,\lambda_2,\lambda_3)|^2 G_{3,\lambda_1,\lambda_2,\lambda_3}(\omega) \tag{3.120}$$

where $G_{2,\lambda_1,\lambda_2}(\omega)$ is given by (3.103) and

$$G_{3,\lambda_1\lambda_2\lambda_3} = \sum_{+-}(2\pm1)\frac{2(\omega_1+\omega_2+\omega_3)}{(\omega_1+\omega_2+\omega_3)^2-(\omega+i\varepsilon)^2}\left[(1+n_1+n_2)(n_3+\frac{1}{2}\pm\frac{1}{2})\pm n_1n_2\right]$$

Only the third and fourth terms in (3.120) contribute to the damping function $\Gamma_R(\omega)$ which has the form

$$\Gamma_R(\omega) = \pi\ \mathrm{sgn}\ \omega\sum_{+-}\{\frac{1}{2}\sum_{\lambda_1\lambda_2}|\Phi_3(R,\lambda_1,\lambda_2)|^2 2(\omega_1+\omega_2)\delta(\omega^2-(\omega_1\pm\omega_2)^2)$$

$$\left[(n_2+\frac{1}{2})\pm(n_1+\frac{1}{2})\right] \tag{3.122}$$

$$+\frac{1}{6}\sum_{\lambda_1\lambda_2\lambda_3}|\Phi_4(R_{\lambda_1\lambda_2\lambda_3})|^2(2\pm1)2(\omega_1+\omega_2\pm\omega_3)\delta(\omega^2-(\omega_1+\omega_2\pm\omega_3)^2)$$

$$\left[(1+n_1+n_2)(n_3+\frac{1}{2}\pm\frac{1}{2})\pm n_1n_2\right]\}$$

where the distribution relation

$$2\omega_i\ \mathrm{sgn}\omega\delta(\omega^2-\omega_i^2) = \delta(\omega-\omega_i)-\delta(\omega+\omega_i) \tag{3.123}$$

has been used. The simplest approximation for $\Gamma_R(\omega)$ would be to replace the matrix elements $|\Phi_3|^2$ and $|\Phi_4|^2$ (generally frequency-dependent) by constants or weakly frequency-dependent functions. The result of this approximation is

$$\Gamma_R(\omega) \simeq c_3 \sum_{+\,-} \rho_2(\omega)_{\omega=\omega_1 \pm \omega_2} \left[(n_2 + \tfrac{1}{2}) \pm (n_1 + \tfrac{1}{2}) \right] \tag{3.124}$$

$$+ c_4 \sum_{+\,-} \rho_3(\omega)_{\omega=\omega_1 + \omega_2 \pm \omega_3} \left[(1+n_1+n_2)(n_3 + \tfrac{1}{2} + \tfrac{1}{2}) \pm n_1 n_2 \right] \tag{3.125}$$

where the n_i are suitable averaged phonon occupation numbers, and the joint densities ρ_i are

$$\rho_2(\omega)_{\omega=\omega_1 \pm \omega_2} = \sum_{q \neq 0} \sum_{j_1 \neq j_2} (\omega_1 \omega_2)^{-1} \, \delta(\omega - \omega_1 \pm \omega_2), \text{ etc.,} \tag{3.126}$$

while the c_i's are adjustable constants or functions. The structure of the damping function is, in this approximation, completely given by the critical points of the joint densities.

The results of this section are only applicable to ionic crystals where non-linear dipole moments are negligible. This is generally not fulfilled even in alkali halides such as LiF, although such an approximation gives a rather satisfactory picture. In general, anharmonicity and non-linear dipole moments govern different parts of the spectra. An analysis of the situation is GaAs will be given in Sect. 4., where both types of couplings are equally important.

The presence of anharmonic couplings and non-linear dipole moments at the same time leads to processes of mixed type which can be analyzed in terms of renormalized dipole moments. With the approximations $\Phi_{4+n}=0, M_{3+n}=0, n \geq 1$, the discussion can be restricted to a renormalization of the linear dipole moment $\underline{M}_1$.

With the help of the functional method, the frequency-dependent renormalized linear dipole moment of a dispersion oscillator is obtained

$$\underline{M}_R(\omega)=\underline{M}_R+\sum_{\lambda}\left[\frac{1}{2}\underline{M}_3(R\lambda\bar{\lambda})-\sum_{R'}\underline{M}_2(RR')\frac{1}{\omega_{R'}}\Phi_3(R'\lambda\bar{\lambda})\right](2n_\lambda+1) \qquad (3.127)$$

$$-\frac{1}{2}\sum_{\lambda_1\lambda_2}\underline{M}_2(\lambda_1\bar{\lambda}_2)G^{(0)}_{2,\lambda_1\lambda_2}(\omega)\Phi_3(\bar{\lambda}_1\lambda_2R)$$

$$-\frac{1}{6}\sum_{\lambda_1\lambda_2\lambda_3}\underline{M}_3(\lambda_1\lambda_2\lambda_3)G^{(0)}_{3,\lambda_1\lambda_2\lambda_3}(\omega)\Phi_4(\bar{\lambda}_1\bar{\lambda}_2\bar{\lambda}_3R)\ .$$

The contribution of the non-linear dipole moments, especially that of $\underline{M}_2$ in the third term, leads to shifts and asymmetries of the resonance infra-red absorption near ω_{TO}.

The results of the foregoing paragraphs are summarized by giving a quite general formula for the susceptibility in crystals which may have several dispersion oscillators R, but with a negligible coupling between them. In this case, the following results are found:

$$\chi^L(\omega)=\frac{\hbar}{V}\left[\sum_i\tilde{\underline{M}}_{R_i}(\omega)\tilde{\underline{M}}_{R_i}(\omega)G_{R_i}(\omega)\right. \qquad (3.128)$$

$$+\frac{1}{2}\sum_{\lambda_1\lambda_2}M_2(\lambda_1\bar{\lambda}_2)M_2(\bar{\lambda}_1\lambda_2)G^{(0)}_{2,\lambda_1\lambda_2}(\omega)$$

$$\left.+\frac{1}{6}\sum_{\lambda_1\lambda_2\lambda_3}M_3(\lambda_1\lambda_2\lambda_3)M_3(\bar{\lambda}_1\bar{\lambda}_2\bar{\lambda}_3)G^{(0)}_{3,\lambda_1\lambda_2\lambda_3}(\omega)\right]\ .$$

In this formula, the renormalized linear dipole moments M_R and anharmonic propagators $G_{Ri}(\omega)$ change the general form of the resonance part of the absorption part of the absorption from delta functions to rather complicated asymmetric profiles.

In crystals with several oscillators with a non-negligible coupling as e.g. in ferro-electrics, the absorption formula (3.128) has to be modified. This modification has its origin in nondiagonal elements of the Dyson equation. For the case of two coupled oscillators, the effective retarded Green function of one of the dispersion oscillators, say (R_1) is found to be

$$G_{R_1}(\omega) = \frac{2\omega_R}{\omega_{R_1}^2 - \omega^2 + 2\omega_{R_1}\left[\Pi_{R_1}(\omega) - \Pi_{R_1R_2}(\omega)G_{R_2}(\omega)\Pi_{R_2R_1}(\omega)\right]} \tag{3.129}$$

where the coupling energy is

$$\Pi_{R_1R_2}(\omega) = \Delta_{R_1R_2}(\omega) - i\Gamma_{R_1R_2}(\omega) \tag{3.130}$$

$$= \sum_{\lambda_1}\left[\frac{1}{2}\Phi_4(R_1R_2\lambda_1\bar{\lambda}) - \sum_{i=1,2}\Phi_3(R_1R_iR_2)\frac{1}{\omega_{R_i}}\Phi_3(R_i\lambda_1\bar{\lambda}_1)\right]$$

$$(2n_1 + 1)$$

$$- \frac{1}{2}\sum_{\lambda_1\lambda_2}\Phi_3(\lambda_1\lambda_2R_1)\Phi_3(R_2\bar{\lambda}_1\lambda_2)G^{(0)}_{2,\lambda_1\lambda_2}(\omega)$$

$$- \frac{1}{6}\sum_{\lambda_1\lambda_2\lambda_3}\Phi_4(\lambda_1\lambda_2\lambda_3R_1)\Phi_4(R_2\bar{\lambda}_1\bar{\lambda}_2\bar{\lambda}_3)G^{(0)}_{3,\lambda_1\lambda_2\lambda_3}(\omega)$$

A corresponding equation holds for the second oscillator (R_2).

The anharmonic expansion parameters in terms of one or multiphonon coordinates are Fourier transforms of the expansion parameters in ordinary space,

$$\Phi_n(\lambda_1\ldots\lambda_n) = \frac{1}{\hbar}\Delta(q_1+\ \ldots\ +q_n)\sum_{L_1\ldots L_n} V_{\alpha_1\ldots\alpha_n}(L_1\ldots L_n) \times$$

$$\times\ \varepsilon^1_{\alpha_1}\ \ldots\ \varepsilon^n_{\alpha n}\ \exp\{i(q_1x_1+\ldots+q_nx_n)\} \tag{3.131}$$

The coupling parameters $V_{\alpha_1 \ldots \alpha_n}$ are usually restricted to cubic and quartic order as explained above and to nearest- and second nearest-neighbor interactions in ordinary space except for anharmonic Coulomb interactions which need a particular treatment. Examples of those coupling parameters are investigated in Sec. 4 for the case of cubic crystals, where the dominating of the short-range cubic nearest-neighbor coupling in infra-red absorption will be exemplified.

The cubic and quartic part of the anharmonic potential play an important role also in the third order and fourth order elastic constants, in the so-called mode Grüneisen parameters and some further anharmonic properties of crystal. Since they are only loosely related to the problem of infra-red absorption, the reader is referred to the extensive literature in the field which includes the work of Leibfried and Ludwig,[40] Ludwig,[8] and Barron and Klein.[41]

3.2.4 Raman Scattering in Terms of Phonons

The general formulae for the Raman scattering cross section has been given in Sec. 3.2.3 (Equation (3.81). The Raman scattering tensor $\underline{i}$ is a fourth-rank tensor with the same symmetry properties as that of the tensor of the elastic stiffness constants C_{ik}. In this section, the electronic polarizabilities P are related to the phonon normal coordinates in order to relate the electronic polarizabilities P to phonon normal coordinates. This allows us to obtain explicit expressions for one- and multi-phonon processes. We develop the polarizabilities $\underline{P}$ in a formal way in nuclear displacements as for the expansion of the crystal moment in the last section. Using the Fourier transformation for the displacement, the following description is obtained.[2]

$$P_{\alpha\beta}(\underline{Q})=P^{o}_{\alpha\beta}+\hbar\left[\sum_{\lambda^{o}} P_{1,\alpha\beta}(\lambda^{o}\underline{Q})A(\lambda^{o}\underline{Q})+\frac{1}{2}\sum_{\lambda\lambda'} P_{2,\alpha\beta}(\lambda\overline{\lambda}')A(\lambda)A(\overline{\lambda}')+\ldots\right] \tag{3.132}$$

The momentum transfer, $\underline{Q}$, may be neglected in those cases, where spatial dispersion plays no role. The first term in Eq. (3.132) is the practically constant part χ_∞ of the electronic crystal polarizability which contributes in only a dispersive way to the light scattering. In this case, the energy transfer, $\hbar\Omega$, is zero and the elastic part of the scattered light, the Rayleigh scattering, is obtained.

The inelastic scattering is due to all higher order terms giving one- and multi-phonon processes. The formal similarity of the expansion of $\underline{P}$ to that of the dipole moment $\underline{M}$ and their analogous roles in the corresponding retarded Green functions for the Raman scattering and the infra-red susceptibility allows for an adaptation of many results obtained for the susceptibility in the foregoing section. Details appear in Birman,[11] Anderson,[42] and Bilz et al.[7]

We note that in cubic crystals three independent symmetric correspondents exist:

$$i_{11} \equiv i_{xxxx} = i(\Gamma_1^+) + 4\, i\,(\Gamma_{12}^+) \tag{3.133}$$

$$i_{12} \equiv i_{xxyy} = i(\Gamma_1^+) - 2\, i\,(\Gamma_{12}^+) \tag{3.134}$$

$$i_{44} \equiv i_{xyxy} = i(\Gamma_{25}^+) \tag{3.135}$$

where the sub-indices x and y denote in an obvious way the scattering geometries. $\Gamma_1^+, \Gamma_{12}^+$, and Γ_{25}^+ are the three irreducible representations which correspond to isotropic (Γ_1^+) and quadrupolar charge deformations. We use here the symmetry notation of Loudon.[43] It is useful to relate the components of the Raman scattering tensor $\underline{i}$ to those of the polarizability tensor $\underline{P}$, especially for the analysis of Raman scattering by a macroscopic sample consisting of randomly oriented scatterers. The second rank tensor $\underline{P}$ possesses two rotational invariants which are independent of the coordinate system; these are the isotropic part of $\underline{P}$,

$$\underline{P}(\Gamma_1^+) \equiv \underline{P}^{is} = \mathrm{Tr}\underline{P} = \frac{1}{3}(P_{xx} + P_{yy} + P_{zz}) \tag{3.136}$$

and the anisotropic part,

$$\gamma^2 \equiv (\underline{P}^{an})^2 = 3\left[(P_{xx} - P_{yy})^2 + (P_{yy} - P_{zz})^2 + (P_{zz} - P_{xx})^2\right.$$

$$\left. + 6(P_{xy}^2 + P_{yz}^2 + P_{zx}^2)\right] = 3\left[P_{xy}^2(\Gamma_{12}^+) + \ldots\right] \tag{3.137}$$

(c.f. Wilson et al.[44])

We obtain for a cubic crystal

$$i_{11} = 3(\mathrm{Tr}\underline{P})^2 = 3|P(\Gamma_1^+)|^2 = 3(\delta\chi_\infty)^2 \tag{3.138}$$

$$= \frac{3}{16\pi^2}(\delta\varepsilon_\infty)^2 = \frac{1}{3}(P_{xx} + P_{yy} + P_{zz})^2$$

Analogously, we may write

$$i_{12} = \frac{1}{3}(P_{xx}P_{yy} + P_{yy}P_{zz} + P_{zz}P_{xx}) \tag{3.139}$$

$$i_{44} = \frac{1}{3}(P_{xy}^2 + P_{yz}^2 + P_{zx}^2) \tag{3.140}$$

and

$$i(\Gamma_1^+) = \underline{P}^2(\Gamma_1^+) \quad . \tag{3.141}$$

The isotropic part of the polarizability is directly connected with the change in the high-frequency dielectric constant ε_∞ .

It should be noted that Eqs. (3.135) and (3.137) are incorrect if the anti-symmetric part of $\underline{P}$ is not negligible. In this case, in these equations P_{xy} has to be replaced by the symmetric combination $\frac{1}{2}(P_{xy} + P_{yx})$, etc. The antisymmetric part is then obtained as

$$i^{as}_{xyxy} = i(\Gamma_{15}^+) = \frac{1}{4}(P_{xy} - P_{yx})^2 \tag{3.142}$$

This contribution is down by a factor $(\Omega/\omega_i)^2$ compared with the symmetric part of $i_{xy,xy}$. Therefore, it should become observable in resonance Raman scattering when this factor is of the order of one.

The crystal's polarizability is formally described by a power series in the ionic displacements. The expansion coefficients of this series are Fourier transforms of the Raman polarizabilities $P_{\alpha\beta}(\{\lambda_i\})$.

Due to the screening of the electron-ion interaction one expects that the Raman coupling parameters constitute a rapidly converging series with respect to an increasing distance of neighbors so that nearest-neighbor and second-nearest neighbor interactions are usually sufficient to describe the spectra. Even for cubic crystals the number of independent parameters for every shell of neighbors is rather high and additional approximations have to be used in order to keep the number of adjustable parameters small. The general notation and symmetry of coupling parameters for some high symmetry crystals may be found in Loudon[43] and Birman.[11]

Of particular interest are the coupling parameters at very long wavelengths, the so-called Pockels or photo-elastic (elasto-optic) coefficients. They can be directly determined with the help of Brillouin scattering and provide therefore an additional check on a particular band or local model of Raman scattering. In this respect they play a role analogous to the elastic constants in lattice dynamics.

In the case of cubic crystals, the situation is simple. As the three elastic constants C_{11}, C_{12}, and C_{44} define the bulk modulus and the two shear constants of crystal, the photoelastic constants p_{11}, p_{12}, and p_{44}, are related to the volume (Γ_1^+) and the two quadrupolar (Γ_{12}^+ and Γ_{25}^+) types of Brillouin scattering.

3.3 Models for infra-red absorption and Raman scattering

3.3.1 General features of infra-red and Raman processes

In the foregoing sections, the formal theory of infra-red absorption and Raman scattering has been described. It yields the lattice susceptibility and the Raman cross section. We are now left with the hard core of the problem, i.e., a discussion of the values of the coupling parameters in different crystals, an understanding of the differences between ionic and covalent crystals and the systematic trends in specific crystal families, such as the alkali halides, etc.

The best approach to the problem would be, of course, a direct quantum-mechanical calculation of the parameters. Here very few first steps have been taken towards explicit calculations mainly for covalent systems. This situation is not surprising since, even for phonons in insulators, microscopic calculation is still in its infancy.[45] It seems to be reasonable, then, to use a model theory with parameters chosen to be as close as possible to the corresponding matrix elements of the microscopic theory.

As mentioned above, a treatment of phonons which initially retains an explicit description of the adiabatic electronic degrees of freedom in the equations of motion before eliminating them in the dynamical matrix, leads to a very satisfactory representation of phonons in different types of crystals. A particular advantage is that it drastically reduces the number of parameters which have to be used in describing dispersion curves of crystals, so that very often a unique relation can be found between the model parameters and independent macroscopic entities. Such a theory obviously has a much greater appeal for a microscopic theory, since then only the macroscopic entities have to be derived microscopically without changing the general structure of the theory. The formalism of the harmonic approximation will be extended in this section to non-linear processes which are still adiabatic in this description. This means that the electronic coordinates can eventually be eliminated. The general idea is to express long-range effective ion-ion forces by short-range electron-ion forces and so not only to reduce the number of independent parameters but also to lead to a deeper insight into the origin of the ion-ion forces.

A non-linear shell model was first used by R. A. Cowley for the description of infrared absorption in alkali halides[46] and in covalent crystals.[47] Satisfactory results were obtained by Knohl[36] and by Bruce[48] for the infrared absorption in alkali halides while the interpretation of Raman scattering by Bruce and Cowley[49] seems to be debatable.[50] The Raman spectra of cubic ionic crystals have been further investigated with a shell model by Haberkorn, et al.,[51] Buchanan et al,[52] and Migoni et al.[53] As mentioned above, there exists an important difference between infrared absorption and Raman scattering. Infrared absorption is based essentially on an induced change in the crystal dipole moment. This means that the effective charges of the crystal ions are responsible for the strength of the infrared absorption. Anharmonic short- and long-range couplings between the ions are then the main factors in obtaining a more refined picture of the situation; Born-Mayer and Coulomb inter-ionic potentials are very good first approximations in that case.

Raman scattering, however, shows a rather different behavior. The change in the crystal polarizability observed in Raman scattering is first of all a change in the electronic polarizability. In a classical picture of an ionic crystal, the main effect of changing the polarizability on an ion is to change its volume, i.e., its ionic radius; therefore, as a typical example, the "breathing" deformability of the polarizable ions in alkali halides should show a clear relation to Raman scattering if it is to be more than a lucky simulation of the interatomic forces. This seems, however, not to be the case for the alkali halides where the "breathing" probably simulates the effect of overlap polarization. We may ask for the general conditions under which differential intra-ionic polarizabilities play an important role in Raman scattering. They are connected with a completely localized change of the polarizability. Such a change requires very flexible electronic polarizabilities of the ions, that is, the electronic wave functions of the ions should be able to readjust themselves in response to some displasive change in the surrounding ions. Wave functions of this type correspond to open-shell configurations which are typical for partially covalent crystals with hybridized wave functions forming bonds. For example, Raman spectra of Zn chalcogenides and Ga pnictides seem to be mainly due to intra-ionic polarizabilities.[54]

Even more impressive examples of intra-ionic Raman polarizabilities may be found in hydrides and oxides. It is well known that the normal polarizability of O^{--} is not a well defined quantity but depends strongly on its ionic radius.[55] This leads to drastic changes of the oxygen polarizability in Raman scattering manifesting themselves in strong second order longitudinal optic overtone spectra in earth alkaline oxides, oxydic perovskites, etc. The spectra of alkali halides, on the other hand, are missing this part of the spectrum and seem to be governed by inter-ionic differential polarizabilities. The alkali halides linear polarizabilities, which are important for the phonons, are nevertheless well defined in terms of individual ions.[55] This means that the ions behave like closed and rather rigid shell electron configurations so that the polarizabilities may be difficult to change without a charge transfer, i.e., inter-ionic processes.[56] In the case of strongly covalent crystals such as diamond and its homologues the concept of bond charges and polarizabilities seems to be useful.

We now go to a more quantitative description of the general features discussed in this introduction. We recall that in infra-red absorption the anharmonic potentials may be described in terms of direct ion-ion potentials so that the discussion of the experimental spectra can be based immediately on the general treatment in the foregoing section without explicit reference to a non-linear extension of the phonon model. In Raman scattering the energy transfer to the crystal is via the polarizable electrons so that an explicit treatment of electronic polarizabilities and deformabilities seems to be worthwhile. We show how to derive a model description of the electron-phonon interaction from the microscopic theory.

3.3.2 Microscopic and model treatment of electron-phonon interaction

The leading term in the electron-phonon interaction in the energy regime of the electrons (i.e., band gap energies $\tilde{-}$eV) is the difference between the adiabatic and the static electron-ion interaction. Here, both terms are defined for some displaced configuration of the ions, but the matrix elements have to be calculated with 'adiabatic' and 'static' electronic wave functions, respectively.

In order to understand this in more detail, we discuss first the case of a single electron interacting with a specific ion in a transition where one or several phonons are created. The adiabatic interaction reads in this case

$$U^{ad}_{e,ph} = \langle S',\underline{R}|U_{ei}|\underline{R},S\rangle - \langle S'_o,\underline{R}_o|U_{ei}|\underline{R}_o S_o\rangle \qquad (3.143)$$

with

$$U_{ei} = -Ze^2/|\underline{r}-\underline{R}| \qquad (3.144)$$

Here $\underline{R}$, $\underline{r}$, s and $\underline{R}_o$, $\underline{r}_o$, s_o denote vectors for the ion, the electron, the quantum numbers of the electronic state for a displaced and the equilibrium configuration, respectively.

We may write Eq. (3.143) in the following form:

$$U^{ad}_{e,ph}\ (\underline{r}-\underline{R}) = -Ze^2 \int \frac{1}{|\underline{r}'-\underline{R}|} \left[\rho^{ad}(\underline{r}'-\underline{R}) - \rho^{st}(\underline{r}'-\underline{R}_o)\right] d^3\underline{r}'$$

with (3.145)

$$\rho^{ad}(\underline{r}'-\underline{R}_o) = \rho^{st}(\underline{r}'-\underline{R}_o)$$

Now we split ρ^{ad} into a rigid and a deformable part where the rigid part is defined by being equal to ρ^{st} if the ion is displaced from R_o to R:

$$\rho^{ad}_{RI}\ (\underline{r}'-\underline{R}) = \rho^{ad}_{RI}\left[\underline{r}'-\underline{R}_o-(\underline{R}-\underline{R}_o)\right] = \rho^{st}(\underline{r}''-\underline{R}_o) \qquad (3.146)$$

Eq. (3.146) means that the rigid part of ρ^{ad} and ρ^{st} cancel. Only the deformable part of ρ^{ad} contributes to the electron-phonon interaction:

$$U^{ad}_{e,ph}(\underline{r}-\underline{R}) = -Ze^2 \int \frac{d^3S'}{|\underline{r}+\underline{S}'-(\underline{R}_o+\underline{U})|}\ \rho^{ad}_{DEF}\left[\underline{r}+\underline{S}'-(\underline{R}_o+\underline{U})\right] \qquad (3.147)$$

with

$$\underline{r}-\underline{R} = \underline{r}_o+\underline{S} - (\underline{R}_o+\underline{U}) \qquad (3.148)$$

Here $\underline{U}$ denotes the displacement of the ion and $\underline{S}$ the corresponding adiabatic shift of the electron. In the rigid ion limit, the electron follows the ion motion completely, i.e. $\underline{S}=\underline{U}$ and $U^{ad}_{e,ph} = 0$.

This reflects the fact that the electron-phonon interaction depends only on the relative displacements $\underline{w} \equiv \underline{S}-\underline{U}$. If we develop now Eq. (3.147) into powers of the relative displacements $\underline{w}$, we obtain with $\underline{a}_o \equiv \underline{r}_o - \underline{R}_o$, using the notation of Eq. (3.143),

$$U^{ad}_{e,ph} = -Ze^2 <S_o' \underline{R}_o \mid \frac{1}{a_o} \Sigma_{m1}^{\infty} \frac{1}{m!} w^m \left[(\hat{w}\cdot\nabla_w)^m U_{ei} \right]_{w=o} \mid S_o R_o>$$

$$= -Ze^2 <\underline{S}_o' R_o \mid \frac{1}{a_o} \underline{w}\cdot \left[\nabla_w \; U_{ei}\right]_{w=o} + \cdots \mid S_o \underline{R}> \qquad (3.149)$$

$$= -Ze^2 (\underline{f}^{ss'} \cdot \underline{w} + \frac{1}{2} \sum_{\alpha\beta} f^{ss'}_{\alpha\beta} w_\alpha w_\beta + \ldots) \qquad (3.150)$$

where $\hat{w} = \underset{\sim}{w}/w$ and $w = |\underline{w}|$.

The matrix elements $f^{ss'}_{\alpha}$, $f^{ss'}_{\alpha\beta}$, etc., define dipolar, quadrupolar, etc., electron-ion coupling parameters with respect to one-, two-, etc., phonon processes assisting transitions between the equilibrium electronic states with quantum numbers s and s'. The center of gravity of the electronic charge is $\underline{R}$ i.e., the position of the displaced ion. This ensures automatically the translational invariance of the treatment and avoids complicated compensation of large terms if the coupling parameters are calculated for the ions in equilibrium positions.

Eq. (3.150) may be extended in an obvious way to the case of many electrons and ions in a periodic crystal. The coupling parameters depend, of course, to some extent on the details of the model. The model dependence is more marked in cases in which a clear separation of the electronic charge density into individual cell or ion contributions is debatable. This problem of model dependence is well-known from band calculations. For example, the electron matrix elements for the band structure of a given crystal from an LCAO calculation differ from those from an APW calculation.[57] We may, nevertheless, state a few general aspects of the theory:

1) the rigid-ion limit leads to vanishing adiabatic electron-phonon coupling. The parameters of this coupling should, therefore, not be confused with the anharmonic phonon-phonon coupling which is well defined in the rigid-ion limit;

2) a model theory should describe the electron-phonon coupling in terms of quantities (vectors and tensors) which correspond to relative displacements between (effective) electron and ion coordinates. This ensures translational invariance in the electron-phonon coupling;

3) the most interesting terms are intra-ionic (or intra-cell) couplings. They describe a strongly localized change of the polarizability;

4) if the local polarizability is well defined (as in alkali halides) and therefore not easily changed, inter-ionic terms should become important. Since often (in particular for weakly polarizable crystals) the relative displacements of ion cores and electrons tightly bound to neighboring ions can be replaced to a good approximation by relative ion-ion displacements, it may be expected that a description in terms of 'formal' expansions into powers of ionic displacements becomes quite successful. The discussion of this section provides a basis for a model theory of Raman scattering.[7]

3.3.3 Bond charge and bond polarizability in infrared and Raman processes

An adiabatic bond charge model[58] gives a very good description of the dispersion curves in diamond and its homologues and even in III-V semiconductors. One might, therefore, look for a non-linear extension of this model for the description of the infrared and Raman processes in covalent crystals.

As a starting point, we use the potentials of the bond charge model. The analysis of the infrared spectra of many ionic crystals suggests that one treat the bond charge Z_b as the effective transverse charge with dominating short-range couplings to the neighboring ions. The simplest possibility would be a cubic complement of the ion-bond charge potential ϕ_{i-bo} which may be renormalized by long range contributions, in analogy to the situation in ionic crystals. This approximation leads to second order phonon absorption and provides an explicit model for the so-called non-linear dipole moment $\underline{M}^{(2)}$. This approach has been discussed in detail by Go, Bilz, Cardona, Rustagi and Weber.[59] The main result is that the non-linear dipole moment may be represented by

$$\underline{M}^{(2)} = \frac{N}{2} Z_b \, \Phi_{ss}^{-1} \, \phi_{i-bo}^{(3)} : \underline{U}^{+} \underline{U}^{-} \tag{3.151}$$

The hypervectors $\underline{U}^{+}$ and $\underline{U}^{-}$ are built of elements $\underline{U}^{+}_{kk'} = \underline{U}_{k} - \underline{U}_{k'}$ and $\underline{U}^{-}_{kk'} = \underline{U}_{k} + \underline{U}_{k'} - 2\underline{U}_{bc}$ respectively. Here the local displacements in a bond are decomposed into those of even parity, $\underline{U}^{+}$ and odd parity, $\underline{U}^{-}$. The rather satisfactory results are discussed in Section 5.

A similar discussion of the second order Raman scattering can be made. Physically, it assumes that the change of polarizability in covalent crystals is mainly caused by the displacement polarizability of the bond charge corresponding to an oscillating electronic charge center of gravity between the ions. A calculation of the spectra shows that this is inconsistent with the observed intensity ratios of quadrupolar (Γ_{12},Γ_{25}) divided by volume (Γ_1) scattering.

Experimentally, the volume scattering is found to be strongest while the theory leads to the contrary result.[59] The failure of the bond charge model for the Raman scattering is not unexpected. Even in the harmonic approximation, the bond charge model gives only 10% of the observed dielectric constant[60] which shows that a rigid point charge approximation for the electronic charge density is insufficient. There are, in principle, two possibilities to overcome these difficulties: the first is a consideration of the missing part of the polarizability by introducing electronic shells at the ion lattice sites. This seems to be a promising idea, but it would imply a somewhat arbitrary division of the electronic charge into bond and shell charges. The second possibility is a dynamical charge transfer of the bond charge in the direction of its neighboring ions. It seems to be difficult to incorporate this charge transfer into the bond model, which would lose its attractive simplicity.

An intermediate step in the right direction may be the introduction of bond polarizabilities which depend in first approximation only on the bond length R^b of two neighboring ions. This description was first used in molecular physics,[61] and for the first order Raman line in covalent crystals by Maradudin and Burstein,[62] and Flytzanis and Ducuing.[63] Its extension to the case of second order spectra is straightforward.[59,64]

The general form of the polarizability reads

$$P_{\alpha\beta}\{R^b\} = \sum_b \hat{R}^b_\alpha \hat{R}^b_\beta \alpha_{||}(R^b) + (\delta_{\alpha\beta} - \hat{R}^b_\alpha \hat{R}^b_\beta)\alpha_{\perp}(R^b) \tag{3.152}$$

where $\alpha_{||}$ and $\alpha_{\perp}$ are the longitudinal and transverse components of the bond polarizability and R^b unit vectors parallel to the bonds. Eq. (3.152) may be used as a starting point for rather different crystals. In the particular case of diamond-type crystals, the general structure of the expansion of P into ionic and bond charge displacements is given by

$$\underline{P} = \underline{P}^o + \underline{P}^{(1)}\underline{U}^+ + \frac{1}{2}\underline{P}^{(2)} : (\underline{U}^+\underline{U}^+ + \underline{U}^-\underline{U}^-) \tag{3.153}$$

with the hypervectors $\underline{U}$ defined after Eq. (3.151). The discussion of the application of the theory to experimental data is given in Section 5. We shall, then, also discuss the relation between bond polarizabilities and the band structure of covalent crystals which leads to an understanding of the systematic trends in the ratio of transverse to longitudinal bond polarizabilities when going from diamond to silicon and germanium.[59]

4. Infrared spectra of crystals

4.1 Qualitative classification of infrared spectra

In simple ionic crystals, the infrared absorption is governed by the frequency-dependent damping of the one dispersion oscillator with frequency $\omega=\omega_{TO}\equiv\omega_R$. Such crystals must be at least diatomic, since a monatomic crystal, with only one atom in a cell, does not have any optical branches. It is, therefore, not surprising that diatomic cubic crystals are the most frequently investigated crystals in infrared optics; it will be shown that a careful analysis of their spectra is still rather complicated. Qualitatively, at least at low temperatures, the structure of reflection and absorption spectra in those crystals is determined by critical points of two-phonon summation processes. In fig. 4-1, the interrelations are shown of dispersion curves $\omega_j(\underset{\sim}{q})$, two phonon combination bands $\omega_1(\underset{\sim}{q})\pm\omega_2(\underset{\sim}{q})$, reflection bands $R = |(\varepsilon-1)/(\varepsilon+1)|$, and the complex dielectric constant $\varepsilon(\omega)$ itself, for a diatomic cubic ionic crystal like LiF, in one of the main symmetry directions (1 0 0) or (1 1 1).

It was suggested more than 60 years ago that the structure in the reflection bands of these crystals might be related to a frequency-dependent damping mechanism of the dispersion oscillator.[69] At low temperatures, the damping function in all crystals is a strongly increasing function of frequency, and very often its values are still small near ω_R, but one order of magnitude higher as ω approaches ω_{LO}. As will be seen in more detail below, the maximum of the damping function in diatomic ionic crystals is given mainly by the critical points of the TA + TO combination bands which are situated in many alkali halides at a frequency $\omega \approx 1.5\ \omega_R$. Usually, this is not very different from ω_{LO} and might, sometimes, give the impression of some "activity" of the longitudinal oscillator at ω_{LO}. It explains, at least at low temperatures, the fact that the lifetime of the transverse optic oscillator at ω_R is usually one order of magnitude higher than that of the longitudinal oscillator, as measured, say, by a Berreman transmission experiment.[70] As indicated in Fig. 4-1, the maximum of the damping function is expressed as a minimum in the reflectivity on the short wavelength side.

The qualitative similarity of reflection and absorption spectra suggests that there might exist an approximate scaling law such that, by reducing the spectra of different crystals with an appropriate measure, the experimental data would be made to look rather the same. This is shown in Fig. 4-2 where the imaginary part of the dielectric function of several alkali halides is represented. The dielectric function is reduced by $\varepsilon_0 - \varepsilon_\infty$ which is proportional to the infrared oscillator strength and, consequently, to the width of the reflection bands. The frequency scale is reduced by the Reststrahlen frequency ω_R in order to bring all maxima into the same position. One sees immediately that the maximum deviation of the ε'' curve from smooth behavior (as expected in a classical theory with constant damping) on the high-frequency side is near $1.5\ \omega_R$ due to the summation bands. On the low-frequency side, the maximum damping appears around $0.5\ \omega_R$. It will be seen that this is only partially due to two-phonon difference bands (like TA-TO). At room temperature, three-phonon

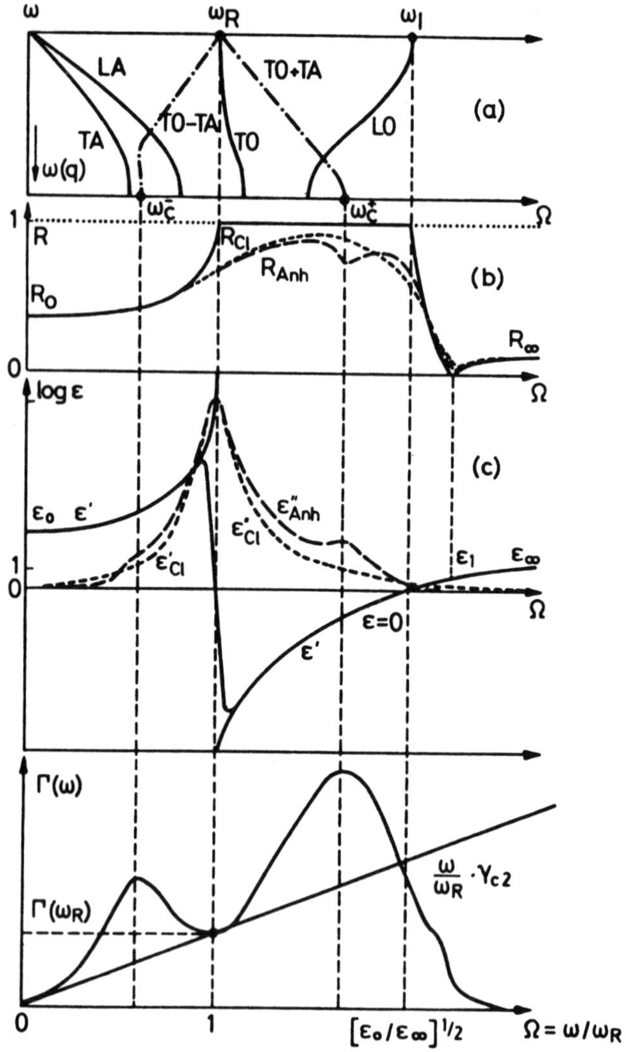

Figure 4-1: (a) Lattice modes of an alkali halide in one of the main symmetry directions (100) or (111). ω_R is the frequency of the infrared active dispersion oscillator; ω_1 is the longitudinal optic mode frequency. Dashed lines are an example of combination bands; summation band TO TA, difference band TO-TA. (b) Reflection R. Full line: without damping; fine dashed line (R_{CL}): with classical constant damping; dashed line (R_{Anh}): with anharmonic frequency dependent damping. (c) Real part ε' and imaginary part ε'' of the dielectric constant. Full ine: ε' with and without damping; fine dashed line (ε'_{CL}) with classical constant damping; dashed line (ε'_{Anh}) with anharmonic frequency dependent damping. The bottom graph in the figure shows the classical and anharmonic damping functions used in part (c) of the figure.

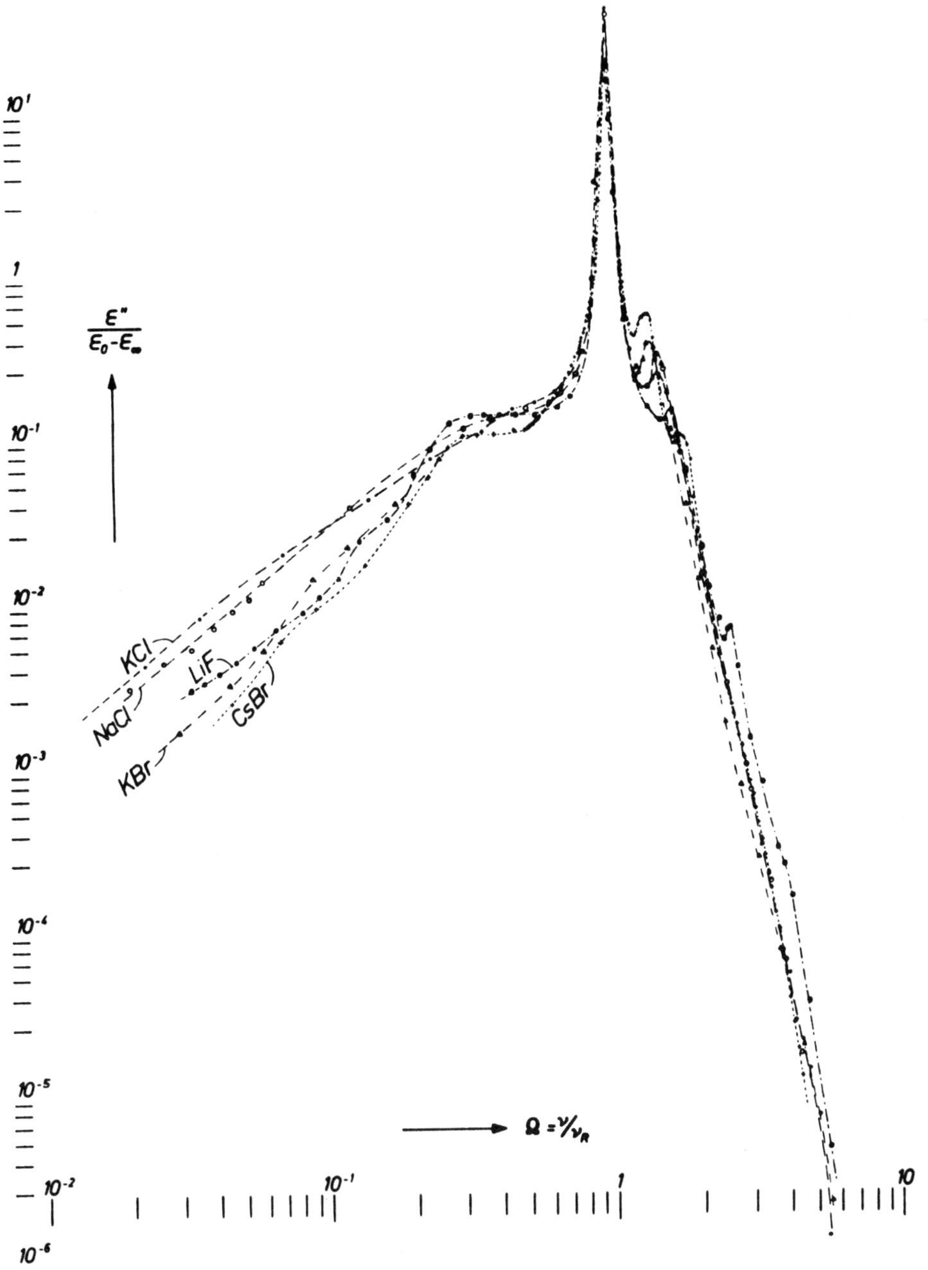

Figure 4-2: Reduced imaginary part of the dielectric constant ε, $\varepsilon''/(\varepsilon_0-\varepsilon_\infty)$ as function of the reduced frequency ω/ω_R in ionic crystals

difference processes begin to dominate the absorption. Furthermore, it is interesting to see that at small frequencies the imaginary part of the dielectric constant tends to become proportional to the reduced frequency. This corresponds to the behavior of a classical oscillator with a damping constant γ, and it means that the damping function itself approaches a linear function of ω. It is, therefore, expected that, at very long wavelengths, a classical picture may be appropriate. This is obviously not the case on the high-frequency side of the resonance where $\varepsilon''(\omega)$ decreases proportional to the seventh power, at least, of ω. This is completely different from classical behavior and is due to the cutoffs of two-, three-, and multi-phonon bands.

The infrared spectra of other diatomic crystals, like the alkali halides with cesium chloride structure or MgO, are similar to those of NaCl, etc. The same holds for crystals with three atoms but only one dispersion oscillator, such as the crystals with CaF_2 structure. There are, however, differences for more complicated crystals with more than one dispersion oscillator. Striking examples are the ferroelectrics, discussed in a forthcoming section.

In the following, we will give a discussion of the processes important for the analysis of infrared spectra of the ionic crystals. It seems to be necessary first to discuss separately the different effects of short- and long-range parts of the anharmonic potential as well as those of the non-linear dipole moments, although in practice they are all connected to one another. This separation is not only to avoid confusion; it is done since the short-range anharmonic couplings (including the corresponding part of the Coulomb potential) seem to govern the main part of the spectra while all other effects are smaller corrections.

The first part of the analysis uses the quasi-harmonic approximation.[39] The renormalization of the quasi-harmonic frequencies due to the self-energy shifts is considered here explicitly for the dispersion oscillators only, which interact with the external electromagnetic field. The other phonon frequencies as well as macroscopic parameters, like ε_0 and ε_∞, are usually regarded as pseudo-harmonic entities, which contain the self-energy effects and can be partially identified with the measured values at a given temperature T and volume V. The approximate thermodynamic potential is, therefore, the Helmholtz free energy F (T,V) from which all important thermodynamic relations can be derived. These conditions allow, within certain perturbation-theoretical approximations, a direct analysis of the experimental spectra with a limited numerical effort and without a complicated theoretical formalism. The validity of the approximations involved always has to be checked by comparing the calculated quantities with those obtained from experiments. A discussion of the temperature-dependence of the line-widths of dispersion oscillators and similar important entities may proceed in this way because, in centro-symmetric crystals such as the alkali halides, the effect of the thermal expansion on the quasi-harmonic parameters can be described at lower temperatures $T<\theta$, by taking into account only the change of volume with temperature. This change leads, in the simplest approximation for the quasiharmonic parameters, to a linear temperature-dependence which allows a determination of purely harmonic values at $T=0$ by an extrapolation of experimental values to zero temperature.[8] The fortunate separation of only volume-dependent quasi-harmonic parameters from dynamical anharmonic effects in alkali halides is based on a strong cancellation of cubic and quartic parts of the anharmonic lattice potential in the thermal expansion and facilitates the analysis of the experimental data greatly.

4.2.1 Anharmonic effects

The analysis of the spectra of alkali halides begins with a discussion of the complex dielectric constant $\varepsilon(\omega)$ where the anharmonic coupling of the dispersion oscillator to the other phonons is considered. The small effect of non-linear dipole moments is neglected in this section but will be discussed afterwards.

The contributions of cubic and quartic anharmonicity to the self-energy of dispersion oscillators are shown by diagrams in Table 4-1. They are related to corresponding expressions for retarded Green functions.

The analytic expressions for the diagrams in Table 4-1 have been found mainly by following diagrammatic perturbation theory used first by Maradudin and Fein[71] and by Cowley.[34,48] For the calculation of the diagrams of Table 4-1 in terms of thermodynamic Green functions, the reader is referred to the corresponding appendix in the handbook article by Cochran and Cowley[3] and to the lectures by W. Halley.

The various diagrams in Table 4-1 are classified according to their temperature-dependence at higher temperatures. The lowest order cubic and quartic terms (diagrams a and b) are explicitly proportional to T and they give contributions of comparable magnitude to the self-energy. This classification is most easily derived[72,34,48] from van Hove's perturbation treatment of the anharmonic potential in terms of a dimensionless parameter λ. We may write:

$$\phi_A = \lambda V_3 + \lambda^2 V_4 + \dots$$

The diagrams a and b both belong to the terms $O(\lambda^2)$ while terms $O(\lambda^4)$ exhibit an explicitly quadratic temperature-dependence at higher temperatures.

Table 4-1: Diagrammatic survey of anharmonic processes important in infrared absorption. Lower index: number of phonon lines, upper index: number of electron lines.

	Diagram	Temp. depend. of occupation numbers	Contrib. to $\pi_R[\omega]$
a)	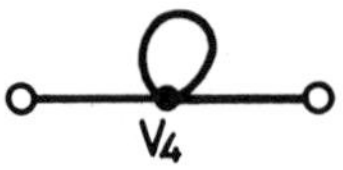	$1 + 2n_1 \quad \propto T$	Δ_R^4
b)		$1+n_1+n_2 \quad \propto T$ $n_2 - n_1$	Δ_R^3 , Γ_R^3
c)		$(\Sigma n_i n_K) \propto T^2$	Δ_R^4 , Γ_R^4
d)	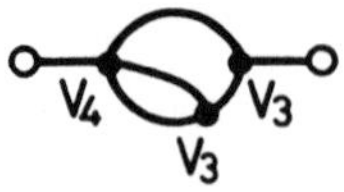	$\propto T^2$	Δ_R , Γ_R
e)		$\propto T^2$	Δ_R^3 , Γ_R^3
f)		$\propto T^2$	Δ_R^3 , Γ_R^3
g)	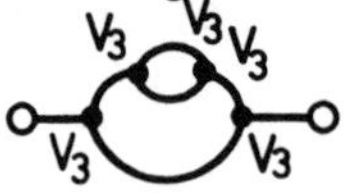	$\propto T^2$	Δ_R^3 , Γ_R^3

The analytic form of the dielectric constant of a single anharmonic dispersion oscillator, Eq. (3.118), is rewritten for the purpose of this section in a slightly different way:[73]

$$\frac{\varepsilon'(\omega) - i\varepsilon''(\omega) - \varepsilon_\infty}{\varepsilon_0 - \varepsilon_\infty} \equiv U(\omega) + iV(\omega) = \frac{\tilde{\omega}_R^2(0)}{\tilde{\omega}^2(\omega) - \omega^2 - i2\omega_R\Gamma_R(\omega)} \tag{4.1}$$

where $\tilde{\omega}(\omega)$ is the pseudo-harmonic frequency

$$\tilde{\omega}^2(\omega) = \omega_R^2 + 2\omega_R\Delta_R(\omega) \tag{4.2}$$

Since $\Delta_R(\omega)$ usually gives a small correction to ω_R only, attention is focused on the damping function $\Gamma_R(\omega)$, which determines the structure of the infrared absorption. The anharmonic approximation of Eq. (4.1) is a satisfactory description in strongly ionic crystals similar to LiF. Furthermore, attention is restricted to a discussion of two-phonon bands.

The first step is the introduction of only one anharmonic coupling parameter of third order between nearest neighbors. As in the Kellermann model, a central short-range two-body potential ϕ is used, that is, ϕ depends only on the absolute value of the distance between nearest neighbors:

$$\phi\left[\,|\underline{U}(L^+) - \underline{U}(L^-)|\,\right] \equiv \phi(r) \tag{4.3}$$

where the indices + and - denote the positive and negative ions in a cell L. There is usually also a contribution from the overlapping negative second nearest neighbors. This contribution is highly dependent on the details of the model (simple shell model, breathing shell model, etc.). It is, therefore, neglected in a first approximation. The anharmonic part of the Coulomb potential is usually neglected as well. The justification for this procedure is found in the fact that the anharmonic derivatives of a potential $\propto r^{-1}$ increase much slower than those of the short-range potential, which vary in the nearest-neighbor region approximately as r^{-9} with comparable harmonic contributions.[34] This argument overlooks the possibility of a constructive interference of long-range parts of the Coulomb potential in the phonon-phonon coupling,[74,36] and it is therefore not sufficient to add to the nearest-neighbor short-range parameter the corresponding part of the Coulomb potential.[48]

In the following, we discuss the damping function in the cubic approximation. This provides a basis for the comparison of experimental data with theoretical calculations and for the discussion of different approximations given later on in this section.

The third derivative V_3^0 of the anharmonic potential ϕ is:[39]

$$\Phi_{\alpha\beta\gamma}(ox,\ ok,\ \ell k') \equiv \Phi_{\alpha\beta\gamma}(L^oL^oL) =$$

$$= \phi^{[3]} x_\alpha x_\beta x_\gamma + \phi^{[2]} (x_\alpha \delta_{\beta\gamma} + x_\beta \delta_{\alpha\gamma} + x_\gamma \delta_{\alpha\beta}) \tag{4.4}$$

where $\phi^{[3]} = r^{-5}(3\phi' - 3r\phi'' + r^2\phi\ )$

$$\phi^{[2]} = r^{-3}(r\phi'' - \phi'),\ r = |\underline{x}|\ .$$

The Fourier transform of this potential is, for the cubic coefficient in the damping function:

$$V(R\lambda_1\overline{\lambda}_2) = \frac{1}{\hbar} \sum_{\alpha\beta\gamma} \{K^3_{\alpha\beta\gamma}(\underset{\sim}{q})(\varepsilon^+_R - \varepsilon^-_R) \left[\varepsilon^+_\beta(\underset{\sim}{q}j_1)\varepsilon^-_\gamma(\underset{\sim}{q}j_2) \right.$$

$$\left. - \varepsilon^-_\beta (qj_1)\varepsilon^+_\gamma(qj_2)\right]\} \tag{4.5}$$

where the tensor $\underline{K}^3$ has the components:

$$K^3_{\alpha\beta\gamma}(\underset{\sim}{q}) = -2i\ \sin(\underset{\sim}{q}_\alpha r_o)\delta_{\beta\gamma}\left[\phi'''\delta_{\alpha\beta} + \phi^{[2]}(1-\delta_{\alpha\beta})\right]$$

i.e.

$$K^3_{\alpha\alpha\alpha}(\underset{\sim}{q}) = -2i\ \sin(\underset{\sim}{q}_\alpha r_o)\phi''' \quad ,$$

$$K^3_{\alpha\beta\beta}(\underset{\sim}{q}) = -2i\ \sin(\underset{\sim}{q}_\alpha r_o)\phi^{[2]} \quad ,$$

$$K^3_{\alpha\beta\gamma}(\underset{\sim}{q}) = 0 \qquad\qquad ,\ \text{otherwise.}$$

The damping function in this cubic approximation is

$$\Gamma^{(3)}_R(\omega) = \pi\omega_R \sum_{+-} \sum_{\lambda_1\lambda_2} \left| V_3(R\lambda_1\lambda_2)\right|^2 \times$$

$$\times\ 2(\omega_1 \pm \omega_2) \left[n_2 + \frac{1}{2} \pm (n_1 + \frac{1}{2})\right] \delta(\omega^2 - (\omega_1 \pm \omega_2)^2) \tag{4.6}$$

$$= \sum_{+-} \sum_{\lambda_1\lambda_2} \sum_{\alpha\beta\gamma} \left| K^3_{\alpha\beta\gamma}(\underset{\sim}{q}) \; f^3_{\alpha\beta\gamma}(\underset{\sim}{q}) \right|^2 \quad \times$$

$$\times \left[\bar{n}_2 + \frac{1}{2} \pm (\bar{n}_1 + \frac{1}{2})\right] \{\delta\left[\omega - (\omega_1 \pm \omega_2)\right] - \delta\left[\omega + (\omega_1 \pm \omega_2)\right] \}$$

with the short notation for the eigenvector combinations:

$$f^{(3)}_{\alpha\beta\gamma}(q) = (\underline{\varepsilon}^+_R - \underline{\varepsilon}^-_R)_\alpha \left[\varepsilon^+_\beta(\lambda_1)\varepsilon^-_\gamma(\lambda_2) - \varepsilon^-_\beta(\lambda_1)\varepsilon^+_\gamma(\lambda_2)\right] =$$

$$= (\frac{\hbar}{2N})^{3/2} \left(\frac{\underline{e}^+_R}{M_+^{1/2}} - \frac{\underline{e}^-_R}{M_-^{1/2}}\right)_\alpha (M_1 M_2)^{1/2} (\omega_R \omega_1 \omega_2)^{-1/2}$$

$$\times \left[\underline{e}^+_1 \, \underline{e}^-_2 - \underline{e}^-_1 \, \underline{e}^+_2\right]_{\beta\gamma} \quad , \qquad (4.7)$$

where

$$\left(\frac{\underline{e}^+_R}{M_+^{1/2}} - \frac{\underline{e}^-_R}{M_-^{1/2}}\right)^2 = \frac{1}{\mu}$$

4.2.2 Critical point analysis

The simple dynamical model described by Eq. (4.5) is, of course, in agreement with the infrared selection rules. For example, two-phonon combinations are not allowed for q at Γ or X, but are possible at L, with the exception of TO + LO and TA + LA, and all overtones are forbidden. These rules for the symmetry points Γ and X can be derived in our approximation form the phase tnesor K $(\underset{\sim}{q})$ since sin sin $(q_\alpha r_o)=0$ at $\Gamma(q_\alpha=0)$, and at x $(q_\alpha r_o=\pi)$. On the other hand, the bilinear tensor of the eigenvectors in Eq. (4.7):

$$\underline{E}^{+-}(\underset{\sim}{q}j_1j_2) \equiv \underline{\varepsilon}^+(\underset{\sim}{q}j_1)\; \underline{\varepsilon}^-(\underset{\sim}{q}j_2) - \underline{\varepsilon}^-(\underset{\sim}{q}j_1)\; \underline{\varepsilon}^+(\underset{\sim}{q}j_2) \qquad (4.8)$$

vanishes if $j_1 = j_2$ because two terms of $\underline{E}^{+-}$ cancel each other in this case. This includes all overtones. For phonon modes at the point $L(2\, r_o q_\alpha = \pi)$ one of the sublattices is always at rest. Assuming for the moment $m_+ < m_-$, the following zeros are established:

$$\varepsilon^+(L|TA) = \varepsilon^+(L|LA) = \varepsilon^-(L|TO) = \varepsilon^-(L|LO) = 0 \tag{4.9}$$

Two phonon combinations are forbidden if in each term of Eq. (4.8) at least one ε is zero. Thus TO + LO and TA + LA are excluded. As this discussion shows, critical point analysis of the two-phonon spectra can be performed on the basis of the present simple dynamical model if the principal behavior of the eigenvectors is known at those critical points. Inspection of a simple rigid ion model easily gives this. From Eq. (4.5), it turns out that the main contributions to the damping function come from regions in the Brillouin zone where $2q_\alpha\, r_o \approx \pi$. This determines a cube in q - space which touches the surface of the Brillouin zone just at the L-points. The four combinations at the L points, TO + TA, TO + LA, LO + TA, LO + LA, are, therefore, especially suitable candidates for a qualitative critical-point analysis of infrared spectra in crystals with sodium chloride structure.

The result of such an analysis is shown for LiF in Fig. 4-3. Here, an experimental damping function for LiF is shown as evaluated from experimental data.[73,77] The reduced frequency scale ω/ω_R practically eliminates the quasi-harmonic shifts between the two curves taken at 70°K and 300°K. The lower temperature data at 70°K are close to zero temperature results since the Debye temperature of LiF is an order of magnitude higher. Therefore, the spectra show hardly any difference bands in contrast to the data obtained at 300°K. The four two-phonon contributions taken at the L point coincide surprisingly well with the four maxima or "kinks" in the experimental summation bands. The small values of the absorption bands beyond the two-phonon cutoff demonstrate the weakness of three-phonon summation bands at these temperatures. This does not hold in the region of difference bands. Figure 4.3 exhibits a very strong increase of the damping function about $\omega = 0.77\omega_R$ which cannot be explained by two-phonon difference bands.

A further point to note is how the frequency dependent damping function differs from classical damping characterized by a constant γ^{cl}. In order to perform an appropriate comparison, it is assumed that $\tilde{\omega}(0) \simeq \tilde{\omega}(\omega_R) \simeq \omega_R$ in Eq. (4.1) and $2\omega_R \Gamma_R(\omega) \equiv \gamma_R(\omega)$ is replaced by $\omega\gamma^{cl}$. Figure 4-3 shows the corresponding linear function as

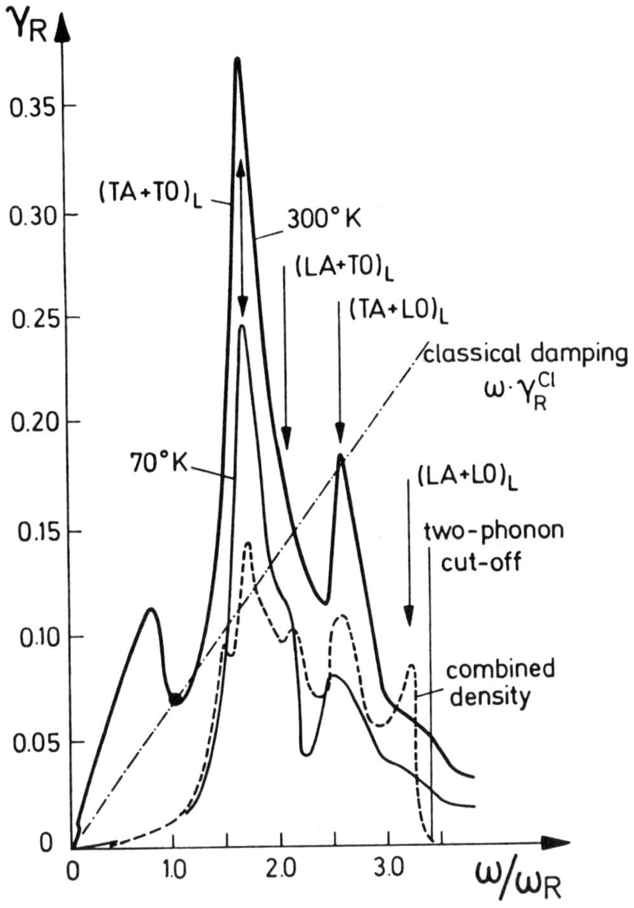

Figure 4-3: Damping function of LiF.

fitted to the experimental data at 300°K. It can be seen from the figure that such a description holds only in a very small frequency region around ω_R and might lead to wrong values of "linewidths," if these are taken from a larger frequency region.

Since the phase tensor $\underline{K}(q)$, (equation following Eq. (4.5)), qualitatively favors regions with larger q-values in the Brillouin zone where the density of states is high, it is expected that a description of the damping function using a combined density approximation should give satisfactory results. Figure 4-3 shows a combined density,

$$\Sigma_{(\tilde{q}j_1j_2)} \delta\left[\omega - (\omega_1 + \omega_2)\right],$$

for LiF as calculated by Smart, et al. at T = 0°K and derived from Hardy's deformation dipole model. The agreement is very good with respect to the positions of maxima, minima and cutoffs, which shows clearly that to obtain a two-phonon critical point it is sufficient to combine two one-phonon critical points, that is, to have $\underline{\nabla}\omega_1(\tilde{q}) = \underline{\nabla}\omega_2(\tilde{q}) = 0$ separately for each branch. Therefore, most of the structure of a combined density can be found by inspection of the one-phonon densities. In particular, cutoffs and gaps are easily determined.

If one extends the expansion of the short-range two-body potential $\phi(r)$, Eq. (4.3), up to the fourth-order, one obtains the coupling parameters $\phi_{\alpha\beta\gamma\rho}(0x, 0x, \ell x', \ell x')$, needed to compute higher order contributions to the self-energy. One parameter, ϕ^{IV}, determines in this approximation the contribution of quartic anharmonicity to the damping function (see diagram c in Table 4-1), which, in the high-temperature limit (where the thermal occupation numbers $n_i + 1/2 = kT/\hbar\omega_i$) becomes proportional to

$$(k/\hbar)^2\, T^2 \left[(\omega_1\omega_2\omega_3)^{-1}\,(\omega_1 + \omega_2 \pm \omega_3)\right] \qquad (4.11)$$

The absolute value of the quartic anharmonic parameter as determined from experimental data depends critically on the details of the theory which is used in the analysis of the data. It seems that often the quartic anharmonicity is overestimated by fitting it to the temperature dependence of the linewidth of the transverse optic eigen-frequency or to Born-Mayer potentials. This leads to values of the calculated absorption which are much too high in the low-frequency regime $\omega<\omega_R$, where the four-phonon difference processes are dominating the absorption at room temperature. We note that the decrease of the absorption constant in the high-frequency regime is mainly due to a corresponding decrease of the damping function. Formula (4.1) reads for $\omega^2>>\omega_R{}^2$

$$\varepsilon'(\omega) \approx \varepsilon_\infty$$

$$\varepsilon''(\omega) \approx (\varepsilon_0 - \varepsilon_\infty)\ 2\omega_R^3\ \Gamma_R(\omega)\omega^{-4} \qquad (4.12)$$

while the experimental values of $\varepsilon''(\omega)$ (Fig. 4-2) show a decrease with at least ω^{-9} and ω^{-10} in the two- and three-phonon regimes, respectively. Therefore, the damping function Γ_R must possess a strong frequency dependency itself. This conclusion is consistent with the result that the non-linear dipole moment, which leads to a weaker frequency-dependency, is not dominating the absorption in alkali halides.

It seems that the description of the damping function of dispersion oscillators in terms of short-range anharmonicity as given in the previous sections is rather satisfactory. In fact, this picture, while remaining qualitatively correct, is modified if long-range parts of the potential, non-linear dipole moments and some other effects are considered. We discuss these next.

The possible importance of the Coulomb potential was discussed by several authors when it turned out that short-range coupling parameters fitted to infrared data were inconsistent with those obtained from a Born-Mayer potential describing lattice compressibility, etc.[75,71,80] It turned out in these papers that a reduction of the value of the cubic parameter ϕ''', Eq. (4.4) by adding its Coulombic nearest-neighbor counterpart improved the agreement between the parameters evaluated from different sources remarkably. It was, therefore, assumed that some "scaling" of the calculated self-energy by an appropriate factor might be enough for the consideration of Coulomb interactions.[71,48]

That this is not quite correct has been shown by Knohl[74,36] who did explicit calculations of the cubic part of the Coulomb potential for LiF. The treatment is very similar to that of the short-range potential with the essential difference that the phase tensor $\underline{K}^3(q)$ has to be replaced by a tensor $\underline{C}^3(q)$ which is the Fourier transform of the anharmonic Coulomb force constant matrix. It turns out that V_3 and V_3^{Coul} interfere destructively so that the complete cubic damping function Γ_3 is reduced by a factor about 2 as compared with its short-range value. Similarly, E. R. Cowley[71] reports a reduction of about 60% in his discussion of thermal properties of NaCl.[81,75]

As Knohl has shown, this "scaling" assumption in LiF is only valid in the frequency regime $\omega \lesssim 2\omega_R$. Similar values may be expected for other alkali halides. Above this frequency, the complete cubic damping function decreases much slower that the (scaled) short-range function and finally gives values even higher than the unreduced short-range contribution. The absorption at long wavelengths has been given special attention by Genzel and his coworkers.[82]

Stolen and Dransfeld[83] discussed the low-temperature behavior of some alkali halides in the submillimeter region while Owens[84] looked into their high-temperature absorption.

From these papers, the clear conclusion can be drawn that the three-phonon combinations dominate the absorption at low frequencies. It seems that the absorption in the nearly static case shows additional contributions from all types of defect scattering, etc. In addition, the assumption of non-interacting phonons in the final states is appropriate at higher frequencies and room temperatures (collision-free regime) but becomes doubtful if both conditions are not fulfilled. The collisions between phonons have then to be taken into account by diagrams of the type shown in Tab. 4-1e,f which are important in phonon transport problems. The reader is referred to the review by Glyde and Klein.[85]

4.2.3 Non-Linear dipole moments

The formal contributions of second- and third-order dipole moments to the infrared susceptibility and their interplay with the anharmonic forces was described in 3.2.3. For the discussion of this section, we represent the different processes in Table 4-1 by diagrams. These diagrams can be derived by replacing the propagator of the dispersion oscillator in Table 4-1 by that of an electronic dipole excitation and the anharmonic vertices V_n by the corresponding non-linear electron phonon couplings V^1_{n-1}

The theory has the structure of the linear response of two coupled active oscillators one of which has a very high resonance frequency (the electronic oscillator) as compared with that of the other, ionic oscillator. The complete system is driven in the frequency regime of the low-frequency ionic oscillator so that the electronic oscillator is always off-resonant and follows the ionic motion adiabatically. The self-energy of the electronic oscillator has a diagonal part corresponding to $\Pi_R(\omega)$ for the dispersion oscillator: this part contributes ε_∞ (the harmonic response) and the susceptibilities $\underline{\chi}^{[2]}$ and $\chi^{[3]}$, due to the 'direct' effects of the second and third order dipole moments. As has been discussed formally in Section 3.2.3, there is also a non-diagonal coupling term in the self-energy matrix of the two oscillators which in the frequency regime of the low-frequency oscillator is conveniently described by a renormalized dipole moment $\tilde{M}_R(\omega)$.

The non-linear dipole moments may be discussed in the framework of a non-linear shell model. This was first done by R. Cowley.[34] Using the shell model expressions for the potential coefficients V^n_m, we obtain the following result for the ratio of the effective non-linear dipole moment M_2 divided by the anharmonic potential V_3:

$$X \equiv \frac{M_2\begin{pmatrix} \underset{\sim}{q} & \underset{\sim}{q} \\ j & j' \end{pmatrix}}{V_3\begin{pmatrix} 0 & \underset{\sim}{q} & -\underset{\sim}{q} \\ & j & j' \end{pmatrix}} = \frac{M^1(0)V^2(00)^{-1}V_2^1(0\lambda\bar{\lambda}')}{V_3\ (0\lambda\lambda')}$$

$$= \frac{Y(I + YCY)_o^{-1}\ \phi_{SCC}(0\lambda\lambda')}{\phi_{CCC}(0\lambda\lambda')} \tag{4.13}$$

where ϕ_{SCC} and ϕ_{CCC} define anharmonic extensions of the shell-core and core-core force constant matrices ϕ_{SC} and ϕ_{CC} respectively. If we assume for simplicity that only the negative ion is polarizable and that anharmonic forces are acting through the shells, then $\phi_{SCC} \approx \phi_{CCC}$ (just as $\phi_{SC} \approx \phi_{CC}$ for the harmonic forces), and one finds

$$eX \approx -\chi_\infty/Y_- e \tag{4.14}$$

Here χ_∞ means the electronic susceptibility due to the polarizability α_- of the negative ions. The same result is derived, in a similar approximation, for the ratio M_3/V_4. Since the shell charge $Y_- e$ is usually found to be near 3e in alkali halides, the relative importance of the non-linear dipole moment is determined by the electronic susceptibility of the crystal. This susceptibility is directly a measure for the difference between the transverse effective (Born) charge Z_T^* and the longitudinal effective (Callen) charge $Z_L^* = \varepsilon_\infty^{-1} Z_T^*$.

To understand the relative importance of the non-linear dipole moment in more detail, we study the wings of the infrared absorption (equations 3.128 and 3.103)

$$\varepsilon''(\omega) = \frac{4\pi\hbar}{V} \operatorname{Im}\{\tilde{M}_R^2(\omega)\, G_R(\omega) + \frac{1}{2} \sum_{\lambda_1 \lambda_2} M_2(\lambda_1\bar{\lambda}_2) M_2(\bar{\lambda}_1\lambda_2) G_{2,\lambda_1\lambda_2}(\omega) + \ldots\} \tag{4.15}$$

$$\simeq \frac{4\pi\hbar}{V} \sum_{\lambda_1\lambda_2} \left| M_2(\lambda_1\bar{\lambda}_2) - \frac{M_R V_3(R\lambda_1\bar{\lambda}_2)}{\omega_R^2 - (\omega_1 \pm \omega_2)^2} \right|^2 \times$$

$$\times 2(\omega_1 \pm \omega_2)\, \delta\left[\omega^2 - (\omega_1 \pm \omega_2)^2\right] \left[n_2 + \frac{1}{2} \pm (n_1 + \frac{1}{2})\right] \tag{4.16}$$

and, with Eq. (4.13):

$$\varepsilon''(\omega) \approx \frac{4\pi\hbar}{V} \frac{M^2}{\omega_R^2} \sum_{\lambda_1\lambda_2} \left| V_3 \left[X \frac{\omega_R}{M_R} - \frac{1}{1 - (\omega_1 \pm \omega_2)^2/\omega_R^2} \right] \right|^2 \tag{4.17}$$

$$\times\; 2(\omega_1 \pm \omega_2)\, \delta\left[\omega^2 - (\omega_1 \pm \omega_2)^2\right] \left[n_2 + \frac{1}{2} \pm (n_1 + \frac{1}{2})\right] .$$

Eq. (4.16) was first derived by Szigeti.[38] As can be seen by inspection of Eq. (4.17), the relative importance is the highest when $\omega = \omega_1 + \omega_2$ is large. In alkali halides, with $\omega_1 + \omega_2 \lessapprox 3\omega_R$, we obtain for the bracket in Eq. (4.17) at this frequency approximately $0.1 - X\omega_R/M_R$.

This means that the second order dipole moment becomes important if the "effective frequency" $M_R/X = \omega^*$ is smaller than 10 ω_R. This frequency was introduced by Borik.[35]

From the shell model, Borik deduced:

$$\omega^* = \left| \frac{1}{2} \omega_R \left(\frac{Z}{Z_p} \frac{\varepsilon_o + 2}{\varepsilon_\infty + 2} - 1 \right) \right| \qquad (4.18)$$

with z_s, the Szigeti effective charge and $Z_p = Z - Z_S$ the polarization charge. In ionic crystals such as the alkali halides, $Z_p \ll Z$, which means that the effective excitation energy $\hbar\omega^*$ for a non-linear dipole moment M_n is rather high as compared with the infrared energy $\hbar\omega_R$, and M_n does not play an important role. The situation is different in III-V compounds where $Z<Z_p$, so that ω^* becomes comparable with ω_R, and then the non-linear dipole moment is essential for the understanding of the infrared absorption.

In ionic crystals with $Z^p \simeq 0.25$ Z, $Z \simeq 1, (\varepsilon_o + 2)/(\varepsilon_\infty + 2) \simeq 2$ it follows that

$$(\omega_R/\omega^*)^2 < 0.01 \qquad (4.19)$$

Since the factor determines the order of magnitude of the relative contribution of the non-linear dipole moment to the absorption in the two-phonon regime, it is now often assumed that this effect is negligibly small.[80,48]

It has been shown by Knohl[36,74] that the situation may be different if the Coulomb anharmonicity is taken into account. This destroys the proportionality of the matrix elements of M_2 to those of V_3.

4.3. Final states interactions of phonons: anharmonic broadening and bound states

In the foregoing section of this chapter, the phonons in the final states were always described as quasi-particles with infinite lifetime. Aside from the dispersion oscillators, the effect of anharmonicity on all the other lattice vibrations was only considered by introducing renormalized eigen-frequencies. This is, of course, an approximation and one may ask for the effect of the finite lifetime of the phonons or more general anharmonic effects on the infrared spectra. Some corresponding diagrams are shown in Table 4-1 and they demonstrate different types of final state interactions.

a) diagram 4-1g: broadening of the spectra by the individual lifetimes of the phonons:

b) diagram 4-1e: dynamical interaction between different two-phonon combinations.

c) diagram 4-1f: Frequency-dependent corrections of the matrix elements (vertex corrections).

d) diagram 4-1d: interaction between two- and three-phonon bands.

Processes (a) seems to be the most important ones leading to a smearing out of sharp features ('erosion' of critical points). There is a striking example in the case of the 1-phonon sideband of the impurity-induced V center mode in alkali halides. Here, the linewidth of the local mode leads to a shift of frequencies in the sidebands and a partial filling of the phonon gap in KBr. Similar effects have to be expected in the phonon spectra of perfect crystals, but, at the present time, evidence is restricted to the fact that calculated spectra quite often show a sharper and stronger oscillating structure than the measured ones exhibit. In practice, the finite lifetime of the phonons can be globally taken into account by keeping ε finite in the Green function. This means a representation of the delta-function by a Lorentzian or a histogram and is usually used to create a smooth curve without spurious features. But it might be regarded as a simple approximation of the phonon lifetime effects and then the absolute value of ε influences the determination of the shift function of the dispersion oscillator. Typical values of ε are near 0.1 THz, i.e., about 2% of the frequency of dispersion oscillators.[48] Process (b) is related to the problem of bound states which was first recognized in connection with certain sharp features in Raman spectra (see Section 5). The problem has been discussed recently by Bruce[48] who investigated the possible influence of diagram 4-1e on the two-phonon spectra of KBr. The contribution to the complex self-energy of the dispersion oscillator is

$$\Delta_R(\omega) - i\Gamma_R(\omega) = \frac{1}{4} \sum_{\lambda i} V_3(R\lambda_1\lambda_2) G_{2,\lambda_1\lambda_2}(\omega) \times$$

$$\times V_4(\lambda_1\lambda_2|\lambda_3\lambda_4) G_{2,\lambda_3\lambda_4}(\omega) V_3(\lambda_3\lambda_4 R) \cdot \qquad (4.20)$$

As has been shown by Bruce, the calculation can be greatly simplified by exploiting the separability of V_4 in this particular case. The calculations are easily extended to take account of diagrams where the single vertex V_4 is replaced by an infinite series of V_4 "Chains." The effect might become visible in the high-frequency regime near ω_L. Bruce's[48] analysis of a thin film experiment in KBr remains, however, doubtful since he did not consider the influence of Coulomb anharmonicity and second order dipole moments which are influential in the same frequency regime.

The vertex corrections of process (c) cannot be expected to give neglig ble contributions if one remembers the results given above for other diagrams of the same order. The calculation is different since V_4 is not separable in this case and has not been carried out so far.

The problem of the temperature dependence of the shift function and the line widths of dispersion oscillators was discussed by R. Cowley,[34] Ipatova, et al.,[79] Lowndes,[86] E. R. Cowley,[76] Bruce,[48] and Fischer.[87] Fischer formulated a consistent extension of the pseudo-harmonic approximation where the renormalized frequencies $\tilde{\omega}_i$ of the phonons are supposed to also contain their imaginary part evaluated at their pole frequencies:

$$\tilde{\omega}_i^{T^2} = \omega_i^{T^2} + 2\omega_i^T \Delta_i(\tilde{\omega}_i^T) - i2\omega_i^T \Gamma_i(\tilde{\omega}_i^T) \tag{4.21}$$

Inserting these complex phonon frequencies into the formalism leads, in the lowest approximation, to the following refinement of the quasi-harmonic two-phonon Green function,

$$\tilde{G}_{2,\lambda_1\lambda_2}(\omega) = -\frac{2(\tilde{\omega}_1^T + \tilde{\omega}_2^T)(1+n_1+n_2)}{[\omega+i(\Gamma_1 + \Gamma_2)]^2 - (\tilde{\omega}_1^T + \tilde{\omega}_2^T)^2} - \frac{2(\tilde{\omega}_1^T - \tilde{\omega}_2^T)(n_2 - n_1)}{[\omega + i(\Gamma_1 + \Gamma_2)]^2 - (\tilde{\omega}_1^T - \tilde{\omega}_2^T)^2} \tag{4.22}$$

Here the entity ε which ensures the proper causal behavior of the free-phonon retarded Green function is replaced by the sum of the line-widths of the phonons. In this way, a further step into a self-consistent treatment of the anharmonic self-energy with respect to all phonons involved has been made. The main approximation in Eq. (4.22) is the neglect of the imaginary parts of the occupation numbers corresponding to correlated fluctuations of the occupation numbers during the absorption process.

Fischer[87] has discussed some qualitative consequences of this approach for the line-width and shift of the resonance frequency. He found, in complete analogy to the simple case of a local mode sideband,[88] a smearing out of features in the self-energy, in particular a 'filling' up of the frequency regions with small damping from the neighboring regions with higher damping. This is exactly the case in the surrounding of the infrared resonance frequency and leads to an additional enhancement of the line-width near ω_R. Though quantitative calculations have still to be carried out, it seems very plausible to conclude that quartic anharmonicity is often overestimated as was suggested by Jooij[89] and E. R. Cowley.[80]

The foregoing discussion also holds in principle for the longitudinal optic mode $\tilde{\omega}_{LO}$. The experimental basis for observing the properties of $\tilde{\omega}_{LO}$ is the thin-film technique first used by Berreman.[70] Physically, we may express the effect of a thin film by stating that the infrared resonance frequency $\tilde{\omega}_R$ with respect to the parallel component of the incident field is shifted to $\tilde{\omega}_{LO}$. As a consequence, the self-energy of the resonance mode in such an experiment is just that of the bulk infrared frequency evaluated at the frequency ω_{LO}, replacing $\tilde{\omega}_R$ by $\tilde{\omega}_{LO}$:

$$\tilde{\omega}_{LO}\pi_{LO}(\omega) = \tilde{\omega}_R\pi_R(\omega) \tag{4.23}$$

Since usually $\tilde{\omega}_{LO}$ is very close to the maximum of the two-phonon damping function, the absorption is stronger near $\tilde{\omega}_{LO}$.

4.4 Other ionic crystals

The infrared absorption of alkali halides seems to be mainly described by an effective nearest-neighbor cubic anharmonicity aside from minor corrections due to Coulomb anharmonicity, quartic terms in the potential and non-linear dipole moments. If we proceed to other cubic ionic crystals, the situation does not change drastically as long as the polarizability of the ions is not very high.

Of special interest are the ionic compounds of silver which exhibit unusual ionic properties, such as extreme diffusivity. If we compare the lattice vibrations of AgCl with those of RbCl, we find that while the acoustic branches of both substances are still rather similar, there are two typical differences in the optical branches. Firstly, the transverse optic branch is lowered especially in the (111) direction as a consequence of d-shell deformabilities of the Ag^+ ion. Secondly, the frequencies in the longitudinal optic branch are increased since the smallness of the Ag^+ ion leads to a contraction of the lattice and, subsequently, to an enhanced coupling between second-nearest neighbor chlorine ions.

The result of the lowering of the TO phonons is a corresponding decrease of the combination branch TO + TA at the L-point which governs the first maximum of the damping function in the cubic approximation. This maximum is situated in alkali halides near $\omega_{LO} = 1.5\omega_R$ but is shifted in AgCl and AgBr to a frequency close to the infrared eigen-frequency ω_R. Therefore, the transmission bands in the silver halides are rather broad as compared with those of the alkali halides even at low temperatures.[90]

Recently, the anharmonic properties of silver halides, in particular those of AgCl, have been discussed by Fischer.[87] He showed that an effective nearest-neighbor cubic anharmonicity is well able to describe the thermal expansion, the temperature dependence of ω_R and, to a certain extent, the infrared spectra in AgCl consistently. Unfortunately, the measured reflection spectra[91] seem to be quite unsatisfactory, probably due to surface and defect scattering.

An interesting aspect of Fischer's calculation is that he considered the influence of the phonon line-width in the two-phonon decay channels Eq. (4.22). As a consequence, the detailed structure in the phonon shift-function is washed out. This might explain the fact that anharmonic renormalization of phonon frequencies as discussed by Cowley and Cowley[92] does NOT lead to easily observable structures in the spectral functions i.e. neutron scattering profiles as one might expect from the above-mentioned calculations.[93,14]

For further information on the experimental spectra and their theoretical analysis, the reader is referred to the infrared bibliography by Pawlik.[94]

5. Raman scattering from crystals

In the last section we have seen that the infrared spectra of ionic crystals are mainly determined by first order resonant absorption from dispersion oscillators with their anharmonic line widths leading to multiple phonon structure in the tails of the absorption. On the other hand, in the strongly covalent crystals of high symmetry such as diamond, etc., genuine second or higher order non-resonant spectra may be observed. The first order resonance, in this case, is situated far above in the optical frequency regime of the energy band transitions. In the model description of these spectra, this means that the role of the effective charge of the nonlinear dipole moments is played by an electronic charge with a mass about 10^{+4} times smaller than the ionic masses are. The infrared spectra of the covalent crystals are low-frequency off-resonant spectra and exhibit simple relations to combined densities, critical points, etc.

This situation is to a certain extent reminiscent of Raman scattering by ionic crystals. In particular, the Raman spectra of cubic ionic crystals with inversion symmetry do not show a first order Raman line as a result of inversion symmetry. Their second order spectra are all far below resonance, i.e., the band gap transition in the UV. Similarities with infrared spectra are nevertheless very superficial and the analysis of infrared spectra does not tell us much about Raman scattering in the same crystal. The Raman spectra are, in addition, of greater complexity than the infrared spectra since they have at least three different parts with symmetries $\Gamma_1^+, \Gamma_{12}^+$, and Γ_{25}^+. The non-resonant Raman phonon spectra only appear as sidebands of electronic transitions and not via direct scattering of the light by an ionic Raman oscillator. This (infrared) ionic Raman effect is very weak due to the ω^4 factor in front of the cross section and has not been observed so far. The possibility of its observation under favorite circumstances has recently been discussed by T. P. Martin.[95] Its realization would be of enormous interest also for the theory since in this case the anharmonic phonon-phonon coupling would play a similar but complementary role to that which it plays in infrared absorption.

On the other hand, the resonance Raman effect in the visible regime of the optical spectrum has been investigated very intensively during the last decade.[96] Its analysis provides interesting details of the electron-phonon coupling connected with particular electronic band-band transitions and allows a determination of the coupling parameters for these specific cases. It provides, therefore, important additional information not given by the off-resonant spectra. Here we will mainly discuss off-resonant spectra.

5.1 Raman spectra of cubic ionic crystals

Fig. 5-1 contains the spectra of three different chlorides. The zero points of the frequency scale are shifted against one another in order to demonstrate the similarities in the general features of the spectra. Corresponding arrangements of spectra can be made for fluorides, bromides and iodides. In all cases, the second order spectra show a strong decrease above the high-frequency limit of the 2 TO regime and exhibit a very weak residual scattering in the following 2 LO frequency regime. This behavior parallels the rather low two-phonon density of states in this latter regime.

The second order spectra of the alkali earth oxides, behave quite differently. In addition to the two-phonon spectra of transverse phonons, they exhibit a very strong scattering in the 2 LO regime. This property seems to be very typical for all oxydic crystals and can even be used to distinguish the oxydic perovskites from those which do not contain oxygen. A similar behavior has been observed in the case of hydrides.[97,98]

The similarities in the spectra suggest a specific mechanism for the change of the polarizability in these crystals. In the case of alkali halides, we shall see that inter-ionic nonlinear polarizabilities are governing the spectra. It seems that this fact is in agreement with a simple inspection of the band structure. While in the case of linear polarizabilities the states below and above the ionization energy are of equal importance,[99] the Raman polarizabilities for second order scattering contain two more energy denominators which favors the relative weight of the low energy transitions. We may, therefore, expect that, in a first approximation, transitions between the (anion p^6) valence band and the first (cation s^2) conduction band should play an important role. In a formal description, these charge transfer transitions correspond to inter-ionic nearest-neighbor polarizabilities which, actually, describe the Raman spectra of alkali halides quite well (see below).

One might argue that the earth alkaline oxides should exhibit a similar behavior since their band structure is similar to that of alkali halides. In particular, the valence and conduction bands originate also from anion p and cation s states, respectively, and the energy gaps show comparable or often higher values (> 10 eV) since the Coulomb splitting for a divalent ionic crystal is higher than for a monovalent one. Surprisingly, the oxides display a strong intra-ionic scattering which corresponds to completely localized transitions at the oxygen sites involving very weak charge transfer processes only. This means that a simple band structure analysis is insufficient for the understanding of the spectra.

We may explain this unexpected behavior by considering the results of Hartree-Fock calculations for the band structure of MgO.[100] Here it turns out that the insulating gap is reduced by a factor of two due to strong correlation effects. Unfortuneately, at this time no calculations exist which explain intra-ionic polarizabilities in ionic crystals explicitly in terms of microscopic matrix elements. Therefore, the foregoing discussion may be considered as a plausible argument only.

Since in the crystals under consideration every ion is a center of inversion symmetry all lattice modes at Γ are Raman inactive. We are therefore concerned with the second order spectra (third order spectra are generally too weak to be observed). As in the case of

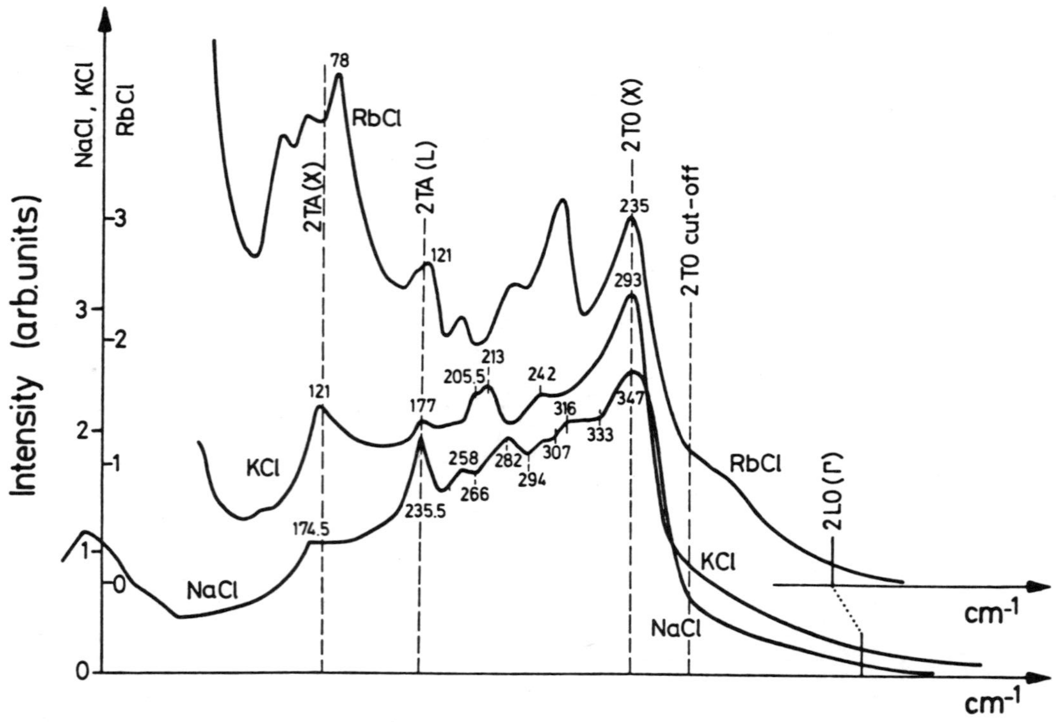

Figure 5-1: Second order Raman spectra ($\Gamma_1^+ + 4\Gamma_{12}^+ + \Gamma_{25}^+$) of alkali chlorides. The spectra are shifted one against another by 55cm^{-1} (KCl and NcCl) and by 58cm^{-1} (RbCl and KCl).

infrared spectra, one might try to relate the main features of the spectra to specific critical points of the joined density of states. The selection rules for the main overtones and combinations for Raman processes in rocksalt crystals given by Birman[11] show that in alkali halides and other cubic diatomic crystals the main peaks and shoulders of IR spectra can be interpreted in terms of combinations of phonons at the L point. This result arises from an interplay of the selection rules, the saddle point character of the phonon dispersion and the q-dependence of the matrix elements determined by a central two-ion potential. A similarly simple and general analysis of the Raman spectra, while very desirable, does not seem to exist although very often overtones and combinations of phonons at the X and the L point have frequencies close to experimental maxima or kinks in the Raman spectra.[101] This does not mean that the main contributions to the scattering intensities actually originate from two phonon processes in the neighborhood of the X or the L point. It rather indicates that there exist unknown correlations between the frequencies of these particular two-phonon processes and the still hidden dominating ones somewhere in the Brillouin zone which are governed by some Raman

'potential.' The knowledge of this potential would be equivalent to an explicit description of the non-linear electron-phonon interaction in ordinary space. To be consistent with experiment, the potential should give dominant volume scattering in the great majority of spectra and should be localized between nearest and second-nearest neighbors.

Simpler Raman spectra are found in the hydrides, oxides and perhaps nitrides where the intra-ionic polarizabilities of the anions dominate the spectra. Here the existence of strongly localized electron-phonon coupling is indicated by intense scattering in the 2LO phonon regime. In favorable cases, such as MgO, as quantitative description of the spectra with only one adjustable parameter can be achieved. A detailed analysis of the origin of the different parts of the spectra in terms of specific two-phonon processes becomes possible. Furthermore, the dominance of the volume scattering leads to the conclusion that the genuinely quartic part of the local electron-ion potential dominates over the integrated cubic process. We note that this statement refers to the displaced ionic configuration and might not be true for the equilibrium configuration.

It is interesting that an analysis of this type is still successful in the case of tetrahedral semiconductors such as the zinc chalcogenides. The reason may be found again in the flexibility of the ground state wave functions which in these crystals are hybridized to some extent with wave functions of neighboring ions. Due to this open-shell behavior, it is impossible to determine a well-defined linear ionic polarizability. The change of the crystal's polarizability is then, as in the case of oxygen, rather well localized at the ion lattice sites and appears as an "intra-ionic" mechanism when described in a model theory.

5.1.1 Raman spectra of alkali halides

After the early measurements of the unpolarized spectra of NaCl by Krishnan[103] and its interpretation by Born and Bradburn[102] in terms of a rigid ion model with weighted two-phonon densities, there were a few investigations mainly using mercury arc lamps.[104,105] Since the advent of lasers about 15 years ago, an increasing number of Raman spectra of ionic and other crystals has been recorded. Here we focus attention on rather new data, in particular those which have been analyzed in terms of formal or model parameters.

An investigation of numerous diatomic ionic crystals has been presented by Krauzman. Among them the spectra of KBr attracted particular attention. The main reason is that the lattice dynamics of KBr is well-known both experimentally and theoretically[107] so that the idea of a quantitative analysis of the spectra is tempting. Of particular interest is the fact the in KBr (just as in NaCl) the longitudinal optic branch near the L point seems to exhibit a strong intra-ionic "breathing" deformability.[108] One might therefore speculate about a non-linear extension of the breathing shell model in terms of non-linear intra-ionic polarizabilities of breathing (Γ_1^+) and dipolar (Γ_{15}^-) type, since both seem to arise in the harmonic lattice dynamics of alkali halides. This idea has been worked out in detail by Bruce and Cowley[49,109,48] with emphasis on the spectra of KBr and NaCl. These authors have fitted three parameters to the experimental data. Since the absolute intensities are unknown, only two ratios have to be determined. In view of this small number of parameters and the experimental uncertainties the results are not unsatisfactory. All important features are reproduced within an intensity factor $\lesssim 2$. Some of the calculated peaks in the upper part of the $\Gamma_1^+(A_{1g})$ spectrum do not seem to have an experimental counterpart. A better result was obtained by Bruce[48] for NcCl with only two parameters and again, agreement seems to be worse in the high frequency part of the Γ_1^+ spectrum. The results of Bruce and Cowley for KBr have been disputed by Krauzman.[50] In his discussion he used the same breathing shell model but calculated the Raman tensor in a more formal way by considering first all types of expansion coefficients of second order into ionic ($\underline{u}$), shell ($\underline{w}$) and breathing ($\underline{v}$) displacements which lead to terms in $\underline{u}\,\underline{u}$, $\underline{u}\,\underline{w}$, $\underline{u}\underline{v}$, etc. He found that even neglecting the breathing mode of the K^+ ion one obtains in a nearest-neighbor approximation 23 parameters contributing to the Γ_1^+ and the Γ_{12}^+ spectra while 13 others contribute to the Γ_{25}^+

spectrum. There is, actually, a further restriction due to the adiabatic condition, but a fit in this parameter space certainly would not give a meaningful result. Krauzman then restricts the analysis to the 5 constants which contribute to the Γ_1^+ and Γ_{12}^+ spectra and involve only core displacements $\underline{u}$. Since the Γ_{12}^+ which depends only on differences of those parameters if rather weak as compared to the Γ_1^+ spectrum he neglects the Γ_{12}^+ part by assuming all 5 parameters to be equal. He then obtains a curve which compares favorably with that of Bruce and Cowley. For the Γ_{25}^+ spectrum, he shows that even a single term in $u_\alpha(K^+)u_\beta(Br^-)$ would give a nice fit to the experimental spectra. A systematic study of the known second order Raman spectra of alkali halides has been carried out by Haberkorn et al.[110] The result of this investigation gives strong support to Krauzman's conjecture in the case of KBr for all alkali halides. With the only exception of NaI, where intra-ionic polarizabilities may contribute about 20% of the spectral intensity, it seems that interionic polarizabilities dominate the spectra. A very strong argument in this direction stems from the fact that in alkali halides with lighter anion masses, no strong Raman scattering in the 2LO regime can be observed. A striking example is RbCl (Fig. 5-1). It is hard to understand that in this case the intra-ionic dipolar Raman polarizability is very weak (according to the scattering above 2TO (Γ)) while in NaCl the corresponding part in the 2LA regime is claimed to originate just from this mechanism. We note investigations of alkali halide spectra in terms of critical points in the two-phonon density of states. Many of them have been carried out by Karo and Hardy,[105,111] and by Krauzman.[50,112] For other investigations of this type in alkali halides and other ionic substances, we refer to the handbook article by Birman,[11] and by Bilz et al.

5.1.2 Alkali hydrides and alkaline earth oxides

The Raman spectra of the alkaline earth oxides (MgO, etc.) and those of LiH and LiD are different from those of the alkali halides by showing a strong scattering intensity in the 2LO frequency regime. We have mentioned above that the anions O^{--} and H^- differ from halogen ions, such as F^-, Cl^-, etc., due to the change of their polarizability with the ionic radius. This real (O^{--}) or near (H^-) instability of the free ions results in a strong intra-ionic nonlinear polarizability, as we now discuss in detail.

A one-parameter calculation of three spectra of MgO has been given by Haberkorn, Buchanan and Bilz. Only the spherical part of the intra-ionic quartic polarizability has been used. The result is surprisingly good in view of the complexity of the shape of the spectra. No inter-ionic (nearest- or second nearest-neighbor) shell model coupling parameters are assumed in contrast to the approach by Pasternak et al.[114] Similar results are obtained for CaO[52] and SrO.[53,115] In the latter case, the importance of inter-ionic polarizabilities in addition to the intra-ionic terms has been demonstrated. The numerical values of the fitted parameters indicate, that a long-range Coulomb-type of interaction may be important. A similar situation seems to exist in LiH and LiD. The spectra have been measured and analyzed in terms of formal parameters by Jaswal et al.,[116] and by Laplace.[98,117] Haberkorn et al.[110] were able to explain the high frequency part of the spectrum with the intra-ionic spherical quartic polarizability. The low-frequency part indicates contributions from inter-ionic terms. It seems that the idea of intra-ionic polarizabilities is quite well established in these cases.

5.1.3. Perovskites

Crystals with a pure or distorted perovskite structure, such as $SrTiO_3$ and $KTaO_3$, are particularly interesting. Many of them are ferroelectrics at lower temperatures and show one or several phase transitions. The phase above the highest transition temperature is often a cubic paraelectric one. These phases are the simplest for a theoretical analysis and, among them, $SrTiO_3$ has attracted the greatest interest since single crystals are easily available and detailed inelastic neutron scattering data are published.[46,118,119] The controversial results of the two latter papers have led to a re-examination of the dispersion curves by Stirling and Currat[120] which has essentially confirmed the earlier data.

Since in this structure the cubic phase has inversion centers at each of the different ions only second-order spectra are allowed. Even in the distorted ferroelectric phases these spectra show very strong intensities comparable to those of the first order lines if oxygen is the anion. Perovskitic halides such as $KMnF_3$, on the other hand, exhibit no measureable second order spectra.[14] It is therefore tempting to extend the intra-ionic treatment of the diatomic oxides to the case of oxydic perovskites.

The first calculation of the unpolarized spectra has been carried out by Bruce.[118] He used a quartic dipolar and a cubic breathing polarizability of the oxygen ion as in his calculation for NaCl,[100] and obtained satisfactory agreement with the experimental data. The use of a breathing deformability, however, seems to be debatable since, as in the case of NaCl and SrO discussed earlier, no evidence for a breathing effect in the lattice vibrations of $SrTiO_3$ could be established.[118] Instead, one might expect that the anisotropic position of the oxygen ion between two neighboring titanium ions in one direction and four strontium ions in the perpendicular plane should lead to an anisotropic polarizability of the oxygen both in lattice dynamics and in Raman scattering.

Rieder et al.[115] have measured the three components of the Raman spectra and have analysed the data in terms of a modified version of one of Stirling's shell models. This version contains a positive shell charge at the Sr ion (overlap shell model) and an anisotropic (elliptic) oxygen polarizability. The results indicate a strong anisotropy but were not very satisfactory with respect to the different positions of experimental and theoretical peaks. Recently, an extended and corrected treatment including the analysis of the $KTaO_3$ spectra has been published.[122] Since here only a few low-lying dispersion curves have been measured, the agreement between theory and experiment is gratifying (Fig. 5-2). It turns out that the harmonic and the quartic intra-ionic polarizability is strongest in the tantalum-oxygen direction which may be related to the strong hybridization of the 2p(O^{--}) wave functions with the 5d (Ta^{+}) wave functions.

5.2. Mainly covalent crystals

Raman spectra of covalent crystals are often analysed in terms of critical points of the one- and two-phonon density of states where the proper selection rules for Raman scattering in a tetra hedral crystal are considered. For a detailed discussion we refer to Birman.[11] Here, we are trying to discuss the situation from a model or microscopic point of view.

5.2.1. Spectra of diamond and its homologues

In section 3, we have discussed the infrared properties of covalent elements C, Si and Ge. It was shown that the dynamical effect of the covalency of these materials could be represented by introducing a 'bond charge' which was allowed to follow the ionic displacements adiabatically. The second order spectra of infrared absorption turned out to be described with very few anharmonic coupling parameters between the bond charge and its neighboring ions. This demonstrates the close analogy between a dynamical treatment of the bond charge and the properties of an effective ionic charge in ionic crystals.

The situation is different for the polarizability and its change in Raman scattering in covalent crystals.[59] This exhibits a one-phonon line with symmetry Γ_{25}^{+} and two-phonon spectra of symmetry Γ_{1}^{+}, Γ_{12}^{+} and Γ_{25}^{+}. The second order Raman tensor is given by Stokes.[12,25]

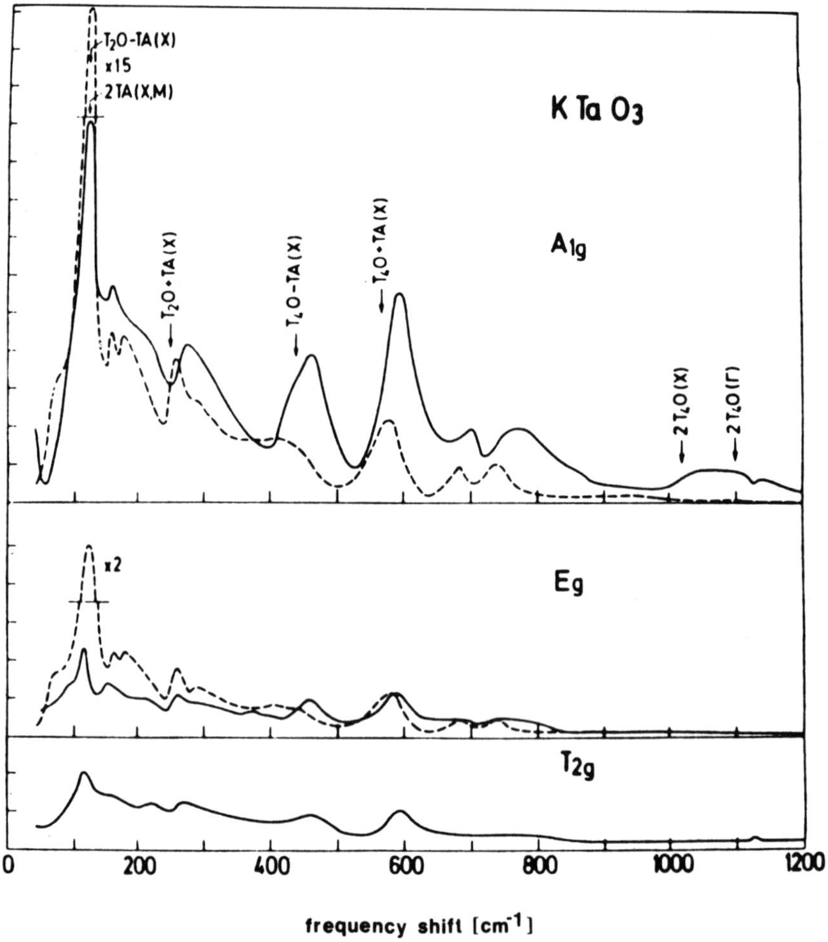

Figure 5-2: Second order Raman spectra of $KTaO_3$. Solid line is theory; dotted line is experiment.

$$i_{\alpha\beta\gamma\delta}(\Omega) = 2 \sum_{\underset{\sim}{q} j_1 j_2} P^*_{\alpha\beta}(\bar{\lambda}_1\lambda_2) P_{\gamma\delta}(\bar{\lambda}_1\lambda_2)(1+n_1)(1+n_2)\delta(\Omega-\omega_1-\omega_2) \tag{5.1}$$

The polarizability tensor $\underline{P}$ may be expanded in powers of the ionic and bond-charge displacements in ordinary space, as was done for the nonlinear dipole moment $\underline{M}^{(2)}$ by Go et al.[59]

$$\underline{P} = \underline{P}^{(o)} + \underline{P}^{(1)} + \frac{1}{2}\underline{P}^{(2)} : (\underline{U}^+\underline{U}^+ + \underline{U}^-\underline{U}^-) \tag{5.2}$$

The contribution of a cubic potential V^3 to the Raman spectra cannot account for the second-order Raman scattering since it leads to a very weak change of the crystal polarizability. Instead, we follow the concept of bond polarizability (BP)[61,62,63] and represent the polarizability of the covalent crystal $\underline{P}$ by a sum of independent BP's each of which is given by:[59]

$$P_{\alpha\beta}\{R^b\} = \sum_b \left[\hat{R}^b_\alpha \hat{R}^b_\beta \alpha_{\parallel}(R^b) + (\delta_{\alpha\beta} - \hat{R}^b_\alpha \hat{R}^b_\beta)\alpha_{\perp}(R^b) \right], \quad \underline{\hat{R}}^b = \frac{\underline{R}^b}{R^b} \tag{5.3}$$

where $\alpha_{\parallel}$ and $\alpha_{\perp}$ are the longitudinal and transverse components of the BP. The expansion coefficients of $\underline{P}$ are then simple linear combinations of $\alpha_{\parallel}$, $\alpha_{\perp}$ and their derivatives:

$$\alpha_v = \frac{4}{3\Omega_o}(\alpha_{\parallel}+2\alpha_{\perp}), \quad \alpha_{\underset{\sim}{q}} = \frac{4}{3\Omega_o}(\alpha_{\parallel}-\alpha_{\perp}), \quad \alpha_1 = r_o \alpha'_v$$

$$\alpha'_1 = r_o^2 \alpha''_v, \quad \alpha_{25} = r_o \alpha_{\underset{\sim}{q}} \left[\ell n(\alpha_{\underset{\sim}{q}}/R^{b^2}) \right]' \quad . \tag{5.4}$$

r_o is the equilibrium bond length. Thus,

$$P^o_{\alpha\beta} = \Omega_o \alpha_v \delta_{\alpha\beta} = \frac{\Omega^o}{4\pi}(\varepsilon_\infty - 1)\delta_{\alpha\beta}$$

$$P^1_{\alpha\beta\mu}(\Gamma^+_{25}) = \frac{\Omega^o}{r_o\sqrt{3}} \alpha_{25}\varepsilon_{\alpha\beta\mu}$$

$$P^2_{\alpha\alpha\mu\nu}(\Gamma^+_1) = \frac{\Omega^o}{12 r_o^2}(-\alpha'_1 + (1-3\delta\mu\nu)\alpha_1)$$

$$P^2_{\alpha\beta\mu\nu}(\Gamma^+_{12}) = \frac{\Omega^o}{24r_o^2}(2\alpha_{25}+3\alpha_{\tilde{q}}\delta^{\mu\nu})(2-\bar{\varepsilon}_{\alpha\beta\mu}-\bar{\varepsilon}_{\alpha\beta\nu}) \quad (5.5)$$

$$P^2_{\alpha\beta\mu\nu}(\Gamma^+_{25}) = \frac{\Omega^o}{12r_o^2}\Big[-\alpha'_{25}-3\alpha_{25}(\delta^{\mu\nu}+1-\varepsilon_{\alpha\beta\nu}-\varepsilon_{\alpha\beta\mu})-9\alpha_{\tilde{q}}(1-\delta^{\mu\nu})(1-\varepsilon_{\alpha\beta\mu}-\varepsilon_{\alpha\beta\nu})\Big]$$

where the subindices α,β refer to the Cartesian coordinates of the dielectric tensor and its derivatives; μ,ν are the coordinates of the differences in the displacement of near-neighbor atoms; $\varepsilon_{\alpha\beta\mu}$, the Levi-Civita tensor, is given by

$$\varepsilon_{\alpha\beta\mu} = (1-\delta_{\alpha\beta})(1-\delta_{\alpha\mu})$$

$$\bar{\varepsilon}_{\alpha\beta\mu} = (1-\delta_{\alpha\beta})(1-\delta_{\alpha\mu})(1-\delta_{\beta\mu}) \quad (5.6)$$

Calculations of the first- and second-order Raman spectra of C, Si, and Ge were carried out with the five fitting parameters $\alpha_q, \alpha_1, \alpha'_1, \alpha_{25}$ and α'_{25}. α_v was obtained from ε_∞. The dominant component of the experimental spectra is that of Γ^+_1 symmetry accompanied by a weaker Γ^+_{25} quadrupole and very weak Γ^+_{12} scattering. These facts are well reproduced by the model in all three substances.

The two prominent spectra with Γ^+_1 and Γ^+_{25} symmetry are practically determined by α'_1 and $\alpha_{25'}$, respectively, which are approximately equal. This corresponds to a very low value of $\alpha_\perp$ as compared to $\alpha_{\parallel}$ in diamond ($\alpha_q \approx \alpha_\nu$) and indicates a simple (longitudinal) charge-transfer mechanism in this crystal. Since the intensity of the measured Raman spectra is only known in arbitrary units, Go et al. determined α_q from the photo-elastic constants $p_{11}+2p_{12}$. The non-fitted photoelastic constants $p_{11}-p_{12}$, p_{44} agree for diamond very well with experimental values.

A feature of particular interest is the peak at the cutoff of two-phonon spectra which often is attributed to a two-phonon bound state.[123] Since the calculation represents the frequency of the experimental peak as well as its intensity, it seems that the result does explain the two-phonon spectrum of diamond without invoking a bound state. A certain "over-bending" of the LO mode in the (100) direction above the Raman frequency with a subsequent extension of the density of states in this frequency region was first proposed by Musgrave and Pople.[124] Thus, within this model, the peak would be due to an overtone volume scattering, in agreement with the conjecture of Uchinokura, Sekine and Matsura,[125] and a recent density of states analysis by Tubino and Birman.[64]

For germanium and silicon a large increase in $\alpha_\perp$ is observed. The calculated first-order Raman tensor agrees quite well with that calculated by Swanson and Maradudin,[126] while for diamond the value has the opposite sign. This discrepancy might be due to the strong dependence of the Raman tensor on δ in Swanson and Maradudin's calculation.

A deeper understanding of the situation may be obtained by focussing attention on the relation between the model parameters and the important features in the energy band structure. We recall that the model assumes that the electronic polarizability of the diamond-like crystals can be expressed as a sum of the electronic polarizabilities of the bonds between nearest neighbor pairs of atoms, each of which depends on the bond length only. Furthermore, it is assumed that the bonds possess axial symmetry with respect to the principle axis of the bond which is preserved throughout the nuclear motion. This model contains six parameters, which describe the electronic polarizability of the crystal up to the second order in nuclear displacements. Firstly, we discuss the connection between the model and the energy bands using the tight binding approximation. The wave functions of valence and conduction bands of diamond-like crystals can be expanded in terms of Bloch sums constructed from the s and p atomic orbitals at each atomic site:

$$\psi_{\underline{k}}^{n} = N^{-\frac{1}{2}} \sum_{j} \sum_{\underline{R}^{i}} e^{i\underline{k}\cdot\underline{R}^{i}} \lambda_{\underline{k}}^{n}\binom{i}{j} u_{j}(\underline{r}-\underline{R}^{i}) \tag{5.7}$$

The static dielectric constant is given by the equation

$$\varepsilon_{\alpha\beta}(0,0) = 1+\frac{8\pi e^2}{v}\sum_{v,c,\underline{k}} \frac{\langle\psi_{\underline{k}}^{\nu}|r_\alpha|\psi_{\underline{k}}^{c}\rangle\langle\psi_{\underline{k}}^{c}|r_\beta|\psi_{\underline{k}}^{\nu}\rangle}{E_{\underline{k}}^{c} - E_{\underline{k}}^{\nu}} \tag{5.8}$$

In the simple bonding-antibonding picture, the energy denominator is replaced by the energy difference between the bonding and antibonding states and thus can be taken out of the summation. The sum is then given by the dipolar matrix element between the bonding and antibonding sp^3 hybrid orbitals which are localized along the four bond directions and lead to a longitudinal bond polarizability, $\alpha_{\parallel}$ only. This situation has been found in diamond. On the other hand, the non-zero transverse bond polarizability, $\alpha_{\perp}$, in Si and Ge indicates a breakdown of the bonding-antibonding picture for these materials. The essential trend in the band structure when going from diamond to germanium is the drastic lowering of the Γ_2' conduction band which results in an increasing contribution of transitions between the highest valence bond Γ_{25}' and this Γ_2' conduction band to $\varepsilon_{\alpha\beta}(0,0)$. The wavefunctions of the Γ_{25}' valence band are essentially given by two $p\pi$ and one $p\sigma$ bonding type atomic orbitals of the nearest neighbor pair of atoms while those of the Γ_2' conduction bond are mainly given by the antibonding type of s atomic orbitals.

The contributions via the E_o gap to $\underline{\varepsilon}$ by the matrix element between the bonding $p\pi$ and the antibonding s functions corresponds to a transverse bond polarizability. As a consequence, the lowering of the E_o gap gives rise to an increasing of the $\alpha_{\perp}$.

To show the relation between the bond polarizabilities and the band structure in more detail it is convenient to use binding orbitals directed along the bond directions, so-called bond orbitals. Without specifying the bond orbitals two orthogonized Bloch functions for the ground states and excited states may be constructed:

$$\psi_{\underline{k}}^{\nu} = N^{-\frac{1}{2}} \sum_{b} \sum_{\underline{R}^i} e^{i\underline{k}\cdot\underline{R}^i} V_{\underline{k}}^{b}(\nu)\Phi_b(\underline{r}-\underline{R}^i)$$

$$\psi_{\underline{k}}^{c} = N^{-\frac{1}{2}} \sum_{b} \sum_{\underline{R}^i} e^{i\underline{k}\cdot\underline{R}^i} C_{\underline{k}}^{b}(c)\phi_b(\underline{r}-\underline{R}^i) \qquad (5.9)$$

b denotes the four bonds in a unit cell, Φ_b and ϕ_b are the bond orbitals in the ground states and the excited states, respectively. Substituting Eq. (5.9) into the expression for $\varepsilon_{\alpha\beta}(0,0_\lambda)$ we obtain

$$\alpha_{\parallel} = (I_1+I_2)P_{\parallel}^2 + 2I_2P_{\perp}^2$$

$$\alpha_{\perp} = (I_1+I_2)P_{\perp}^2 - 2I_2P_{\parallel}P_{\perp} \qquad (5.10)$$

with $I_1 = 2e^2 \sum_{cv\underline{k}} A_{cv}^{11}(\underline{k})/(E_{\underline{k}}^{c}-E_{\underline{k}}^{v})$

$$I_2 = \frac{2}{3} e^2 \sum_{cv\underline{k}} A_{cv}^{12}(\underline{k}) / (E_{\underline{k}}^c - E_{\underline{k}}^v)$$

and $$A_{cv}^{bb'}(\underline{k}) = v_{\underline{k}}^{b*}(v) c_{\underline{k}}^{b}(c) c_{\underline{k}}^{b'*}(c) v_{\underline{k}}^{b'}(v)$$

$$P_{\parallel,\perp} = \langle \Phi_b | r_{\parallel,\perp}^{b} | \phi_b \rangle$$

Again, $\alpha_\perp$ becomes zero if we replace Φ_b and ϕ_b by bonding and antibonding sp^3 orbitals. For the E_o gap, Φ_b is a $p\pi$ bonding type and ϕ_b an s antibonding type orbital and, therefore, the matrix element P_1 is different from zero, so $\alpha_\perp$ becomes even more important as the E_o gap decreases.

For an ab initio calculation of the bond polarizabilities a variational perturbation procedure can be used as proposed by Flytzanis.[127] Using only the ground state Slater type sp^3 bond orbitals (which in the lowest approximation corresponds to a closure approximation) Go et al. are able to find a systematic increase of $\alpha_\perp$ relative to $\alpha_\parallel$ from diamond to germanium. However, a more accurate wave function seems to be necessary in order to make a comparison with the fitted values of $\alpha_\parallel$ and $\alpha_\perp$.

It has to be noted that the consideration of local field corrections [128] results in a quantitative modification of the results without changing the qualitative picture.

One should remark that Tubino et al.[64] have independently derived similar results using the same concept of bond polarizabilities. In particular, the peak at the two-phonon cut off observed in diamond has also been attributed by these authors to overtone scattering. Their work differs from that by Go et al. by using only the zero and first order derivatives of the bond polarizabilities. One notes that Cowley,[47] in his first attempt to calculate the Raman spectra of diamond, used only one parameter which is equivalent to the second derivative of the longitudinal polarizability and obtained a rather good result for the unpolarized Raman spectrum which mainly consists of Γ_1 scattering.

The problem of two-phonon bound states is discussed in papers by J. Ruvalds.[123] At the present state of the art, it seems that the problem is still open to discussion.

To investigate the Raman spectra in crystals with partially ionic character, we recall that the Raman spectra of the cubic oxides can be successfully described mainly with intra-ionic non-linear polarizabilities. It seems, therefore, to be appealing to attempt a similar analysis of the tetra-hedral II-VI and III-V compounds, which are partially ionic systems with increasing covalency. Kunc and Bilz[54] introduced non-linear polarizabilities such as used by Bruce and Cowley and by Buchanan et al. for the ionic crystals. The concept of the inter-ionic polarizabilities means that the ions contribute individually to the spectra. Generally, one may expect that lighter or more polarizable ions show stronger scattering than heavier or less polarizable ions. The second order Raman cross section is proportional to the fourth power of the core displacements i.e., in the harmonic approximation, to the inverse square of ionic masses. The mass factors in the polarizabilities $P_{\alpha\beta}(\lambda_1\overline{\lambda}_2)$ dominate over the influence of ionic polarizabilities. We note that the heavier atom is responsible for the lower-frequency part of the spectra; the lighter for the higher frequency part.

From this analysis, it turns out that the spherical quartic part of the non-linear potential, leads to a satisfactory description of all second order Raman spectra.

A comparison of this treatment with the bond polarizability approach shows that both models contain a longitudinal bond polarizability which is equal to the sum of the two intra-ionic polarizabilities for equal ions. The statement that the second derivative of the longitudinal bond polarizability governs the situation and leads to overtone spectra may be shown to be equivalent to the importance of the quartic term.

We note that the bond charge model is missing any correlations between different bonds. This causes difficulties for an extension of this model to the more ionic crystals such as the II-VI compounds. The shell model on the contrary, contains bond-bond correlations at the very beginning since a shell displacement at the ion automatically affects all four bonds connected with that ion. This may be a reason why the shell model with intra-ionic differential polarizabilities works for the tetrahedral II-VI compounds and even for the rocksalt-type oxides. In this respect the oxygen behaves more like an open-shell ion even in the rocksalt structure. It can not be compared with the closed-shell ions in alkali halides where the ionic radius seems to be a well-defined constant. The corresponding rigidity of the charge density distribution around the ions may explain the fact that inter-ionic differential polarizabilities govern the Raman spectra of the alkali halides.[50] This means that for these crystals a change in the polarizability requires a charge transfer between cations and anions.

We arrive at the conclusion that a simple shell model is able to describe the trends in the lattice dynamics and Raman spectra from the ionic to the covalent crystals mainly in terms of intra-ionic linear and non-linear polarizabilities. This may provide a basis for a further detailed comparison with the corresponding trends in the band structures.

5.3. Resonant Raman scattering

In the foregoing section, it was assumed that the energies of the incoming or scattered light were always sufficiently below the optical gap of the solids under consideration. This condition allowed us, in many cases, to neglect the detailed nature of the excited electronic states and to focus attention on the effect of the modulation of the electronic charge density by phonons in a low-frequency external field. This "classical" regime turned out to be very appropriate for the use of model theories which are powerful in making explicit contact between the lattice dynamics of a certain family of solids and their light scattering properties. The excited states in the crystal could be represented by an "effective" gap of the order of the lowest ionization energy or the corresponding model electron ion forces.

In the case of semiconductors with gaps between 1 to 3 eV, many laser frequencies are close to or even above the lowest band gaps and resonant Raman scattering (RRS) becomes possible. Here, the specific features of one or few electronic transitions and their particular coupling to the phonons of the system are dominating. Many new and interesting physical aspects, then, come into play such as the lifetime of intermediate states, etc. We are not going to give a detailed account of RRS in this article. The reader who is interested in a detailed discussion of RRS is referred to the Proceedings of the three last International Conferences on Light Scattering.[11,96,129,130]

REFERENCES

1. R. Stinchcombe, this volume.

2. M. Born and K. Huang, Dynamical Theory of Crystal Lattices (Oxford University Press, Oxford, 1973).

3. W. Cochran and R. A. Cowley, Handb. Phys., 15/2a, 59 (Springer Verlag, Berlin, 1967).

4. A. A. Maradudin, et. al., Phys. Rev. B, 6,1106 (1972).

5. G. K. Horton and A. A. Maradudin (eds.), Dynamical Properties of Solids, Vol. 1 (North-Holland Publishing Co., Amesterdam, 1974); G. K. Horton and A. A. Maradudin (eds.), Dynamical Properties of Solids, Vol. 2. (North-Holland Publishing Co., Amsterdam, 1974).

6. J. L. Birman, Handb. Phys., 25/2b, (Springer Verlag, Berlin, 1974).

7. H. Bilz, et.al., to be published, 1978.

8. W. Ludwig, Springer: Tracts in Mod. Phys., 43 (Springer Verlag, Berlin, 1967).

9. W. Cochran, Crit. Rev. in Solid State Sciences, 2, 1 (1971).

10. S. K. Sinha, Crit. Rev. in Solid State Sciences, 3, 273 (1973).

11. J. L. Birman, Handb. Phys., 25/2b (Springer Verlag, Berlin, 1974).

12. R. A. Cowley, this volume.

13. L. J. Sham, in Dynamical Properties of Solids, Vol. 1, Horton and Maradudin eds. (North-Holland Publishing Co., Amsterdam, 1974), 301.

14. H. Bilz et. al., in Dynamical Properties of Solids, Vol. 1, Horton & Maradudin eds. (North-Holland Publishing Co., Amsterdam, 1974), 343.

15. L. C. Pauling, The Nature of the Chemical Bond (Cornell University Press, Ithaca, N. Y., 1940).

16. J. C. Phillips, Covalent in Crystals, Molecules and Polymers (University of Chicago Press, Chicago, 1969).

17. M. Hass and H. B. Rosenstock, Phys. Rev., 153, 962 (1967).

18. H. Fröhlich, Theory of Dielectrics, 2nd ed. (Clarendon Press, Oxford, 1968).

19. B. Szigeti, in Cooperative Phenomena, H. Haken and M. Wagner eds. (Springer Verlag, Berlin, 1973), 147.

20. J. C. Phillips, Rev. Mod. Phys., 42, 317 (1970); J. C. Phillips, Phys. Rev. Lett., 29, 1551 (1972).

21. J. A. vanVechten, Phys. Rev., 187, 1007 (1969).

22. W. A. Harrison, Phys. Rev. B, 8, 4487 (1973); W. A. Harrison and J. C. Phillips, Phys. Rev. Lett., 33, 410 (1974); W. A. Harrison, Phys. Rev. Lett., 34, 1198 (1975).

23. G. G. Hall, Phil. Mag., 43, 338 (1952).

24. C. A. Coulson et. al., Proc. R. Soc. Lond., 270, 357 (1962).

25. G. Lehman and J. Friedel, J. Appl. Phys., 33, 281 (1962).

26. G. Baym, Lectures on Quantum Mechanics (Benjamin, New York, (1969)).

27. J. Fiutak, Canad. J. Phys., 41, 12 (1963).

28. L. P. Kadanoff and G. Baym, Quantum Statistical Mechanics (Benjamin, New York, 1962).

29. B. Wilhelmi, Z. Phys. Chem. (Germany), 238, 305 (1968); D. L. Greenaway with G. Harbeke, Optical Properties and Band Structure of Semiconductors (Permagon Press, New York, 1968).

30. L. P. Kadanoff and P. C. Martin, Ann. Phys. (New York), 24, 419 (1963).

31. G. Platzek, in Handbuch der Radiologie, E. Marx ed. (Akadamische Verlagsgesellschaft, Leipzig, 1934).

32. N. H. March et. al., Many Body Problems in Quantum Mechanics (Cambridge U.P., London, 1967).

33. R. Wehner, Phys. Status. Solidi, 15, 725 (1966); B. Szigeti, J. Phys. Chem. Solids, 24, 225 (1963).

34. R. A. Cowley, Advan. Phys., 12, 421 (1963).

35. H. Borik, Phys. Stat. Solidi B, 39, 145 (1970).

36. U. Knohl, Phys. Status Solidi, 53, 295 (1972).

37. M. Lax and E. Burstein, Phys. Rev., 97, 39 (1955).

38. B. Szigeti, Proc. Roy. Soc. A, 258, 377 (1960).

39. G. Leibfried and W. Ludwig, Solid State Physics, Vol. 12, F. Seitz and D. Turnbull eds. (Academic Press, New York, 1961).

40. G. Leibfried, A. Phys., 171, 1 (1963).

41. T. H. K. Barron and M. L. Klein, in Dynamical Properties of Solids, Vol. 1, G. K. Horton and A. A. Maradudin eds. (North-Holland, Amsterdam, 1974).

42. A. Anderson ed., The Raman Effect, 2 Vols. (Dekker, New York, 1971, 1973).

43. R. Loudon, Adv. Phys., 13, 423 (1964).

44. E. B. Wilson et. al., Molecular Vibrations (McGraw-Hill, New York, 1955).

45. R. Zeyher, in Light Scattering in Solids, Proceedings of the Third International Conf. on Light Scattering, Campinas, Brazil, 1975, M. Balkanski et. al. eds. (Flammarion, Paris, 1976), 87.

46. R. A. Cowley, Phys. Rev., 134, A981 (1964).

47. R. A. Cowley, J. Phys., 26, 659 (1965).

48. A. D. Bruce, J. Phys. C, 5, 2909 (1973).

49. A. D. Bruce and R. A. Cowley, J. Phys. C, 5, 595 (1972).

50. M. Krauzman, Solid State Commun., 12, 157 (1973).

51. R. Haberkorn et. al., Solid State Commun., 12, 681 (1973).

52. M. Buchanan et. al., J. Phys. C, 7, 439 (1974).

53. R. Migoni et al., Phys. Rev. Lett., 37, 1155, (1976).

54. K. Kunc and H. Bilz, Solid State Commun., 19, 1027 (1976).

55. J. R. Tessman et al., Phys. Rev., 92, 890 (1953).

56. S. T. Pantelides, Phys. Rev. Lett., 35, 250 (1975).

57. J. Callaway, Quantum Theory of the Solid State (Academic Press, New York, 1974).

58. W. Weber, Phys. Rev. B, 8, 5082 (1973).

59. W. Weber et. al., Proc. of the Twelveth Ann. Conf. on the Physics of Semiconductors, Stuttgart, M. Pilkuhn ed. (1974); S. Go, H. Bilz, M. Cardona, Phys. Rev. Lett., 34, 580 (1975); S. Go et. al. in, Light Scattering Solids, M. Balkanski et. al. eds. (Flammarion, Paris, 1976).

60. W. Weber et. al., Proc. of the Twelveth Ann. Conf. on The Physics of Semiconductors, Stuttgart, M. Pilkuhn, ed. (1974).

61. M. Wolkenstein, Compt. Rend. Acad. Sci., 32, 185 (1941).

62. A. A. Maradudin and E. Burstein, Phys. Rev., 164, 1081 (1967).

63. C. Flytzanis and J. Ducuing, Phys. Rev. A, 8, 1218 (1969); C. Flytzanis, Phys. Rev. B, 6, 1264 (1972).

64. R. Tubino and J. L. Birman, in Light Scattering in Solids, M. Balkanski et. al. eds. (Flammarion, Paris, 1976).

65. R. Wallis ed., International Conference on Lattice Dynamics, Copenhagen, 1963 (Permagon, New York, 1965).

66. D. H. Martin, Adv. Phys., 14, 39 (1965).

67. S. S. Mitra and S. Nudelman eds., NATO Advanced Study Inst., 1968 Proceedings (Plenum Press, New York, 1970); S. Nudelman and S. S. Mitra, Optical Properties of Solids (Plenum Press, New York, 1969).

68. M. Balkanski, J. Lumin., 7, 451 (1973).

69. E. Madelung, Gesell. Wiss. Göttingen Nachr., Math-Phys., Klasse 1, 100 (1909).

70. D. W. Berreman, Phys. Rev., 130, 2193 (1963).

71. A. A. Maradudin and A. E. Fein, Phys. Rev., 128, 2589 (1962).

72. L. Van Hove et. al., Problems in Quantum Theory of Many Particle Systems (Benjamin, New York, 1961).

73. R. Wehner, Phys. Status Solidi, 15, 725 (1966).

74. U. Knohl, Phys. Stat. Solidi, 53, 295 (1972).

75. J. I. Berg and E. E. Bell, Phys. Rev. B, 4, 3572 (1971).

76. E. R. Cowley, Phys. Rev. B, 3, 2743 (1971).

77. H. Bilz, Phonons in Perfect Lattices and Lattices with Point Imperfections, R. W. H. Stevenson ed. (Plenum Press, New York, 1966).

78. C. Smart et al., International Conference on Lattice Dynamics, R. Wallis ed. (Permagon, New York, 1965).

79. I. P. Ipatova et. al., Phys. Rev., 155, 882 (1967).

80. E. R. Cowley, J. Phys. C, 5, 1345 (1972).

81. K. Hisano et. al., J. Phys. C, 5, 2511 (1972).

82. M. Klier, Z. Physik, 150, 49 (1958); L. Genzel, et. al., Z. Physik, 154, 13 (1958); G. Seger and L. Genzel, Z. Physik, 169, 66 (1962).

83. R. H. Stolen and K. Dransfeld, Phys. Rev., 139, A1295 (1965).

84. J. C. Owens, Phys. Rev., 181, 1228 (1969).

85. H. R. Glyde and M. L. Klein, Crit. Rev. in Solid State Sciences, 2, 181 (1971).

86. R. P. Lowndes, Phys. Rev. B, 1, 2754 (1970); R. P. Lowndes, Phys. Rev. B, 6, 1490 (1972).

87. K. Fischer, Phys. Stat. Solidi B, 66, 295 (1974).

88. H. Bilz and R. Zeyher, Phys. Status Solidi, 20, K167 (1967).

89. J. E. Mooij, Phys. Lett. (Netherlands), 28a, 573 (1969).

90. G. L. Bottger and A. L. Geddes, J. Chem. Phys., 46, 3000 (1967).

91. A. Hadni et. al., Appl. Optics, 7, 161 (1968).

92. E. R. Cowley and R. A. Cowley, Proc. Roy. Soc. A, 292, 209 (1966).

93. H. J. Maris, Phys. Lett., 17, 228 (1965).

94. E. D. Pawlick, Far-Infrared Spectroscopy, K. D. Möller and W. G. Rothschild, ed. (Wiley, New York, 1971).

95. T. P. Martin and L. Genzel, Phys. Stat. Solidi B, 61, 493 (1974).

96. M. Cardona ed., Light Scattering in Solids (Springer Verlag, Berlin, 1975).

97. S. L. Cunningham et al., Phys. Rev. B, 10, 3500 (1974).

98. D. LaPlaze, J. Phys. (France), 37, 1051 (1976).

99. A. Dalgarno, Adv. in Phys., 11, 281 (1962).

100. S. H. Lin, et. al., Chem. Phys. Lett., 29, 389 (1974).

101. M. Krauzman, thesis, University of Paris (1969), unpublished.

102. M. Born and M. Bradburn, Proc. Roy. Soc. A, 188, 161 (1947).

103. R. S. Krishnan, Proc. Ind. Acad. Sci., 19, 216 (1944).

104. H. C. Welsh, et. al., Nature, 164, 737 (1949).

105. A. M. Karo and J. R. Hardy, Phys. Rev., 141, 696 (1966); A. M. Karo and J. R. Hardy, Phys. Rev., 160, 702 (1966).

106. M. Krauzman, Solid State Common., 12, 157 (1973); M. Krauzman, in Light Scattering in Solids, G. B. Wright ed. (Springer, New York, 1969), 109.

107. R. A. Cowley, et. al., Phys. Rev., 131, 1030 (1963); W. Cochran, Adv. Phys., 9, 387 (1960).

108. V. Nuesslein and U. Schroeder, Phys. Status Solidi, 21, 309 (1967); U. Schroeder, Solid State Common., 4, 347 (1966).

109. A. D. Bruce and R. A. Cowley, J. Phys. C, 6, 2422 (1973).

110. R. Haberkorn et al., International Conference on the Physics and Chemistry of Semiconductor Hetrojunctions and Layer Structures, 1970, G. Szigeti ed. (Akademiai Kiado, Budapest, 1971).

111. J. R. Hardy, et. al., Phys. Rev., 179, 837 (1969).

112. M. Krauzman, in Light Scattering in Solids, G. B. Wright ed. (Springer, New York, 1969), 109.

113. R. Haberkorn, et. al., Solid State Commun, 12, 681 (1973); M. Buchanan, et. al., J. Phys. C, 7, 439 (1974).

114. A. Pasternak, et. al., Phys. Rev. B, 9, 4584 (1974).

115. K. H. Rieder, et. al., Phys. Rev. B, 12, 3374 (1975).

116. S. S. Jaswal, et. al., J. Phys. Chem. Solids, 35, 571 (1974).

117. D. LaPlaze, C. R. Hebd. Seances Acad. Sci. B (France), 276, 619 (1973).

118. W. G. Stirling, J. Phys. C (GB), 5, 2711 (1972).

119. M. Iizumi, et. al., J. Phys. C, 6, 3021 (1973).

120. W. G. Stirling and R. Currat, J. Phys. C (GB), 9, L579 (1976).

121. C. H. Perry, et. al., in Light Scattering in Solids, M. Balkanski, et. al. eds. (Flammarion, Paris, 1976).

122. R. Migoni et al., Phys. Rev. Lett., 37, 1155 (1970).

123. J. Ruvalds and A. Zawadowski, Phys. Rev. B, 2, 1172 (1970).

124. M. J. P. Musgrave and J. A. Pople, Proc. Roy. Soc. (London), A268, 474 (1962).

125. K. Uchinokura, et. al., J. Phys. and Chem. Solids, 35, 171 (1974).

126. L. R. Swanson and A. A. Maradudin, Solid State Commun., 8, 859 (1970).

127. C. Flytzanis, Phys. Rev. Lett., 23, 1336 (1969).

128. W. Hanke and L. J. Sham, Phys. Rev. Lett., 33, 582 (1974).

129. G. B. Wright (ed.), Light Scattering in Solids, Proc. 1968 International Conf. on Light Scattering in Solids (Springer, New York, 1969); M. Balkanski (ed.), Light Scattering in Solids, Proc. 1970 International Conf. on Light Scattering in Solids (Flammarion, Paris, 1971); M. Balkanski, et. al. (eds.), Light Scattering in Solids, Proc. 1975 International Conf. on Light Scattering in Solids (Flammarion, Paris, 1976).

130. W. Richter, in Springer: Tracts in Modern Physics, G. Höhler (ed.), 78 (Springer Verlag, Berlin, 1976).

NEUTRON SCATTERING AND INTERACTIONS BETWEEN EXCITATIONS

R. A. Cowley

Department of Physics
University of Edinburgh
Mayfield Road
Edinburgh EH9 3JZ, Scotland

ABSTRACT

The use of neutron scattering to determine the interactions between excitations is reviewed. A brief summary is given of the use of the triple axis crystal spectrometer and of the theory of neutron scattering. The effects on the neutron scattering cross-section of a frequency dependent self-energy of an excitation, and of the interactions between the modes are described, and illustrated. The interference between a single excitation and multiple excitation processes is reviewed, and the way in which these different effects may be distinguished is described. Neutron scattering may also be used to study the effects of the interactions between the excitations on two-excitation response functions. This is illustrated by examples from magnon-magnon scattering, from liquid helium, and from structural phase transitions.

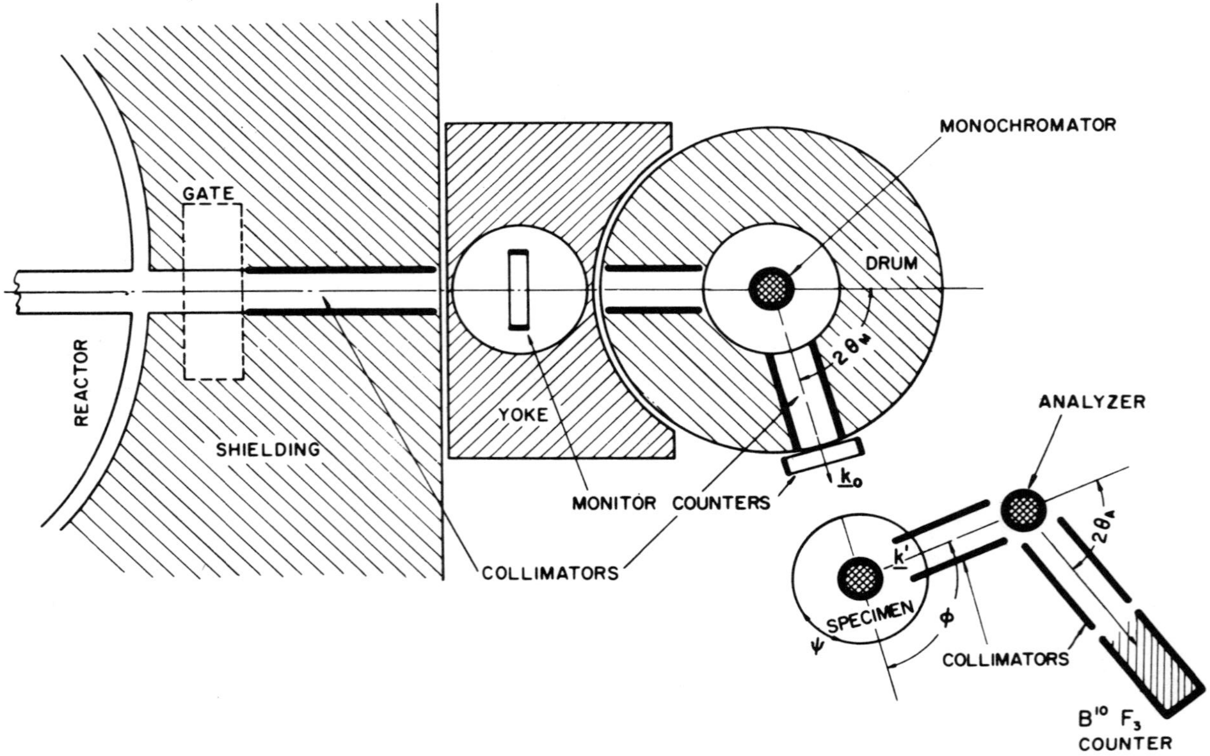

Fig. 1. A triple axis crystal spectrometer for neutron scattering. The incident energy is determined by Bragg reflection at the monocrhomator and the scattered energy by Bragg reflection at the analyser.

INTRODUCTION

The energy and wavelength of thermal neutrons make them uniquely suitable for the study of the elementary excitations in condensed matter. The experiments are performed by producing a beam of monoenergetic neutrons and then determining the energy and wavevector of the scattered neutrons. The changes in the energy and wavevector then enable the energy and wavevector of the elementary excitations to be determined. The determination of the dispersion relations of the normal modes of vibration in crystals, or of the spin waves in magnetic materials, has now been accomplished for many materials and is a standard technique. In these lectures I wish to focus attention on the use of neutron scattering to obtain more detailed information about the elementary excitations and their interactions than is provided by the dispersion relation.

In the next section there is a brief account of one of the most useful types of neutron scattering equipment, and the way in which it is used in practice. This is followed by a review of the relevant parts of the theory of neutron scattering. In section 3 the neutron scattering from a weakly anharmonic crystal is discussed in some detail as it provides an excellent means of illustrating different types of effects. Neutron scattering can be used to provide information about two-excitation correlation functions as well as the more usual one-excitation response. In section 4 we show that information about the effect of the interactions between the excitations on the two-particle correlation functions can be obtained using neutron scattering techniques and illustrate this with examples taken from magnetism, liquid helium and anharmonic crystals.

NEUTRON SCATTERING

(i) Triple Axis Crystal Spectrometers

One of the most useful spectrometers for determining the neutron scattering from a sample is the triple axis crystal spectrometer[1] illustrated schematically in fig. 1. A monochromatic beam of neutrons is obtained by Bragg reflection from a single crystal monochromator. The plane spacing of the monochromator and the Bragg angle, Θ_M, then determine the incident neutron energy, E_o, and wavevector, $\underline{k}_o$. The neutrons scattered through an angle ϕ, at the specimen are analysed by Bragg reflection at the analyser single crystal. The Bragg angle, Θ_A, determines the scattered neutron energy, E', and wavevector, $\underline{k}'$. The frequency transfer, ω, and wavevector transfer, $\underline{Q}$, are given by:

$$\begin{aligned} \hbar\omega &= E_o - E' \\ \underline{Q} &= \underline{k}_o = \underline{k}' \end{aligned} \qquad (2.1)$$

The spectrometer is initially set up by adjustment of Θ_M, Θ_A, ϕ and the specimen crystal orientation ψ, so that the scattered neutron intensity is recorded for some preselected point in $(\underline{Q},\omega)$ space. The spectrometer is then readjusted so that the intensity can be recorded for a neighbouring point in $(\underline{Q},\omega)$ space. Since there are 4 adjustable angles, it is possible to measure the scattering intensity along any predetermined line in $(\underline{Q},\omega)$ space, and then to repeat the process along the next required line. Most frequently the spectrometer is programmed to perform scans in which the wavevector transfer, $\underline{Q}$, is held fixed[1] while the frequency transfer is varied, and this is known as a constant $\underline{Q}$ scan.

Since even the best nuclear reactors are very weak sources of neutrons compared with readily available X-ray or light sources, the scattered intensity in a neutron scattering experiment is frequently very low. Considerable skill is therefore required to choose the optimum monochromator and analyser crystals, and in the choice of collimation. It is not appropriate to discuss these challenging problems here, but it is essential to point out the two main ways of using the spectrometer, and the way in which the resultant raw data is related differently to theoretical expressions. This difference arises because in a constant $\underline{Q}$ scan, for example, each point in $(\underline{Q},\omega)$ space specifies three conditions in eqns. (2.1). (Scattering is recorded only in a plane so that $\underline{Q}$ is two-dimensional.) Since there are 4 adjustable angles, Θ_A, Θ_M, ϕ and ψ, one of these may be held fixed throughout the scan. In the fixed analyser mode of operation, Θ_A, is held fixed and so the analyser system has a constant energy acceptance throughout the scan. Since the monitor counter in the incident beam usually has an efficiency proportional to $1/k_o$, the observed scattering intensity is proportional to $k_o \frac{d^2\sigma}{d\Omega dE'}$, where the differential cross-section is defined in eqn. 10.6 of ref. 2.

The alternative mode of operation is to keep Θ_M fixed. Since the angular collimation of the analyser system is invariably kept fixed throughout the scan, the energy acceptance of the analyser varies as Θ_A varies. The observed intensity is then proportional to $\frac{d^2\sigma}{d\Omega d\Theta_A}$, which using Bragg's Law becomes

$(k')^3 \cos \Theta_A \frac{d^2\sigma}{d\Omega dE'}$. Since the factor $(k')^3 \cos \Theta_A$ may easily vary by a factor 4 across a scan, it is clearly essential to include it before theoretical calculations can be compared with these experimental results. Alas it is frequently difficult in neutron scattering papers to ascertain whether or not this factor has been included in the data presented. From this description it would appear that all experiments should be performed in the constant $2\Theta_A$ mode. This is an incorrect impression, there are frequently excellent experimental reasons to prefer the latter configuration.

The other main class of neutron spectrometers is based on the time-of-flight technique. I shall not discuss this technique here as it is usually not appropriate for the study of the details of the correlation functions with which we shall be concerned below.

(ii) Theory of Neutron Scattering

The theory of neutron scattering was reviewed in ref. 2 and 3. Here we collect together a few of the most important results. Neutrons are scattered by nuclear scattering and by magnetic scattering. Initially we consider only the former, and also neglect the randomness in the neutron scattering length which may arise from the different isotopes and different spin states of the neutron and nucleus intermediate state. The nuclear coherent scattering is given (cf eqn. 10.11 of ref. 2) as:

$$\frac{d^2\sigma}{d\Omega dE'} = \frac{k'}{k_o} S(\underline{Q},\omega), \qquad (2.2)$$

where for crystals $S(\underline{Q},\omega)$ is the frequency transform of the thermally averaged correlation function

$$S(\underline{Q},t) = \sum_{\substack{\ell\ell' \\ KK'}} b_K b_{K'} \langle \exp(-i\underline{Q}.\underline{r}_{\ell K}(0)) \exp(i\underline{Q}.\underline{r}_{\ell' K'}(t))\rangle \qquad (2.3)$$

where b_K is the coherent neutron scattering length of nuclei of type K, while $\underline{r}_{\ell K}(t)$ is the position of the atom of type K in the ℓth unit cell at time t. In crystals it is nearly always useful to make a phonon expansion:

$$\underline{r}_{\ell K}(t) = \underline{R}_\ell + \underline{R}_K + \underline{U}_{\ell K}(t), \qquad (2.4)$$

where $\underline{R}_\ell$ is the position of the ℓth unit cell, $\underline{R}_K$ the mean position of the Kth type of atom within the unit cell, and $\underline{U}_{\ell K}$ is the displacement. When eqn. (2.4) is substituted into eqn. (2.3), two terms result. Bragg scattering which arises as $t \to \infty$, and inelastic scattering. If the crystal is harmonic the inelastic scattering is further usefully divided into one-phonon, two-phonon, etc. dependent upon the number of phonons which carry the correlation between (ℓK) and ($\ell' K'$). The displacements, $U_{\ell K}$, may be written in terms of the normal modes[4] ($\underline{q}j$) as:

$$\underline{U}_{\ell K} = \sum_{\underline{q}j} \left(\frac{\hbar}{2\omega(\underline{q}j)N}\right)^{\frac{1}{2}} e_K(\underline{q}j) \exp\left(i\, \underline{q}\cdot(\underline{R}_\ell+\underline{R}_K)\right) A(\underline{q}j),$$

where $\underline{e}_K(\underline{q}j)$ is the phonon eigenvector, $\omega(\underline{q}j)$ the frequency, and $A(\underline{q}j)$ is the sum of a phonon creation and annihilation operators. The one-phonon part of the inelastic scattering is then given by:

$$S^I(\underline{Q},\omega) = \sum_{\underline{q}jj'} \mathrm{Im}\left[F(\underline{Q}|\underline{q}j)F(-\underline{Q}|-\underline{q}j')(n(\omega)+1)G_{jj'}(\underline{q},\ \omega-i\varepsilon)\right] \quad (2.5)$$

where the phonon structure factor is given by:

$$F(\underline{Q}|\underline{q}j) = \sum_K b_K \exp(-W_K) \exp\left(i(\underline{Q}+\underline{q})\cdot\underline{R}_K\right)\left(\underline{Q}\cdot\underline{e}_K(\underline{q}j)\right)$$

$$\Delta(\underline{Q}+\underline{q}) \left(\frac{\hbar}{2\omega(\underline{q}j)}\right)^{\frac{1}{2}}, \quad (2.6)$$

where $\exp(-W_K)$ is the Debye-Waller factor (eqn. 10.24 of ref. 2) and $\Delta(\underline{Q}+\underline{q}) = 1$ if $(\underline{Q}+\underline{q})$ is a reciprocal lattice vector, $\underline{\tau}$, and zero otherwise, while

$$n(\omega) = \left[\exp(\hbar\omega/k_B T) - 1\right]^{-1},$$

and

$$G_{jj'}(\underline{q},\ \omega-i\varepsilon) = \langle\langle A(\underline{q}j);\ A(\underline{q}j')\rangle\rangle\ . \quad (2.7)$$

In the harmonic approximation

$$G_{jj'}(\underline{q},\ \omega-i\varepsilon) = \left[\delta(\omega-\omega(\underline{q}j)) - \delta(\omega+\omega(\underline{q}j))\right]\delta_{jj'} \quad (2.8)$$

Measurement of the neutron scattering cross-section then enables the frequencies of the phonons to be determined, through the positions in frequency of the peaks in the scattered intensity. It also enables information about the eigenvectors to be obtained from the intensity of the neutron groups through eqn. (2.5).

The magnetic scattering of neutrons by unpolarised neutron beams may be obtained by a similar development to that outlined above. The result[2,3] is, eqn. 10.31, that

$$\frac{d^2\sigma}{d\Omega dE'} = \frac{k'}{k_o} \sum_{\mu\nu} F_{\mu\nu}(Q) \frac{1}{2\pi} \int dt \, \exp(-i\omega t) \quad \Gamma_{\mu\nu}(Q,t) \tag{2.9}$$

where for a system containing only one type of ion, the constants, dipole interaction and magnetic form factor, $f(\underline{Q})$, give

$$F_{\mu\nu}(\underline{Q}) = \left(\frac{\gamma e^2}{m_e c^2}\right)^2 \left|\tfrac{1}{2}gf(\underline{Q})\right|^2 \exp(-2W)\left(\delta_{\mu\nu} - \frac{Q_\mu Q_\nu}{|\underline{Q}|^2}\right) \tag{2.10}$$

while the spin dependence is contained in the correlation function:

$$\Gamma_{\mu\nu}(\underline{Q},t) = \langle S_\mu(\underline{Q},0)\, S_\nu(-\underline{Q},t)\rangle \tag{2.11}$$

and

$$S_\mu(\underline{Q}) = \sum_\ell \exp(i\, \underline{Q}.\underline{R}_\ell) S_\mu(\ell),$$

is the $\underline{Q}$th Fourier component of the spin density. Further development in terms of spin waves then shows that the positions of the neutron groups give the frequency of the spin waves, and the intensities provide information about the nature of the spin motions.

THE ANHARMONIC CRYSTAL

(i) Phonon Self-Energy

In the last section the neutron scattering cross-section from a harmonic crystal was reviewed. In real crystals the interatomic forces are not harmonic and give rise to interactions between the different normal modes of vibration. Provided, however, that these anharmonic interactions are weak their effect on the neutron scattering can be evaluated using low-order perturbation theory so that the anharmonic crystal provides an ideal system with which to study the effects of the interactions between excitations.

The anharmonic terms in the Hamiltonian of the crystal may be written[4] in terms of a power series in the phonon coordinates,

$A(\underline{q}j)$:

$$H_A = \sum_{1,2,3} V(1\ 2\ 3)\ A(1)\ A(2)\ A(3) + \sum_{1,2,3,4} V(1\ 2\ 3\ 4)\ A(1)\ A(2)\ A(3)\ A(4) + \ldots\ldots \tag{3.1}$$

where we have written the suffix 1 to represent $(\underline{q}_1 j_1)$ etc. and the coefficients in the expansion may be obtained from the anharmonic terms in the interatomic potential, and the phonon eigenvectors.

The anharmonic terms allow the decay of one phonon into pairs of other phonons. This influences the neutron scattering from a particular normal mode because these processes may give rise to a finite lifetime of that mode, and to a shift in its frequency. Since the particular normal mode is an intermediate state between the neutron and the final state of two phonons, it is not necessary to conserve frequency (energy) in this state. Consequently it is necessary to examine the decay probability for all energy transfers, $\hbar\omega$, in the experiment. The decay probability and shift in frequency may be obtained either by using the golden rule of perturbation theory and second-order perturbation theory, or by using diagrammatic techniques. In the case when there is only a single mode $(1 = \underline{q}j)$ of a particular symmetry the Green's function for the mode becomes[4,5]:

$$G_j(\underline{q}, \omega - i\varepsilon) = \frac{2\,\omega(\underline{q}j)}{\omega(\underline{q}j)^2 - \omega^2 + 2\omega(qj)\ \Sigma_{jj}(q, \omega - i\ )} \tag{3.2}$$

where

$$\Sigma_{jj}(\underline{q}, \omega - i\varepsilon) = -\frac{18}{\hbar^2} \sum_{2,3} |V(1\ 2\ 3)|^2 (R(\omega) - i\ S(\omega)), \tag{3.3}$$

where

$$R(\omega) = \frac{n_2+n_3+1}{(\omega_2+\omega_3+\omega)_p} + \frac{n_2+n_3+1}{(\omega_2+\omega_3-\omega)_p} + \frac{n_3-n_2}{(\omega_2-\omega_3+\omega)_p} + \frac{n_3-n_2}{(\omega_2-\omega_3-\omega)_p},$$

and

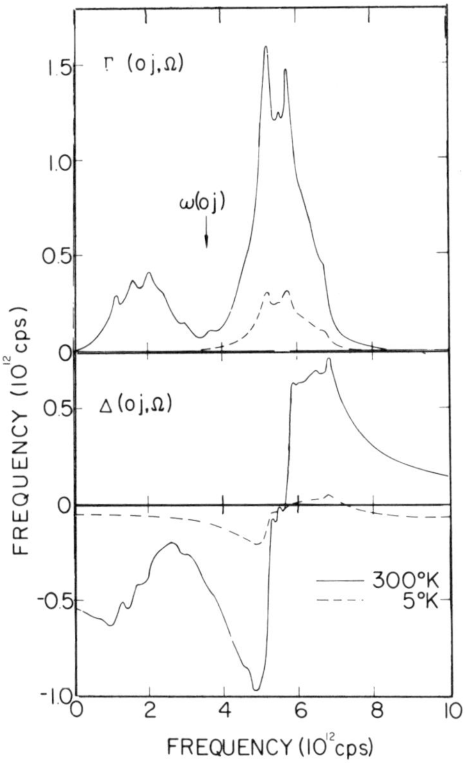

Fig. 2. The Width $\Gamma(oj, \omega)$ and shift $\Delta(oj, \omega)$ functions in KBr at two different temperatures[6].

$$S(\omega) = \pi(n_2+n_3+1)\left[\delta(\omega_2+\omega_3-\omega) - \delta(\omega_2+\omega_3+\omega)\right]$$

$$+ \pi(n_3-n_2)\left[\delta(\omega_2-\omega_3-\omega) - \delta(\omega_2-\omega_3+\omega)\right] \qquad (3.4)$$

Other terms in the self-energy arise from the thermal expansion and from the quartic term in eqn. (3.1), but they are not of importance for our present discussions.

Both the real and imaginary parts of the phonon self-energy are frequency dependent so that in principle the phonon in an anharmonic crystal cannot be specified by a single frequency and damping. Examples of the shift (real part, $\Delta(\underline{q}j,\omega)$ and width (imaginary part, $\Gamma(\underline{q}j,\omega)$) of $\Sigma_{jj}(\underline{q},\omega)/2\omega(\underline{q}j)$ are shown in fig. 2 for the q = 0 TO mode in KBr. These shift functions have been used to calculate the imaginary part of the Green's function which determines the scattering cross-section (fig.3). Clearly there is considerable structure and it would be quite unreasonable to try to characterise this line shape by a single frequency and width. Such a description can be valid only if $\Gamma(\underline{q}j, \omega)$ is much smaller than $\omega(\underline{q}j)$ when the neutron scattering line shape becomes Lorentzian with a half-width given by $\Gamma(\underline{q}j, \omega)$.

I am unaware of any well documented examples of cases in which the frequency dependent phonon self-energy has given rise to an anomalous neutron scattering line shape. Infra-red absorption measurements at oblique angles of incidence in alkali halides have, however, demonstrated[7] that the structure does occur as shown in fig. 4.

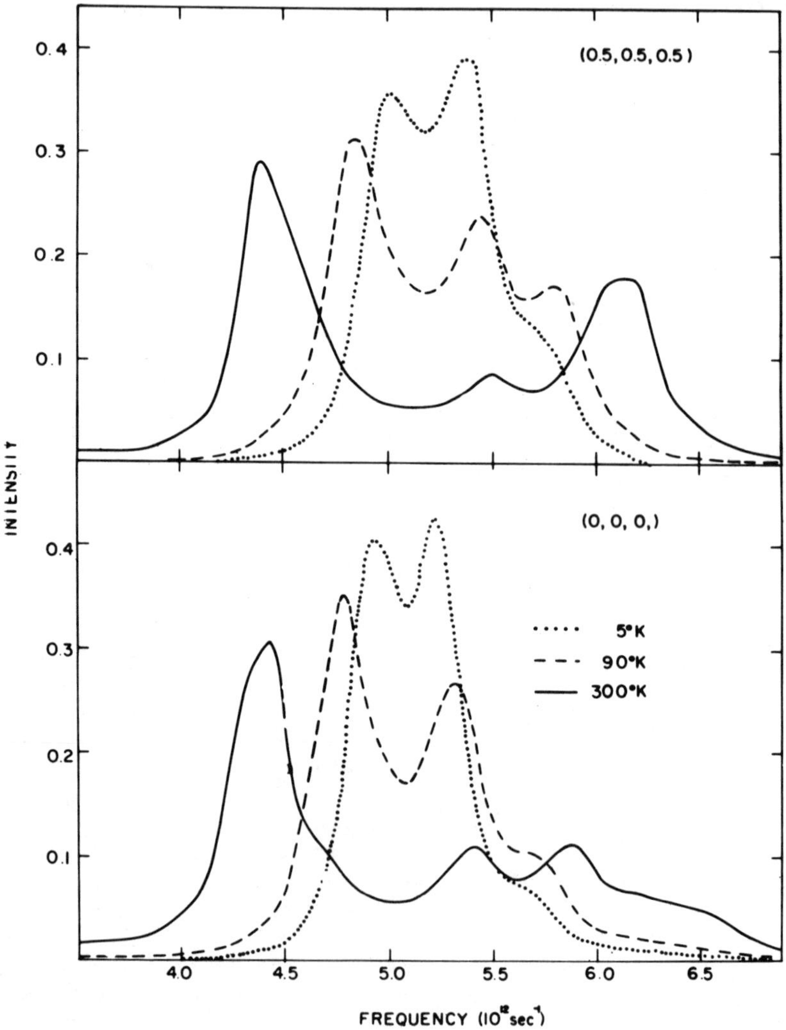

Fig. 3. The line shape of two LO modes in NaI[6]

(ii) Interference between Modes

If there is more than one mode with a particular symmetry the Greens function, eqn. 2.7, becomes a matrix in the branch indices j and j'. Under these circumstances it satisfies the matrix equation[4,5]:

$$\sum_{j'} \left[(\omega(qj)^2 - \omega^2)\delta_{jj'} + 2\omega(qj)\Sigma_{jj'}(\underline{q},\omega)\right] G_{j'j''}(\underline{q},\omega)$$

$$= 2\omega(\underline{q}j)\,\delta_{jj''} \quad , \qquad (3.5)$$

where the off-diagonal parts of the self-energy matrix are given by similar expressions to eqn's (3.3) and (3.4) but with $|V(1\ 2\ 3)|^2$ replaced by $V(1\ 2\ 3)\ V(1'\ 2\ 3)$.

Although it is not difficult to evaluate this expression in detail, at least formally, it is instructive to obtain the expressions for the scattering in the neighbourhood of $\omega(\underline{q}j)$ assuming that there are only two modes, j. Under these conditions

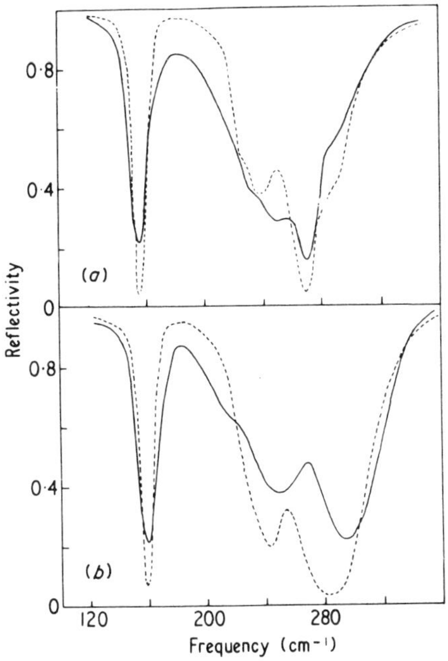

Fig. 4. The infra-red reflectivity[7] of NaCℓ at 300 K (a) for a film thickness of 2.5 μm and an angle of incidence of 45° and (b) an angle of incidence of 65°. The solid line shows the experimental results and the dotted line calculations using a frequency dependent self-energy.

we can write

$$G_{jj'}(\underline{q},\omega) \approx G_j(\underline{q},\omega)\, \Sigma_{jj'}(\underline{q},\omega)\, G_{j'}(\underline{q},\omega), \tag{3.6}$$

where the $G_j(\underline{q},\omega)$ is the diagonal part of the Green's function evaluated in section 3(i). It is then convenient to divide the intensity scattered near $\omega(\underline{q}j)$ into two parts. One contribution is

$$S_1 + S^a_{12} = (n(\omega)+1)\Big[|F(\underline{Q}|\underline{q}j)|^2 - 2\,\mathrm{Re}\big[F(\underline{Q}|\underline{q}j)\, F(-\underline{Q}|-\underline{q}j')\, G_{j'}(\underline{q},\omega)\, \Sigma_{jj'}(\underline{q},\omega)\big]\Big]\, \mathrm{Im}\big[G_j(\underline{q},\omega)\big], \tag{3.7}$$

and the other

$$S^b_{12} = (n(\omega)+1)\Big[-2\,\mathrm{Im}\big[F(\underline{Q}|\underline{q}j)\, F(-\underline{Q}|-\underline{q}j)\, G_j'(\underline{q},\omega)\, \Sigma_{jj'}(\underline{q},\omega)\big]\Big]\mathrm{Re}\big[G_j(\underline{q},\omega)\big]. \tag{3.8}$$

The former contribution (3.7) has roughly the same shape as $\mathrm{Im}[Gj(\underline{q},\omega)]$, but the intensity is different from that given by the normal one-phonon expression because of the term S^a_{12}. The latter is asymmetric about $\omega(\underline{q}j)$ as shown schematically in fig. 5.

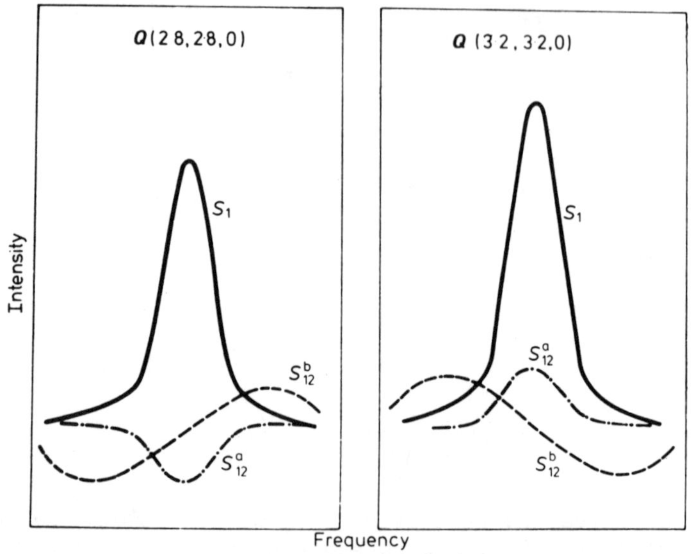

Fig. 5. Schematic representation of the different contributions discussed in the text. The conditions under which S^a_{12} and S^b_{12} change sign between $\underline{Q}$ = (2.8, 2.8, 0) and (3.2, 3.2, 0) are discussed in the text.

Clearly the former term makes it more difficult to use the intensities of neutron groups to determine the eigenvectors of the normal modes, while the latter term may alter the apparent frequencies of the groups. One of the interesting aspects of these terms is that their effects alter from one-lattice point to another. Since the $F(\underline{Q}|\underline{q}j)$ are in general different for $\underline{Q}_1 = \underline{\tau}_1 - \underline{q}$ and $\underline{Q}_2 = \underline{\tau}_2 - \underline{q}$, the detailed shape of the scattering will be different for $\underline{Q}_1$ and $\underline{Q}_2$. This feature is illustrated in figs. 6 and 7 which show the different neutron groups observed for different $\underline{Q}$ but the same phonon wavevector $\underline{q}$ in $BaTiO_3$[8] and lead germanate.[9] In both cases the interference is between a heavily damped ferroelectric optic mode and a well defined acoustic mode.

These terms may also lead to a difference in the scattering for $\underline{Q}_1 = \underline{\tau} + \underline{q}$ and $\underline{Q}_2 = \underline{\tau} - \underline{q}$. In non-centro symmetric crystals, $\Sigma_{jj'}(\underline{q},\omega)$ may be odd in $\underline{q}$ so that the interference terms may change sign in the two measurements at $\underline{Q}_1$ and $\underline{Q}_2$. An example of this

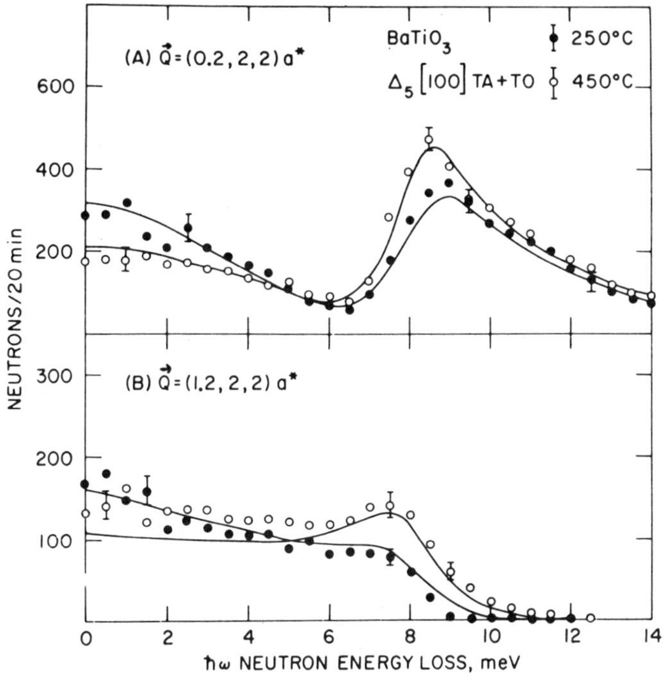

Fig. 6. The neutron scattering observed[8] for a particular $\underline{q}$ = (0.2, 0, 0) around two different lattice points (0,2,2) and (1, 2, 2) in $BaTiO_3$. Note the very different shapes observed.

occurs in KD_2PO_4[10,11] where the ferroelectric critical scattering, shown in fig. 8, is clearly not symmetric about the lattice point (5, 1, 0). This arises from the asymmetry introduced by the piezoelectric coupling between the ferroelectric mode and the acoustic modes.

(iii) Interference with the Multi-phonon Scattering

A similar effect from that discussed above may occur in an anharmonic crystal due to the interference between the one-phonon and multi-phonon scattering[12]. The neutron interacts with a phonon which decays into a pair of phonons which then interact again with the neutron. The detailed expressions for this term in

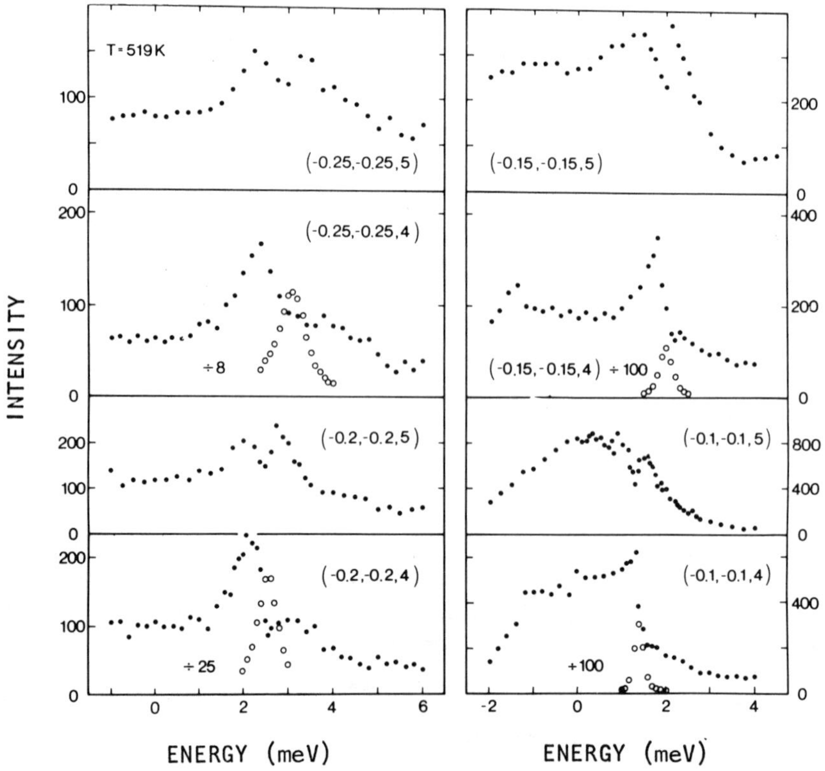

Fig. 7. The neutron scattering[9] for $\underline{q}$ = (0.25, 0.25, 0), (0.2, 0.2, 0), (0.15, 0.15, 0) and (0.1, 0.1, 0) in lead germanate at 519 K, near the lattice points (0 0 4), (0 0 5) and (0 0 6) (open circles).

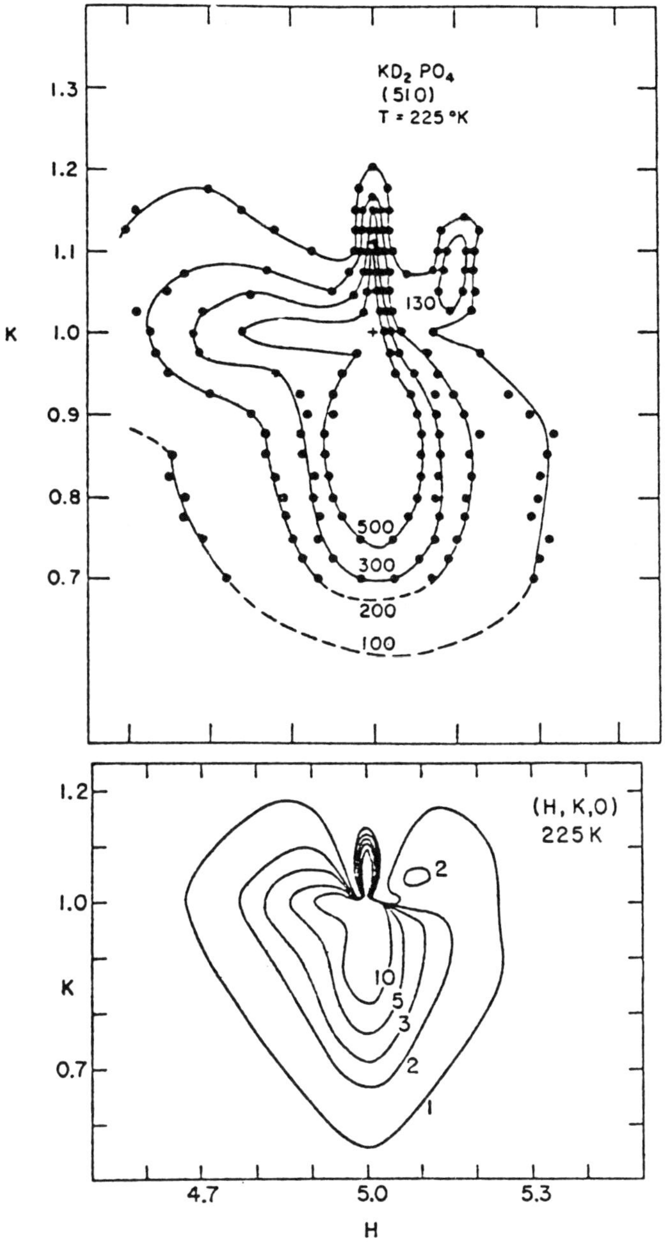

Fig. 8. Observed[10] and calculated[11] critical scattering in the ferroelectric KD_2PO_4. The unusual asymmetry arises from the piezoelectric interference between the ferroelectric mode and the acoustic modes.

the scattering cross-section[12] are cumbersome and of interest only to those unfortunate enough to perform detailed calculations. The qualitative results are similar to those discussed in the previous section. One term, S^a_{12}, is roughly proportional to $\mathrm{Im}[G_j(\underline{q},\omega)]$ and the other term, S^b_{12}, to $\mathrm{Re}[G_j(\underline{q},\omega)]$. The former then alters the intensity of the peaks and the latter is asymmetric about the peaks as illustrated in fig. 5.

The effect of this interference with the multi-phonon background has now been demonstrated in KBr[13], fig. 9, K metal[14] fig. 10, and solid He[15]. One of the most elegant demonstrations of its effect has been performed using X-ray scattering from alkali halides[12,16] as shown in fig. 11.

(iv) Separation of the Different Processes

The effects described in sections (i) - (iii) are all ways in which the one-phonon cross-section may deviate appreciably from that expected from a harmonic crystal. In all cases the intensity of the neutron group may be modified suggesting that in anharmonic crystals the one-phonon sum-rule[17] is of little practical use. More seriously these effects may make uncertain whether the intensities and possibly the frequencies of the neutron groups can be used directly to determine eigenvectors and frequencies of the normal modes.

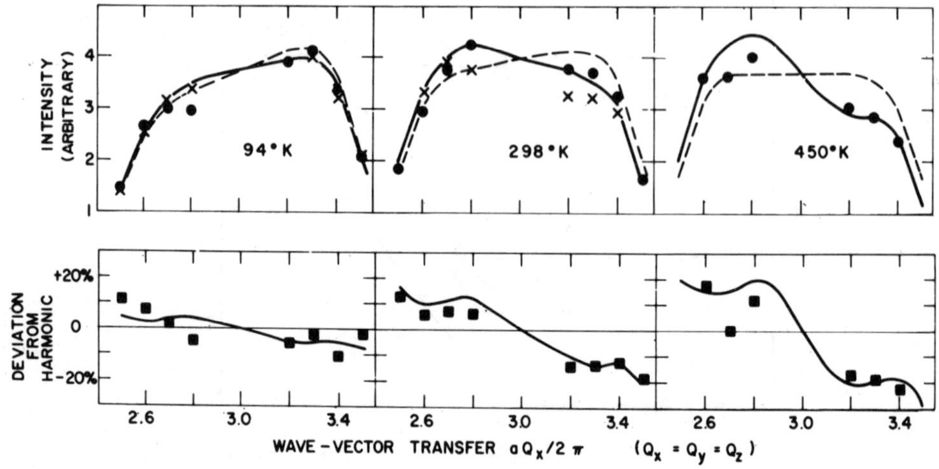

Fig. 9. The observed intensity[13] as a function of Q for LO modes in KBr. The solid lines are theoretical calculations including the multi-phonon asymmetry and the dotted line shows the intensity calculated using the harmonic approximation.

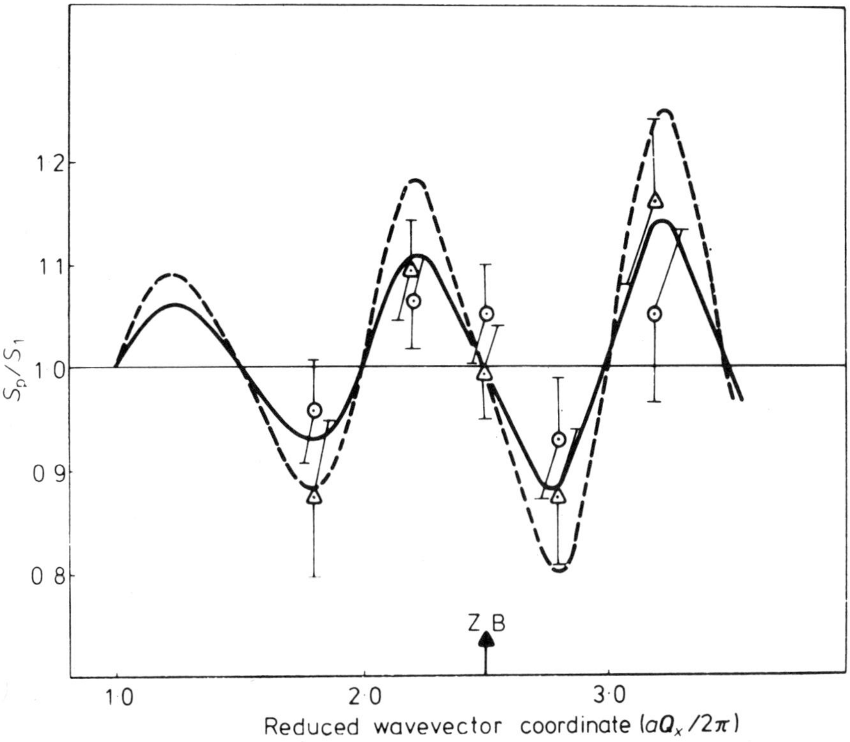

Fig. 10. The fraction of the harmonic intensity observed[14] and calculated for K metal at 99 K, O and solid line and 150 K, Δ and broken lines.

It is of interest to be able to distinguish between the different processes experimentally. Alas this is almost impossible using optical techniques but can be performed readily using neutrons. If measurements are performed at different wavevectors $\underline{Q}$ but the same phonon wavevector, $\underline{q}$, then spectra of identical shape will result from the self-energy effects described in (i). If different shapes are observed then the interference effects (ii) and (iii) are important. In centrosymmetric crystals these can be distinguished by performing experiments at $\underline{\tau} + \underline{q}$ and $\underline{\tau} - \underline{q}$ when the interference effects are largely unchanged in case (ii), but reversed in sign in case (iii). Alas in non-centro-

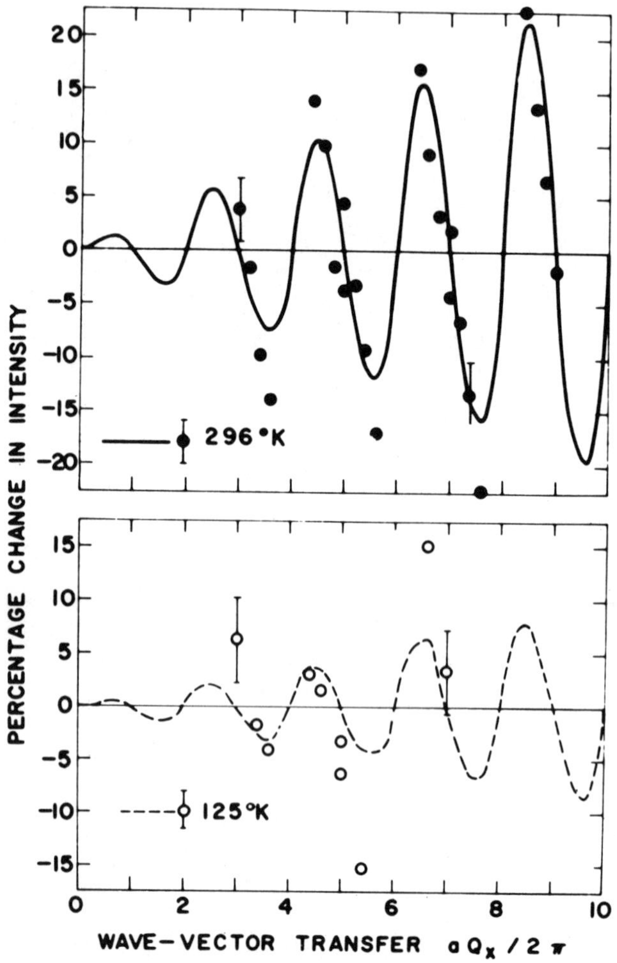

Fig. 11. The difference between the observed and harmonic intensity expressed as a fraction of the harmonic intensity for X-ray scattering[16] in NaCℓ as compared with calculation[12].

symmetric crystals it is not possible to distinguish readily between these two interference effects, but it might be expected that the size of the effects will tend to increase with increasing Q for the multi-phonon processes (iii) but not for the single-phonon interference (ii).

4. INTERACTIONS BETWEEN PAIRS OF EXCITATIONS

(i) Two-Magnon Scattering

Although neutrons are scattered by two-phonon processes in crystals there have to my knowledge been no serious attempts to measure their spectra. In part this is because it would be necessary to subtract off the three-phonon scattering and four-phonon scattering so that reliable measurements would be difficult, but also because it is much easier to interpret one-phonon scattering and so this has been the primary object of study. In this section a review is given of these experiments which have or may have indicated that the two-excitation spectra is not that which would be expected if the excitations did not interact with one another.

The simplest of these measurements is the two-magnon scattering in antiferromagnets. This arises from the $\Gamma_{zz}(\underline{Q},t)$ part of the spin-spin correlation function, eqn. 2.11. In terms of creation operators for spin deviations at site ℓ, $S_z(\ell) = S - a(\ell)\, a^+(\ell)$. The $a(\ell)$ and $a^+(\ell)$ may then be transformed into magnon creation operators $\alpha^+(q)$ so that for a simple two sublattice antiferromagnet at T = 0 K,

$$\Gamma_{zz}(\underline{Q},\omega) = \sum_{\underline{q}} \sinh\,(\Theta_{\underline{q}}-\Theta_{\underline{Q},\underline{q}})\ \delta(\omega-\omega(\underline{q})-\omega(\underline{Q}-\underline{q})), \qquad (4.1)$$

where $\omega(\underline{q})$ is the magnon frequency and the $\Theta_{\underline{q}}$ are the phase factors which enter into the transformation to spin-wave variables. Measurement of the $\Gamma_{zz}(\underline{Q},\omega)$ then provides a direct measure of a two-magnon correlation function. It is of interest to point out that the correlation function is quite different from that measured in two-magnon Raman scattering, in which the two-magnons are created on neighbouring sites while for two-magnon neutron scattering they are created at the same site. This leads to a very different two-magnon spectrum with the neutron scattering spectrum in particular being very different from a two-magnon density of states. The second point of interest is that the two-magnon scattering is zero at T = 0 in a simple ferromagnet because the ground state has $S_z = S$. The scattering arises in antiferromagnets because the magnon creation operators are a linear combination of site spin deviation creation and annihilation operators. It is therefore a direct manifestation of the fact that the molecular field ground state is not the true ground state in an antiferromagnet.

Measurements of the two-magnon scattering have been made in CoF_2[18] and MnF_2[19]. The results in the former case are illustrated in fig. 12. Clearly the scattering can be observed and is very different in shape from that of the two-magnon density of states.

One of the most interesting aspects of the two-magnon Raman scattering has been the insight which it has provided into the effect of the interactions between the magnons in modifying the spectral shape. Magnon-magnon interactions will also modify the

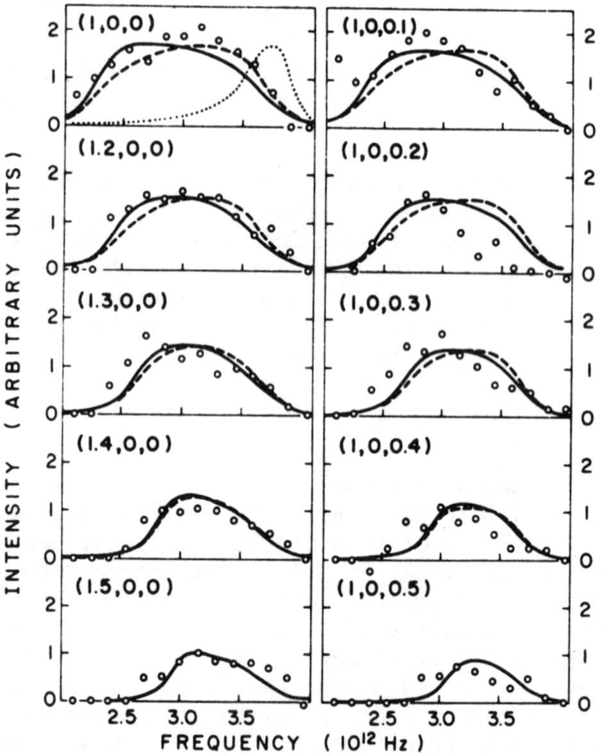

Fig. 12. Two-magnon scattering[18] from CoF_2. The dotted line shows the two-magnon density of states, the dashed line was calculated using eqn. (4.1) and the solid line includes the interactions between the excitations approximately.

shape of the spectra in neutron scattering experiments, but since the neutron couples to different pairs of magnons the details of the theory are different, and more difficult. Initially the theory of the interactions was evaluated[18] by including only the Ising terms in the exchange Hamiltonian, but more recently a more comprehensive theory has been developed[20]. Alas the experimental results, fig. 12, are sufficiently uncertain to provide any detailed information about the magnon-magnon interactions.

(ii) Roton-Roton Interactions in Liquid Helium

The macroscopic properties of superfluid liquid helium can be understood in terms of the elementary excitations, phonons and rotons, and their interactions. It is convenient to describe the neutron scattering from superfluid liquid helium in terms of two components: one component is well defined in frequency and corresponds to neutron scattering from the excitations whereas the other component has a broad distribution in frequency and from its similarity with multi-phonon scattering in crystals is known as the multi-phonon scattering. In fig. 13 the distribution of the scattering in frequency (energy) and wavevector is illustrated[21] for these two components.

The feature of the liquid helium which makes it unique for our purposes is the isotropy of the dispersion relation: the frequency depends on $|\underline{Q}|$. As first pointed out by Ruvalds and Zawadowski[22] this isotropy leads to several important consequences. Firstly the density of two non-interacting roton states has a discontinuity when $\omega = 2\Delta$, where Δ is the roton energy. Secondly if there is any attractive roton-roton interaction there will be a two roton bound state formed with an energy $\omega = 2\Delta - \varepsilon_b$, just below 2Δ. The two-roton density of states is then unusually sensitive to the roton-roton interactions. Furthermore Raman scattering measurements[23] have shown that a two roton bound state does exist at least for the pairs of rotons, with $\underline{Q} = 0$, excited in a Raman scattering experiment.

In principle neutron scattering could be used to measure the two-roton density of states for a range of $\underline{Q}$, but as yet there is not a microscopic theory of liquid helium and so the coupling between the neutrons and the rotons is unknown. Instead it is possible to deduce some information about the two-roton density of states from the properties of the single excitations. In particular a single excitation will decay into pairs of rotons if the decay satisfies conservation of momentum and frequency. The resultant process will give rise to a self-energy expression similar to that of eqn. 3.3. Now Pitaevskii[24] showed that if eqns. 3.4 are evaluated for the isotropic roton spectrum shown in fig. 13, that close to 2Δ

$$\Sigma(\underline{Q},\omega) = \frac{A}{Q} \ln \left|\frac{\omega - 2\Delta}{D}\right| \qquad \omega < 2\Delta \qquad (4.2)$$

where A and D are constants. As a result of the singularity in this expression as $\omega \rightarrow 2\Delta$, the frequency of the single excitation cannot exceed 2Δ if the excitations are non-interacting. If the interactions between the excitations are attractive then Zawadowski et al.[25] have shown that the frequency cannot exceed $2\Delta - \varepsilon_b$.

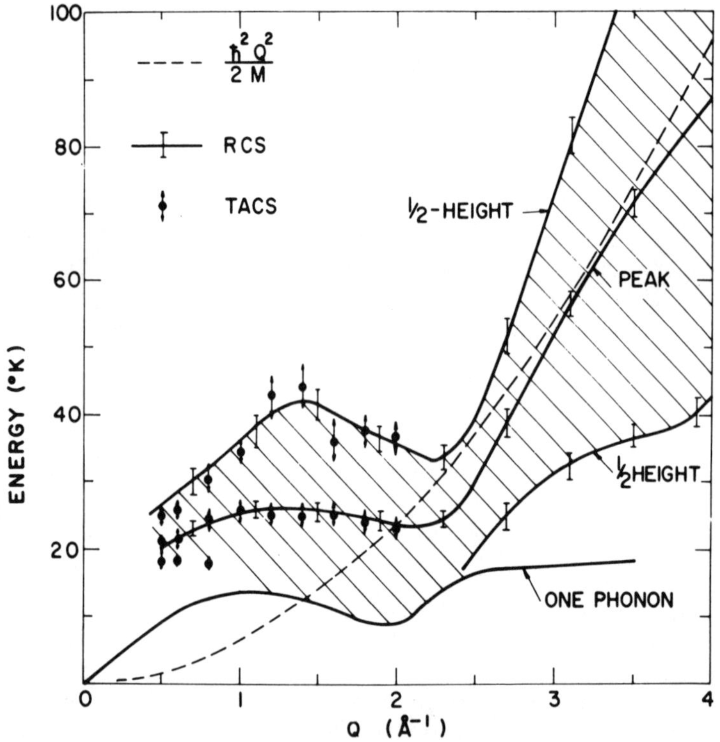

Fig. 13. Neutron scattering[21] from liquid helium at 1.1 K. The one-phonon part is scattering by a well-defined excitation and the hatched region gives the extent of the diffuse multi-phonon scattering.

Measurements have been made for the single excitation spectrum for frequencies $\omega \approx 2\Delta$ both for $Q \sim 1.1$ Å^{-1}, ref. 26, and for $Q \sim 3.0$ Å^{-1}, ref. 27, as shown in fig. 14. In both cases at elevated pressure the peak of the neutron groups are at frequencies in excess of 2Δ. The interactions between the rotons must therefore be repulsive for these wavevectors. A detailed fit to the lineshapes shown in fig.14 enables the roton-roton interactions to

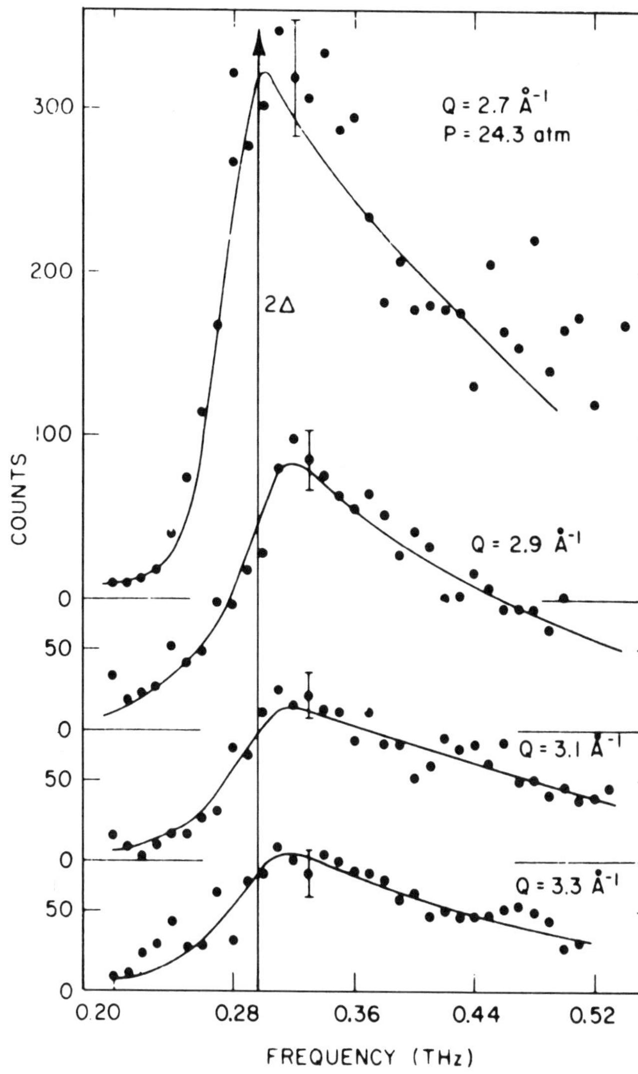

Fig. 14. Neutron groups[27] in liquid helium at 24.3 atm and 1.1 K.

be obtained, and the interaction has the wavevector dependence shown in fig. 15. Also shown in fig. 15 is a calculation of the roton-roton interactions using an expression of Lee[28] based on the Jackson-Feenberg theory of liquid helium.

(iii) Zero and First Sound

The self-energy of an excitation due to decay into pairs of excitations was discussed in section 3(i) using low order perturbation theory. As discussed above this expression fails in the case of decay into pairs of rotons in isotropic liquid helium due to the extreme sensitivity of the non-interacting density of roton states. In all materials it fails in the limit as $q \to 0$ and $\omega \to 0$ because thermodynamic processes of necessity involve multiple scattering effects. The difficulty arises in eqns. (3.4) from those terms for which n_2 and n_3 belong to the same branch so that as $\underline{q}_1 \to 0$ $\underline{q}_2 \to -\underline{q}_3$ and $\omega_2 \to \omega_3$. Clearly under these circumstances eqn. (3.4) cannot be valid when $\omega < \Gamma_2$ where Γ_2 is

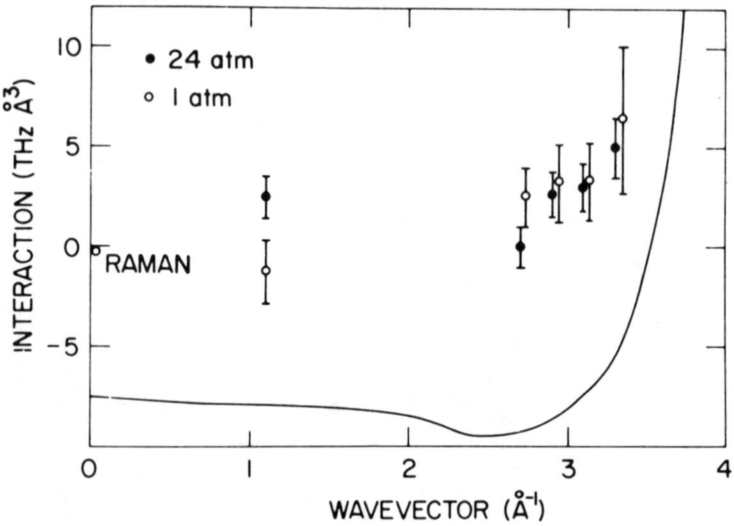

Fig. 15. The roton-roton interaction[27] as a function of the total wavevector of the rotons compared with a theoretical calculation.

the width of mode 2. Collisionless or zero sound[29] is then that measured in a neutron scattering experiment, for $\omega > \Gamma$, and for which eqns. (3.4) give a well defined result.

First sound or collision dominated sound occurs when $\omega < \Gamma$ when a detailed treatment of the thermodynamic processes shows that the isothermal elastic constants are obtained when the limit of eqn. (3.4) is taken so that $\omega \to 0$ followed by $q \to 0$. The full expression also recovers the well known macroscopic result for the difference between the adiabatic and isothermal elastic constants. The first experiment to demonstrate the difference between the $\omega > \Gamma$ and $\omega < \Gamma$ sound was performed by Svensson and Buyers[30] in KBr. The results are shown in fig. 16 where they are compared with theoretical calculations. Although the theoretical results are too large by a factor of 2, there is clearly a significant effect, and now the same type of experiment has been performed in a number of different materials.

(iv) Central Peaks at Structural Phase Transitions

The occurrence of quasi-elastic scattering in the spectral response close to a structural phase transition must be associated with some relatively long time scale[31]. There have been several proposals for the origin of this new time-scale. The three principal types of explanations are: (a) the quasi-elastic scattering is associated with some type of defect[32], (b) it arises from scattering by dynamic or static domain walls[33] and (c) it arises from a coupling of the soft mode to phonon density fluctuations in a very similar way to the acoustic modes discussed in section 4(iii)[34]. It seems unlikely that any one of the explanations will be able to account for all the central peaks which have now been observed, but in this section we discuss only alternative (c) as it is another example of the effect of interactions between elementary excitations.

As discussed above in section 4(iii), if the soft mode associated with a structural phase transition can couple to the phonon density fluctuations, then there will be a difference between the behaviour when $\omega > \Gamma$ and when $\omega < \Gamma$. An immediate consequence of this difference is that there will be a central peak of width Γ (a typical phonon inverse lifetime).

Now below T_c, when the soft mode has condensed, the soft mode has the same symmetry as the crystal structure. The soft mode must therefore be able to couple to temperature fluctuations and there will always be a central peak resulting from this coupling. Above T_c, the soft mode must have a different symmetry from the crystal and so it cannot couple linearly with temperature

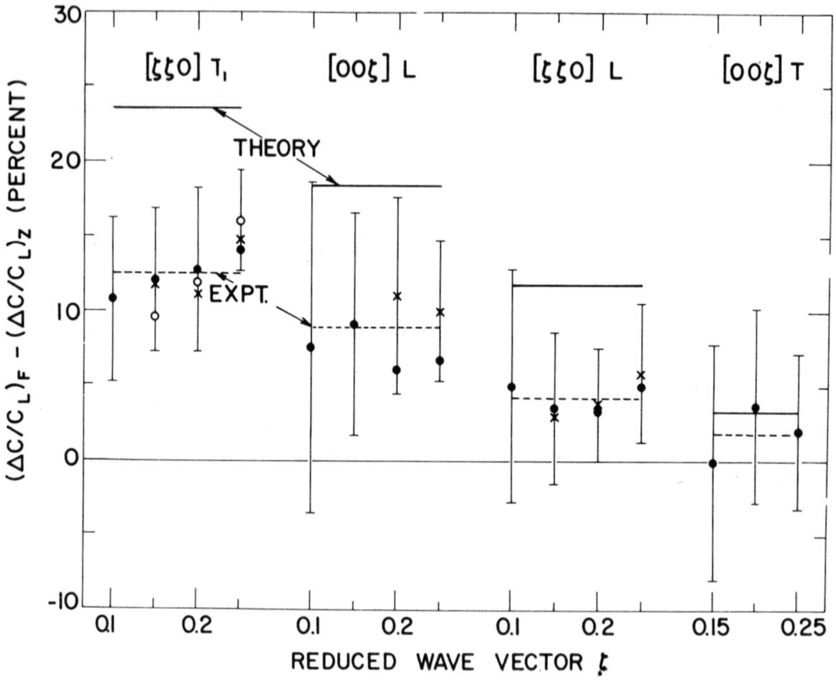

Fig. 16. Difference between the first and zero sound elastic constants[30] in KBr compared with theory[29].

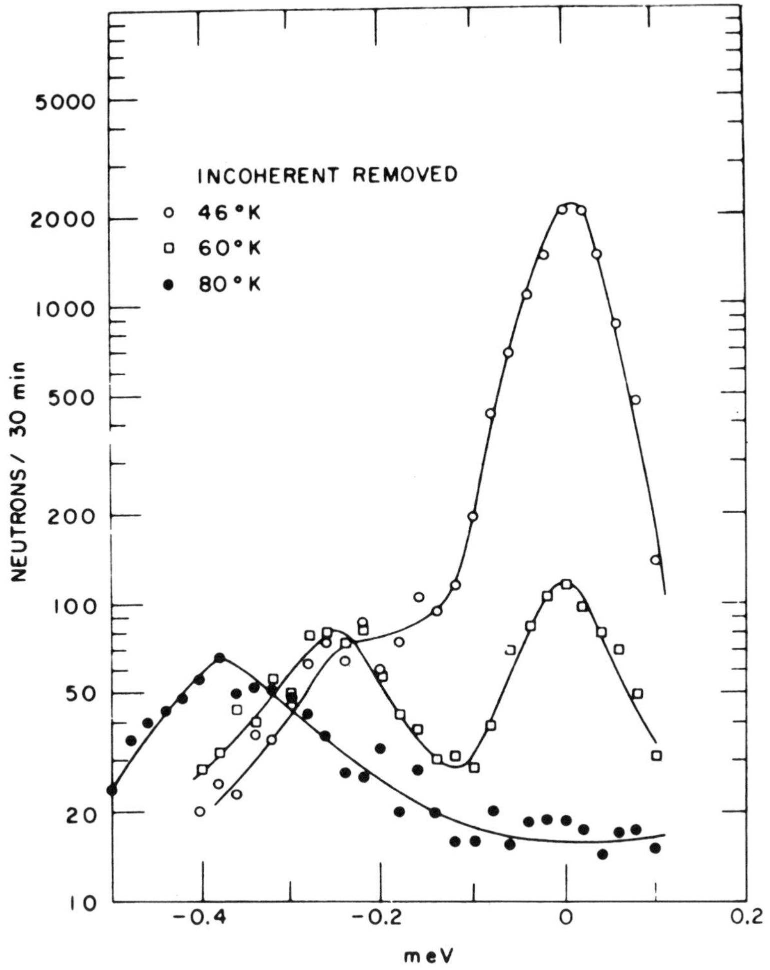

Fig. 17. The neutron scattering[35] from the soft acoustic mode in Nb_3Sn. Note the increase in quasi-elastic scattering as $T \to T_c$.

fluctuations. There will be no central peak due to this mechanism. It may, however, be possible for the soft mode to couple to phonon density fluctuations which do not alter the local temperature. This occurs only if the soft mode wavevector has $\underline{q} = 0$, and furthermore if the coupling coefficients between the soft mode and two modes belonging to the same branch of the dispersion relation, $V\begin{pmatrix} 0 & \underline{q} & -\underline{q} \\ s & j & j \end{pmatrix}$, is non-zero. For ferroelectric transitions this implies that the material is piezoelectric in the paraelectric phase, like KDP. In these circumstances a central peak is expected to occur both above and below T_c.

Unfortunately there are few if any of the central peaks which have been unambiguously shown to result from this mechanism. One possible example is the central peak in Nb_3Sn, as shown in fig. 17, for which the soft mode is an acoustic mode which may couple to the phonon density fluctuations.

Recently there has been particular interest in incommensurate phase transitions at which the soft mode has a wavevector, $\underline{q}_{inc}$, which is incommensurate with the lattice of the high temperature structure. The low temperature distortion is then described by a wave which has the same energy irrespective of the phase of this wave. Consequently there is a new set of 'acoustic' modes associated with this invariance of the energy of the system and these modes are known as 'phasons'. A further unusual property then arises because of these phasons. The interaction[36] between these phase modes and the amplitude modes, which describe the changes in the amplitude of the distortion, then renormalise the self-energy of the amplitude modes in an exactly analogous way to that described in section 4(iii) so that at all $T < T_c$, the frequency of the amplitude modes for $\underline{q} = \underline{q}_{inc}$ is zero. There is infinitely strong quasi-elastic scattering in an experiment if its resolution is ideally good. The dispersion relation for the amplitude mode central peak is also of an unexpected form. The intensity varies as $[\underline{q} - \underline{q}_{inc}]^{-1}$.

This type of behaviour also occurs in ideal Heisenberg ferromagnets[37] where the spin waves play the role of the phase modes and the longitudinal susceptibility diverges for $\underline{q} = 0$. This predicted divergence of the longitudinal susceptibility has never been observed experimentally; possibly because real ferromagnets are not ideal but have some anisotropy so that the spin waves have an anisotropy gap which then removes the divergence in the susceptibility. Maybe incommensurate phase transitions provide an example at which this divergence could be observed, and so test an extreme example of the effects of the interactions between excitations.

REFERENCES

1 B.N. Brockhouse, 1961, Inelastic Scattering of Neutrons in Solids and Liquids (IAEA, Vienna) p.113.

2 R.B. Stinchcombe, 1977, this volume.

3 W. Marshall and S.W. Lovesey, 1971, Theory of Thermal Neutron Scattering (Oxford Univ. Press).

4 for example R.A. Cowley, 1968, Rept. Prog. Physics, 31, 123.

5 A.A. Maradudin and A.E. Fein, 1962, Phys. Rev. 128, 2589.

6 E.R. Cowley and R.A. Cowley, 1965, Proc. Roy. Soc. A237, 259.

7 K. Hisano, F. Placido, A.D. Bruce and G.D. Holah, 1972, J. Phys. C, 5, 2511.

8 J. Harada, J.D. Axe and G. Shirane, 1971, Phys. Rev. B4, 155.

9 R.A. Cowley, J.D. Axe, M. Iizumi (unpublished).

10 J. Shalyo Jr., B.C. Frazer and G. Shirane, 1970, Phys. Rev., B2, 4603.

11 R.A. Cowley, 1976, Phys. Rev. Lett., 36, 744.

12 R.A. Cowley and W.J.L. Buyers, 1969, J. Phys. C, 2, 2262.

13 R.A. Cowley, E.C. Svensson and W.J.L. Buyers, 1971, Phys. Rev. Letters, 23, 525.

14 J. Meyer, G. Dolling, R. Scherm and H.R. Glyde, 1976, J. Phys. F, 6, 943.

15 see H.R. Glyde, 1974, Can. J. Phys., 52, 2281.

16 W.J.L. Buyers, J.D. Pirie, and T. Smith, 1968, Phys. Rev., r., 165, 999.

17 V. Ambegaoker, J. Conway and G. Baym, 1965, In Lattice Dynamics, Pergamon Press, ed. R.F. Wallis, p. 261.

18 R.A. Cowley, W.J.L. Buyers, P. Martel and R.W.H. Stevenson, 1969, Phys. Rev. Lett., 23, 86.

19 T.M. Holden, E.C. Svensson, R.A. Cowley, W.J.L. Buyers and R.W.H. Stevenson, 1970, J. Appl. Phys., 41, 896.

20 C.R. Natoli and J. Ranninger, 1973, J. Phys. C, 6, 370.

21 R.A. Cowley and A.D.B. Woods, 1971, Can. J. Phys., 49, 177.

22 J. Ruvalds and A. Zawadowski, 1970, Phys. Rev. Lett., 25, 333.

23 C.A. Murray, R.L. Woerner and T.J. Greytak, 1975, J. Phys. C, 8, L90.

24 L.P. Pitaevskii, 1959, Sov. Phys. JETP 9, 830.

25 A. Zawadowski, J. Ruvalds and J. Solana, 1972, Phys. Rev., 15, 399.

26 E.H. Graf, V.J. Minkiewiez, H. Bjerrum Møller and L. Passell, 1974, Phys. Rev. A10, 1748.

27 A.J. Smith, R.A. Cowley, A.D.B. Woods, W.G. Stirling and P. Martel, 1977, J. Phys. C, 10, 543.

28 D.K. Lee, 1967, Phys. Rev. 162, 134.

29 R.A. Cowley, 1967, Proc. Phys. Soc., 90, 1127.

30 E.C. Svensson and W.J.L. Buyers, 1968, Phys. Rev., 665, 1063.

31 for an outdated review see R.A. Cowley, 1974, Ferroelectrics, 6, 163.

32 B.I. Halperin and C.M. Varma, 1976, Phys. Rev. B, 14, 4030.

33 J. Krumhansl and J.R. Schrieffer, 1975, Phys. Rev. B, 11, 3535, S. Aubry, 1975, J. Chem. Phys., 62, 3217.

34 G.J. Coombs and R.A. Cowley, 1973, J. Phys. C, 6, 121, 143.

35 G. Shirane and J.D. Axe, 1971, Phys. Rev. Lett., 27, 1803.

36 A.D. Bruce and R.A. Cowley (to be published).

37 E. Brezin and D.J. Wallace, 1973, Phys. Rev. B, 7, 232; D.R. Nelson, 1976, Phys. Rev. B, 13, 2222.

OPTICAL RESPONSE OF QUANTUM CRYSTALS

Isaac F. Silvera

Natuurkundig Laboratorium der Universiteit van Amsterdam
Valckenierstraat 65, Ansterdam-C, The Netherlands

INTRODUCTION

Quantum solids are crystals in which, even at T = 0 K, the atoms of the ordered array (or molecules) undergo large rms displacements or zero-point motion (ZPM) about their equilibrium lattice sites[1]. The ZPM in these crystals, in comparison to normal crystals, is not small relative to the nearest neighbor distance, R_o. The dynamical aspects of ordinary crystals can be treated classically or quantum mechanically in the quasi-harmonic approximation. Quantum crystals must be treated quantum mechanically because the zero-point energy is comparable to the static lattice energy. This in itself would present no difficulty if potentials were harmonic, however real potentials are highly anharmonic and for quantum crystals the usual perturbative treatment of anharmonicity breaks down. New theoretical approaches which have been developed in the last decade to handle the dynamical problems will be discussed in the first section.

To qualify as a quantum solid the atoms must have a weak interaction potential and a light mass. The most extreme stable quantum solids are ^{3}He and ^{4}He. Molecular hydrogen and its isotopes are quantum crystals with regard to the translational motions, just as the heliums. However H_2 has an added aspect, that of rotation of the homonuclear diatomic molecules in the solid state. For the rotational motions, quantum properties become important if the anisotropic interactions are weak and the moment of inertia is small[2]. This is the case for H_2 which remains an almost free rotor in the solid at T = 0 K, whereas in normal molecular solids such as N_2, CO_2 etc., the molecules are orientationally localized. An added feature is that because of the rotational and internal vibrational motions of the molecules a much richer spectrum of excitations is accessible to optical studies than is found in the atomic quantum solids. In both the translational and the rotational motions, the large amplitude ZPM is responsible for important changes (from that of harmonic solids) in the spectral function arising from nuclear motions of the solids.

These lectures are organized in the following way. First a description and review will be given of the ground state properties of translational and orientational quantum solids as well as a discussion of interactions and solid phases. Next the infrared and Raman interaction with light will be discussed, followed by a description of experimental techniques. Finally the Raman and infrared quasi-particle spectra of helium and hydrogen and their isotopes will be presented.

TRANSLATIONAL QUANTUM SOLIDS

The ground state and excitations of a quantum crystal can be studied in terms of the Hamiltonian

$$H = K + V = \sum_i \frac{P_i^2}{2m} + \frac{1}{2} \sum_{i \neq j} V_I(ij) \qquad (1)$$

where m is the mass of the lattice particles and the potential energy term V is approximated by a sum of isotropic pairwise intermolecular potentials $V_I(ij)$; the notation ij refers to the separation r_{ij} between particles i and j. For quantum solids at low pressure the thermodynamic and dynamical properties can be described reasonably well by a potential of the Lennard-Jones (LJ) form:

$$V_I(r) = 4\varepsilon \left[\left(\frac{\sigma}{r}\right)^{12} - \left(\frac{\sigma}{r}\right)^{6} \right] . \qquad (2)$$

Although realistic potentials have a softer exponential hard core, the LJ potential is quite adequate for our purpose except at high pressures. In this potential σ represents the distance at which the potential is zero; ε is the well minimum which occurs at $r_m = 2^{1/6}\sigma$. In Fig. 1 we show the LJ potentials for He, H_2 and Argon. To compare these solids it is useful to rewrite the Hamiltonian (1) in terms of reduced energy units ε and with $r \to r/\sigma$:

$$\mathcal{H} = H/\varepsilon = -\frac{1}{2}\lambda^2 \sum_i \nabla_i^2 + \frac{1}{2} \sum_{i \neq j} v(ij) \qquad (3)$$

Here $\lambda = \hbar/\sigma\sqrt{m\varepsilon}$ is the quantum parameter which differs from that defined by de Boer by a factor $1/2\pi$. For heavy masses or deep potentials wells λ^2 is small and the T = 0 K equilibrium structure of the solid will be determined by the potential energy term. In the case of helium and hydrogen λ^2 is large and the kinetic energy term plays an important role in determining the equilibrium lattice parameter. In Table I we list some values of λ^2.

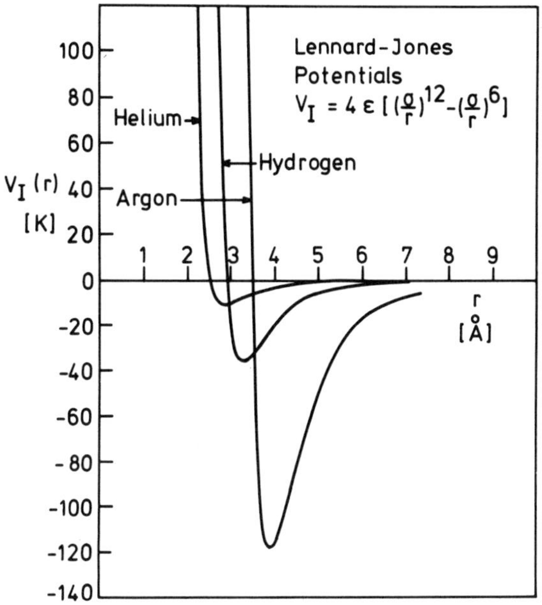

Fig.1. The Lennard-Jones isotropic potential using parameters given in Table I.

Table I. Lennard-Jones parameters and the quantum parameter, λ^2.

Atom or Molecule	$\varepsilon(^oK)$	$\sigma(\AA)$	λ^2
^3He	10.2	2.56	.241
^4He	10.2	2.56	.182
H_2	37	2.92	.076
D_2	37	2.92	.038
Ar	119.3	3.45	.0027

It is illuminating to study the problem of the quantum solid as approached from a simple model of a harmonic solid. We follow an example from Guyer[1] of an atom in an fcc crystal. We take a single particle model where each atom sits in the potential minimum of its z (z = 12 for fcc) stationary neighbors. The harmonic hamiltonian (per atom) is

$$H = P^2/2m + z\,v(r_o) + \frac{z}{2}\,v''(r_o)u^2 \tag{4}$$

where $v(r_o) \simeq \varepsilon$, $v''(r_o)$ is the second derivative of the potential, $\underline{u} = \underline{r}-\underline{r}_o$ is the displacement from equilibrium, and the factor 1/2 arises because the energy is shared with the z neighbors. The ground state solution to $(H - z\,v(r_o))\Phi = E\Phi$ is the gaussian wavefunction

$$\Phi(u) = (\frac{\alpha}{\pi})^{1/4} \exp(-\alpha\, u^2/2) \tag{5}$$

with $\alpha = \sqrt{mzv''}/\hbar$ and zero-point energy $\hbar\omega/2$ where $\omega = \sqrt{zv''/m}$. The dynamic energy of the particle is $\varepsilon_o = 9\hbar\omega/8$ and the total ground state energy (static plus dynamic)

$$\langle H\rangle = z\,v(r_o)/2 + \frac{9}{8}\,\hbar\,\left[zv''(r_o)/m\right]^{1/2} \tag{6}$$

For argon the dynamic term is about an order of magnitude smaller than the static term, but for helium the positive dynamic term is greater than the static term. A Hartree calculation of $\langle H\rangle$ yields 66 K compared with an experimental value of −1 K. The situation for this model, however, is even more extreme. The particle occupies a volume of space given by the rms displacement:

$$\langle u^2\rangle^{1/2} = \sqrt{3/2\alpha} = \sqrt{3\hbar/2}\,\left[2/mzv''\right]^{1/4} \tag{7}$$

For helium this is so large that there is strong overlap of the hard cores. The crystal can lower its total energy by expanding. This reduces the kinetic energy of localization at the expense of static potential energy. The mean position of the particle is no longer at the well of its neighbors. This result is a catastrophe for this simple model. In Fig.2 we show the potential that a helium atom would experience using the pairwise LJ potential and the nearest neighbor distance of helium at the pressure required for solidification (∿ 25 Atm.). The single particle well has a hump in the center. The second derivative of the potential, $v''(r_o)$, is negative and the oscillator frequencies are imaginary. Stable solutions do not exist. The same result is found for H_2.

There are a number of difficulties with this simple model. First the derivatives such as v'' are evaluated at a point in space. Evidently, for a broad distribution they should be averaged over the motion. Secondly, the neighbors were fixed; a more appealing solution would allow the neighbors the same motion as the

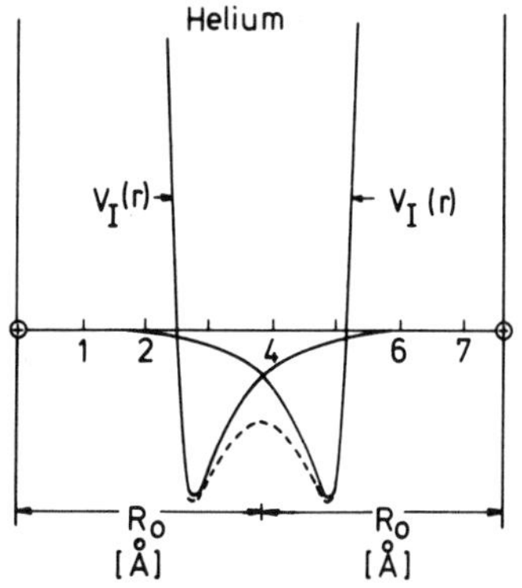

Fig.2. The potential that a helium atom would experience in the well of rigidly fixed nearest neighbors using the Lennard-Jones potential.

central atom, or the problem requires self consistency. Thirdly, there is no correlation in the problem. In a real crystal the motions of the atom are correlated. In particular short-range correlations are required such that the relative motions of neighboring atoms minimizes overlap.

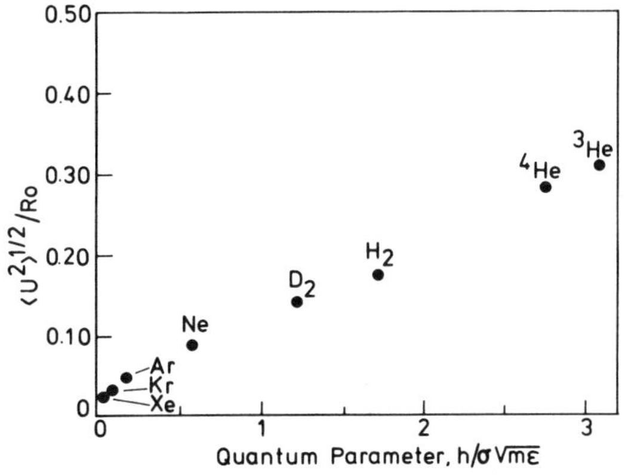

Fig.3. The localization as a function of quantum parameter for solids at about zero pressure.

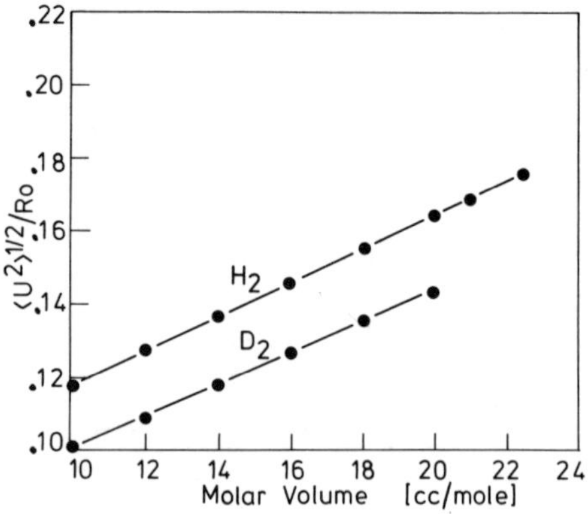

Fig.4. The dependence of localization on density and mass for H_2. The corresponding pressures can be found in Fig.12.

Before discussing improved theories, let us consider some systematics involving the quantum parameter, $2\pi\lambda$, and some results using self-consistent theories. In Fig.3 we show the localization of the rare gas atoms, $\langle u\rangle^{1/2}/R_o$, as a function of the quantum parameter[3]. We see that for ^3He the rms displacement is 30% of the nearest neighbor distance. The 18% value for H_2 has been measured experimentally by neutron scattering[4]; the 10% for neon results in important anharmonic effects; these are much less for argon. In Fig.4 we show the effect of pressure and mass on the localization. Pressure compresses the lattice, which localizes the particles (the particle in a well picture starts to become applicable). Even though the zero-point energy increases due to the increase of v'' (Eq.6), $\langle u^2\rangle^{1/2}$ decreases, Eq.(7). The dependence on mass is also shown in the latter equation.

In order to understand the quantum solids we would like a theory that can accurately predict the ground state properties, including the energy, the single particle distribution function (the probability of finding a particle at a point in space), the pair correlation function, PCF, (the probability of finding two particles at a given relative separation) and the translational excitation spectrum. We shall not probe too deeply into the theory as detailed treatments exist in recent reviews[1] and our aim here is to present the prominent features of these theories and the relation to optical properties. We deviate from the historical development and note that experimental measurements show phonons to be well defined excitations. Inelastic neutron scattering in He and H_2 have shown well defined phonon groups, which are however severely broadened in certain regions of the Brillouin zone. Dispersion relations exist just as in ordinary solids; an experimental one-phonon dispersion curve for He is shown in Fig.5.

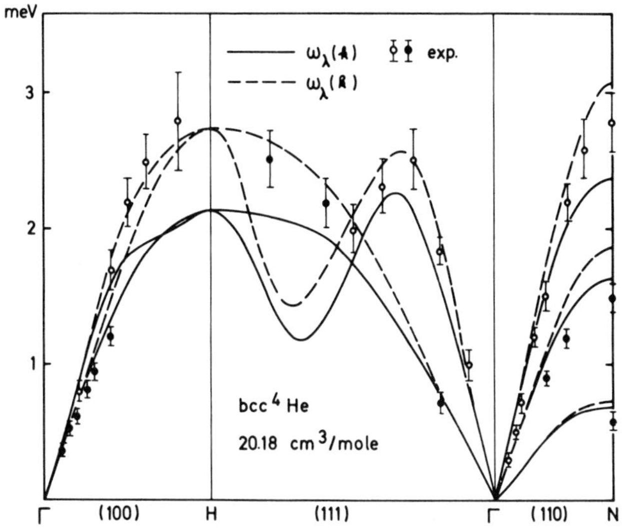

Fig.5. Dispersion curves for bcc ^{4}He determined by neutron scattering along with calculated curves (after ref.6).

A theory which explicitly contains phonons is the self consistent phonon theory in which one approximates the Hamiltonian, Eq.(1), by

$$H_o = -\frac{\lambda^2}{2}\sum_i \nabla_i^2 + \frac{1}{2}\sum_{i\neq j} \underline{u}_i \underline{\underline{\Phi}}_{ij} \cdot \underline{u}_j \qquad (8)$$

The spring constants $\underline{\underline{\Phi}}_{ij}$ are not those found by a Taylor series expansion as in the quasi-harmonic theory but are determined by a variational treatment to bring the energy levels of Eq. (8) as close as possible to those of Eq.(1). The basis functions are correlated gaussian states ($\psi_o = \langle \underline{u}_1, \ldots, u_N|0\rangle$)

$$\psi_o = A \exp\left[-\frac{1}{2}\sum_{i,j} \underline{u}_i \cdot \underline{\underline{G}}_{ij} \cdot u_j\right] \qquad (9)$$

The result of the variational treatment is that the force constants are given by

$$\Phi_{\alpha\beta}(ij) = \langle 0|\ \nabla_\alpha(i)\ \nabla_\beta(j)\ V\ |0\rangle\ , \qquad (10)$$

that is the force constants are the average over the distribution rather than the value of the second derivative evaluated at the equilibrium position. This is in fact an average over the pair correlation function defined by

$$g(12) = \int \psi_o^* \psi_o \, dr_3 \ldots\ldots dr_n / \langle 0|0\rangle \qquad (11)$$

where $\langle 0|0\rangle$ gives the normalization. The pair correlation function is a generalized gaussian distribution. The self-consistency results because the states $|0\rangle$, i.e. G_{ij} of Eq.(9), depend on Φ. This is seen by substitution of Eq.(9) in Eq.(8) which yields $(\lambda G)^2 = \underline{\underline{\Phi}}$. The phonon frequencies generated by this theory are called self consistent phonons.

One difficulty with the self-consistent phonon theory is that it does not properly describe the short range behavior of the wave function on the PCF. Here it is a pure generalized Gaussian function and allows too much overlap of neighboring particles. The asymtotic $r_{ij} \to 0$ pair wave function should obey the equation

$$\left[\frac{1}{2m}(p_i^2 + p_j^2) + v(ij)\right]\psi_{ij} = E\psi_{ij} \qquad (12)$$

For an LJ potential the WKB solution in the limit $r_{ij} \to 0$ is

$$\psi_{ij} \sim \exp\left[-K/r_{ij}^5\right] \qquad (13)$$

This is a possible form for a "Jastrow" short-range cut-off function, f_{ij}. Improved variational functions are found by multiplying the trial function, Eq.(9) by such a Jastrow function

$$\psi_{0J} = \psi_o \prod_{i>j} f_{ij}(r_{ij}) \qquad (14)$$

which further complicates the self-consistent equation and generates new force constants.

Most recent developments have been the theory of Horner[1] which brings the PCF into the theory in a more fundamental way and imposes certain moment restrictions on the distributions. Finally we point out that the self-consistent theories do not fully account for the anharmonicity. Important corrections to phonon frequencies and lifetimes are found from diagramatic perturbation theory of cubic and quartic anharmonic terms[5]. In Fig.6 we show the phonon spectral functions for bcc ^{4}He calculated by Horner[6]. These functions, which would be delta functions for a harmonic solid, are highly structured. Reasonably good agreement between theory and the experimental dispersion relations are found. In Fig.7 we show the pair distribution functions calculated by Horner[6], with and without short-range correlations. Large differences are found for the nearest neighbor, whereas for the third shell, the differences between the gaussian and the full PCF cannot be seen here.

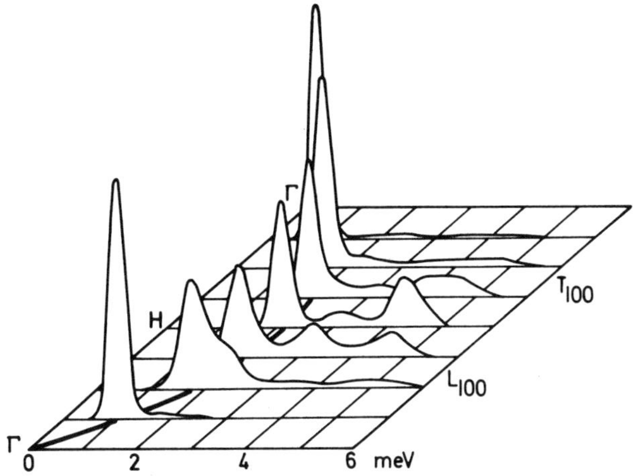

Fig.6. Phonon line shapes in bcc ^{4}He at 20.92 cc/mole. L and T refer to longitudinal and transverse phonon branches (after ref.6).

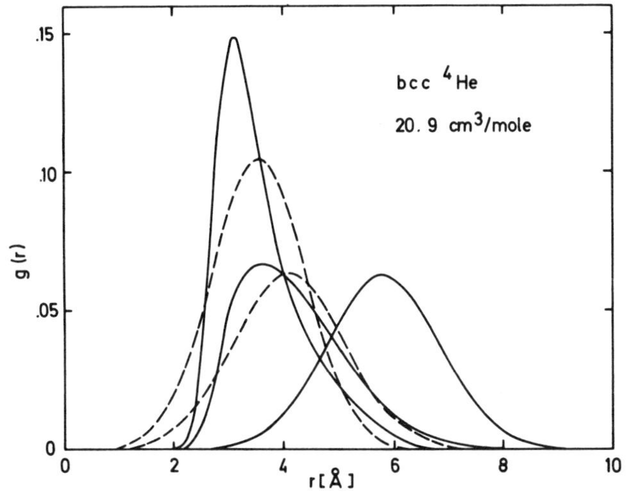

Fig.7. Calculated pair correlation functions in bcc ^{4}He for the first three shells of neighbors. The dashed lines are without short range correlations (after ref.6).

ORIENTATIONAL QUANTUM SOLIDS

Orientational quantum solids are molecular solids in which, even at T = 0, the angular distribution is broad, or non-localized. Molecular hydrogen and its isotopes are the most extreme (and only) examples. H_2 is an almost free rotor in the solid and in the ground orientational state has a fully spherical distribution; by contrast the symmetry axes of N_2 molecules are well localized at T = 0 K in the solid state.

Let us confine our attention to linear molecules with pairwise interactions. Then we must extend V of Eq.(1) to include an anisotropic potential, V_A, and the wave function must depend upon the orientational coordinates of all molecules

$$H = \sum_i \frac{-\hbar^2\nabla_i^2}{2m} + \frac{1}{2}\sum_{i\neq j} V_I(ij) + \frac{1}{2}\sum_{i\neq j} V_A(ij) \qquad (15)$$

where we use the set of polar coordinates $\Omega_i \equiv \theta_i,\phi_i$, to specify the orientation of the molecular axes with respect to some coordinate system. To demonstrate why H_2 is a free rotor solid let us set $V_A = 0$ and study two neighboring molecules 1 and 2. We solve the wave equation

$$\left\{ \frac{-\hbar^2}{2m}\left[\nabla_1^2 + \nabla_2^2\right] + V(r_{12}) \right\}\psi = E\psi \qquad (16)$$

To a good approximation the wave function is separable,

$$\psi = \psi(\Omega_1)\,\psi(\Omega_2)\,\psi(r_1, r_2) \qquad (17)$$

and the solution of the angular part is that of the well known free rotor with $E = BJ_i(J_i+1)$ where $B = \hbar^2/2I$ is the rotational constant and I is the moment of inertia. J_i is the rotational quantum number and the rotational wave function is the spherical harmonic $Y_{J_iM_i}(\theta_i,\phi_i)$ which we shall also specify as $|JM\rangle$ with M the projection quantum number. We can now rewrite the orientational kinetic energy as a spin operator $B\underline{J}^2$. Returning to the full Hamiltonian, Eq.(15) with $V_A = 0$ we can write this as a rotational kinetic energy part and a translational kinetic energy part which depends only on the motion of the molecular centers of mass. Thus we have free rotors which can (and do) also behave as translational quantum solids. For the lowest rotational state, with all molecules in the J = 0 state, the solid wave function is a product state of $Y_{oo}(\Omega_i)$'s. Since Y_{oo} is a constant the angular distribution is spherically symmetric i.e. the molecules are not at all angularly localized. For each level Y_{JM}, all (2J+1) of the M sublevels will be equally populated and again no localization.

To see how valid this picture is we now remove the restriction $V_A = 0$. Then if V_A is small we can use first order perturbation theory to correct the wave function. We again consider a pair of molecules, with the result

$$|J_1M_1J_2M_2>' = |J_1M_1J_2M_2> + \tag{18}$$

$$+ \sum_{\substack{J_1'J_2' \\ M_1'M_2'}} \frac{<J_1'M_1'\ J_2'M_2'\ |V_A(\Omega_1\Omega_2)|J_1M_1J_2M_2>}{B\left[J_1(J_1+1)+J_2(J_2+1)-J_1'(J_1'+1)-J_2'(J_2'+1)\right]} |J_1'M_1'J_2'M_2'>$$

where $|J_1M_1J_2M_2> = |J_1M_1>|J_2M_2>$ is a product of free rotor states. If the mean value of V_A, $<V_A>$ is small compared to B, the correction will be small and J can be considered to be a good quantum number. For H_2, $B \simeq 60\ cm^{-1}$ and $<V_A>/B \simeq 0.07$. By contrast, for N_2, $B \simeq 2\ cm^{-1}$ and $<V_A>/B \simeq 8$, thus the free rotor states are no longer valid and the molecules will be angularly localized by the anisotropy barriers at T = 0 K.

Since H_2 behaves as an almost free rotor in the solid, it is useful to discuss the rotational properties of the isolated molecule. Molecular hydrogen is a homonuclear diatomic molecule with spin $I_N = \frac{1}{2}$ on each proton. Because the protons are indistinguishable and fermions, the total molecular wave function involving the nuclear coordinates must be antisymmetric with respect to particle permutation. The relevant part of the wave function can be written as a product of vibrational, rotational and nuclear spin wave functions. Inversion symmetry provides parity as a good quantum number; the nuclear and rotational wave functions are either symmetric (S) or antisymmetric (AS). The allowed combinations for an AS total wave function require consideration of only the latter two, and are given in Table II, along with the nuclear weight and ortho-para designation. We also give the combinations for tritium which is similar to H_2 (i.e. $I_N = \frac{1}{2}$), and deuterium for which the total wave function must be symmetric because the deuterons have spin 1 and are bosons. The non-homonuclear molecules such as HD do not have the ortho-para distinction since the nucleons are distinguishable. The classification of states is of great physical importance as ortho-para transitions (conversion) are forbidden for isolated molecules and the two species have different properties, as we shall see. The crystal structure of the solid will be most strongly influenced by the angular distribution of the molecules characterized by the rotational quantum number, thus from Table I, we would expect similar behavior for the even J species, p-H_2 and o-D_2, or odd J species, o-H_2 and p-D_2.

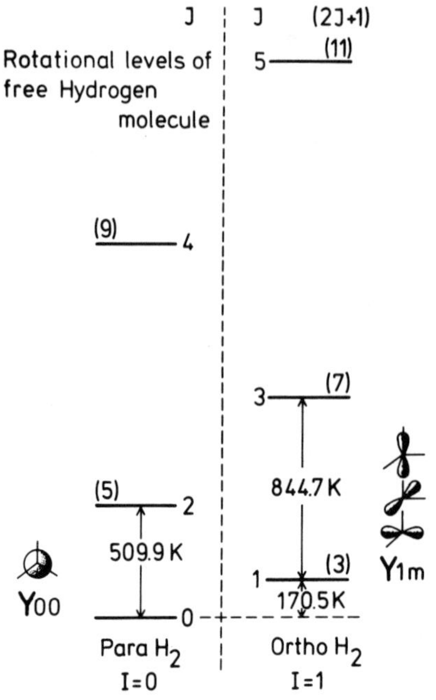

Fig.8. Free rotor energy levels for molecular hydrogens. The degeneracy is given by numbers in parenthesis.

The reason for the large physical distinction between the species can be seen by considering the rotational energy levels given by $E = BJ(J+1)$ and shown in Fig.8 for H_2. Due to the very large splittings of the rotational levels, at the low temperature of the solid hydrogens ($\lesssim$ 20 K), only the J = 0 and 1 levels are thermally populated. A J = 1 molecule will remain metastably in that level, as conversion to the J = 0 state is very slow in the solid.

The anisotropic pair potential of Eq.(15) can be expressed in terms of a small number of terms if only the J = 0 or 1 rotational states are populated, as is the case in the low temperature solid. Because the molecular orientational wave functions are the spherical harmonics, the potential is expanded in the $Y_{LM}(\theta,\phi)$. This expansion is limited to $L \leq 2$, since matrix elements of spherical harmonics with $L > 2$ vanish within the J = 0,1 manifold of states. The most general form of the potential is[7]

Table II. Allowed combinations of nuclear-spin states and rotational states for hydrogen, deuterium and tritium and the ortho-para designations. Antisymmetric is abreviated by AS and symmetric by S; I_{mol} is the total molecular nuclear spin and J the rotational quantum number.

Molecule and spin of nucleon		I_{mol}	J	$\psi(I_{mol})\psi(J)$	Nuclear weight	Designation
Hydrogen Tritium $I_N=\frac{1}{2}$	state	0	even			para
	symmetry	AS	S	AS	1	
	state	1	odd			ortho
	symmetry	S	AS	AS	3	
Deuterium $I_N=1$	state	1	odd			para
	symmetry	AS	AS	S	3	
	state	0, 2	even			ortho
	symmetry	S	S	S	6	

$$V_A(R) = (16\pi/5)^{1/2}\, B(R)\, \left[Y_{20}(\Omega_1) + Y_{20}(\Omega_2)\right]$$

$$+ 4\pi \sum_{\substack{j=0,2,4\\ \mu}} \varepsilon_j(R)\alpha_j C(22j;\mu,-\mu)\, Y_{2\mu}(\Omega_1)Y_{2-\mu}(\Omega_2) \quad (19)$$

where Ω_1 and Ω_2 specify the angles of the molecular axes with respect to the vector R connecting the molecular centers, and $C(22j;\mu,-\mu)$ is a Clebsch-Gordon coefficient. The coefficients $\alpha_o = \sqrt{5}$, $\alpha_2 = \sqrt{\frac{1}{2}}$ and $\alpha_4 = \sqrt{70}$; coefficients ε_o, ε_2 and ε_4 arise in part from valence forces, anisotropic van der Waals forces, and electric quadrupole-quadrupole (EQQ) interactions. In 1955 Nakamura[8] treated V_{anis} theoretically and showed that the EQQ interaction corresponding to j = 4 in Eq.(19) was the dominant term in the zero pressure solid, a result which has been subsequently confirmed by experiment and will be discussed later. To a good approximation, due to the symmetry and the large nearest neighbor distance, the anisotropic interaction in the zero pressure solid can be taken to be pure EQQ which is often reexpressed in the form

$$V_{EQQ}(R) = 4\pi \sum_{\mu} \frac{5}{6} \sqrt{70}\, C(224;\mu,-\mu)\, \Gamma_{12} Y_{2\mu}(\Omega_1)\, Y_{2-\mu}(\Omega_2) \quad (20)$$

where $\Gamma_{12} = (6/25)e^2Q^2/R_{12}^5$ is the EQQ coupling parameter and Q is the EQ moment. In the zero pressure rigid lattice $\Gamma = \Gamma_o = 1.101$ K for H_2 and 1.21 K for D_2. The quadrupolar interaction energy of a pair of classical molecules is lowest when the molecular axes are perpendicular to each other in a "T" configuration and highest when parallel in an "H" configuration.

The effect of the rotational state on the ground state translational properties is the following. The isotropic pair potential, V_I is essentially the same for J = 0 and J = 1 molecules. For the translational properties we average the ground state over the populated rotational states, $\langle\ \rangle_J$. For para $\langle V\rangle_J = \langle V_I\rangle_J + \langle V_A\rangle_J = V_I + 0$; whereas for ortho $\langle V\rangle_J = V_I + \langle V_A\rangle_J$. If all three $|JM\rangle = |1M\rangle$ states are equally populated $\langle V_A\rangle_J = 0$; if not the main effect is to slightly change the equilibrium volume of the solid since V_A is attractive. As a result we can consider both species as interacting via the same isotropic potential.

RENORMALIZED INTERACTIONS

We have already seen that the orientational motions have a minor effect on the translational motions and isotropic interactions. The effective isotropic interactions are however strongly effected by the ZPM as shown by Eq.(10) in which the force constants are determined by averaging over the PCF. The ZPM also renormalizes the anisotropic interactions. For pair interactions, such renormalizations also correspond to averages over the PCF

$$\langle 0|V_A(1,2)|0\rangle = \int g(1,2)\, V_A(1,2) dr_1 dr_2 \quad (21)$$

These averages are performed by expressing V_A in a crystal frame (see Eq.31) rather than in a frame with respect to the intermolecular axis as in Eq.(19). Harris[7] has shown that the averaging of V_{EQQ} does not effect the angular functional dependence of the interaction and the only effect is to modify the coupling parameter Γ such that $\Gamma_{eff} = \xi\Gamma_o$ where Γ_o is the rigid lattice value $6e^2Q^2/25R_o^5$. A PCF such as shown in Fig.7 will evidently put a higher weight on large R values for V_{EQQ} so that $\xi < 1$. For H_2 $\xi \simeq .90$ which is significant in terms of the accuracy of measurements and theory. The averaging will not, in general preserve the functional dependence of functional expressions.

CRYSTAL STRUCTURES AND PHASES

Knowledge of the phase diagram and crystal structures is vital for optical studies. In Fig.9 we show this for ^{3}He and ^{4}He. Because of the large zero-point energy they are both liquid at T = 0 K and zero pressure; ^{3}He solidifies under a pressure of 25 Atm, ^{4}He, 30 Atm. The equilibrium structure depends on a very delicate balance of energy; the difference in the static lattice energies for hexagonal close packed (hcp) and face centered cubic (fcc) is of the order 10^{-3} K. Kinetic energy plays an important role in the structural determination. Both heliums have a body centered cubic (bcc) structure at low pressure. The structure has 8 nearest neighbors and is less closely packed than fcc and hcp. Although this reduces the attractive potential, evidently the kinetic energy is more strongly affected since there is more atomic volume available to the atoms. Both isotopes have hcp and fcc phases accessible either by varying the density (pressure) and/or temperature. In Fig.10 we show the phase diagram of H_2 and D_2. H_2 solidifies at 13.96 K and zero pressure into an hcp lattice and remains so down to T = 0 K. Ortho-H_2 also solidifies in the hcp phase but has an order disorder phase transition at 2.8K in which the orbitals of the J = 1 molecules become orientationally ordered along certain crystalline directions. This transition lowers the free energy due to the EQQ anisotropic interactions. The structure is the Pa3 space group shown in Fig.11. The molecular centers are on the sites of an fcc lattice whereas the p-like orbitals orient along the four body-diagonals forming the four interpretating simple cubic lattices shown in the figure. The phase diagrams of p-D_2 and o-D_2 are essentially identical except for scaling of T and P. The phase diagrams have not been extensively studied at higher pressures and mixed ortho para concentrations.

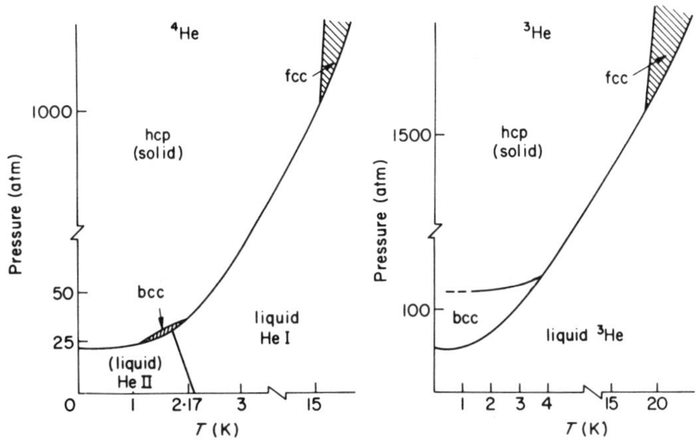

Fig.9. Phase diagrams for ^{3}He and ^{4}He.

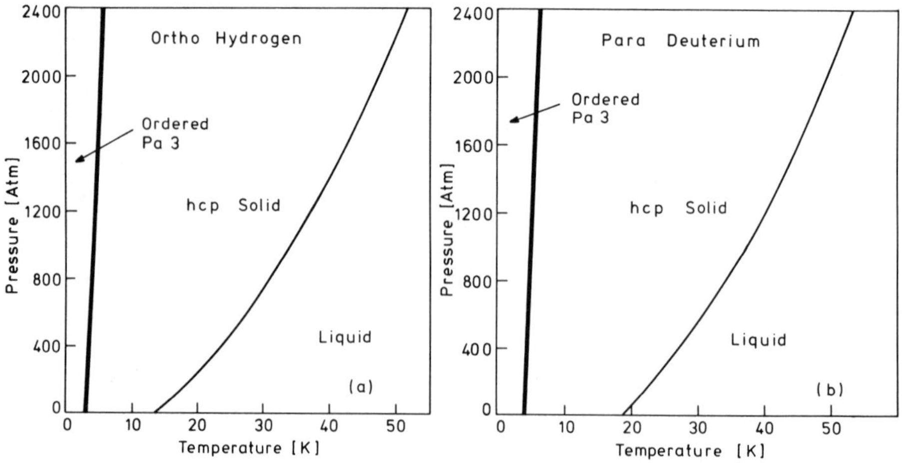

Fig.10. The phase diagrams for ortho-H_2 and para-D_2. Hysteresis between the Pa3 and hcp phase is indicated by the heavy line. Para-H_2 and ortho-D_2 have no Pa3 phase in this pressure region, remaining hcp to T = 0 K.

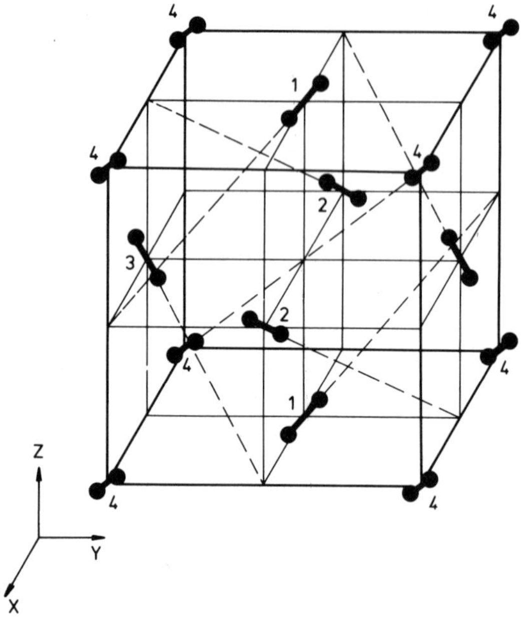

Fig.11. The Pa3 space group. For H_2 the molecular axes are oriented along the body diagonals.

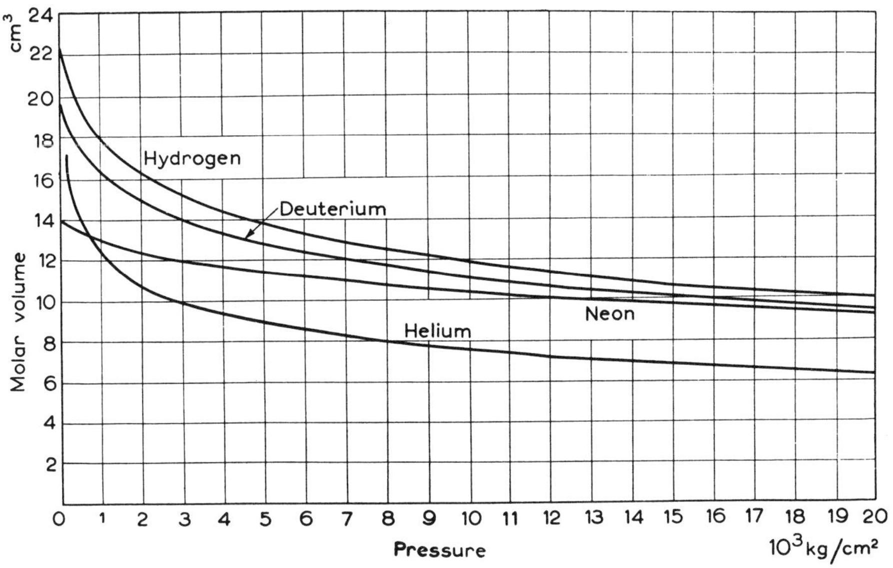

Fig.12.The isotherms of H_2, D_2, ^4He and Neon at 4^oK (after ref.9).

The various phases are easily accessible as the necessary pressures are readily generated in the laboratory. The density can also be easily varied by a factor of two as seen from the P-V curves measured by Stewart[9] shown in Fig.12.

INTERACTION WITH LIGHT

The interaction of light with the solids which provides the optical response is adequately described in the electric dipole approximation. There are two important optical techniques which have been used for studying these solids :infrared spectroscopy and Raman spectroscopy. Because these are weakly absorbing materials, transmission techniques can be used in the infrared . Measurements are made of

$$T = \frac{I}{I_o} = e^{-a(\omega)d} \qquad (22)$$

where I_o is the light flux incident on a sample with absorption coefficient a and I is the flux remaining after passing a distance d into the sample. The absorption coefficient at T = 0 K found from Fermi's golden rule is

$$a_{if}(\omega) = \frac{4\pi^2\omega\rho_i}{\hbar cnV}|<f|\underline{P}.\underline{\lambda}\ |i>|^2\ [\delta(E_{fi}-\hbar\omega) + \delta(E_{fi}+\hbar\omega)] \qquad (23)$$

where $\underline{P}$ is the total dipole operator, $\sum_i e\underline{r}_i$, $<f|$ and $<i|$ are the final and initial states $\underline{\lambda}$ is the polarization vector of the electric field, ρ_i is the probability that state $|i>$ is populated, E_{fi} is the energy of the transition, n is the index of refraction which we shall take as 1, and V is the volume. To obtain the total absorption for polarization $\underline{\lambda}$ at a temperature T we sum over all final and initial states. This can be manipulated to the following form[10]

$$a(\omega) = \frac{4\pi^2\omega}{\hbar cnV}(1-e^{-\beta\hbar\omega}) \int dt\ e^{i\omega t} <\underline{\lambda}.\underline{P}(t)\underline{\lambda}.\underline{P}(o)>_o \qquad (24)$$

where $\beta = 1/KT$. The bracketed quantity under the integral is the dipole-dipole correlation function. The brackets mean the thermal average of the included quantity i.e. $\sum_i \rho_i <i|\ |i>$ .The fourier transform of the correlation function, including the thermal factor, is the spectral function which is a measure of the absorption of radiation as a function of frequency.

Raman scattering is the process in which an incident photon interacts with a substance inducing a transition from state $|i>$ to $|f>$ with a shift in frequency of the photon corresponding to the energy $\hbar\omega_{fi}$. The scattering efficiency from state $|i>$ to $|f>$ is given by

$$S_{\gamma\beta}(if) = \left(\frac{4\omega\omega'^3}{c^4}\right)\frac{\rho_i}{V}|<f|\alpha_{\gamma\beta}|i>|^2\ \delta(\omega_s-\omega_{if}) \qquad (25)$$

in the polarizability approximation. Here $\alpha_{\gamma\beta}$ is the $\gamma\beta$ Cartesian component of the total polarizability of the system and $\omega_s = \omega-\omega'$ is the frequency shift of the incident and scattered light. S is the number of photons scattered per incident photon per unit solid angle, per unit frequecy shift per unit sample length. It can also be written as the integral of a correlation function.

OPTICAL RESPONSE

In this section we consider the optical response of helium and hydrogen. We shall concentrate on the distinguishing aspects of these quantum solids, in particular the relationship between the large amplitude ZPM and the single and multi quasi-particle excitations. Although optical techniques have certain limitations, in particular conservation of momentum confines excitations to total transfer of wave vector $\underline{K} \simeq 0$, which is not the case for inelastic neutron scattering, there are a number of advantages. Optical measurements can be made rapidly relative to neutron measurements. In cases such as measurements on "pure" o-H_2 where the sample is destroying itself at ∿ 2%/hr due to conversion to p-H_2, most neutron studies would be excluded, as is also the case for ^{3}He due to its large neutron-capture cross section. Furthermore optical studies of collective excitations are not limited by "incoherent scattering". Finally optical techniques enable higher precision in the determination of excitation energies.

Experimental Techniques

In Raman scattering experiments one measures the intensity of the light as a function of the shift from the laser excitation frequency which is a measure of the quasi-particle frequency. The experimental system used by Slusher and Surko[11] for measurements on solid helium is shown in Fig.13. Crystals were grown in the pressure cell C which could be used up to 150 Atm. The growth technique was to hold the pressure constant as the temperature was lowered. Presumably, single crystals can be grown in this technique. The light from an Ar^+ laser is focussed in the sample and right angle scattered light is collected and analyzed with a double spectrometer. The light was detected with a photo multiplier and counter. A computer was used for control and to store intensity and frequency information. Automatization is generally

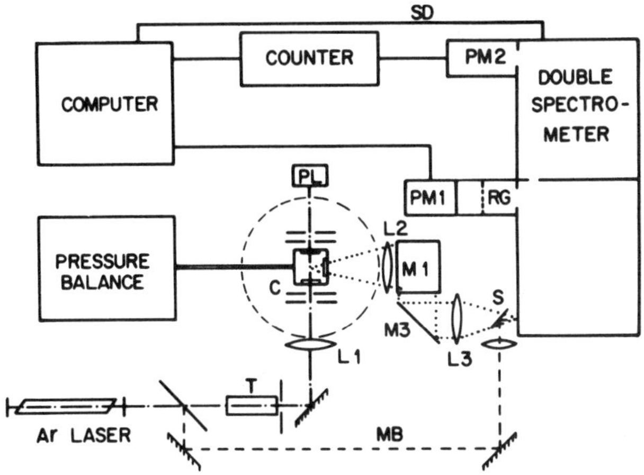

Fig.13. An experimental system for laser Raman scattering: PL is a laser power monitor, PM are photo multipliers, SD is a coupling to the spectrometer drive. Frequency calibration was done with a Ronchi grating (RG) (after ref.11).

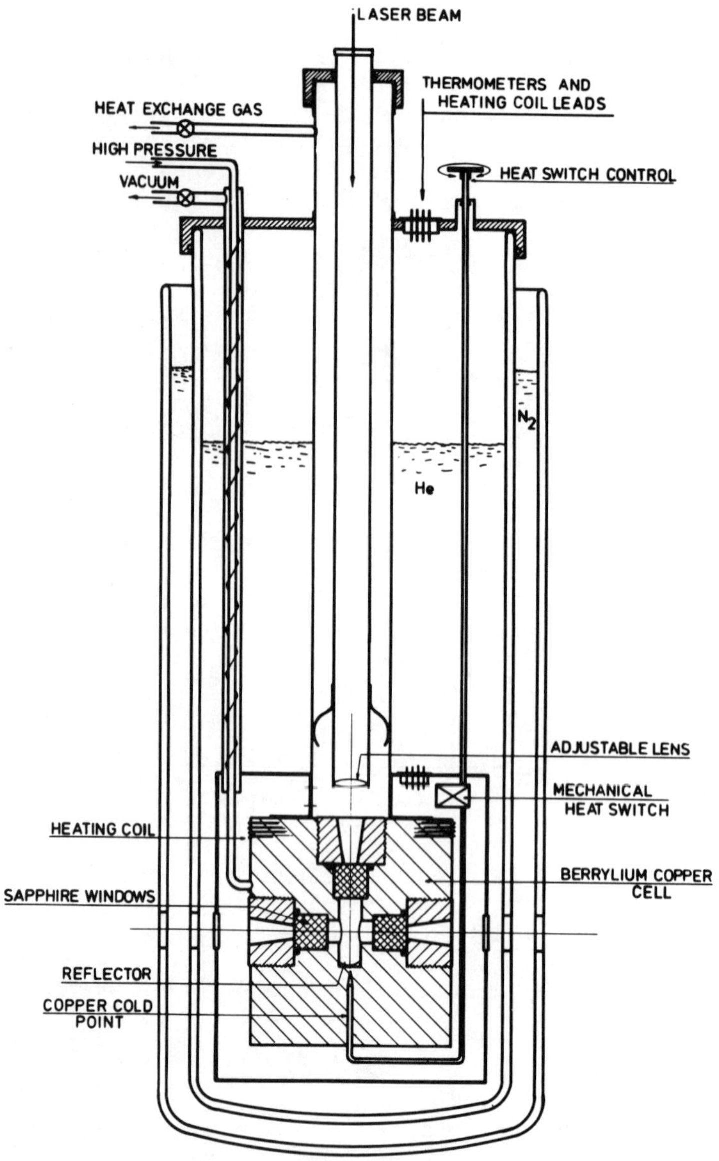

Fig.14. A high pressure cell and cryogenic system for Raman scattering in condensable gases. This system can be used for pressures of several thousand atmospheres and in the temperature range 1 to 300 K.

not of great advantage in Raman scattering, but in this case it was used for signal averaging necessary because of the weak scattering crossection of He.

A high pressure cell that we have used is shown in Fig.14. The berylium-copper cell with three sapphire windows has been tested and used to $\sim$6000 Atm. Crystals are grown either at constant temperature (with a gradient) by raising the pressure or at constant pressure with a lowering of temperature. Samples are sealed in by freezing the feed capillary. This has the advantage of enabling isochoric measurements. The laser beam is reflected back out of the cell by a spherical mirror at the bottom to prevent heating due to dissipation of the $\sim$1 watt of power. An on-line frequency calibration system (not shown) enables measurement accuracy of $\simeq 0.05\ cm^{-1}$. The frequency shift range for studies by these techniques in quantum solids is $\sim 2\ cm^{-1}$–$10^3\ cm^{-1}$.

Far infrared techniques have only been used for studies in the hydrogens as helium has no known (first order) absorption spectrum. Far infrared spectroscopy is characterized by weak broad band radiation sources (generally high pressure mercury discharge lamps). The low power is compensated for by use of sensitive bolometer detectors and in the last decade the displacement of the grating spectrometer by Fourier interference spectroscopy. The latter makes better use of the weak source because of its high light gathering capability and the multiplex advantage. A computer controlled system that we have used is shown in Fig.15. The computer is vital in this technique as the measured interferogram must be Fourier transformed to give the power spectrum. The technique is described in detail in a number of places[12]. The low frequency limit is $\simeq 5\ cm^{-1}$.

A cryostat for studying the FIR absorption spectrum of condensable gases at zero-pressure is shown in Fig.16. The sample is condensed in a sealed tube of brass and slowly cooled to crystallize. The sample length is determined by the quantity of gas condensed. Radiation from the spectrometer passes through the sample and is detected by a cooled doped silicon bolometer. We have also built and operated a high pressure ($\sim$6 KBar) berylium copper cell which uses the same general principles as in Fig.16.

The Use of Symmetry

In general, before embarking on a somewhat difficult development of experimental apparatus and techniques involving low temperature, pressure, and the ability to grow (single) crystals in situo which are destroyed upon warming up, it is useful to be able to anticipate the observation of first order (single quasi-particle) spectra. This requires a determination of selection rules and an intensity estimate. The latter will be discussed in the following sections. Selection rules for collective excitations can be determined most readily by use of group theoretical techni-

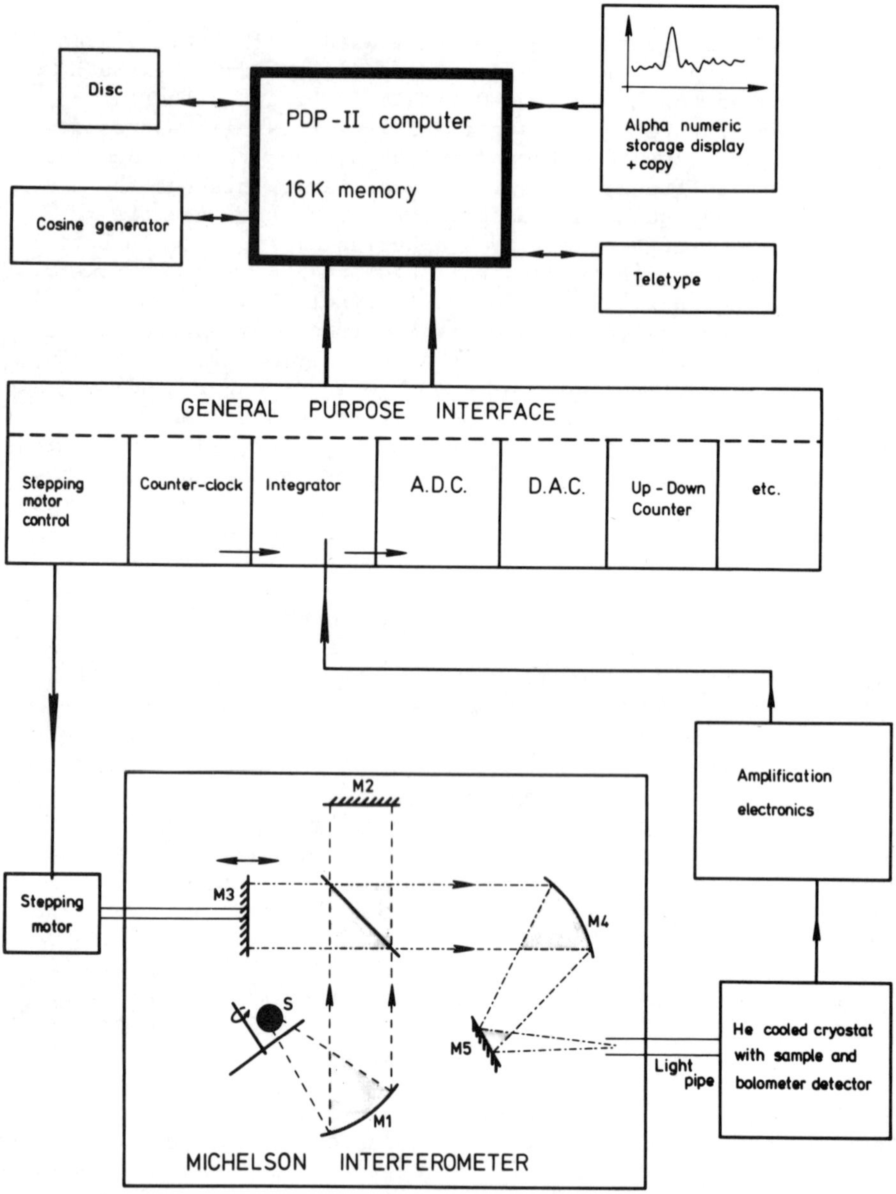

Fig.15. A computer controlled fourier interference spectroscopy system for far infrared studies.

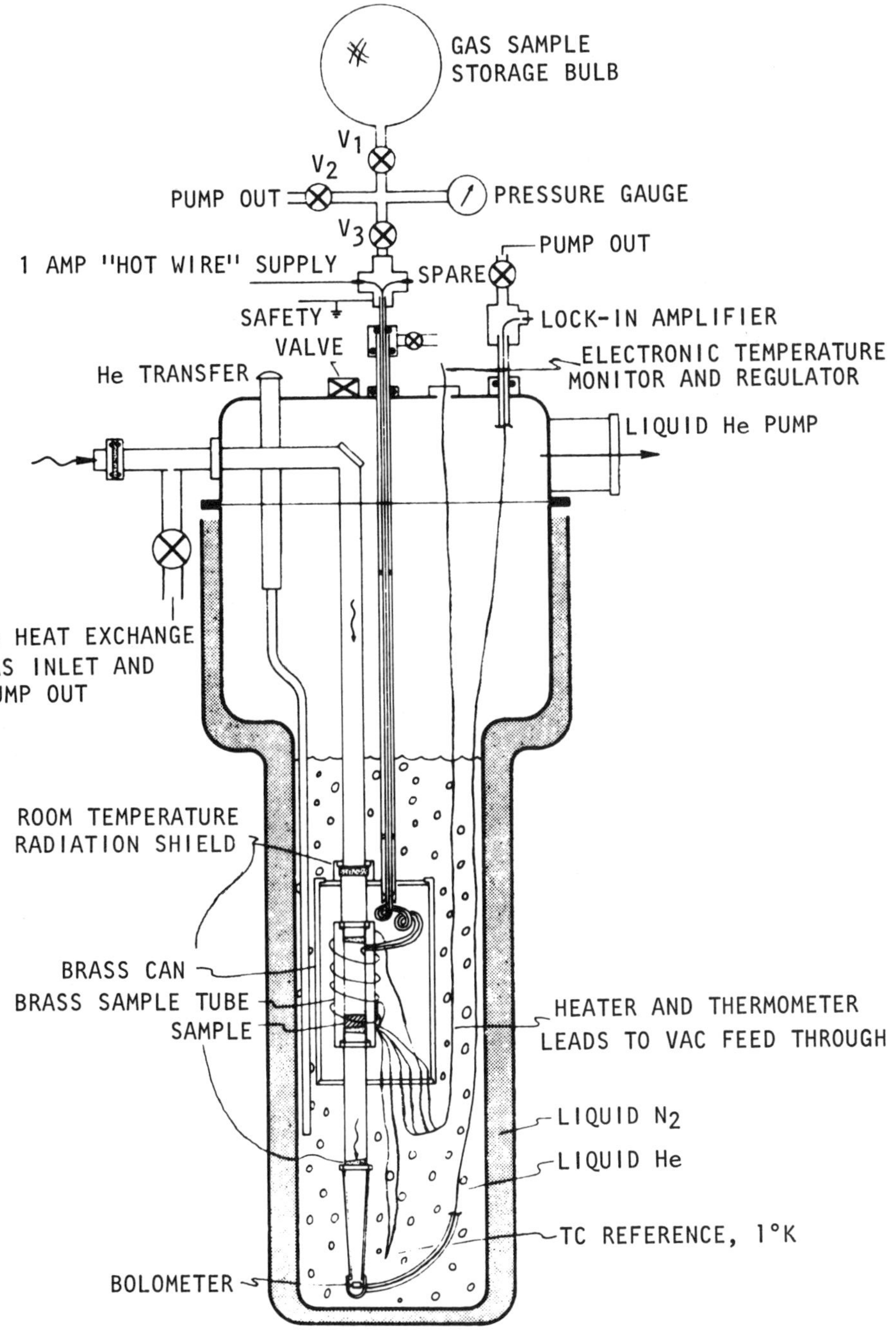

Fig.16. A cryostat for far infrared studies on condensable gases (after ref.13).

ques, if the crystal space group is known. Intuitive techniques can also be used if the normal mode structure is known. Basically, to be infrared active there must be an oscillating total electric dipole moment associated with the mode (Eq.23) and to be Raman active the total polarizability must have a variation at the frequency of the mode (Eq.25).

For the various structures of helium and hydrogen (fcc, hcp, bcc, and Pa3), a group theoretical analyses shows that the only first order allowed infrared active modes are the phonons in the orientationally ordered Pa3 structure. For Raman scattering, single phonon transitions are allowed only in the hcp lattice. In addition the collective reorientational excitations (librons) of the Pa3 structure are allowed Raman processes.

Raman Spectra: Phonons

Phonon spectra have been observed in the hcp phases of the hydrogens and heliums by Raman scattering. The first measurements were made by Silvera, Hardy and McTague[14] on H_2 and its isotopes. A spectrum for HD at zero pressure is presented in Fig.17 showing a sharp peak at $\sim$36 cm^{-1} and a broad higher frequency spectrum. First order spectra are only allowed for zone center optical modes (ZCOM) since the light wave vector $\underline{K} \simeq 0$. The hcp structure has a longitudinal and transverse ZCOM. Only the latter is active. This mode corresponds to an opposed motion, transverse to the c axis, of the two atoms in the hcp unit cell. The sharp feature in the spectrum was identified as the transverse mode based on its sharpness, polarization properties,

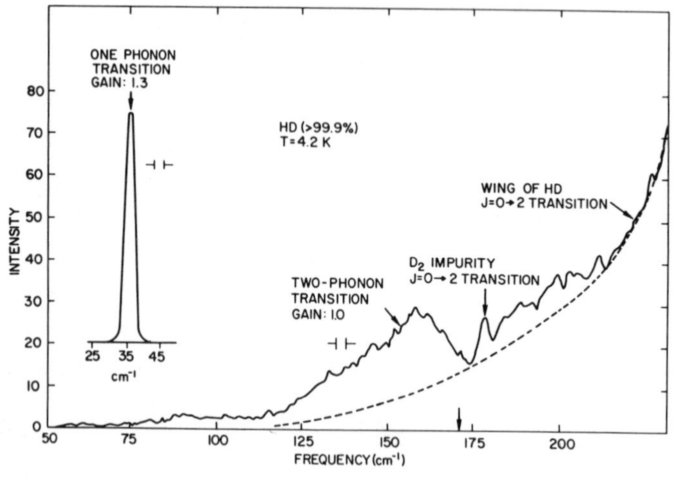

Fig.17. Raman spectrum of solid hcp HD. The sharp peak in the inset is an optical phonon. The weak 2-phonon spectrum is shown on the rising wing of the higher frequency J = 0 → 2 rotational transition (after ref.14).

comparison with neutron measurements for the hydrogens, and predictions from self consistent theory. The frequencies of the sharp peak ranged between 35.5 to 38.6 cm^{-1} for D_2 and H_2 which suggests that hardly any isotope effect exists (for a harmonic solid the frequencies should scale as the square route of the masses, 1.41, for H_2-D_2). However this is explained by the fact that all samples were measured at zero pressure for which H_2 and D_2 have different molar volumes as the expansion due to zero-point kinetic energy differs. As a consequence the renormalized forces of H_2 are softer than those of D_2 compensating in part for the mass difference. Differences of the frequencies for disordered ortho-para species of ~ 2 cm^{-1} have also been measured. This surprising result is explained by an indirect effect. The attractive EQQ moments of the $J = 1$ species have short-range order at low temperature which causes a lattice contraction and thus a change of the renormalized force constants.

The broad feature in Fig.(17) was attributed to a two-phonon band. The frequency of this band should extend from zero to twice the maximum zone boundary frequency indicated by the arrow on the frequency shift axis. This two-phonon band has been further studied by Fleury and McTague[15] in H_2 who demonstrate reasonable agreement with a density of states scaled from that of helium.

The light scattering intensity mechanism requires a variation of the crystal polarizability with the relative displacement of the atoms. The main contribution to the crystal polarizability, α_c, is given by the independent polarizability approximation as a sum of the polarizabilities of the atoms

$$\alpha_c = \sum_i \alpha_o(i) = N\alpha_o \tag{27}$$

where α_o is the polarizability of an isolated atom. This expression does not depend on phonon coordinates; two mechanisms provide deviations from Eq.(27) required for Raman scattering intensities. At long ranges the leading term arises from an induced dipole mechanism; at shorter ranges, already in the well region of the potential, overlap of charge clouds will modify the polarizability. Let us consider the first mechanism which has been handled in detail by Werthamer[16]. The polarization of atom is given by

$$P(r_i, t) = \alpha_o \underline{E}_o \cdot \left[\hat{\lambda}\hat{\lambda} + \alpha_o \sum_{j\neq i} (3\hat{r}_{ij}\hat{r}_{ij} - \hat{\lambda}\hat{\lambda}) / |\underline{r}_{ij}|^3 \right] \tag{28}$$

where $\underline{E}_o = |\underline{E}_o|\hat{\lambda}$ is the applied electric field, $\hat{\lambda}$ is a unit vector in the direction of E_o, and $\underline{r}_{ij}$ is the vector between atoms i and j. The first term in the brackets is the dipole moment on i induced by $\underline{E}_o$. $\underline{E}_o$ also induces dipoles on all neighbors j whose dipolar electric fields act on i to induce an additional moment, giving rise to the second term which depends on the separation of atoms. The coefficient of $\underline{E}_o$ in Eq.(28) is the polarizability of atom i. This can be suitably used in Eq.(25) to calculate the intensity. The overlap mechanism has been shown to reduce this correction to the polarizability[17]. Note that the intensity will be proportional to α_o^4. An estimate of the scattering efficiency from Eq.(25) yields

$$S \simeq \frac{N}{V} \left(\frac{\omega}{c}\right)^4 \left(\frac{\langle u^2 \rangle}{R_o^2}\right) \frac{\alpha_o^4}{R_o^6} \qquad \text{(29a)}$$

where N/V is the number density and $\left[\langle u^2 \rangle / R_o^2\right]^{1/2}$ is the localization discussed in the first lecture. The measured intensities in H_2 were found to agree quite well with this theory. Predictions could be made of the intensity expected for helium for which α_o is about 0.25 that of H_2. With all factors, the intensity for He should be down by $\sim 10^2$.

The Raman spectra of ^{3}He , ^{4}He and mixtures have been studied by Slusher and Surko[11] as a function of density (pressure) in the range ≃ 17–21 cc/mole. (The 25 atm. density of ^{4}He is ∿ 21 cc/mole). A typical spectrum for an hcp ^{3}He–^{4}He mixture is shown in Fig.18 Because of the weak Raman crossection of helium these measurements required about 1 minute measuring time for each point. The sharp peak was identified as the ZCOM. In these experiments, the isotope ratios could be measured at the same density yielding values between 1.17 and 1.18 which should be compared to $\sqrt{4/3} = 1.155$ for a harmonic solid. Evidently the deviations arise from the difference in ZPM and short range correlations in ^{3}He and ^{4}He.

The most profound difference from the H_2 spectra is the extent and strength of the broad feature which is attributed to multi-

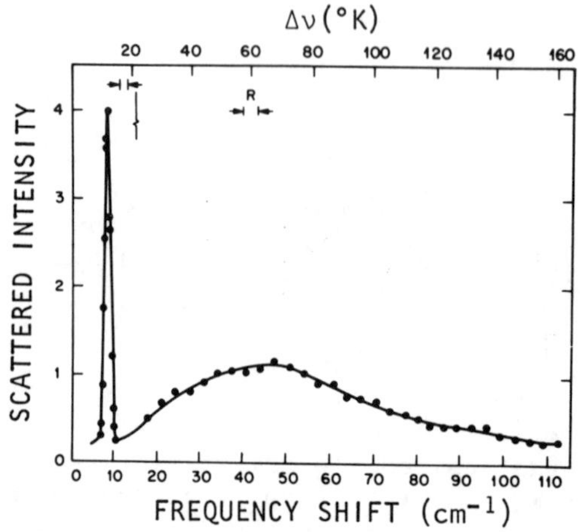

Fig.18. The Raman spectrum of a solid helium mixture ($^4He_{0.8}{}^3He_{0.2}$) at 1.4 K and 48 Atm. (after ref.11).

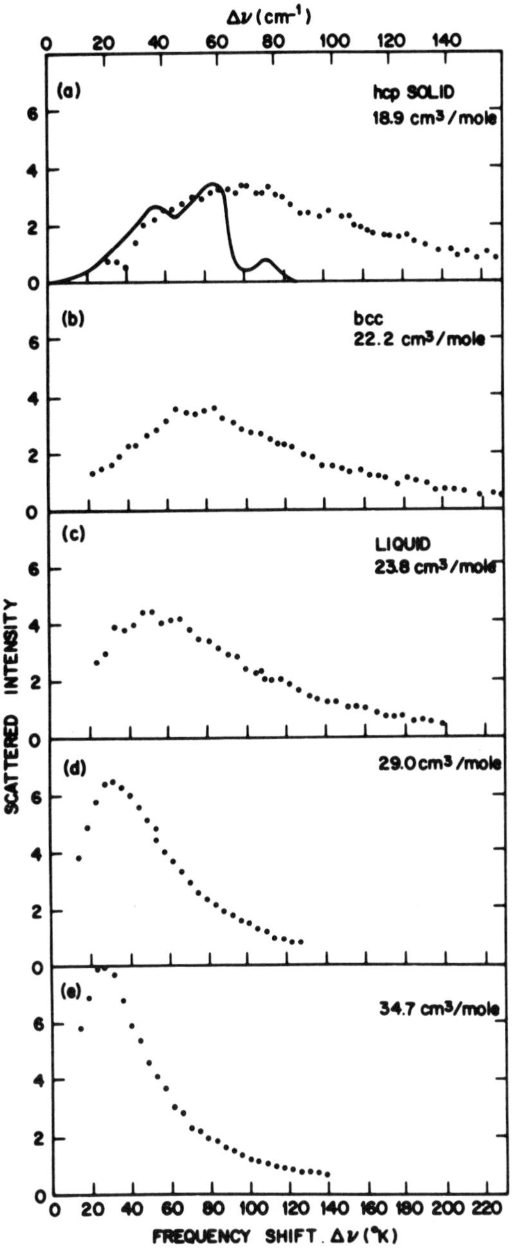

Fig.19. Raman spectra in liquid and solid ^{3}He. The ZCOM in the hcp phase has not been drawn in the spectrum (after ref.11).

phonon processes. In Fig.19 we show this feature along with the joint density of states function shown by the solid line. Surko and Slusher argue convincingly that the residual intensity arises from a process in which a photon creates a pair of nearly-free-particle excitations with a dispersion $\varepsilon_q = \hbar^2 q^2/2m^*$ where m* is an effective mass. Such "free-particle" excitations have previously been observed by neutron scattering. This process has a much higher probability in He than in H_2 or other rare gas solids because the He atom is so weakly bound to its lattice site (binding energy $\simeq$ 40 K). They use the anisotropic polarizability (corrected for overlap) found from Eq.(28) to calculate the Raman intensities. A quantum solid wave function is used for the initial state and plane waves for the final state. Excellent agreement is found in a fit to the data. We suggest here an alternate approach of analyses that explicitly shows the dependence on ZPM. A sum rule for the total integrated scattering in the Raman spectrum is found from Eq.(25):

$$\int S_{\gamma\beta} d\omega = \left(\frac{4\omega\omega'^3}{c^4}\right) \frac{\rho_o}{\hbar V} < 0 \,|\, \alpha^*_{\gamma\beta}\, \alpha_{\gamma\beta} \,|\, 0 > \qquad (29b)$$

where we have taken the initial state $|i>$ as the ground state $|0>$, summed over all final states, and used $\sum_f |f><f| = \underline{\underline{1}}$. The brackets in Eq.(29b) imply averaging of $\underline{\underline{\alpha}}^* \underline{\underline{\alpha}}$ (see Eq.21) over the ground state particle distributions. Thus the integrated intensity will depend on the width and cut-off of the PCF. We have not carried out these averages, but a similar approach is used in analyzing the IR spectrum of H_2; quantitative results will be discussed in a later section. If we expand $\alpha_{\gamma\beta}$ of Eq.(29b) in a phonon expansion i.e. $\alpha_2 = \alpha_o + (\nabla\alpha)_o \cdot \underline{u} + \cdot$, we see that the first order spectrum goes as $<u^2>$, and multiphonons as $<u^4>$, etc. In the limit that the PCF becomes very narrow, the multi quasi-particle scattering intensity will be reduced in comparison to first order spectra.

Raman Spectra: Librons

The ground state of the pure J = 1 H_2 and D_2 has the four sublattice structure of Fig.11 with the axis of the molecular orbitals aligned along the <111> directions. The low-lying excited reorientational states are collective normal modes of the solid. For a classical solid[18] one would expect these modes to correspond to small amplitude orientational oscillations about the <111> directions, with the restoring forces arising from the torque constant τ exerted by the neighbors, and frequencies being proportional to $\sqrt{\tau/I}$ where I is the moment of inertia. In H_2 the reorientations can take place within the J = 1 manifold of states (M_J = 0, ±1) so the BJ^2 term is constant in the Hamiltonian and the moment of inertia does not enter into the excitation frequencies. The collective librational modes are called librons.

The ground state of the Pa3 structure has all molecules in the $M_J = 0$ state where the axes of quantization are the four <111> ordering directions. There are two possible states per molecule (±1). The modes correspond to coherent excitations of the $M_J = \pm 1$ states on all molecules with a well defined wave vector $\underline{k}$ (as a result of the translational invariance). One expects 8 modes at each point in $\underline{k}$ space, however symmetry considerations show that at $\underline{k} = 0$ there are one two-fold (Eg) and two three-fold degenerate (Tg) modes, all of which are Raman active. The Raman spectrum was observed by Hardy, Silvera, and McTague[19] (HSM) and is shown in Fig.20. These measurements were made in large single crystals of D_2. When first observed in earlier work, the spectra were rather surprising and unexpected. Calculations[20] based on the predicted Pa3 structure, using the well known EQQ interaction, indicated that three peaks should be observed at about twice the frequency of the sharp peaks in the figure. Coll and Harris[21] and Berlinsky and Harris[22] (HBC) were able to show that the large shift to lower frequencies arose from a combination of strong anharmonic anisotropic interactions, which were perturbatively treated, and renormalization of the EQQ interaction[7]. They could also show that large cubic anharmonic interactions were responsible for the broad 2-libron band identified in the figure. HSM provided definitive identification of the single-libron modes from polarization measurements in single crystals. We outline the calculations. The interaction Hamiltonian is

$$H = \frac{1}{2} \sum_{i,j} V_{EQQ}(i,j) \tag{30}$$

where the interaction of Eq.(20) is expressed in the crystal coordinate system

$$V_{EQQ}(i,j) = \frac{20\pi}{9} (70\pi)^{\frac{1}{2}} \Gamma_{ij} \sum_{M,N} C(224;\, M,N) \times Y_2^M(\Omega_i)\, Y_2^N(\Omega_j)\, Y_4^{M+N*}(\Omega_{ij}) \tag{31}$$

Here Ω_{ij} defines the orientation of the vector $\underline{R}_{ij}$ between sites i and j with respect to a crystal coordinate system, Ω_i specifies the orientation of molecule i in the same coordinate system, and C(224; M,N) is a Clebsch-Gordan coefficient.

This is transformed further so that the orientation of each molecule (ω_i) is specified relative to orientations from its local equilibrium axis

$$Y_2^M(\Omega_i) = \sum_N D_{MN}^{(2)*}(\chi_i)\, Y_2^N(\omega_i) \tag{32}$$

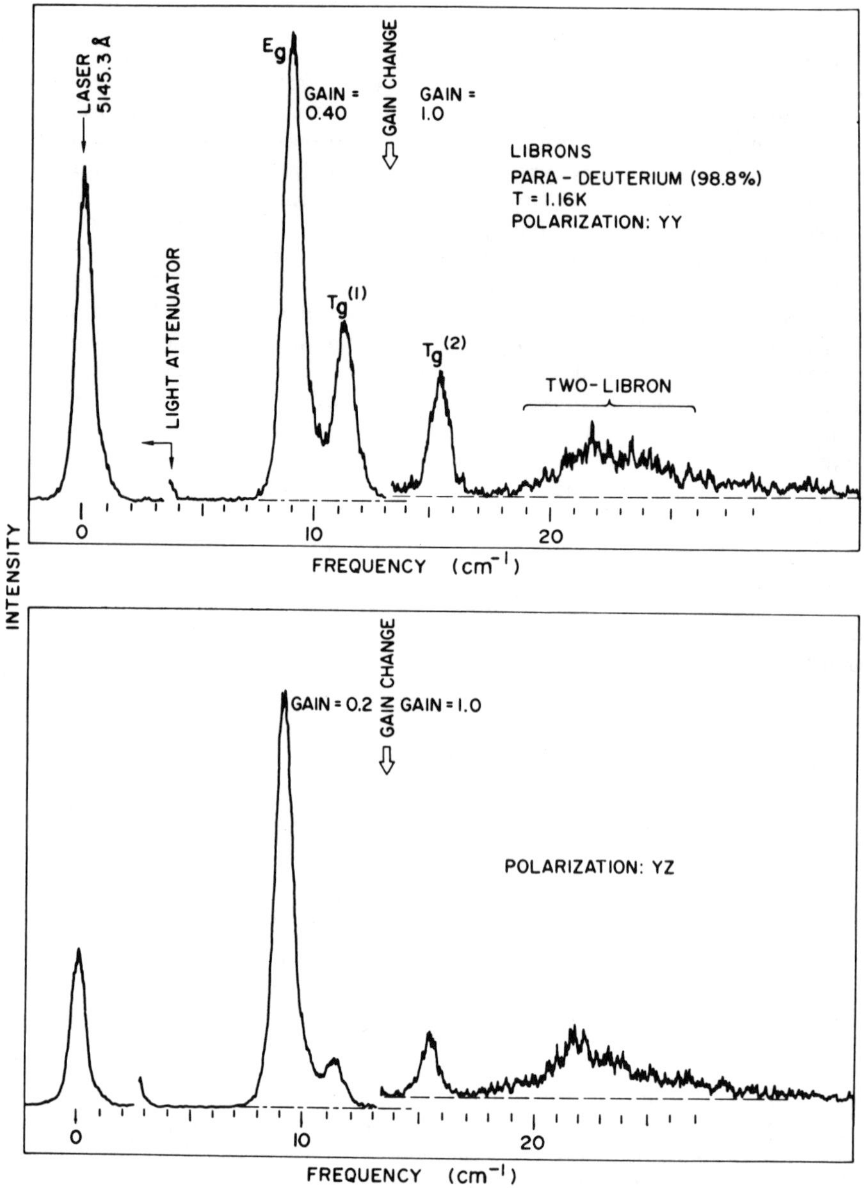

Fig.20. Raman spectra of 1 and 2-libron transitions in single crystals of D_2 for two polarizations. All structure disappears above the ordering temperature (after ref.19).

where $D^{(2)}_{MN}$ is a rotation matrix and χ_i are the Euler angles specifying the orientation of the local axes with respect to the crystal axes. In this set of coordinate systems in the ground state all molecules are in the $|JM\rangle = |10\rangle$ states. It is convenient to replace the spherical harmonics by operator equivalents

$$Y_2^M(\omega_i) = A_M\, O_i^M(J_i) \tag{33}$$

where $\underline{J}_i$ is the angular momentum operator. Eq.(30) can now be written in the form

$$V_{EQQ} = \frac{1}{2} \sum_{i,j} \sum_{M,N} \zeta_{ij}^{M,N}\, O_i^M\, O_j^N \tag{34}$$

The O_i^M can then be expressed in terms of boson operators c_{i+} and c_{i-} which create excited states $M = \pm 1$ on a site i when applied to the ground state. Further transformations can be carried out by Fourier transforming the c_i and details are given in ref.21. V_{EQQ} can be separated into a constant part, a part quadratic in c_{iM}, and higher order terms. Diagonalization of the quadratic part gives the modes and dispersion relations of Raich and Etters[20] and others. The important contribution of HBC[21,22] was to note that the higher order terms, in particular the cubic anharmonicity, are large and important. They were able to correct the frequencies, in close agreement with experiment, and to show that the cubic anharmonicity allowed for intense 2-libron scattering.

The intensities of the Raman libron transitions are calculated by evaluating Eq.(25) in which the initial state is the ground state with all molecules in the $|JM\rangle = |10\rangle$ state and the final state is a libron state. The polarizability of the crystal is the sum of the molecular polarizabilities given in the independent polarizability approximation, Eq.(27). In the case of phonons Eq.(27) was inadequate for explaining the scattering mechanism because it did not depend on phonon coordinates. However for the crystal polarizability of molecular hydrogen we have

$$\alpha_\mu^{(2)} = \sum_i \alpha_\mu^{(2)}(\Omega_i) \tag{35}$$

where $\alpha_\mu^{(2)}$ is a second rank spherical tensor with

$$\alpha_\mu^{(2)}(\Omega_i) = \left(\frac{8\pi}{15}\right)^{\frac{1}{2}} (\alpha_\parallel - \alpha_\perp)\, Y_{2\mu}(\Omega_i) \tag{36}$$

Here $(\alpha_\parallel - \alpha_\perp)$ is the anisotropy in the molecular polarizability. $\alpha_\mu^{(2)}$ can easily be reexpressed in terms of Cartesian coordinates; of the XYZ crystal coordinate system[23,19]. Since $\alpha_u^{(2)}$ depends directly on the angular coordinates, this will be the most important part of the crystal polarizability for rotational Raman scattering[24]. By evaluating Eq.(27) in this manner for various polarizations of incident and scattered light, as shown in Fig.20, HSM[19] were able to identify the 1-libron modes, and indeed this was the first strong experimental confirmation of the Pa3 space group as the ground state of ordered H_2.

FAR INFRARED SPECTRA: PHONONS AND LIBRONS

Far IR phonons are first order active only in the orientationally ordered state of H_2 or D_2. These spectra were first

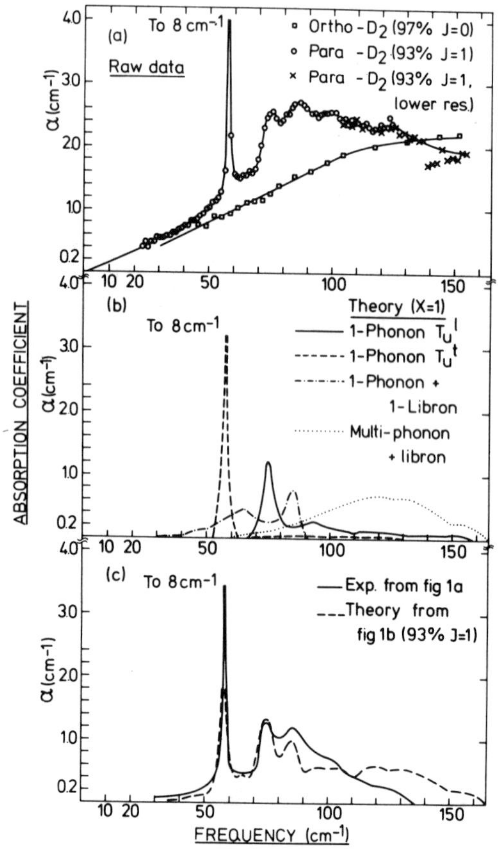

Fig.21. Experimental and theoretical far infrared spectra of solid D_2. See text for explanation (after ref.26).

observed by Hardy, Silvera, Klump and Schnepp[25] (HSKS) and have been remeasured and further analyzed more recently by Jochemsen, Berlinsky, Goldman and Silvera[26] (JBGS). The spectrum for D_2 from the latter work is shown in Fig.21; similar results are found for H_2. The Pa3 space group has four molecules per unit cell, yielding several optical phonon modes labeled T_u^l, T_u^t, E_u and A_u. Only the transverse, T_u^t, and the longitudinal, T_u^l, are infrared active. For a harmonic lattice one would expect two sharp peaks. In Fig.21 we see an intense broad structured spectrum with only one sharp peak. The p-D_2 spectrum appears to have a large unstructured background. HSKS attempted to subtract this off by warming the sample into the hcp disordered phase which has no first order activity and should provide an adequate background. Although this background is present in the disordered phase, unfortunately a second order IR absorption spectrum also occurs preventing a proper background determination. JBGS solved this problem by using an equivalent length sample of pure J = 0 ortho-D_2 which should not have this second order spectrum as the J = 0 molecules have no EQ moment which is required for the intensity mechanism. One sees (square data points) that there remains a large unexplained absorption which can now, nevertheless, be subtracted off to give the spectrum in Fig.21 (solid line). This should be representative of the dynamical processes in p-D_2. The spectrum has been sorted out into several components shown in Fig.21. First the T_u^t and T_u^l frequencies and spectral functions (see Eq.24) were calculated within the self-consistent phonon approximation with anharmonic corrections. The T_u^t phonon has a sharp peak (in the figure it is broadened by the instrumental width) with about 15% of its spectral weight spread out to higher frequencies. The T_u^l phonon is severely broadened by the anharmonicity. A 1-phonon + 1-libron band adds further structure. Finally a multi-phonon + 1-libron band adds a broad unstructured component. When all of these calculated components are added up they yield the dashed curve in Fig.21, in reasonable agreement with experiment.

We now show how all of these terms arise by developing a formalism that incorporates the translational and orientational motions of the solid. JBGS[26] generalize the approach of Schnepp[27] who first calculated the infrared absorption intensity for α-N_2 which also has the Pa3 structure. Essentially we must develop an expression for the dipole moment of Eq.(23). The dipole moment arises from the electric field due to an EQ moment on a J = 1 molecule which induces a dipole moment on neighboring molecules; this is summed to give the total crystal moment. The dipole moment is zero in the ground state since molecules sit at sites of inversion symmetry, however translational or angular displacements break this symmetry. Calculations are simplified if we work with spherical tensors. The potential at a site j due to a J = 1 molecule at i is

$$\phi_{(ij)} = \frac{4\pi eQ}{5} \sum_{\nu} Y_2^{\nu}(\Omega_i)\, Y_2^{\nu^*}(\Omega_{ij})/R_{ij}^3 \qquad (37)$$

The quantities Ω_i, Ω_{ij} and $\underline{R}_{ij}$ are the same as defined after Eq.(31). The $m^{\underline{th}}$ component of the electric field is found from the gradient of ϕ:[28]

$$E^m(j) = 4\pi eQ\sqrt{3/7} \sum_{\nu} (-1)^{\nu}\; C(123;m\nu)\, \frac{Y_2^{-\nu}(\Omega_i)Y_3^{m+\nu}(\Omega_{ij})}{R_{ij}^4} \qquad (38)$$

where $C(123;m\nu)$ is a Clebsch-Gordan coefficient. This will induce a dipole at site j by interacting with the polarizability which has an isotropic part

$$\bar{\alpha} = (\alpha_{\parallel} + 2\alpha_{\perp})/3 \qquad (39)$$

and an anisotropic part depending on its orientation

$$\alpha_2^m(j) = \sqrt{8\pi/15}\; 3\kappa\bar{\alpha}\; Y_2^m(\Omega_j) \qquad (40)$$

where the anisotropy $\kappa = (\alpha_{\parallel} - \alpha_{\perp})/3\bar{\alpha}$. The dipole moment has a scalar part due to Eq.(39) and a tensor part due to Eq.(40). To maintain simplicity we treat only the scalar part (complete details can be found in ref.29).

$$P^{\mu}(j) = 4\pi\bar{\alpha}eQ\sqrt{3/7} \sum_{i\neq j} \sum_{\nu} C(123;\mu\nu)\, \frac{Y_2^{\nu}(\Omega_i)Y_3^{\mu+\nu^*}(\Omega_{ij})}{R_{ij}^4} \qquad (41)$$

which when summed over all sites j gives the dipole moment of the crystal. We now want to evaluate the matrix element of Eq.(23) between the ground state and various phonon and libron excited states. To do this we expand P^m in phonon and libron creation annihilation operators. The phonon coordinates are brought out explicitly by the expansion for the $i^{\underline{th}}$ molecule

$$P_i^m = (P_i^m)_o + \Sigma u_{ij}(\nabla P_i^m)_o + \ldots \qquad (42)$$

The dilation operator $\underline{u}_{ij} = \underline{R}_{ij} - (\underline{R}_{ij})_o$ can then be expanded in phonon creation operators. The first term in Eq.(42) is a constant, the second will represent 1-phonon creation, the third 2-phonon etc. The $Y_2(\Omega_i)$ term in Eqs.(41) or (42), can be expanded in terms of libron creation operators c_{im}, giving a no-libron and a 1 and 2-libron term. The no-libron 1-phonon term provides the intensity of the T_u^t and T_u^l modes. Pure 1-libron absorption is not allowed by symmetry. The 1-phonon + 1 libron term is also calculated from Eq.(42); the line shape can be approximated by a phonon density of states shifted upward by the mean libron frequency ($\sim 12\ cm^{-1}$)

The higher order terms of Eq.(42) are difficult to evaluate. These can be determined by a sum rule which also explicitly shows the effect of the zero-point motion on the absorption processes. The total integrated absorption due to lattice creation processes at T = 0 is the integral of Eq.(23)

$$A = \int \alpha(\omega) d\omega = \frac{4\pi^2 \rho_o}{3\hbar^2 cnV} \sum_f |\langle f|\underline{P}|00\rangle|^2 (E_f - E_o) \qquad (43)$$

where we now take the initial state $|i\rangle$ to be the ground orientational and translational state $|00\rangle$ and average over the X,Y,Z components of $\underline{P}$. Eq.(43) can be rewritten as

$$A = \frac{4\pi^2 \rho_o}{3\hbar^2 cnV} \langle 00 \mid [H, \underline{P}^*] \cdot \underline{P} \mid 00 \rangle \qquad (44)$$

so that the total absorption is the indicated average over the ground state. For H we use the total Hamiltonian given by Eq.(15), with $V_A = V_{EQQ}$. After some algebra, Eq.(44) reduces to the evaluation of quantities of the form

$$\langle 0 \mid \frac{Y_4^M(\Omega_{ij})\, Y_4^{N*}(\Omega_{kl})}{R_{ij}^5\, R_{kl}^5} \mid 0 \rangle \qquad (45)$$

where the brackets mean averages over the phonon ground state of the lattice such as indicated in Eq.(21) (note that in Eq.(45) the evaluation of three particle correlations are also required). The sum rule can be broken up into three parts all of which include multi-phonon absorption : a no-libron term, a 1-libron term, and a 2-libron term. By taking the limit as the zero-point motion goes to zero, one finds that only 1-phonon processes (T_u^l and T_u^t) have intensity. Using correlated Gaussians with a short range cut-off, for D_2 one finds the total integrated absorption to be enhanced by about a factor of two , with about half in the no-libron transitions and half in the 1-libron transitions (the 2-libron part is negligible). Thus the ZPM of the quantum solids has a profound effect on the interaction with light. From the sum rule it can be shown that as the width of the PCF goes to zero, the total absorption reduces to that arising from the 1-phonon allowed processes.

CONCLUSIONS AND TRENDS

In these lectures we have attempted to present the basic underlying and distinguishing features of quantum solids and to show the profound effects that the large ZPM has on the quasi particle spectrum. A number of experimental results of the optical response have been presented to demonstrate the properties of real systems. We have confined the examples to the fundamental low-lying excitations. A large number of optical properties, phenomena, and measurements have been omitted. There still remains a great deal of exciting research to be carried out. The current trends are toward studies as a function of density. As the pressure of a solid is increased the lattice particles are localized. These studies will give insight into the transition from quantum to harmonic behavior. Finally the most quantum of all atomic systems, spin aligned atomic hydrogen with a quantum parameter $\lambda^2 = 0.55$, more than twice that of ^{3}He, has never been stabilized. Achievement of this would provide an exotic new system with many fascinating aspects.

REFERENCES

1. A number of excellent reviews on quantum solids exist, for example: R.A.Guyer "The Physics of Quantum Solids", Sol.State Physics, Vol.23, Academic Press New York 1969;
H.R.Glyde "Solid Helium" in Rare Gas Solids ed. M.L.Klein and J.A.Venables, Academic Press, New York 1976;
also see the articles by H.Horner and by T.Koehler in "Dynamical Properties of Solids" Vols. I and II, G.K.Horton and A.A.Maradudin, eds., North Holland Publ. Co. Amsterdam,1974.
2. For a review of solid hydrogen see I.F.Silvera "The Solid Hydrogens: Properties and Excitations" Proceedings of Low Temp. Conf. 14, Vol.5, North Holland Publ. Co., Amsterdam 1975.
3. Data for helium is from R.D.Etters and R.L.Danilowicz, Phys.Rev. A, 1698(1973); for rare gas solids and H_2 and D_2, V.Goldman, Phys.Rev. 174, 1041 (1968) and private communication.
4. M.Nielsen, Phys.Rev. B7, 1626 (1973).
5. See for example V.V.Goldman, G.K.Horton and M.L.Klein, Phys.Rev. Lett. 24, 1424 (1970).
6. H.Horner, J.Low Temp. Phys. 8, 511 (1972).
7. See for example A.B.Harris, Phys.Rev. B1, 1881 (1970) and references therein.
8. T.Nakamura, Prog.Theor. Phys. (Kyoto) 14, 135 (1955).
9. J.W.Stewart, J.Phys.Chem.Sol. 1, 146 (1956).

10. See for example, R.G.Gordon, Adv.Mag. Resonance 3, 1 (1968).
11. R.E.Slusher and C.M.Surko, Phys.Rev. B13, 1086 (1976); C.M.Surko and R.E.Slusher, ibid p.1095.
12. See for example G.W.Chantry, "Submillimetre spectroscopy", Academic Press, New York, 1971.
13. I.F.Silvera, Rev.Sci. Inst. 41, 1592 (1970).
14. I.F.Silvera, W.N.Hardy, and J.P. McTague, Phys.Rev. B5, 1578, (1972).
15. P.A.Fleury and J.P.McTague, Phys.Rev.Lett. 31, 914 (1973).
16. N.R.Werthamer, Phys.Rev. 185, 348 (1969).
17. D.W.Oxtoby and W.M.Gelbart, J.Mol.Phys., 29, 1569 (1975).
18. S.H.Walmsley and J.A.Pople, Mol.Phys. 8, 345 (1964).
19. W.N.Hardy, I.F.Silvera and J.P.McTague, Phys.Rev. 12, 753 (1975).
20. See for example, J.C.Raich and R.D.Etters, Phys.Rev. 168, 425 (1968).
21. C.F.Coll, III and A.B.Harris, Phys.Rev. B4, 2781 (1971).
22. A.J.Berlinsky and A.B.Harris, Phys.Rev. B4, 2808 (1971).
23. C.F.Coll and A.B.Harris, Phys.Rev. B2, 1176 (1970).
24. P.J.Berkhout and I.F.Silvera, Comm. on Physics 2, 109 (1977).
25. W.N.Hardy, I.F.Silvera, K.N.Klump and O.Schnepp, Phys.Rev. Lett. 21, 291 (1968).
26. R.Jochemsen, A.J.Berlinsky, V.V.Goldman and I.F.Silvera, submitted for publication.
27. O.Schnepp, J.Chem.Phys. 46, 3983 (1967).
28. M.E.Rose, "Elementary Theory of Angular Momentum", John Wiley and Sons, New York 1957.
29. R.Jochemsen, F.Verspaandonk, A.J.Berlinsky and I.F.Silvera, to be published.

ANHARMONIC INTERFERENCE IN SCATTERING EXPERIMENTS

W. J. L. Buyers

Atomic Energy of Canada Limited

Chalk River, Ontario, Canada, K0J 1J0

Although deeply entrenched in the folklore of condensed matter physics the phonon concept is of limited validity. Even with the most direct technique for studying phonons, neutron inelastic scattering, it is not possible in principle to distinguish between the one-phonon response and the multiphonon background.

In 1963 Ambegaokar, Conway and Baym[1] showed theoretically that one-phonon scattering of neutrons could interfere with two- or more-phonon scattering. They proved that the one-phonon part of the spectrum, $S_P(Q,\omega)$, satisfied a sum rule, now called the ACB sum rule,

$$\int \omega \, S_P(\vec{Q},\omega)d\omega = d(Q)^2 Q^2/2m, \qquad (1)$$

in which d(Q) is the Debye-Waller factor. Although the lineshape of the one-phonon response is changed by the anharmonic interference, the sum rule is not.

The interference effect was first observed[2,3] in the diffuse scattering of X-rays from NaCℓ. The X-ray one-phonon intensities[4], from which $\omega_{\vec{q}}$ is conventionally obtained, were found to contain a component antisymmetric in $\vec{q}$ about a reciprocal lattice point, $\vec{\tau}$, when $\vec{q}\,||\,\vec{\tau}$. Initially the effect of form factor changes arising from ionic deformation was calculated[3]. This calculation yielded a term in the scattering antisymmetric in $\vec{q}$ but gave too small a dependence on $\vec{Q}$ and T. Moreover the ionic deformations required for the observed asymmetry were much larger than those predicted by the simple[3] or breathing[5] shell model.

At Chalk River Roger Cowley suggested that the X-ray asymmetries might be anharmonic in origin. We calculated[6] the interference between the one-phonon and two-phonon scattering and obtained excellent agreement[4] with the X-ray results[3,5] on NaCℓ and NaF. To eliminate any possible influence of the X-ray form factor we searched for and found[7] similar asymmetries between $+\vec{q}$ and $-\vec{q}$ in the scattering of neutrons by phonons in KBr.

The picture that emerged of the lowest order anharmonic interference was that the probe (neutron or X-ray photon) is scattered by a phonon which subsequently decays into a pair of phonons. The one-phonon response, $G(\vec{q}\,j\,\omega)$, is coupled by the third order anharmonic potential, $V^{(3)}(\vec{q}\,j\,12)$, where $1 \equiv \vec{q}_1 j_1$, to the two-phonon response, $K(12\omega)$. The scattering is proportional to the imaginary part of

$$\vec{Q}\cdot\vec{e}(\vec{q}\;j)\;G(\vec{q}\,j\,\omega)\sum_{12} iV^{(3)}(\vec{q}\,j\,12)\;K(12\omega)\;\vec{Q}\cdot\vec{e}(1)\vec{Q}\cdot\vec{e}(2)$$

$$\equiv \vec{Q}\cdot\vec{e}(\vec{q}\;j)\;G(\vec{q}\,j\,\omega)\;H(\vec{Q},\vec{q}\;j\;\omega). \tag{2}$$

If each atom is on a centre of symmetry the real quantity $iV^{(3)}$ is antisymmetric in $\vec{q}$. Because H is a smooth two-phonon continuum the effect of interference consists of a symmetric contribution proportional to Im G Re H (Fig. 1, curve(a)) and an asymmetric contribution to the lineshape proportional to Re G Im H (Fig. 1, curve(b)) that extends well out into the wings.

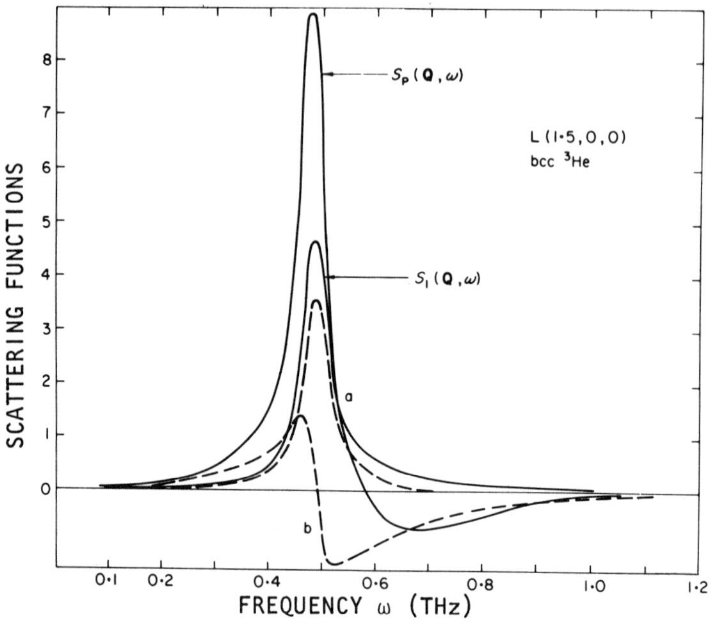

Fig. 1. The one-phonon-like scattering, S_P, is the sum of the pure one-phonon component, S_1, and the interference terms a,b described in the text (from Ref. 8).

EXPERIMENTS ON SOLID AND LIQUID HELIUM

Neutron experiments on hcp[9,10] and bcc[11,12] helium illustrate well the limitations of the phonon concept in highly anharmonic solids. In bcc ^4He the one-phonon-like intensity did not scale with $Q^2\ d(Q)^2$. The experimenters evaluated the first-moment, M^1, over the region of the peak, assuming the multiphonon scattering could be subtracted off as a background. Via the ACB sum rule (Eq.1) they found a Debye-Waller factor that appeared to oscillate in Q (Fig. 2). The harmonic Debye-Waller factor should give the broken curve.

It was rapidly shown[13,14], however, that the amount of non-gaussian behaviour in the vibrational distribution required to account for large oscillations in the Debye-Waller factor was much larger than could reasonably be expected even for a solid as anharmonic as ^4He. As the ACB sum rule is rigorous, it was suggested[13] that incorrect backgrounds had been used to evaluate the first moment.

Werthamer[15] pointed out that the Q-dependence of the "one-phonon" contribution to the first-moment was similar in liquid[16] (Fig. 3) and solid ^4He. The loss in one-phonon strength near 1 $Å^{-1}$ is accompanied by a transfer to the multiphonon scattering. This behaviour and its pressure dependence suggest[17] that anharmonic interference effects are present in the liquid; if so, the third-order interaction in the liquid must be comparable to that in the solid.

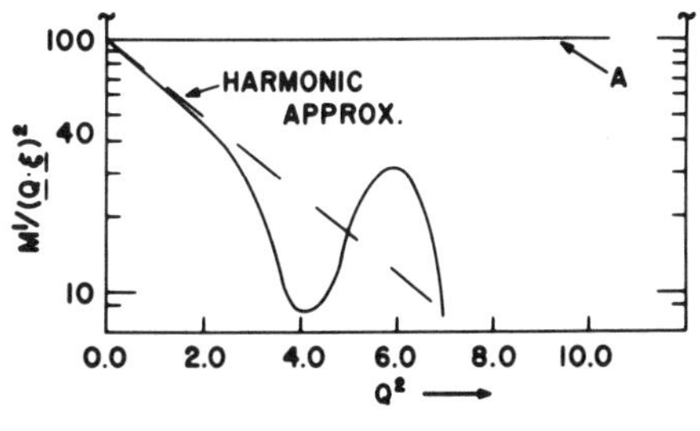

Fig. 2. The anomalous first-moment of the one-phonon-like scattering in bcc ^4He (Ref. 12).

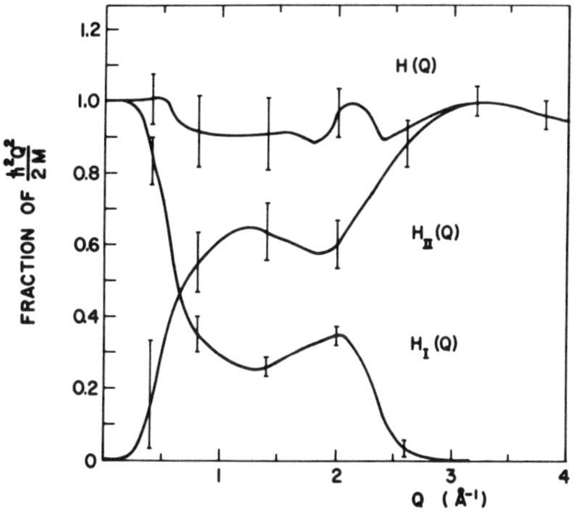

Fig. 3. The one-phonon, H_I, and multiphonon, H_{II}, contributions to the first moment of the scattering from liquid ^4He (Ref. 16).

In view of the results for liquid ^{4}He and the alkali halides it should have been no surprise that the first moment of solid helium appeared to oscillate, but it was not until the numerical work of Horner[18] and Glyde[19] that the importance of anharmonic interference came to be appreciated for the helium solids. The one-phonon response, $G(\vec{q}\, j\, \omega)$, was calculated by Horner from self-consistent potentials obtained by averaging the interatomic potential with an approximate form of the pair distribution that emerges at the end of the iterative loop. The third-order correction to the self-energy is inside the iterative loop. The interference scattering between the resultant $G(\vec{q}\, j\, \omega)$, which exhibits subsidiary peaks[20], and the two-phonon scattering is then calculated. The total scattering[21] shows (Fig. 4) how the interference at equivalent wave vectors (a) has modified the overall intensity and (b) enhanced or diminished some of the subpeaks. Glyde's calculation differs in that self-consistent phonon theory is used with the third-order correction to the self-energy as a perturbation. His lineshapes are much smoother than Horner's (Fig. 1). As expected, both authors found that the ACB sum rule would appear to be violated if the first moment was evaluated over only the main peak of $S(\vec{Q},\omega)$. If a complete integration were performed the symmetric scattering $\propto$ Im $G(\vec{q}\, j\, \omega)$ (Fig. 1, curve a) would give a positive contribution which is cancelled by a negative contribution from the asymmetric scattering (Fig. 1, curve b). An exact proof follows from Kramers-Kronig relations. In an experiment, however, the multiphonon scattering is modified by the wings of the asymmetric scattering, which extend far out beyond the apparent peak region. Consequently an incorrect background and hence first moment will be obtained.

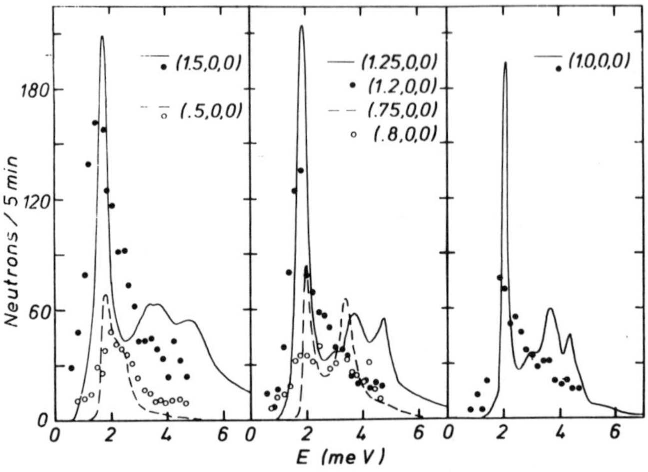

Fig. 4. Neutron scattering in bcc ^{4}He (Ref. 21), theory (lines) and experiment (points).

The conclusion is that the ACB sum rule, although valid within its specified limits, is of little use to experimenters. It is impossible in practice to distinguish between the one-phonon scattering and the multiphonon scattering because of their mutual interference. This history of anharmonic interferences will, it is hoped, make clear that any future comparison of theory and experiment for anharmonic systems should be based on the total density-density response, $S(\vec{Q},\omega)$, and not on any one-phonon-like feature of the spectrum.

REFERENCES

1. V. Ambegaokar, J. Conway and G. Baym, "Lattice Dynamics" (Ed. R.F. Wallis) Pergamon, 1965, p. 261.
2. W.J.L. Buyers and T. Smith, Phys. Rev. 150, 758 (1966).
3. W.J.L. Buyers, J.D. Pirie and T. Smith, Phys. Rev. 165, 999 (1968).
4. For figures see chapter by R.A. Cowley, or refs. 3,5,6,7.
5. J. Melvin, J.D. Pirie and T. Smith, Phys. Rev. 175, 1085 (1968).
6. R.A. Cowley and W.J.L. Buyers, J. Phys. C 2, 2262 (1969).
7. R.A. Cowley, E.C. Svensson and W.J.L. Buyers, Phys. Rev. Lett. 23, 525 (1969).
8. H.R. Glyde, "Rare Gas Solids" (Eds. M.L. Klein, J.A. Venables) Academic, 1976, Vol. I, p. 382.
9. R.A. Reese, S.K. Sinha, T.O. Brun and C.R. Tilford, Phys. Rev. A 3, 1688 (1971).
10. V.J. Minkiewicz, T.A. Kitchens, F.P. Lipschultz, R. Nathans and G. Shirane, Phys. Rev. 174, 267 (1968).
11. E.B. Osgood, V.J. Minkiewicz, T.A. Kitchens and G. Shirane, Phys. Rev. A 5, 1537 (1972); errata A 6, 526 (1972).
12. V.J. Minkiewicz, T.A. Kitchens, G. Shirane and E.B. Osgood, Phys. Rev. A 8, 1513 (1973).
13. V.F. Sears and F.C. Khanna, Phys. Rev. Lett. 29, 549 (1972).
14. A.K. McMahan and R.A. Guyer, "Low Temperature Physics - LT13", Plenum, 1972, Vol. I, p. 110.
15. N.R. Werthamer, Phys. Rev. Lett. 28, 1102 (1972).
16. R.A. Cowley and A.D.B. Woods, Can. J. Phys. 49, 177 (1971).
17. A.D.B. Woods, E.C. Svensson and P. Martel, Phys. Lett. 43A, 223 (1973). See also E.C. Svensson, P. Martel, V.F. Sears and A.D.B. Woods, Can. J. Phys. 54, 2178 (1976).
18. H. Horner, Phys. Rev. Lett. 29, 537 (1972).
19. H.R. Glyde, Can. J. Phys. 52, 2881 (1974).
20. A.K. McMahan and H. Beck, Phys. Rev. A 8, 3247 (1973) show that Horner's subsidiary peaks are enhanced by the sparseness of his mesh used for $\vec{q}$-space sums, and by his inclusion of $V^{(3)}$ within the iterative loop. The iterative procedure diverges for a $V^{(3)}/V^{(2)}$ ratio slightly larger than that used by Horner.
21. H. Horner, "Low Temperature Physics - LT13", Plenum, 1972, Vol. 2, p. 3.

CONTRIBUTORS

Umberto Balucani
Consiglio Nazionale delle Richerche
Laboratorio di Elettronica Quantistica
Via Panciatichi, 56/30 - 50127 Firenze, Italy

Abraham Ben Reuven
Department of Chemistry
Tel Aviv University
Tel Aviv, Israel

Heinz Bilz
Max-Planck Institut für Festkörperforschung
Büsnauer strasse 171
7000 Stuttgart 80, West Germany

W.J.L. Buyers
Chalk River Nuclear Laboratories
Chalk River, Ontario Canada K0J 1J0

M. G. Cottam
Department of Physics
University of Essex
Box 23, Wivenhoe Park, Colchester
England CO4 3SQ

Roger Cowley
Department of Physics
University of Edinburgh
James Clerk Maxwell Building
The King's Buildings, Mayfield Road
Edinburgh EH9 3JZ

P. Fleury
Bell Telephone Laboratories
Murray Hill, New Jersey

Dean Frenkel
Universiteit van Amsterdam
Laboratorium voor Fysische Chemie
Nieuwe Prinsengracht 126
Amsterdam-C.

William M. Gelbart
Department of Chemistry
University of California at Los Angeles,
Los Angeles, California 90024

Allen Goldman
School of Physics and Astronomy
Tate Laboratory of Physics
University of Minnesota
116 Church Street S.E.
Minneapolis, Minnesota 55455

J. W. Halley
School of Physics and Astronomy
Tate Laboratory of Physics
University of Minnesota
116 Church Street S.E.
Minneapolis, Minnesota 55455

Roger Hastings
Department of Physics
North Dakota State University
Fargo, North Dakota 58102

Raymond Kapral
80 St. George St.
Toronto, Canada M5S 1A1

P. D. Loly
Department of Physics
The University of Manitoba
Winnipeg, Manitoba R3T 2N2

Gene F. Mazenko
The James Franck Institute
The University of Chicago
5640 Ellis Avenue
Chicago, Illinois 60637

Horia I. Metiu
Department of Chemistry
University of California at Santa Barbara
Santa Barbara, California 93106

Cherry Murray
Department of Physics
Massachusetts Institute of Technology
Cambridge, Massachusetts 02139

Irwin Oppenheim
Chemistry Department
M. I. T.
Cambridge, Massachusetts 02193

Francis Pinski
Physics Department
State University of New York
Stony Brook, New York 11794

Aneesur Rahmann
Argonne National Laboratory
Argonne, Illinois 60439

R. B. Stinchcombe
Oxford University
Department of Theoretical Physics
1 Keble Road
Oxford, OX1 3NP

I. F. Silvera
Natuurkundig Laboratorium
Der Universiteit van Amsterdam
Valckenierstraat 65 - Amsterdam-C.

M. Thorpe
Physics Department
Michigan State University
East Lansing, Michigan 48824

SUBJECT INDEX